Accelerating Fossil Energy Technology Development Through Integrated Computation and Experimentation

Topical Conference at the 2012 AIChE Annual Meeting

Pittsburgh, Pennsylvania, USA
28 October - 2 November 2012

ISBN: 978-1-62276-750-2

Printed from e-media with permission by:

Curran Associates, Inc.
57 Morehouse Lane
Red Hook, NY 12571

Some format issues inherent in the e-media version may also appear in this print version.

Printed by Curran Associates, Inc. (2013)

For permission requests, please contact AIChE
at the address below.

AIChE
3 Park Avenue
New York, NY 10016-5991

Phone: (203) 702-7660
Fax: (203) 775-5177

www.aiche.org

Additional copies of this publication are available from:

Curran Associates, Inc.
57 Morehouse Lane
Red Hook, NY 12571 USA
Phone: 845-758-0400
Fax: 845-758-2634
Email: curran@proceedings.com
Web: www.proceedings.com

TABLE OF CONTENTS

Practical Membranes for CO_2 Capture 1
Christina Myers, David Luebke, Wei Shi, Shan Wickramanayake, Brian Kail, Robert L. Thompson, Hunaid Nulwala

Theoretical and Experimental Approach to Nitrogen Selective Membranes 2
Ekin Ozdogan, Panithita Rochana, Jennifer Wilcox

Poly(ionic liquid) Membranes for CO_2/N_2 Separation: Insight From Atomistic Simulation 4
Weijie Fang, Zhonglin Luo, Jianwen Jiang

Multifold CO_2 Transport Property Enhancement of Interfacially Constrained Nanocomposite Membranes 5
Lakshmi Suhasini Kocherlakota, Tiep Hoang Pham, René M. Overney

Nanocomposite Membrane of a Polymer of Intrinsic Microporosity and Zeolitic Imidazolate Frameworks for Gas Separation 6
Qilei Song, Shaheen A. Al-Muhtaseb, Easan Sivaniah

Steam Stability of Layered Zeolites Suitable for Hydrogen Transport 8
Bahman Elyassi, Xueyi Zhang, Michael Tsapatsis

Physical Aging and Carbon Dioxide Plasticization of Thin Polyetherimide Films 9
Jianzhong Xia, Tai Shung Chung, Pei Li, Norman R. Horn, Donald R. Paul

The Design and Modeling of a Controlled Atmosphere Food Storage System Based On Membrane Technology 10
Hang Yin, Na Peng, Kok Lin Chong, Glenn Lipscomb, Tai-Shung Chung

Effect of Confinement On Ionic Liquid Properties for Carbon Capture Processes 11
Ruth E. Baltus, Laila Banu

Metallic Based Aqueous Amine Membranes for the Separation of Carbon Dioxide From Fossil Fuel Flue Gas 12
Brian L. Bischoff, Lawrence E. Powell

A Comprehensive Multi-Step Description of the Permeation of Hydrogen in a Dense Metal Membrane Over a Broad Temperature Range 13
Nicholas Deveau, Yi Hua Ma, Ravindra Datta

High Temperature Metal Membranes for Hydrogen Permeation without Platinum Group Metals 14
J. Douglas Way, Colin A. Wolden, Kehinde Adeyemo, Mayur Ostwal

Long Term H_2 Permeance and Selectivity Stability of Pd and Pd-Au Large Scale Membranes in Actual Syngas Atmospheres 15
Federico Guazzone, Jacopo Catalano, Ivan Mardilovich, Nikolas Kazantzis, Yi Hua Ma

Composite Catalytic Membranes: Enhancing H_2 Recovery Membranes by Appropriate Coupling with Water-Gas-Shift Reaction 16
Elva Romero, Daejin Kim, Benjamin Wilhite

Toward a Common Method of Cost Estimation for CO_2 Capture and Storage At Fossil Fuel Power Plants 17
Edward S. Rubin

Proposed Baselines and Standard Methods for Systems Analyses of Power Generation Technologies 18
Michael Matuszewski

Methodology for Evaluating the Cost of Emerging CCS Technologies 19
Abhoyjit S. Bhown, George Booras, Ronald Schoff

A Common Modeling Framework to Evaluate and Compare the Economics of CCUS Technologies 20
Edward S. Rubin

NETL CO_2 Injection and Storage Cost Model 21
Tim Grant, Dave Morgan, Jason Valenstein, Andrea Poe, Marta Milan, Richard Lawrence

A Monte Carlo Approach for Estimating and Validating Contingencies for CCS 22
Paul Worhach, Mark Woods, John Wimer

CO_2 Mineralisation Through Hydrometallurgical Leaching of Brown Coal Fly Ash 23
Lian Zhang, Teck Kwang Choo

Statistical Evaluation of Mineral Carbonation of a Fly Ash/Brine System 24
Grace Muriithi, Leslie F. Petrik, Wilson M. Gitari, Frédéric J. Doucet

Ex situ CO_2 Mineralization Using Reactive Species 25
Juan Ma, Roe-Hoan Yoon

Base Metal Production From CO_2 Capture and Storage Via the Ics Integrated Mineral Carbonation Process 30
Richard Hunwick
Coupling Industrial Carbon Mineralisation and Ocean Alkalinity Enhancement 31
Phil Renforth, Tim Kruger
Evaluating the Viability of CO_2 Mineralization Via Reaction of Caustic Waste Materials 33
Robert Dilmore, D. Allen, S. W. Hedges, Yee Soong
The NETL Aqueous Process: History and Recent Developments 34
William O'Connor, G. E. Rush, Circe Verba
Summary Results of Epri's Engineering and Economic Case Studies of Post Combustion Capture Retrofit Applied to Various North American Host Sites 35
Des Dillon
Electrochemical Membrane for Carbon Dioxide Separation and Power Generation 37
Stephen Jolly, Hossein Ghezel-Ayagh, Jennifer Hunt, Dilip Patel, William A. Steen, Carl F. Richardson, Olga A. Marina
Modeling, Simulation and Optimization of Absorption, Membrane and Adsorption-Based CO_2 Capture for Variable Feed CO_2 Concentration and Feed Flow 44
M. M. Faruque Hasan, Richard C. Baliban, Josephine A. Elia, Christodoulos A. Floudas
Pressure Swing Adsoprtion: Design and Optimization for Pre-Combustion Carbon Capture 45
Alexander W. Dowling, Lorenz T. Biegler, David C. Miller
Assessment and Analysis of Post-Combustion Carbon Capture with Membrane Separation Technology 46
Xiangping Zhang, Xuezhong He, Truls Gundersen
Dynamic Modelling and Analysis of Post-Combustion CO_2 Capture Processes 49
Niall Mac Dowell, Nilay Shah
Comparisons of Solvents for Post-Combustion CO_2 Capture Using a Multi-Objective Analysis Approach 50
Anita S. Lee, John R. Kitchin
Post-Combustion Gas Permeation Carbon Capture System Models 51
Juan E. Morinelly, David C. Miller
Fuel Processing: Renaissance of the Second Decade (Keynote) 52
Brant. A. Peppley, Rajesh D. Parmar, Kunal Karan
Desulfurization of JP-8 Fuel Using Ag-Loaded Silica Supports Under Ambient Conditions 55
Dat T. Tran, Jessica St. John, Deryn Chu, Scott R. J. Oliver
Desulfurization of Commercial Fuels Using Mixed Oxide Supported Silver Adsorbents At Room Temperature for Fuel Cell Applications 56
A. H. M. Shahadat Hussain, Hongyun Yang, Bruce J. Tatarchuk
Multi-Physics Modeling of Auto-Thermal Diesel Surrogate Reforming 57
Rajesh D. Parmar, Amrit Jalan, Dushyant Shekhawat, William H. Green Jr., Brant A. Peppley, Kunal Karan
Heat-Integrated Steam Reformer for Hydrogen Production 58
Saurabh A. Vilekar, Christian Junaedi, Kyle Hawley, Dennis Walsh, Subir Roychoudhury
Diesel Reforming Over Pyrochlore-Based Monolith 59
Daniel J. Haynes, David Berry, Dushyant Shekhawat, Mark W. Smith, Matthew M. Seabaugh, Richard Long, James J. Spivey
Carbon Deposition Over Supported Rh-Ni Catalysts in Steam Reforming of Liquid Hydrocarbon Fuel for H_2 Production: Influence of Support, Temperature, and Sulfur 60
Chao Xie, Yongsheng Chen, Chunshan Song
Fuel Processing with Diesel Fuel and Kerosene for APU Applications 61
Joachim Pasel, Jan Meißner, Remzi Can Samsun, Ralf Peters, Detlef Stolten
Reactor Design for Diesel Reforming Using a Graded-Composition Monolith 68
Mark W. Smith, David A. Berry, Dushyant Shekhawat, Daniel J. Haynes, James J. Spivey, John Zondlo
Carbon Molecular Sieve Membranes for N_2/CH_4 Separation 69
Xue Ning, William J. Koros
High Performance Carbon Molecular Sieves (CMS) Hollow Fiber Membranes for Natural Gas Separation 70
Nitesh Bhuwania, William J. Koros
Evaluation of Facilitated Transport Membrane for Olefin/Paraffin Separation 71
Majid Keyvani, Glenn Towe, Ali Hamza
Gas Transport of Paraffins and Olefins in Liquid Crystal Membranes 86
Feras Rabie, Stephen M. Martin

Synthesis and Characterization of ZIF-8 Membrane with Exceptionally High Propylene / Propane Separation Performance 87
Hyuk Taek Kwon, Sagar Lonkar, Hae-Kwon Jeong

Design of ZIF/Polyimide Mixed Matrix Membrane for Olefin/Paraffin Separations 89
Chen Zhang, William J. Koros

TR Polymers for Olefin/Paraffin Separation 90
Zachary P. Smith, David F. Sanders, Ruilan Guo, Benny Freeman, Donald R. Paul, James E. McGrath

Asymmetric Hollow Fiber Membranes for Aggressive Sour Gas Purification 91
Brian E. Kraftschik, William J. Koros

Carbonate Ceramics and Concrete: A Disruptive Technology for CO_2 Utilization and the Construction Business 92
Richard Riman, S. Gupta, Q. Li, V. Atakan, Ç. Vakifahmetoglu, J. Azurdia, J. Krishnan, M. A. Bitteto, L. Tang, J. Czerepinski, Larry E. McCandlish, Nick Decristofaro

Synthesis of $CaSiO_3$ for Use As a Cementitious Material for Carbonate Concrete 93
Nick Decristofaro, V. Atakan, C. Vakifahmetoglu, L. McCandlish, J. Krishnan, L. Tang, Q. Li, R. E. Riman

CO_2 Utilization in Concrete Block Production 94
Yixin Shao, Liang Hu, Hilal El-Hassan

Catalyzed Activation of CO_2 by Lewis Base Site in Hybrid Metal Organic Frameworks 97
Liang Chen

Utilization Technologies for CO_2 Derived From North Dakota Lignite 98
Jason D. Laumb, Melanie D. Jensen, Robert M. Cowan, Alexander Azenkeng, Loreal V. Heebink, Laura J. Raymond, Peter A. Letvin

Harnessing Applied Potential for Sunlight Driven CO_2 Utilization 108
Benjamin R. Buckley

Accelerating Progress Toward Operational Excellence of Fossil Energy Plants with CO_2 Capture 109
Stephen E. Zitney, Eric A. Liese, Priyadarshi Mahapatra, Richard Turton, Debangsu Bhattacharyya

Characterization and CO_2 Sorption Properties of Materials for CCS Applications 164
Jiajun He, Erik C. Rupp, Jennifer Wilcox

Reducing Uncertainties in Model Predictions Via History Matching of CO_2 Plume Migration At the Sleipner Project, Norwegian North Sea 165
Peng Lu, Chen Zhu, Per Aaggard

Gas Hydrates Modeling: Spanning Multiple Scales 166
Nagasree Garapati, Srinath Velaga, Brian Anderson

High-Resolution Numerical Simulation of CO_2 Migration in Saline Aquifers 167
Ruslan Iskhakov, Denis Voskov, Hamdi Tchelepi

Pore Scale CO_2 Displacement in Sandstone with Comparison to Core Scale Dynamics 168
Dustin Crandall, Robert P. Warzinski, William O'Connor, Aslam M. Kabir, Grant Bromhal

Modeling of Pressure and Displacement Signatures for Detecting A Fracture in the Caprock During Geologic Sequestration 169
Hema Siriwardane, Raj Gondle, Grant Bromhal

Effect of Calcite Dissolution On the Leakage Through Abandoned Wells in Geological CO_2 Storage 172
Ezzeddine Laabidi, Rachida Bouhlila

Hydrogen Production From Glycerol Reforming Over Hydrogen Production From Glycerol Reforming Over Ni Based Catalysts 173
Dong-Ju Moon, Su Hyun Cho, Seung-Hoon Kim, Seung-Hwan Lee, Bang Hee Kim, Sang Woo Kim, Arunabha Datta, Kwan-Young Lee

Computer Simulation of Water and Ammonia Electrolysis 175
Ali Estejab, Damilola A. Daramola, Gerardine G. Botte

Desulfurization of JP-8 Fuel Using Metal-Loaded Meso/Macroporous Aluminosilicates 177
Jessica St. John, Dat T. Tran, Deryn Chu, Scott R. J. Oliver

A Novel KMgAl Sorbent for CO_2 Capture in a Fluidized Bed 178
Fukui Xiao, Wei Wei, Lei Li, Xia Wen, Feng Wang, Ning Zhao

Microscopic Behavior of Chemical Demulsifiers At the Water-in-Brazilian Asphaltene Model Oil Emulsions: Langmuir Monolayers and Their Corresponding Langmuir-Blodgett Films 182
Juliana S. A. Carneiro, Montserrat Fortuny, Alexandre F. Santos, Cláudio Dariva, Elton Franceschi, Bianca M. S. Ferreira, Sandro R. P. Da Rocha

New Reservoir Opportunities for CCUS: The Bio-Geochemistry of Mobile Water Intervals with Residual Oil: A Permian Basin Case History 183
David Vance, Steve Melzer, Robert Trentham

Evolution of Carbonate Dissolution Features Produced Under Variable pCO_2 Conditions Relevant to CO_2-EOR and Geologic CO_2 Storage 185
Susan Carroll, Megan Smith, Yue Hao

Opportunities for Utilizing Anthropogenic CO_2 for Enhanced Oil Recovery (EOR) 186
Michael L. Godec
Potential for CO_2 Ehanced Oil Recovery in the Appalachian Basin 187
Timothy Carr
Carbon Dioxide Pipeline Infrastructure Dependence On Reservoir Risk 188
Jeffrey Bielicki, Richard Middleton, Melisa Pollak, Elizabeth Wilson, Jeffrey Fitts, Catherine Peters
A Phased Approach to Developing A Pipeline Network for CO_2 Transport During CCUS 190
Melanie D. Jensen, Peng Pei, Anthony C. Snyder, Loreal V. Heebink
Micro-Channel Fuel Processors As Hydrogen Source for Fuel Cells in Distributed Energy Supply Systems (Keynote) 196
Gunther Kolb, S. Keller, M. O'Connell, Jochen Schuerer, David Tiemann, Martin Wichert
Tri-Generation of Hydrogen, Heat and Power From a High Temperature Fuel Cell 198
Jack Brouwer, Roxana Bekemohammadi, Pere Margalef
Effect of Storage Time On the Catalytic Activity of Gold for Selective CO Oxidation 201
Fabio B. Passos, Thiago S. Mozer
Hydrogen and Methane Rich Syngas From a Catalytic Coal Gasifier with in Situ Capture of H_2S/CO_2 and Ex Situ Regeneration of Catalysts 202
Nicholas Siefert, Dushyant Shekhawat, Shawn Litster, David Berry
Molten Salt Gasification (MSG) of Refinery Wastes to Produce Hydrogen 207
Lyman Frost
Highly Efficient Integrated Chemical Looping Gasification Fuel Cell System 208
Liang Zeng, Mandar Kathe, Siwei Luo, Fanxing Li, Liang-Shih Fan
Experimental Investigation On the Effects of Trace Contaminants in Coal Derived Syngas On Performance of SOFC Ni-YSZ Anode 209
Koji Kuramoto
A Microchannel Water-Gas Shift Reactor Simulation and Optimization for a Mobile Application 219
M. H. Akbari, S. A. Asaee
Microplasma Reforming for Fuel Cell Feed 236
Peter J. Lindner, R. S. Besser
Dry Reforming Assisted Nonthermal Plasma Catalytic Reactor 243
Mahammadunnisa Shaik, P. Manoj Kumar Reddy, B. Ramaraju, Ch. Surbahmanyam
Dry (CO_2) Reforming of Methane Using Rh Substituted Pyrochlores ($La_2Zr_2O_7$) 244
Devendra Pakhare, Victor Abdelsayed, Daniel J. Haynes, Dushyant Shekhawat, James J. Spivey
Simulation of Methane Steam Reforming Enhanced by in Situ CO_2 Sorption Utilizing K_2CO_3 Promoted Hydrotalcites for H_2 Production 262
Naruewan Chanburanasiri, Ana M. Ribeiro, Alirio E. Rodrigues, Navadol Laosiripojana, Suttichai Assabumrungrat
Bi-Reforming of Methane and Natural Gas Under High Pressure 270
Alain Goeppert, Miklos Czaun, Robert May, Surya Prakash, George Olah
Production of Hydrogen and Nanocarbon From Catalytic Decomposition of Methane Over a Ni-Fe/Al_2O_3 Catalyst 272
Yi Jin, Gaowei Wang, Yongdan Li
Natural Gas Direct Carbon Fuel Cell for Hydrogen and Electricity Production (NGDCFC) 273
Meyer Steinberg
CFD Simulation of Pressurized, Staged Oxyfuel Combustion 274
Benjamin M. Kumfer, Fei Xia, Bhupesh Dhungel, Richard L. Axelbaum
Double Perovskite Materials for Applications in Chemical Looping and Oxygen Storage 276
Jonathan W. Lekse, Sittichai Natesakhawat, John P. Baltrus, Christopher Matranga
Using $Ca_xLa_{1-x}Mn_{1-y}M'_yO_{3-\delta}$ As Oxygen Carriers for Chemical-Looping Combustion 277
Tobias Mattisson, Ali Hedayati, Abdul-Majeed Azad, Magnus Ryden, Henrik Leion
Mechanism of Methane Chemical Looping Combustion with Fe_2O_3 Promoted with CeO_2 285
Duane Miller, Ranjani V. Siriwardane
Oxygen Carriers for Chemical-Looping Combustion with Solid Fuels 286
Tobias Mattisson, Henrik Leion, Magnus Ryden, Anders Lyngfelt
Development and Performance Evaluation of a Cost-Effective Oxygen Carriers 294
Liangyong Chen, Yi Zhang, Fang Liu, Chunli Han, James K. Neathery, Kunlei Liu
Study of Wyoming PRB Coal for Application of Solid Fueled Chemical Looping Combustion with TGA-MS 295
Yi Zhang, Jeffrey Slone, Aurora M. Rubel, James K. Neathery, Fang Liu, Chunli Han, Liangyong Chen, Kunlei Liu
Techno-Economic Evaluation of Chemical Looping Combustion for Industrial-Scale Applications 296
John T. Ramsey, Michael J Gallagher Jr., Stephen Carpenter, George Richards

CFD Modeling of Commercial-Scale Entrained-Flow Coal Gasifiers 297
Jinliang Ma, Stephen E. Zitney

Ultrasound Measurements of Temperature Profile Across Gasifier Refractories 298
Yunlu Jia, Mikhail Skliar

Recent Developments in Carbonaceous Chemistry for Computational Modeling (C3M) with Uncertainty Quantifications (UQ) 302
Kiran Chaudhari, Richard Turton, Chris Guenther, Mehrdad Shahnam, Aytekin Gel, Philip Nicolleti, Tingwen Li, Dirk Vanessendelft, Nate Weiland

A One-Dimensional Transient Model of a Single-Stage, Downward-Firing, Entrained-Flow Gasifier 303
Job Kasule, Richard Turton, Debangsu Bhattacharyya, Stephen E. Zitney

Co-Pyrolysis of Low Rank Coals and Biomass At Transport Gasifier Conditions 304
Nathan T. Weiland, Nicholas C. Means, Chris Guenther

Dynamic Modeling of Detachment, Transport, and Flow of Slag in an Entrained-Flow Gasifier 311
Pratik Pednekar, Debangsu Bhattacharyya, Richard Turton, Raghunathan Rengaswamy

Hydrogen Production From Renewable Sources: R&D Challenges, Targets and Pathways (Keynote) 313
Sara Dillich, Eric Miller, Erika Sutherland, Sarah Studer

Auto-Thermal Reforming of Biodiesel-Ethanol-Diesel Blends for Solid Oxide Fuel Cell Applications 315
Jiefeng Lin, Thomas Trabold, Mark Walluk, Daniel Smith

Development of a Butanol Fuel Processor for Person-Portable Fuel Cell Power Systems 323
Stephen Jolly, Hossein Ghezel-Ayagh

Cartridge-Based Ceramic Microreactor Networks for Process Intensification in Fuels Reforming: Demonstration and Modeling of Complex 2-D Thermal Integration 324
Bhanu Vardhan Reddy Kuncharam, Shalini Damodharan, Benjamin Wilhite

Protonic Membrane Reactors: Converting Hydrocarbon Resources and CO_2 to Fuels 325
Eric D. Wachsman

One-Stage Process for Water Gas Shift in a Pd-Based Membrane Reactor 326
Giuseppe Barbieri, Adele Brunetti, Alessio Caravella, Enrico Drioli

Water Gas Shift Reaction in Thin Pd-Layer Supported On PSS Membrane Reactor 328
Angelo Basile, Pietro Pinacci, Francesca Drago, Simona Liguori, Pietro Paolo Morrone, Adolfo Iulianelli

Techno-Economic Assessment of Membrane Reactor Technologies for Pure Hydrogen Production for Fuel Cell Vehicle Fleets 329
Leonardo Roses, Giampaolo Manzolini, Stefano Campanari, Ellart De Wit, Michael Walter

Experimental and Simulation Studies of Effects of Operating Conditions and Membrane Properties On the Performance of Zeolite Membrane Reactors for High Temperature Water Gas Shift Reaction 345
Seok-Jhin Kim, Shaowei Yang, Junhang Dong

Pore Scale Models for Imbibition of CO_2 Analogue Fluids in Etched Micro-Model Junctions Using Micro-Fluidic Experiments and Direct Flow Calculations 346
Emily Chapman, Jianhui Yang, John Crawshaw, Edo Boek

Selection of Domain Size for Lattice Boltzmann Simulation to Calculate Flow Properties of Porous Media 347
Md Rakibul Sarker, Shameem Siddiqui, Dustin McIntyre, Grant Bromhal, Mohamed Soliman

A Lattice Boltzmann Model with Realistic Equations of State for Modeling Pore-Scale Effects On Capillary Trapping of CO_2 During Geological Carbon Sequestration 348
Marcel Schaap, Dorthe Wildenschild

Insights From Image-Based Modeling of Fluid Flow in Bulk Proppant Packs and Propped Fractures Under Varying Confining Stress 350
Yijie Shen, Paula Sanematsu, Seth M. Bradley, Clinton S. Willson, Ali Takbiri Borujeni, Karsten E Thompson, Mayank Tyagi

Recent Trends of Streamline Simulation Technology Applications in Oil Industry 351
Abdullah Al-Najem, Shameem Siddiqui, Mohamed Soliman

Hydrogeochemical Modelling in Coastal Carbonate Aquifer: Porosity and Permeability Evolution 366
Rachida Bouhlila, Ezzeddine Laabidi

Structure Sensitivity of Methanol Synthesis From Carbon Dioxide On Copper-Based Catalysts 377
Sittichai Natesakhawat, Victor Abdelsayed, Jonathan W. Lekse, John P. Baltrus, Paul R. Ohodnicki Jr., Bret H. Howard, Xingyi Deng, Christopher Matranga

Fluorine-Modified Cu/Zn/Al/Zr Catalysts Via Hydrotalcite-Like Precursors for the CO_2 Hydrogenation to Methanol 378
Fukui Xiao, Peng Gao, Li Feng, Ning Zhao, Wei Wei, Yuhan Sun

Dehydrogenation of Ethylbenzene to Styrene with CO_2 Over V_2O_5/Al_2O_3-ZrO_2 Catalyst 385
Shujiao Zhang, Xiaohong Li, Jie Feng, Wenying Li, Hongxia Fan, Qian Wang

A Comparative Study of Activated CO_2 Adsorption and Protonation On the γ-Al_2O_3, β-Ga_2O_3, In_2O_3 Surfaces 397
Jingyun Ye, Changjun Liu, Qingfeng Ge
C-H Activating Carboxylation of Terminal Alkynes with CO_2 and Its Application in Synthesis 398
Yugen Zhang, Jackie Y. Ying, Dingyi Yu, Siti Riduan
Photochemical Reduction of CO_2 Using Delafossite Oxides 400
Jonathan W. Lekse, James P. Lewis, M. Kylee Underwood, Christopher Matranga
Two-Step Solar Thermochemical Cycle for Splitting CO_2 Via Ceria Redox Reactions - Experimental Investigation with a 3.8 Kw Solar Reactor 401
Philipp Furler, Jonathan Scheffe, Aldo Steinfeld
Applicability of Perovskite-Based Oxides As Redox Materials for Thermochemical CO_2 Dissociation 402
Jonathan Scheffe, David Weibel, Aldo Steinfeld
Bench-Scale Electrochemical Production of Synthesis Gas 403
Eric Dufek, Tedd Lister, Michael McIlwain
Electrode and Catalyst Design for Electrochemical Reduction of CO_2 411
Molly Jhong, Sichao Ma, Paul J. A. Kenis
CO_2 - Ionic Liquids Synergies: Reactive Capture and Conversion of CO_2 to Chemicals 413
Valerie Eta, Pasi Virtanen, Ikenna Anugwom, Päivi Mäki-Arvela, Jyri-Pekka Mikkola
Numerical Simulation of Reservoir-Wellbore Fluid Flow Simulations At Geologic CO_2 Sequestration Sites: Coupled or De-Coupled? 415
Rajesh Pawar, Haruko Wainwright, Robert Dilmore, Jens Birkholzer, Jim Houseworth, Grant Bromhal
Investigating the Role of Coupled Thermal-Hydrological-Mechanical (THM) Processes During Geological Sequestration of CO_2 416
Sharad Kelkar, Rajesh Pawar, George A. Zyvoloski, Philip Stauffer, Hari Viswanathan
Reservoir Model Optimization Using the Seismic Response to CO_2 Injection and Stochastic Inversion 417
Abelardo Ramirez, Kathy Dyer, Donald J. White, Yue Hao, James Johnson
Assessing the Role of Thermal Stresses in Caprock Integrity During CO_2 Storage 418
Gennady Gor, Jean-Herve Prevost
Cement Degradation Under Conditions Relevant to Geological Carbon Sequestration 419
Leopold Brunet, Li Li, Zuleima Karpyn, Brian Strazisar, Barbara Kutchko
Laboratory and Numerical Simulation Based Investigation of Two-Phase Brine-CO_2 Displacement Processes Using Cores From Three CO_2-Sequestration Candidate Formations 426
Shameem Siddiqui, Talal Gamadi, Otman Algadi, Abdulrahman A. Alquraishi, Dustin McIntyre
Modification of Fracture Hydraulic Properties by CO_2-Acidified Brine Flow 437
Hang Deng, Brian R. Ellis, Catherine Peters
New Experimental Data and Physico-Chemical Model for the Solubility of Carbon Dioxide in Aqueous Solutions of Monoethanolamine 438
Michael Wagner, Inga Von Harbou, Jaeik Kim, Gerd Maurer, Hans Hasse
Mass Transfer Evaluation of CO_2 Absorption in a Wetted Wall Column 439
Reynolds A. Frimpong, Joseph E. Remias, James K. Neathery, Kunlei Liu
Kinetic Study of Carbon Dioxide Absorption by Aqueous Solutions of N-Methyldiethanolamine and Piperazine 440
Alberto Servia, Nicolas Laloue, Julien Grandjean, Sabine Rode, Christine Roizard
Aqueous Morpholine As a Potential Post-Combustion CO_2 Absorption Solution 446
Krishna Mohan Jinka, Taylor Kuhn, Joseph E. Remias, James K. Neathery, Kunlei Liu
Novel Liquid-Like Nanoscale Organic Hybrid Materials for CO_2 Capture 447
Camille Petit, Youngjune Park, Kunyi Andrew Lin, Ah-Hyung Alissa Park
Development of Reaxff Force Field for Carbon Dioxide Capture with Ionic Liquids: A Combined First Principles and Classical Simulation 448
Bo Zhang, Adri C. T. Van Duin, Karl Johnson
Vibrational Spectroscopy Characterization of CO_2-Ionic Liquid Systems for Pre-Combustion CO_2 449
Anita S. Lee, John R. Kitchin
Evaluation of Mercury Capture Technology Using Various Sorbents Injection 450
Xiaofei Wang, Pratim Biswas
Overview on Mercury Control Options for Coal-Burning Power Plants 451
Evan J. Granite, Henry W. Pennline
Palladium Sorbents for High Temperature Capture of Mercury, Arsenic, Selenium and Phosphorus from Fuel Gas 452
Evan J. Granite, Henry W. Pennline, Erik C. Rupp, Dennis Stanko, John P. Baltrus, Hugh Hamilton, Liz Rowsell, Stephen Poulston, Andrew Smith, Wilson Chu, Tony Wu
Author Index

(149a) Practical Membranes for CO2 Capture

Christina Myers[1], David Luebke[1], Wei Shi[2], Shan Wickramanayake[3], Brian Kail[3], Robert L. Thompson[3] and Hunaid Nulwala[4], (1)US DOE/NETL, Pittsburgh, PA, (2)URS , South Park, PA, (3)URS, South Park, PA, (4)Department of Chemistry, Carnegie Mellon University, Pittsburgh, PA

Membrane technology has advanced greatly in the last 10 years, and due to its potential cost advantages, membranes are now being applied to a wide variety of processes. One of the major focal points for research is in the utility of membranes in CO_2 separations, particularly in coal gasification plants.

Our research examines supported ionic liquid membranes (SILMs) and mixed matrix membranes (MMMs) in both flat and hollow fiber configurations. NETL's membrane development team includes researchers from a wide variety of backgrounds including computational modeling, materials synthesis, membrane fabrication, performance testing, and systems analysis.

SILMs were chosen for their improved transport properties. Gas diffusivities in the IL transport media are higher than those observed in polymers leading to greater permeabilities. More importantly, the materials are stable at elevated temperatures and have negligible vapor pressure, making them suitable for gasification applications. Earlier testing of the various SILMs demonstrated their stability for gas separations at 300°C, and other modifications increased permeability and selectivity. While the studies have demonstrated the potential for laboratory gas separation, the work is expanding to include performance studies under more realistic gas compositions. The presentation will detail evolution of the SILM technology from inception toward practicality.

(149b) Theoretical and Experimental Approach to Nitrogen Selective Membranes

Ekin Ozdogan, Panithita Rochana and Jennifer Wilcox, Department of Energy Resources Engineering, Stanford University, Stanford, CA

The main objective of the proposed research project is to develop N_2-selective catalytic membrane technology with potential sustainability benefits of indirect CO_2 capture and ammonia synthesis. The N_2-selective membrane technology benefits from the driving force of N_2 flue gas (73 wt.%) stream for indirect CO_2 capture as it provides atomic nitrogen on the permeate side of the membrane during separation. With these properties if the H_2 is provided as a sweep gas alongside this novel technology, additional benefits, such as formation of ammonia as a byproduct of the N_2 separation process would be observed.

Within this study, N_2 in particular, has been investigated to determine the mechanism of surface adsorption, dissociation and subsequent atomic diffusion into the bulk crystal structure. Theoretical calculations have been carried out using Vienna *ab initio* Simulation Package (VASP)[i],[ii]. The electron exchange correlation effects are described by a generalized-gradient approximation (GGA) using the Perdew-Burke-Ernzerhof functional[iii] with a plane-wave expansion with a cutoff of 600 eV. The surface Brillouin zone integration has been calculated using a 7x7x7 Monkhorst-Pack mesh[iv]. For absorption energy estimates, binding energy calculations of atomic N in the BCC structure of bulk V and V-Ru alloys have been simulated using a unit cell of 16 atoms with an optimized lattice constant of 2.98Å (3.03Å experiment). Atomic N was found to be stable in the octahedral, face-octahedral and tetrahedral interstitial crystal sites (Figure 1.). It was determined that the octahedral site is the most favorable binding site for N within bulk V, with a binding energy of -2.132 eV. N binding in V is nearly two orders of magnitude stronger compared to the well-known H binding[v] case (-0.076 eV). These strong N binding energies indicate that N may have difficulty diffusing through the material. We found through alloying V with Ru, that this binding energy could be tuned so that it approaches that of H. Alloying V with 6.25 at.% Ru yields a weaker N binding energy of -0.889. It is anticipated that increasing the content of Ru in the alloy will lead to a further reduction in the stability of N within the crystal lattice. Optimal N permeability includes a careful balance between N solubility and diffusivity within a given material.

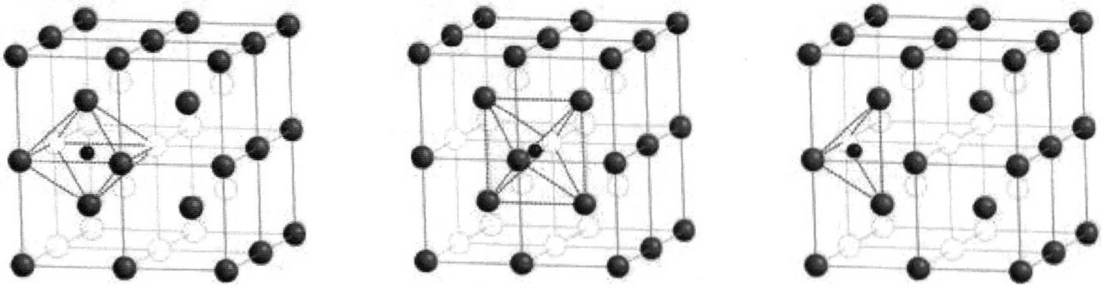

(a) O-site (b) Face-O site (c) T-site

Figure 1 Schematic representation of atomic nitrogen in the interstitial crystal sites

(a) octahedral, (b) face-octahedral and (c) tetrahedral.

Binding energy calculations are carried at different interstitial sites of Vanadium bulk structure. Nitrogen was found to be stable primarily at O-sites within the bulk V lattice. Bader charge analysis has been carried out to investigate the mechanism associated with bulk absorption phenomena. Additionally, alloys of ruthenium (Ru)-V have been studied indicating that Ru can be used as a dopant to tune the electronic structure of the bulk to enhance atomic diffusion. Surface calculations on the 3 low-index surfaces, V(110), V(100) and V(111) based on DFT indicated that the V(111) surface binds N_2 the strongest at the fcc site, with an adsorption energy of 1.4 eV.

Experimental efforts involve pure and mixed gas permeation to determine the N_2 and CO_2 permeabilities of membrane foils comprised of Group V metals and their different composition alloys with Ru over a range of temperatures from 500 to 1000 °C. The schematic of the permeation test set-up is given in Figure 2. In this study, linear regression analyses of the flux isotherms were performed to obtain Sieverts' Constant (n) and it was shown to be 0.5 for N_2 permeation through V and Nb membranes. Unlike the lighter gases such as H_2 and N_2 permeating through the metallic membranes by solution-diffusion mechanism, CO_2 was shown to diffuse only through the grain boundaries or defects on the dense metal surfaces. As the CO_2 fluxes were compared with N_2 fluxes for the pure foil membranes, N_2 fluxes were found to be 2 to 3 orders of magnitudes higher than the CO_2 fluxes. Characterization of the foils before and after the permeation tests was performed by XRD, XPS and SEM techniques to investigate the surface and subsurface structural changes in the membranes. In total, characterization and flux measurements were compared alongside electronic structure predictions of DFT to determine the performance of Group V metals and their alloys for N_2-selective membrane application.

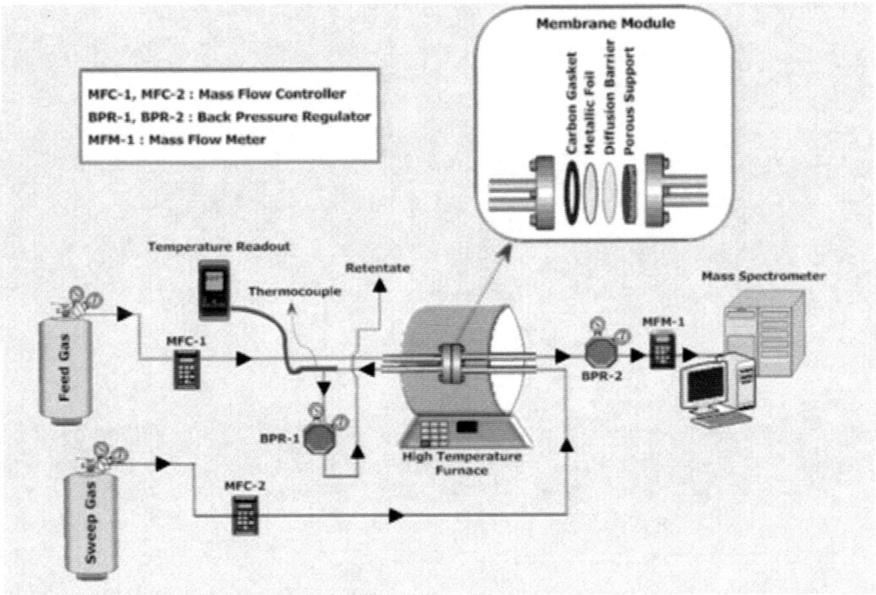

Figure 2.Schematic of Experimental Set-Up

References:

[i] G. Kresse and J. Hafner, *Phys. Rev. B* **48** 13115 (1933).

[ii] G. Kresse and J. Furthmuller, *Comput. Mater. Sci.* ***6*** *15 (1996).*

[iii] P. Perdew, K. Burke, and M. Ernzerhof, *Phys. Rev. Lett.* **77** 3865 (1996).

[iv] H.J. Monkhorst and J.D. Pack, *Phys. Rev. B* **13** 5188 (1976).

[v] S. Aboud and J. Wilcox, *J. Phys. Chem. C* **114** 10978-10985.

(149c) Poly(ionic liquid) Membranes for CO2/N2 Separation: Insight From Atomistic Simulation

Weijie Fang[1], Zhonglin Luo[2] and Jianwen Jiang[1], (1)Chemical and Biomolecular Engineering, National University of Singapore, Singapore, Singapore, (2)National University of Singapore, Singapore, Singapore

Polymeric ionic liquids (poly(IL)s) are a new class of membrane materials for gas separation. In this study, we conduct atomistic simulation to investigate the membrane structures of poly(IL)s and their capability for CO_2/N_2 separation. The poly(IL)s examined are poly[VBIM][TF_2N] and poly[VBIM][SCN]; for comparison, ionic liquids ([BMIM][TF_2N] and [BMIM][SCN]) are also considered. The densities, solubility parameters, ion mobilities and heats of vaporization for the poly(IL)s and ILs are predicted and compared with available experimental data. The fractional free volume increases in the order of poly[VBIM][TF_2N] < poly[VBIM][SCN] < [BMIM][TF_2N] < [BMIM][SCN]. The poly(IL)s and ILs with [TF_2N]$^-$ have a high percentage of large voids (> 3 Å). The predicted solubility and diffusion coefficients CO_2 and N_2 are consistent with experimental data. The analysis of radial distribution functions reveals that gas molecules in poly(IL)s have strong interactions with the end groups of poly-cation chains, but with both cations and anions in ILs. This study provides an atomistic insight into experimental observation, in which the anions of a poly(IL)s do not play an important role in CO_2 sorption while the anions of ILs do.

(149d) Multifold CO2 Transport Property Enhancement of Interfacially Constrained Nanocomposite Membranes

Lakshmi Suhasini Kocherlakota[1], Tiep Hoang Pham[2] and René M. Overney[1], (1)Chemical Engineering, University of Washington, Seattle, WA, (2)Chemical Engineering, University of Washington, Seattle

Multifold CO_2 Transport Property Enhancement of Interfacially Constrained Nanocomposite Membranes

Lakshmi S. Kocherlakota, Tiep Hoang Pham, René M. Overney*

AIChE 2012-Abstract

Polymeric membranes are gaining wide attention in separation processes owing to their economic viability and operational flexibility. Poly(l-trimethylsilyl-1-propyne)(PTMSP), a high free volume glassy polymer, exhibits extraordinarily high gas permeability coefficients and high organic-vapor/permanent-gas selectivities. These unusual transport properties in PTMSP are attributed to its high fractional free volume. In this paper, we will address the impact on interfacial constraints on PTMSP's free volume and transport properties. In particular, we found that within the submicrometer regime towards silicon-oxide surfaces, PTMSP membranes reveal CO_2/helium selectivities that exceed bulk PTMSP transport properties by up to three-fold.[1] This finding was attributed to a further increase in free volume, as verified by molecular energetic mobility analysis[2] (intrinsic friction analysis, IFA) that showed an enhancement of the methyl side groups in ultrathin membranes compared to bulk membranes[1]. For scale-up purposes, "inverse" nanocomposites, made up of nanorods (100 to 200 nm diameter) of PTMSP within anodic alumina oxides sieve membranes were produced, yielding a two-fold increase of CO_2 permeability over virgin bulk membranes, and about a 20% improvement of the selectivity of CO_2extraction from helium and nitrogen. While PTMSP membrane are notorious for aging under ambient conditions, the embedded inverse nanocomposite systems show improved longevity.

[1] Kocherlakota LS, Daniel B. Knorr Jr. , Laura Foster, René M. Overney, Enhanced gas transport properties and molecular mobilities in nano-constrained poly[1-(trimethylsilyl)-1-propyne] membranes, Polymer 2012; doi:10.1016/j.polymer.2012.03.067

[2] Knorr DB, Kocherlakota LS, and Overney RM. Journal of Membrane Science 2010;346(2):302-309.

(149e) Nanocomposite Membrane of a Polymer of Intrinsic Microporosity and Zeolitic Imidazolate Frameworks for Gas Separation

Qilei Song[1], Shaheen A. Al-Muhtaseb[2] and Easan Sivaniah[1], (1)Cavendish Laboratory, Physics, University of Cambridge, Cambridge, United Kingdom, (2)Chemical Engineering, Qatar University, Doha, Qatar

Developing energy-efficient and environmentally-friendly separation processes has become an important research topic in dealing with global issues, such as carbon capture, natural gas production, and water purification. Membrane separation technology has great potential for solving these challenges as an alternative to conventional industrial processes. Nanocomposite membranes, also known as mixed matrix membranes, are promising materials for gas separation with potentially enhanced permeability and selectivity. In this study, we report the preparation and gas separation performance of nanocomposite membranes combined of zeolitic imidazolate frameworks (ZIF-8) nanocrystals in a polymer of intrinsic microporosity (PIMs).

The PIMs polymers present microporosity due to their unique contorted and rigid molecular structure, and have shown great potential for a wide range of applications, such as membrane for gas separation, gas storage, and catalysis. As membrane for gas separation, PIMs have shown high gas permeability but have selectivity that need to be enhanced to a satisfactory level for industrial applications. Zeolitic imidazolate frameworks (ZIFs), a sub-family of MOFs, have attracted significant attention with tuneable pore sizes, exceptional chemical stability, and versatile structures analogous to that of inorganic zeolites. The ZIFs nanoparticles with sub-nanometer channels could present synergetic effects as fillers for mixed matrix membranes: (1) disruption of the packing of polymer chains; and (2) molecular sieving effect. The PIMs/ZIFs composite membrane is a representative example of combination of *amorphous microporous* polymer and *crystalline microporous* MOFs. Therefore, understanding the physical and chemical properties of this composite, such as porosity, free volume, thermal stability, and gas sorption properties, is not only important for developing high performance composite membrane for gas separation, but also useful for many other applications, such as adsorbent and catalysis.

In this study, the as synthesised ZIF-8 nanoparticles, dispersed as colloids in solvent solution, were incorporated into the PIM-1 polymer matrix forming a nanocomposite membrane, which gives uniform dispersion of ZIF-8 in the polymer matrix. A series of composite membrane were fabricated from solution casting of the PIM-1/ZIF-8 solution with loading of ZIF-8 up to the significantly high values of ~40 wt%. The nanocomposite membranes were further treated with various techniques, such as methanol treatment, thermal treatment, and photochemical modification, aiming to improve the permeability or selectivity. The pure gas permeation properties were evaluated in a constant-volume variable-pressure apparatus (time lag method). Mixed gas permeation was tested in a similar membrane cell equipped with a gas chromatography.

For the nanocomposite membrane without methanol treatment, the gas permeability decreased with the loading of ZIF-8 nanoparticles, while the selectivity was maintained without significant loss. After methanol treatment, the permeability of nanocomposite membrane increased with the loading of ZIF-8 nanocrystals and surpassed the Robeson's upper bound. The effects of annealing temperature (100-300°C) on the gas permeation performance of pure polymer and nanocomposite membrane were monitored. For the pure polymer membranes annealed at high temperature (up to 300°C), the gas permeability slightly decreased while the selectivity was found to be stable. For the PIM-1/ZIF-8 nanocomposite membranes, the gas permeability decreased with the annealing temperature, interestingly, the selectivity could be significantly enhanced. Particularly, for the composite membrane with 20 wt% loading of ZIF-8 annealed at 300°C, the CO_2 permeability could reach approximately 1000 Barrer with selectivity of both CO_2/N_2 and CO_2/CH_4 as high as 30. The ideal selectivity of other gas pairs, such as O_2/N_2, H_2/N_2 and H_2/CH_4, were all enhanced surpassing the Robeson's upper bound.

Gas sorption properties of these nanocomposites were also examined with various types of gases, so their potential as adsorbents or gas storage materials were also evaluated.

Acknowledgements

We acknowledge the NPRP grant from the QNRF, the Engineering and Physical Sciences Research Council (EPSRC, UK), and the China Scholarship Council.

Notes: Corresponding author: es10009@cam.ac.uk (Dr. Easan Sivaniah)

References

1. N. B. McKeown and P. M. Budd, *Chem. Soc. Rev.*, 2006, **35**, 675-683.

2. T.-H. Bae, J. S. Lee, W. Qiu, W. J. Koros, C. W. Jones and S. Nair, *Angew. Chem., Int. Ed.*, 2010, 49, 9863-9866.

3. T. C. Merkel, B. D. Freeman, R. J. Spontak, Z. He, I. Pinnau, P. Meakin and A. J. Hill, *Science*, 2002, **296**, 519-522.

(149f) Steam Stability of Layered Zeolites Suitable for Hydrogen Transport

Bahman Elyassi, Xueyi Zhang and Michael Tsapatsis, Chemical Engineering and Materials Science, University of Minnesota, Minneapolis, MN

In this study, we demonstrate potential of zeolite nanosheets with 6-member ring (6MR) apertures for preparation of hydrogen permselective membranes. The preparation technique is based on exfoliation of layered zeolites into their constituent nanosheets, formation of closely packed monolayers on porous substrates, and then performing secondary growth. Steam stability of these zeolites in conditions relevant to steam methane reforming and water gas shift reactions over a period of about 3 months using X-ray diffraction (XRD), scanning electron microscopy (SEM), transmission electron microscopy (TEM), multinuclear solid state NMR, and Ar adsorption analyses was investigated. It was found that zeolites with lower concentrations of structural defects perform better in steam. In line with this observation, enhanced steam stability of the zeolites treated with silicon tetrachloride ($SiCl_4$) was seen. This work highlights the importance of zeolite synthesis or post-synthesis methods suitable to eliminate structural defects for applications where high temperature steam is present.

(149g) Physical Aging and Carbon Dioxide Plasticization of Thin Polyetherimide Films

Jianzhong Xia[1,2], Tai Shung Chung[2], Pei Li[2], Norman R. Horn[1] and Donald R. Paul[1,2], (1)Chemical Engineering, University of Texas at Austin, Austin, TX, (2)Chemical and Biomolecular Engineering, National University of Singapore, Singapore, Singapore

Industrial gas separation membranes have selective dense layers with thicknesses around 100 nm. It has long been assumed that these thin layers have the same properties as thick (bulk) films. However, recent research has shown that thin films with such thickness experience accelerated physical aging relative to bulk films and, thus, their permeation properties can differ significantly from the bulk. This paper explores the aging and carbon dioxide plasticization behavior of this new polyethermide and compares the results with those for PSF and Matrimid® 5218 from the previous work. Particular attention is devoted to the permeation response of thin films for a variety of CO_2 conditions due to the current strong interest in CO_2 related separations, such as the removal of CO_2 from natural gas and flue gases. The CO_2 plasticization of Extem thin films is proven to be a function of film thickness, aging time, exposure time, pressure and prior history.

(149h) The Design and Modeling of a Controlled Atmosphere Food Storage System Based On Membrane Technology

Hang Yin, Chemical and Biomolecular Engineering, National University of Singapore, singapore, Singapore, Na Peng, Chemical and Biomolecular Engineering, National University of Singapore, Singapore, Singapore, Kok Lin Chong, National University of Singapore, singapore, Singapore, Glenn Lipscomb, Department of Chemical and Environmental Engineering, University of Toledo, Toledo, OH and Tai-Shung Chung, Department of Chemical & Biomolecular Engineering, National University of Singapore, Singapore, Singapore

Fruits and vegetables continue to respire after harvesting. In addition to refrigeration, controlled atmosphere storage technology can prolong the shelf-life of fruits and vegetables by creating a specific atmosphere. In this study, an innovative storage system based on green membrane technology is evaluated for food preservation. This is a novel study of hollow fiber membrane modules in a controlled atmosphere storage (CAS) system to maintain an environment that prolongs the lifespan of fruits and vegetables during transportation and storage. Compared to conventional CAS systems that utilize nitrogen flushing, the proposed membrane controlled atmosphere storage (MCAS) system provides a nitrogen-enriched air stream to control the concentration of both oxygen and carbon dioxide in the storage container. Detailed mathematical models to describe mixed gas permeation in a hollow fiber membrane module and guide system design are presented. The performance of MCAS systems under on-off and continuous operation modes is evaluated and a case study of the newly developed MCAS system in food transportation presented.

(223a) Effect of Confinement On Ionic Liquid Properties for Carbon Capture Processes

Ruth E. Baltus, Chemical and Biomolecular Engineering, Clarkson University, Potsdam, NY and Laila Banu, Chemical & Biomolecular Engineering, Clarkson University, Potsdam, NY

Work in our laboratory has focused for a number of years on examining the potential of room temperature ionic liquids for post combustion carbon capture processes. Experimental work has involved measurements of carbon dioxide solubility and diffusivity in different ionic liquids as well as on gas permeation measurements through supported ionic liquid membranes (SILMs) prepared using anodic alumina films as the support membrane.

Results from these studies have raised questions about the impact of a solid interface on the properties of ionic liquids. Gas permeation through SILM systems with different pore size indicates both carbon dioxide and nitrogen permeation is larger when the ionic liquid is confined within a 20 nm pore size membrane when compared to a 100 nm pore size membrane. Ideal selectivity values are generally the same in the two membrane systems. A comparison of measured membrane permeation to values predicted from bulk solubility and diffusivity values shows higher carbon dioxide permeance in the membranes compared to values predicted from bulk properties of the ionic liquids.

Gas uptake into confined ionic liquids have also been experimentally measured, with solubility and diffusivity of carbon dioxide in the confined ionic liquids compared to bulk phase characteristics. To date, results have been mixed with results from some ionic liquids showing no impact of the solid interface whereas others indicate a significant difference in ionic liquid properties when confined. In this presentation. Results from these studies will be presented and discussed.

(223b) Metallic Based Aqueous Amine Membranes for the Separation of Carbon Dioxide From Fossil Fuel Flue Gas

Brian L. Bischoff and Lawrence E. Powell, Oak Ridge National Laboratory, Oak Ridge, TN

According to the US Department of Energy, there is scientific evidence that CO_2 emissions from fossil fuels have caused climate change and that a reduction of these emissions is essential for reducing the risk of future devastating effects. At present, the leading technology for the post combustion capture of CO_2 relies on the separation of CO_2 from the flue gas stream by employing chemical absorption with an aqueous monoethanolamine (MEA) solution. The separation is followed by compression of the gas to a liquid or supercritical state for transportation, and ultimately long term storage in geological formations. However, there is little experience with large scale amine separation systems especially in oxidizing environments typical of a fossil fuel flue gas stream. Furthermore, there are concerns about the degradation rate and evaporative losses of the amine solution and the amount of energy required for regeneration.

We propose using metallic supported liquid amine membranes to overcome the shortcomings of the large energy intensive amine process. Because the amine solution is immobilized in the pores of a thin nanoporous layer there is a lower vapor pressure thereby reducing the evaporation rate and the thin layer requires a small fraction of the amine compared to the absorption method. The membrane can be operated at a constant temperature, lower than the absorption process regeneration temperature, which reduces the degradation rate of the amine. We prepared a series of robust membranes based on a stainless steel support structure and a very thin nanoporous aluminum oxide layer that holds the aqueous amine solution. We will present laboratory results showing flux data and selectivity data for a range of amines under varying temperatures. We will also present stability data which demonstrates a strong promise for long-term application.

(223c) A Comprehensive Multi-Step Description of the Permeation of Hydrogen in a Dense Metal Membrane Over a Broad Temperature Range

Nicholas Deveau[1], **Yi Hua Ma**[2] and Ravindra Datta[1], (1)Chemical Engineering, Worcester Polytechnic Institute, Worcester, MA, (2)Worcester Polytechnique Institute, Worcester, MA

Development of cheaper and more durable dense metal membranes are being sought for generation of pure hydrogen suitable for fuel cells from coal. Toward this end, we have developed a comprehensive multi-step model for describing permeation of hydrogen in such membranes. The various steps include: s_1) dissociative adsorption on the surface, followed by: s_2) solution and then: s_3) diffusion of hydrogen atoms within the bulk metal matrix, and finally: s_4) evolution from the bulk to the surface, and: s_5) associative desorption on the permeate side.

The efficacy of the model is ascertained by first applying it to Pd and Pd-Cu membranes, for which much literature exists. Of course, for these membranes, typically Sieverts' law is assumed, which is limited to high temperatures (> 300 °C) and low hydrogen pressures, when diffusion (step, s_3) is controlling, and hydrogen concentration in bulk metal is low. Over a broader range of temperatures spanning the crystallographic phase-transition in the Pd-H and Pd-Cu-H systems, in fact, the slopes are very different at low and high temperatures, and also there is a pronounced peak in permeability in the intermediate temperature range. The latter behavior has not yet been theoretically explained.

Our approach considers the 5-step network in analogy to an electrical circuit and describes well the permeation behavior of hydrogen in Pd and Pd-Cu membranes, including the peak at intermediate temperatures. Further, it can adequately describe the absorption isotherms of the metal-hydrogen system. A comparison of step resistances also allows a systematic reduction to a three-step (adsorption, diffusion, and desorption control) model that is adequate over a broad range of temperatures and pressures.

(223d) High Temperature Metal Membranes for Hydrogen Permeation without Platinum Group Metals

J. Douglas Way, Colin A. Wolden, Kehinde Adeyemo and Mayur Ostwal, Chemical and Biological Engineering, Colorado School of Mines, Golden, CO

Group V metal membranes (tantalum, vanadium, niobium and their alloys) have extremely high theoretical permeability for dissociated hydrogen[1], but are unable to self-catalyze the hydrogen dissociation reaction. Such membranes need to be coated with an active catalyst layer, typically palladium or a Pd-alloy, increasing the materials cost of the membrane. Additionally, the use of a metallic surface catalyst layer limits the upper operating temperature of the membrane, as temperatures higher than 623 K tend to promote alloying, which causes loss of flux. As group V metals are also prone to low-temperature hydrogen embrittlement, this limits their applicability.

Transition metal carbides, nitrides, and sulfides have been studied as substitutes for platinum group catalysts in applications requiring hydrogen dissociation, such as the water-gas shift reaction. The goal of this study was to determine if these lower cost materials could be used as catalyst layers for group V metal membranes.

Molybdenum carbide (Mo_2C) was chosen as the test transition metal catalyst, and vanadium as the group V bulk metal. Magnetron sputtering was used to apply layers of Mo_2C to both sides of commercial cold-rolled vanadium foils with thicknesses ranging from 50 to 125 microns. Membranes sputtered at temperatures of 473 K or below had a cubic structure, and were the most permeable to hydrogen, while higher-temperature (873 K) sputtering produced an orthorhombic lattice[2].

These variations in catalyst structure produced corresponding variations in performance. All membranes were single-gas tested in hydrogen and helium at temperatures from 823 to 1073 K and feed gas pressures up to 690 kPa gauge (100 psig). No membranes had any detectable helium leakage through defects, and membranes without Mo_2C layers had no hydrogen flux, demonstrating that all hydrogen dissociation activity was caused by the catalyst.

At temperatures below 1023 K, the structure and thickness of the carbide catalyst layer had the most influence on the hydrogen permeability. As temperature increased, the permeability reached a maximum at 1023 K, and then decreased as predicted by Steward[1] due to decreasing hydrogen solubility in the vanadium foil. The highest H_2 permeability was 5.9×10^{-8} $mol.m/m^2.s.Pa^{0.5}$ at 1023 K, about 70% higher than pure Pd. We believe these are the first measurements of hydrogen permeability for vanadium at high temperature. The hydrogen flux for this 50 micron thick membrane at a differential feed pressure of 690 kPa was 0.7 $mol/m^2.s$. Membranes display long term stability (> 150 hrs) at high temperature with no loss of flux and TEM analysis after testing confirms that the Mo_2C did not reduce or otherwise alloy with the vanadium foil. Effects of process parameters such as V foil thickness, catalyst layer thickness, sputter conditions, and permeation temperature will be discussed and compared to Steward's predictions[1].

1. Steward, S.A., *Review of Hydrogen Isotope Permeability Through Materials*, 1983, Lawrence Livermore National Laboratory.

2. Gade, S.K., S.J. Chmelka, S. Parks, J.D. Way, and C.A. Wolden, *Dense Carbide/Metal Composite Membranes for Hydrogen Separations Without Platinum Group Metals.* Advanced Materials, 2011. **23**(31): p. 3585–3589.

(223e) Long Term H2 Permeance and Selectivity Stability of Pd and Pd-Au Large Scale Membranes in Actual Syngas Atmospheres

Federico Guazzone[1], Jacopo Catalano[2], Ivan Mardilovich[2], Nikolas Kazantzis[3] and Yi Hua Ma[4], (1)Chemical Engineering, Worcester Polytechnic Institute, Worcester, MA, (2)Worcester Polytechnic Institute, (3)Chemical Engineering, WPI, Worcester, MA, (4)Worcester Polytechnique Institute, Worcester, MA

The integration of composite Pd based membrane reactors (WGS or MSR reactions) into large scale H_2 producing units entails valuable cost savings for pre-combustion CO_2 sequestration. In addition, it was recently demonstrated that composite Pd and Pd-Au membranes were stable for more than 250 h in actual syngas atmosphere (450°C, 13 bara) and that H_2 permeate purity was higher than 99.9%. The Pd-Au membranes tested earlier were prepared by the deposition of a thin (0.15-0.2 μm) layer of Au on top of the dense Pd layer. The membranes showed some initial flux decline attributed to poisoning of the layer by unknown contaminants and a maximum recoverability of 77% after 300 h in pure H_2 at 450°C. The objective of the present work was to further investigate the stability of permeance and purity and the of Pd/Au membranes with a higher Au content on the Pd surface of 0.5 μm thickness.

Composite Pd and Pd/Au porous stainless steel membranes having an area of more than 200 cm^2 (2.54 cm OD, 25.4 cm L) were prepared by WPI's patented technology and tested at the National Carbon Capture Center (NCCC) in Wilsonville, Alabama. The Syngas composition was slightly different from the previous study since in this second run, the syngas was shifted before entering the membrane module. Hence, CO concentrations were lower than 1% and H_2O/CO ratio was higher than 1. The operation conditions and H_2S concentration levels were kept the same as in the previous study. A Pd-0.5 μm/Au membrane showed a remarkably stable hydrogen flux for over 475 h albeit an initial hydrogen flux decline of ~40% after 20 hours of exposure to the syngas stream. The hydrogen purity, as high as 99.9%, was also stable for over 475 hours. Upon test at NCCC, the permeance of the Pd-0.5 μm/Au membrane was totally recovered after 160 h at 450°C in a pure H_2 atmosphere. Hence, a thicker Au layer enhanced drastically the recoverability of membranes exposed to actual syngas atmospheres which represents a breakthrough accomplishment in the development of composite Pd membranes for WGS-CMR applications.

In an attempt to understand the causes of the initial permeance flux decline, a preliminary XPS study of the surface of Pd coupons exposed to actual syngas was performed and showed the presence of carbon contamination in the Pd bulk. Further analysis would be needed to confirm this finding.

(223f) Composite Catalytic Membranes: Enhancing H2 Recovery Membranes by Appropriate Coupling with Water-Gas-Shift Reaction

Elva Romero, Daejin Kim and **Benjamin Wilhite**, Artie McFerrin Department of Chemical Engineering, Texas A&M University, College Station, TX

Membranes for gas purification are a promising alternative to existing cryogenic, absorbtion and pressure-swing adsorption systems. However, permselective membrane materials are often limited by surface inhibition (e.g., CO inhibition of palladium surfaces) or low gas permselectivities (~50:1 for most polymeric membranes). To address these limitations, our research group has developed a composite catalytic-permselective (CCP) membrane design approach, in which a permselective film (e.g., polymeric, metallic) is integrated with a porous catalytic film such that reaction and diffusion within the catalytic layer directly modifies the gas composition at the surface of the underlying permselective barrier [1]. This modification of gas composition has been predicted to enhance permeation rates, membrane permselectivities and reduce surface corrosion/inhibition.

In this talk, we will review general design rules for implementing this strategy, and demonstrate their extension to the specific application of hydrogen purification from diesel reformate mixtures using water-gas-shift as the enhancing catalytic reaction. Detailed two-dimensional simulations will be presented comparing the resulting CCP design against reference cases of (i) an untreated membrane, and (ii) an untreated membrane employed in a packed-bed membrane reactor configuration. Experimental results for the case of an electroless-plated 30-micron Pd film coupled with water-gas-shift will be presented for comparison with model predictions. Lastly, a discussion of how this approach may be extended to polymeric systems in order to enhance permselectivity will be presented.

[1] B.A. Wilhite. Composite Catalytic-Permselective Membranes: A Novel Strategy for Enhancing Selectivity and Permeation Rates via Reaction and Diffusion. Ind. Eng. Chem. Res., 2011, 50(17), 10185-10193.

(248a) Toward a Common Method of Cost Estimation for CO2 Capture and Storage At Fossil Fuel Power Plants

Edward S. Rubin, Engineering and Public Policy, Carnegie Mellon University, Pittsburgh, PA

There are significant differences in the methods employed by various organizations to estimate the cost of carbon capture and storage (CCS) systems for fossil fuel power plants. Such differences often are not readily apparent in publicly reported CCS cost estimates. As a consequence, there is a significant degree of misunderstanding, confusion, and mis-representation of CCS cost information, especially among audiences not familiar with the details of CCS costing. Given the international importance of CCS as an option for climate change mitigation, efforts to improve and systematize the methodology and communication of CCS costs are especially urgent and timely. This paper recommends a path forward to achieve that goal based on the recommendations of an international Task Force on CCS Costing Methods, which will be discussed in the proposed presentation. They include recommendations for a standardized nomenclature for cost estimates plus guidelines for reporting CCUS cost-related information in technical reports, journal or conference papers, and oral presentations.

(248b) Proposed Baselines and Standard Methods for Systems Analyses of Power Generation Technologies

Michael Matuszewski, Performance Division - Office of Program Planning & Analysis, National Energy Technology Laboratory, Pittsburgh, PA

The Department of Energy's National Energy Technology Laboratory (NETL) routinely conducts analyses of power generation systems in order to guide responsible expenditure of public funding used for the development and demonstration of advanced power generation technologies, especially in the field of carbon capture, utilization and storage. Many of these NETL analyses ultimately need to be compared to one another or to third-party studies to exercise due diligence in confirming and comparing results. As a result, NETL has found great value in standardizing methodologies for capital cost estimation, operating cost estimation, calculation of required product pricing as well as methodologies for calculating the performance of power and fuel generation systems.

This presentation will cover the details of standard methods NETL uses for: calculation of required electricity selling price, computing Nth-of-a-kind plant costs based on learning curve assumptions, coal & feedstock selection and composition, and recommended specifications for sequestered CO2.

In addition to the standard methodologies discussed above, there are many more assumptions that must be made when simulating a "typical" power generation system. Often times, these assumptions are made in the absence of any specific requirements and are made at the sole discretion of the engineering team performing the analysis. Due to the many choices that can be made, rarely do the assumptions made by engineers performing systems analyses of these facilities converge. This makes comparison difficult and requires a great deal of reconciliation amongst other public studies in order to provide valuable conclusions by way of comparison. Therefore, this presentation will also cover the standardized representations of baseline power generation facilities, covering coal feed to busbar, to which the effects of implementing advanced power generation technologies and CO2 capture can be compared. Overviews of NETL's baseline Pulverized Coal, Integrated Gasification Combined Cycle, and Natural Gas Combined Cycle power plants will be presented in this context.

(248c) Methodology for Evaluating the Cost of Emerging CCS Technologies

Abhoyjit S. Bhown[1], George Booras[1] and Ronald Schoff[2], (1)Electric Power Research Institute, Palo Alto, CA, (2)Electric Power Research Institute, Charlotte, NC

The potential for carbon capture, utilization, and storage to help mitigate the effects of carbon emissions from power plants on the global climate has resulted in a boom in technology research and development in the area. Stakeholders outside of the R&D community include the financial, government, environmental, and electric utility sectors. Due to the exponential increase in the development of carbon capture technologies over the last decade, it has become increasingly important to differentiate the technologies. One metric that is commonly used is the cost of those technologies. In this paper, we present an approach to using different cost estimating methods and analysis based on a metric of Technology Readiness Level (TRL). TRL offers a convenient and consistent way to rank a technology's stage of development from initial concept to commercial launch. In some cases, for early-stage technologies with low TRL, a rigorous cost estimating methodology may not be as meaningful or useful as other metrics such as energy consumption, but it does provide insight into key challenges and helps focus research goals. As a technology progresses in its TRL ranking, increasingly stringent and increasingly detailed estimating methods should used to predict a process's economic viability, though the preparation of detailed capital cost estimates becomes increasingly more expensive. Ultimate deployment of a new CCS technology will depend heavily on its economics, and bracketing its cost will continue to be an important factor in assessing its commercial viability.

(248d) A Common Modeling Framework to Evaluate and Compare the Economics of Ccus Technologies

Edward S. Rubin, Engineering and Public Policy, Carnegie Mellon University, Pittsburgh, PA

The current status of performance and cost estimates for CCUS in power plant applications will be reviewed in this presentation, along with a discussion of the factors that most strongly influence these costs. Because of the large number of variables involved in assessing the economics of CCS options for particular applications—or in assessing the opportunities and risks of R&D investments in new technology—tools are needed that can help managers, researchers, analysts, and technology developers systematically evaluate alternative options in the context of their situation. This presentation will describe and illustrate the use of such a tool developed by Carnegie Mellon University for the U.S. Department of Energy's National Energy Technology Laboratory (DOE/NETL). The IECM (for Integrated Environmental Control Model) is a publicly available model now used by roughly 2000 people at over 800 organizations in more than 50 countries worldwide to evaluate alternative CCUS options for PC, IGCC and NGCC power plants. Examples of how this modeling framework can be employed to systematically (and transparently) evaluate and compare CCUS options for a variety of situations (including new plant vs. retrofit applications, first-of-a-kind vs. Nth-of-a-kind costs, etc.) will be discussed and illustrated. The limitations of a standardized approach also will be discussed.

(248e) NETL CO2 Injection and Storage Cost Model

Tim Grant[1], Dave Morgan[2], Jason Valenstein[3], Andrea Poe[3], Marta Milan[3] and Richard Lawrence[4],
(1)DOE/National Energy Technology Laboratory, Pittsurgh, PA, (2)DOE/National Energy Technology Laboratory, (3)Booz Allen Hamilton, Inc., USA, (4)Advanced Resources International, Inc., USA

The U.S. Department of Energy's National Energy Technology Laboratory (NETL) has developed a model to estimate the costs of storing captured CO_2. This model includes the range of costs from initial regional geologic evaluation through site characterization, permitting, injection/MVA operations, post-injection site care to final site closure and transfer to long-term stewardship. Differences in storage costs across different geologic formations are driven by two basic factors: (1) injectivity which determines the number of injection wells drilled to accommodate a given rate of CO_2 injection, and (2) the volume of CO_2 to be stored which determines, per in-situ reservoir parameters, the areal extent of the plume and hence the Area of Review (AoR) of a Class VI well permit. The AoR defines the areal extent of MVA activities which dominates costs during injection and post-injection operations. The basic framework for this model provides costs for compliance with various sections of EPA's Class VI regulation and Subpart RR of the GHG Reporting Program. Cost analysis at two levels is provided by this model: (1) site specific where the modeler can enter their own reservoir data, and (2) regional costs derived from the model's geologic database and presented in the form of cost supply curve output data. Within the financial module, the model will solve for the first year breakeven cost to store a tonne of CO_2. Another potential CO_2 storge reservoir incorporated in this model are oil and gas reservoirs utilized in Enhanced Oil Recovery (EOR) operations. EOR reservoirs are important for their utilization of captured CO_2 for additional recovery of original oil in place (OOIP). A test matrix was developed to evaluate this cost model across several simple scenario variations to test for cost sensitivities. For saline reservoirs, scenario variations include: change in time span for site characterization, permitting and post-injection site care; change in application of seismic data acquisition and density of monitoring wells, as well as a change in the number of old wells that require remediation (corrective action). Important cost considerations and modeling scenarios for EOR are the cost of compliance with Subpart RR, conversion to Class VI permits, or concurrent saline storage operations. Both saline and EOR operations will be modeled for change in Financial Responsibility and saline operations will be modeled for change in capital structure. Transportation is the link between capture and sequestration. The test matrix will utilize NETL's Capture-Transport-Storage (CTS) model to analyze the use of dedicated and network pipelines in various intra and inter-regional CCS scenarios. An important goal here is to assess the cost to store captured CO_2 in various basins or regions across the lower 48 states over the next century.

(248f) A Monte Carlo Approach for Estimating and Validating Contingencies for CCS

Paul Worhach, Nexant, Inc., San Francisco, CA, Mark Woods, NETL, Booze Allen, Pittsburgh, PA and John Wimer, Department of Energy, National Energy Technology Laboratory, Morgantown, WV

The AACE International recommended practice for estimating process and project contingencies provides a set of contingency factors to be applied to total process capital cost, as a function of process technology development and class of estimate. Process contingency ranges from 0 to 10 percent for commercial technologies to 40 percent plus for new concepts with limited data, and project contingency ranges from 15 to 30 percent, with a recommended factor of 25 percent.

Monte Carlo simulation can be a useful technique to validate and to quantify key sensitivities in the application of these factors. A Monte Carlo simulation model was developed based upon a cost estimation case for an IGCC plant from the NETL draft report "Supplemental Analysis for Baseline Studies – GEE IGCC and SP PC Zero Liquid Discharge Plants." Probability ranges of P10/P90 and P1/P99 were developed for project account costs, and a correlation factor matrix was constructed based upon a set of key risk drivers. Triangular, Double Triangular, and PERT distributions were specified for the cost ranges. The simulation calculates the total contingency required to cover 95% of the simulated cost range, and additional cases were run to compare the alternative distributions and probability ranges, confidence targets, and the use of and sensitivity to the correlation matrix factors.

While the Monte Carlo simulation approach for contingency estimation has limitations, it can be a useful technique to validate factor-based methodologies by building up cost account-based probability ranges, specifying correlations, and quantifying sensitivities. The simulations can be used as an indication of where the factor-based contingencies might be adjusted to better estimate total project costs.

(290a) CO2 Mineralisation Through Hydrometallurgical Leaching of Brown Coal Fly Ash

Lian Zhang and Teck Kwang Choo, Chemical Engineering, Monash University, Melbourne, Australia

The use of an industry waste, brown coal fly ash collected from the Latrobe Valley, Victoria, Australia, has been tested for the post-combustion CO_2 capture through indirect minersalisation in acetic acid solution. After the optimum leaching, the majority of calcium and magnesium in fly ash were dissolved into the solution, the carbonation potential of which have been investigated intensively through the use of a continuously stirred high-pressure autoclave reactor and an advanced analytical protocol for the carbonation precipitate chracterisation. A large CO_2 capture capacity for fly ash under mild conditions has been confirmed. The CO_2 was fixed in both carbonate precipitates and water-soluble bicarbonate, and the conversion between these two species were achievable at approximately 60°C and a CO_2 partial pressure above 3 bar. The kinetic analysis confirmed a fast reaction rate for the carbonation of the brown coal ash-driven acetic acid leachate at a global activation energy of 12.7 kJ/mol. It is much lower than that for natural minerals and is also very close to the potassium carbonate/Piperzine system. The CO_2 capture capacity of this system was also proven to reach maximum 264 kg CO_2/t which is comparable to the natural minerals tested in the literature. As the fly ash is a valueless waste and requires no comminution prior to use, the technology developed here is highly efficient and cost-effective, the resulting carbonate products of which are also invaluable for the use as additive to cement and in the paper and pulp industry.

(290b) Statistical Evaluation of Mineral Carbonation of a Fly Ash/Brine System

Grace Muriithi, Chemistry, University of the Western Cape, Cape Town, South Africa

Muriithi N. Grace[1][1] Petrik F. Leslie[1], Gitari M. Wilson[2] and Doucet J. Frédéric [3],

[1]Environmental Nano Sciences Group, Chemistry Department, University of the Western Cape, South Africa.

[2]Department of Ecology and Resource Management, School of Environmental Studies, University of Venda, South Africa.

[3]Industrial Mineralogy Laboratory, Council for Geoscience, Pretoria, South Africa.

Statistical evaluation of mineral carbonation of a fly ash/brine system

Abstract

As countries continue on their development path, coal is bound to play a vital role as the main source of energy especially in BRICS (Brazil, Russia, India, China and South Africa) where coal is available in considerable quantities. International regulations, specifically the 2nd Kyoto agreement reached in Durban, South Africa (28th November to 9th December 2011) binds developing countries to reduce their carbon emissions even as they industrialize. Finding alternatives to reduce CO_2 emissions is going to be paramount to achieve the targets set. The effect of four parameters on the mineralization potential of class F fly ash and a $NaSO_4$ brine type was evaluated in a statistically designed protocol. Main effects of temperature, particle size and solid to liquid (S/L) ratio were the most significant while the interactions of temperature and particle size as well as temperature and S/L ratio were also significant. Between 2.75 % to 6.5 % conversion of the CaO in fly ash into $CaCO_3$ was achieved depending on the factor combinations. Mineral carbonation at 90 C and 4 Mpa using bulk ash and S/L ratio of 1 gave the highest degree of carbonation of 6.5 %. Carbonation conducted at 30 C and 1 Mpa using the >150 µm particle size and S/L ratio of 0.1 gave the lowest degree of carbonation of 2.75 %. Application of brine instead of ultra-pure water resulted in higher carbonation efficiency due to the utilization of the additional Ca^{2+} and Mg^{2+} present in brine. Moreover the carbonation effluents after the experiments are cleaner with regards to most elements (Pb, Zn, Cu, NO_3, Na, SO_4, As, Se, K, Al, Mg, Ca, Si, Sr and Ni) while Mo, B, V and Cl increased with carbonation. McNemar's test (p-value of 0.01) of the factor combinations reveals that carbonation carried out at 90 C and 4 Mpa using bulk ash and S/L ratio of 1 has the best factor combinations as the highest removal of contaminants was achieved (≥70 %).

---

[1]Corresponding author; Muriithi N. Grace; Email: gmuriithi@uwc.ac.za; Tel: +27219593878;

Fax: +27219593838

Ex situ CO_2 Mineralization Using Reactive Species

Juan Ma, Roe-Hoan Yoon
Department of Mining and Minerals Engineering
Virginia Tech, Blacksburg, VA, USA

Abstract

While geologic sequestration method may be used to store large volumes of CO_2, it involves two distinct steps, *i.e.*, capture and storage, with the former being a more costly step. Also, finding suitable storage sites, pumping supercritical CO_2, safekeeping it underground, and handling of saline water discharges could impede its deployment. CO_2 mineralization, on the other hand, can be a single-step process and the product can be stored permanently. In some cases, the product can be utilized as construction materials which will defray the cost for CO_2 mineralization.

The idea of CO_2 mineralization was first proposed by Seifritz in 1990[1]. After that, many researchers investigated ex situ CO_2 mineralization by reaction of CO_2 with natural silicate minerals such as serpentine, olivine and wollastonite, forming geologically stable carbonate minerals. Lackner et al. were the first to conduct significant studies. They firstly studied direct dry carbonation and found that the reaction was too slow[2, 3]. Then they investigated a well-known acid extraction route, where magnesium hydroxide ($Mg(OH)_2$) was extracted from silicate minerals using hydrochloric acid (HCl) The energy cost in the dehydration step needed for HCl recovery was too high [4-6]. Researchers at Albany Research Center (ARC) studied an alternative process[7-11], namely direct aqueous mineral carbonation, where gaseous CO_2 directly reacts with fine particle-sized minerals in suspension in an autoclave. This process was uneconomical due to the substantial energy required for the activation of silicate minerals and the large scale of associated mining operations to produce the needed ore.

Twenty years of research and development (R&D) work addressing CO_2 mineralization using silicate minerals has failed to produce an economical technology that could be applied on a large-scale. The main limitation was the high energy cost required for the extraction or activation of silicate minerals. To get around this problem, reactive species such as Mg^{2+}, Ca^{2+} and Fe^{2+} ions could be used in place of silicate minerals to react with CO_2, which is the subject of the present work. The precipitation of carbonates from reactive cations offers much faster kinetics than reacting with silica minerals where cations are bound in a silicate structure. The present work focused on using Mg^{2+} ions because Mg-carbonate has a higher proportion of CO_2 than Ca- or Fe- carbonate due to the lower molecular weight of Mg and seawater contains a much higher concentration of Mg^{2+} ion than Ca^{2+} and Fe^{2+} ions.

In the present work, we have explored the possibility of reacting CO_2 with reactive species that are already present in waters such as produced water from oil and gas industries and sea water. Many of the species present in these waters (*e.g.*, Ca^{2+} and Mg^{2+} ions) are reactive without or with minimal activation energies, thereby reducing both capital and operating costs. Recognizing that sea water represents the single largest reservoir of the reactive species, we have conducted mineralization experiments using 1,400 mg/L of Mg^{2+} and 400 mg/L of Ca^{2+} ions which represent average concentrations in sea water. The amounts of these species present in sea water alone are such that we can use these two reactive species for the next 100,000 years at the present level of global CO_2 emission (35 Gt/year). Additionally, approximately 70 billion barrels of produced water are currently generated every year all over the world[12], of which the U.S. contributes 20 to 30 billion barrels[13]. About 35% of this water is treated and discharged into surface water bodies—which could be costly. Some of the discharged produced waters have high levels of Mg^{2+} ions, e.g., brine sample from Oriskany, Pennsylvania contains 2490 mg/L Mg^{2+} ions[14] and that from Guernsey, Ohio contains 3440 mg/L Mg^{2+} ions[15]. The process of CO_2 mineralization using reactive species can be used to clean-up the produced water while achieving both capture and storage of CO_2, all simultaneously.

It was found that Ca^{2+} ions react more readily than Mg^{2+} ions. A series of mineralization experiments were conducted at temperatures in the range of 10 to 40°C and the product was characterized to be nesquehonite ($MgCO_3 \cdot 3H_2O$). During the precipitation process, the Mg^{2+} ion concentration decreased and solution pH increased as shown in figure 1. The kinetics of mineralization and the %Mg^{2+} ion utilization varied depending on energy dissipation rate, temperature, pH, and NaCl concentration. The maximum Mg^{2+} ion utilization we achieved was 86% as shown in figure 1.The activation energy for the formation of nesquehonite was 64.6 kJ/mol, which was determined by using the rate constants at different temperatures shown in figure 2 and Arrhenius plot shown in figure 3. It was found that the activation energy barrier can be readily overcome by simple agitation and heating at slightly elevated temperatures, *e.g.*, 40°C. We also conducted thermodynamic calculations to construct the species distribution diagrams, predict the pH of CO_2 mineralization, and to predict %Mg ion utilization (or extraction) from sea water. The CO_2 mineralization method developed in the present work may lead to a process of extracting magnesium, which is listed as a "near-critical" material in the 2011 DOE Critical Materials Strategy Report.

One of the concerns in CO_2 mineralization with reactive species may be pH lowering. We have addressed this issue by contacting spent solutions with basic minerals such as limestone and olivine. It was found that in the presence of these minerals the pH rises to the pH of minimum solubility of the buffering mineral. The pH of minimum solubility of limestone is 8.3 and that of olivine is 8.6. Other means of pH neutralization will be discussed.

Key words: CO_2 sequestration, mineralization, Mg^{2+} ions, kinetics, temperature, pH

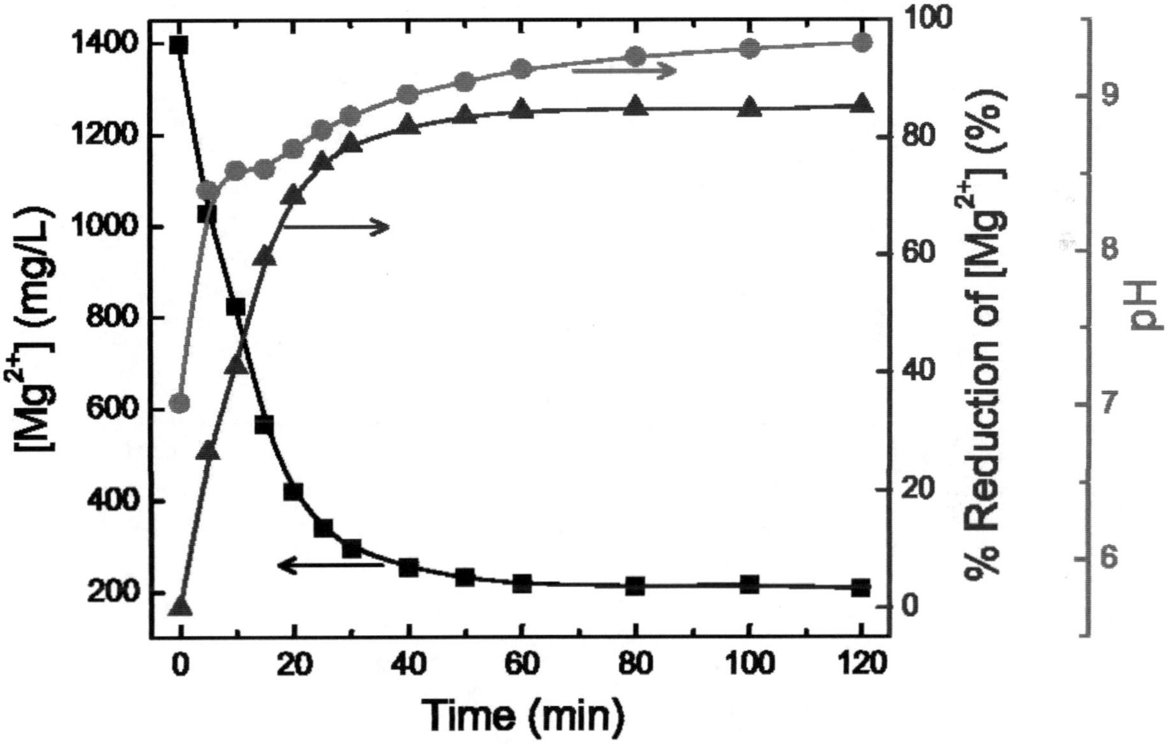

Figure 1 Changes in Mg^{2+} ion concentration ([Mg^{2+}]), reduction of [Mg^{2+}] and pH of the solution during the formation of nesquehonite.

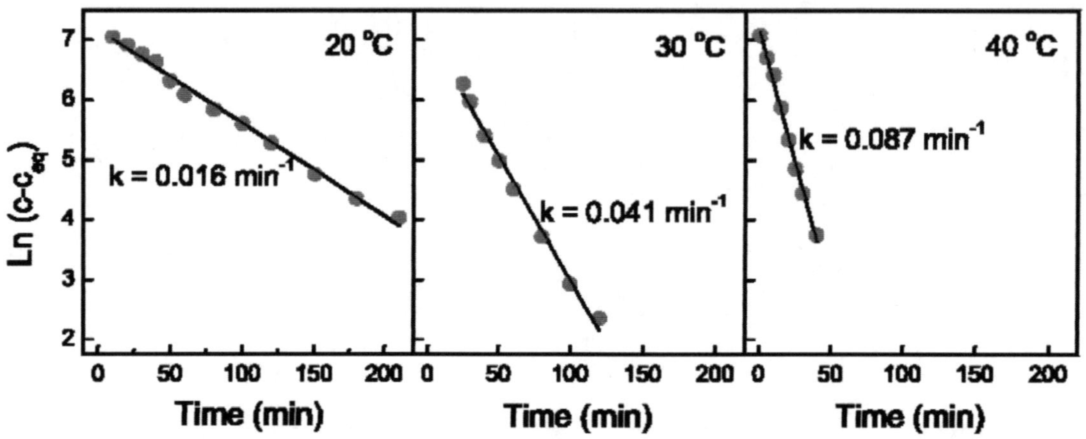

Figure 2 The first-order kinetic plots for nesquehonite precipitation at 20, 30, 40 °C.

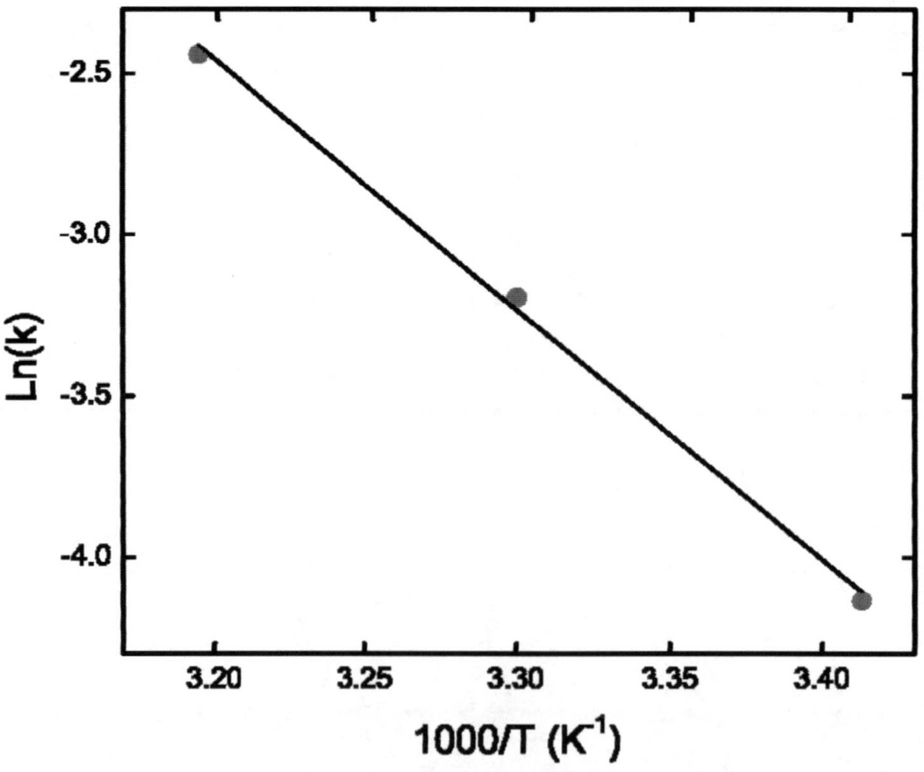

Figure 3 Arrhenius plot for nesquehonite precipitation.

References

1. Seifritz, W., "*CO2 disposal by means of silicates*". Nature, 1990. **345**.
2. Lackner, K.S., et al., "*Carbon dioxide disposal in carbonate minerals*". Energy, 1995. **20**(11): p. 1153-1170.
3. Lackner, K.S., D.P. Butt, and C.H. Wendt, "*Progress on binding CO2 in mineral substrates*". Energy Conversion and Management, 1997. **38**: p. S259-S264.
4. Wendt, C., et al., "*Thermodynamic Calculations for Acid Decomposition of Serpentine and Olivine in MgCl2 Melts, III. Heat Consumption in Process Design*", 1998, Tech. Report No. LA-UR-98.
5. Nilson, D. and G. Hundley, "*Preliminary feasibility study of the sequestration of carbon dioxide gas with minerals: a study of the LANL aqueous process*". Albany Research Center, US DoE, 1999.
6. Newall, P., et al., "*CO2 storage as carbonate minerals*"1999: CSMA Consultants Ltd.
7. O'Connor, W., et al., "*Carbon dioxide sequestration by ex-situ mineral carbonation*", 2000, Albany Research Center (ARC), Albany, OR.
8. O'Connor, W., et al. "*Continuing studies on direct aqueous mineral carbonation for CO2 sequestration*". 2002.
9. O'Connor, W.K., et al., "*Carbon dioxide sequestration by direct mineral carbonation: results from recent studies and current status*", 2001, Albany Research Center (ARC), Albany, OR.
10. Gerdemann, S.J., et al., "*Ex situ aqueous mineral carbonation*". Environmental science & technology, 2007. **41**(7): p. 2587-2593.
11. O'Connor, W., et al., "*Aqueous mineral carbonation*". Final Report, DOE/ARC-TR-04, 2005. **2**.
12. Veil, J. and M. Puder, "*Regulatory considerations in the management of produced water—A US perspective*". Produced Water Management-Gas TIPS, 2005: p. 25-28.
13. Kharaka, Y., et al. "*Can produced water be Reclaimed? Experience with placerita oil field, California*", 1998.
14. Soong, Y., et al., "*CO_2 sequestration with brine solution and fly ashes*". Energy Conversion and Management, 2006. **47**(13): p. 1676-1685.
15. Druckenmiller, M.L. and M.M. Maroto-Valer, "*Carbon sequestration using brine of adjusted pH to form mineral carbonates*". Fuel processing technology, 2005. **86**(14-15): p. 1599-1614.

(290d) Base Metal Production From CO2 Capture and Storage Via the Ics Integrated Mineral Carbonation Process

Richard Hunwick, Integrated Carbon Sequestration Pty Ltd, Sydney, Australia

The ICS Process is a new, patented hydrometallurgy-based process being developed for carbon capture and storage by mineral carbonation. The ultramafic rocks such as serpentinites preferred for mineral carbonation are also the source rocks for many economically significant base and precious metals including iron ore, nickel, cobalt, chromium and platinum group metals (PGMs). These are inherently liberated into solution or otherwise rendered more accessible by the Process. But instead of requiring large quantities of expensive and environmentally challenging mineral acids as in the preferred processes for recovering these metals: high-pressure acid leaching (HPAL), the only chemical consumed in large quantities by the ICS Process is CO_2. In a carbon-constrained world, this gas represents a negative cost to mineral processors, because emitters such as coal-fired power stations will pay them to take it.

The ICS Process (for Integrated Carbon Sequestration) as its name suggests, seamlessly converts CO_2 emitted from power stations and other point sources to mineral carbonates. Uniquely amongst CCS processes, it avoids any need to produce or handle pure carbon dioxide gas.

The presentation will discuss the key features of the ICS Process, its development status and outline possible scenarios for the development of ultramafic rock-based mineral resources, starting with the most amenable weathered nickel-rich laterites and saprolites, then extending mining operations into the bulk of the unweathered ultramafics as markets for carbon trading develop in coming years.

(290e) Coupling Industrial Carbon Mineralisation and Ocean Alkalinity Enhancement

Phil Renforth, Earth Sciences, University of Oxford, Oxford, United Kingdom and Tim Kruger, University of Oxford, Oxford, United Kingdom

Coupling industrial carbon mineralisation and ocean alkalinity enhancement

Introduction

The ocean carbonate system has the potential to store ~10^2 - 10^3 billions of tonnes of CO_2as a stable bicarbonate solution. Some propose that this system can be manipulated by the addition of oxide and hydroxide minerals to the surface ocean (Kheshgi 1995; Renforth et al., in review), or through 'enhanced weathering' of silicate and carbonate minerals on the land surface (Schuiling and Krijgsman, 2006, Manning, 2008, Renforth et al., 2009, Renforth, Submitted). Dissolution products of silicate/carbonate weathering are transported to the surface ocean where they remain in solution.

Mineral carbonation has exclusively considered the precipitation of solid carbonates as the repository for sequestered CO_2(e.g. O'Connor et al., 2005). In this presentation, a case will be made for considering the coupling of 'conventional' mineral carbonation with ocean alkalinity enhancement. The presentation will touch upon recent enhanced weathering and ocean alkalinity modification research (Renforth, Submitted), and previous work investigating industrial materials (Renforth et al., 2009, Renforth et al., 2011).

Ocean Alkalinity Enhancement

Increasing ocean alkalinity was first proposed by Kheshgi (1995), in which particulate hydroxide minerals are added to the surface ocean. Dissolution of the particles promotes a number of reactions in the carbonate system which ultimately results in a drawdown of atmospheric CO_2 (Equation 1). The surface ocean (<150m) is relatively well equilibrated with the atmosphere, whereas, the deep ocean is isolated. Therefore, to effect change in the CO_2 concentration of the atmosphere, alkalinity modification would need to be limited to the surface ocean (modifying the alkalinity in the deep ocean would have little effect). Other researchers have described similar schemes such as CO_2stripping of flue gas using limestone (Rau, 2011), and direct application of limestone into upwelling regions (Harvey, 2008). Generally, the aim of these techniques is to promote the reaction in Equation 2

$$CaO + 2CO_2 + H_2O \text{ ---> } Ca^{2+} + 2HCO_3^- \text{Equation 1}$$

$$CaCO_3 + CO_2 + H_2O \text{ ---> } Ca^{2+} + 2HCO_3^- \text{ Equation 2}$$

Although, as Harvey (2008) points out, the saturation state of seawater results in very slow carbonate dissolution rates such that any particles added to the surface would sink out of the surface layer of the ocean before they dissolved. Therefore, material pre-treatment is required to enhance dissolution rate.

The carbon dioxide cycle in the ocean is complex and is influenced by variables such as temperature, pressure, and composition of the overlying air. For a detailed discussion of the thermodynamics of carbon dioxide in seawater the reader is referred to Millero (1995), and Zeebe and Wolf-Gladrow (2001). All atmospheric gases dissolve in seawater, but CO_2also hydrates to form carbonic acid. This can be described by the sequence of equilibrium reactions.

The ratio between atmospheric and dissolved CO_2 is constant at a given temperature and salinity. Dissolution is driven by the difference of partial pressure of CO_2 in the atmosphere and surface seawater. The concentration of total dissolved CO_2 (including CO_2, HCO_3^- and CO_3^{2-}) is dependent on seawater pH. In a closed system, if pH increases, the proportion of dissolved carbon dioxide falls while bicarbonate reaches a maximum at a pH around 7-8. A further rise in pH results in decreasing bicarbonate proportion and therefore carbonate ions reach their maximum. If dissolved CO_2 is buffered by atmospheric CO_2 (as it is in the surface ocean) a positive pH change will draw CO_2into solution.

The basic concept of is to increase ocean pH by adding alkalinity in the form of calcium or magnesium oxide/hydroxide. 1 mole of added calcium or magnesium oxide neutralises 2 moles of carbonic acid, as 2 moles of alkalinity are added. In order to re-establish equilibrium, 2 moles of atmospheric carbon dioxide have to dissolve in the ocean. However, this theoretical level of CO_2 uptake is not reached in reality. In reality, between 1.6-1.8 moles of carbon dioxide are absorbed (depending on temperature, and atmospheric CO_2concentration) due to equilibrium reactions in the carbonate system.

The surface ocean is already supersaturated with respect to calcium carbonate by 4-5 times. Therefore, it may be intuitive to expect carbonate minerals to precipitate if dissolved Ca and pH were to increase, which would reverse the effect of ocean alkalinity enhancement. However, the presence of magnesium (and phosphate) in the surface ocean inhibits the abiotic precipitation of calcium carbonate minerals below approximately 19-25 times supersaturated (Berner, 1975, Morse and He, 1993). Additional work is required to understand this limitation for seawater perturbed by calcium and magnesium oxides and the biological response to such a proposal. However, if all oceanic surface waters (1.81×10^{16} m^3) were able to store this additional alkalinity, approximately 1400 Gt of CO_2 could theoretically be sequestered (equivalent to a 200 ppmv reduction in CO_2concentration in the atmosphere).

Coupling Mineral Carbonation and Ocean Alkalinity Enhancement

Coupling mineral carbonation and ocean alkalinity may be a mechanism for sequestering atmospheric carbon dioxide. While there is potentially numerous ways to achieve this coupling, a direct combination would involve the introduction of ground silicate and water into a high temperature/pCO_2 reactor. After solid separation, the high temperature liquor would be circulated back into the reactor and the solid carbonate precipitate would be fed into a calciner, where it is dried and decarbonated. The evolved gases would be returned to the reactor and the solid lime would be distributed into the surface ocean. Such a scheme 1) removes the need to purify and transport CO_2 from flue gas streams, 2) CO_2 is captured directly from the air which would mitigate non-point source CO_2 emissions, 3) 1.7 moles of CO_2 are drawn-down into the ocean for every mole of carbonate used (rather than the 1:1 ratio in carbonate precipitation), and 4) the independence of the scheme from point source CO_2 emissions means that it could be located in areas with cheap isolated energy. However, such schemes are required to be adjacent to the coast to aid in material distribution, and the CO_2 emissions from fuel combustion in the calciner would need to be captured and stored. Furthermore, the thermal treatment of reactor precipitates would likely produce meta-stable calcium silicates (similar to cement production), the surface ocean geochemistry of which has not been considered.

(290f) Evaluating the Viability of CO2 Mineralization Via Reaction of Caustic Waste Materials

Robert Dilmore[1], D. Allen[2], S. W. Hedges[1] and Yee Soong[1], (1)National Energy Technology Laboratory, U.S. Department of Energy, Pittsburgh, PA, (2)Geological Sciences, Salem State College, Salem, MA

Consensus in the scientific community is that anthropogenic greenhouse gas (GHG) emissions are a dominant factor contributing to global climate change. While CO_2 is not the strongest greenhouse gas with respect to radiative forcing, it is, by far, the largest volume GHG, and is recognized by most researchers to be the greatest contributor to observed increases in mean atmospheric temperatures. As such, separation of anthropogenic CO_2 from coal fired power plants and other large industrial sources, and storage of that CO_2 out of the atmosphere has been proposed as a means to slow the rise of, and possibly stabilize atmospheric concentrations of CO_2. Proposed sinks for anthropogenic CO_2 include storage in geologic formations (saline formations, coal seams, depleted oil- and gas-bearing formations, basalts, and organic-rich shales), uptake of CO_2 into plants with terrestrial storage in soil and biomass, and reaction of CO_2 *ex situ* with naturally occurring minerals or byproduct streams from industrial processes.

Reacting CO_2 with caustic industrial byproduct streams (solids, slurries, or solutions) has been proposed by many as a possible viable approach for *ex situ* CO_2 sequestration. Cited advantages of this paradigm include proximity (often co-location) of CO_2-bearing industrial byproduct gas streams to caustic byproduct materials, possibility of reacting whole flue gas streams with caustic byproduct – thereby obviating CO_2 separation from mixed byproduct gas streams, concomitant neutralization of caustic waste materials that would otherwise need to be neutralized with acidic reagents, ability to realize significant reaction efficiency at very modest reaction conditions, storage of anthropogenic CO_2 in a stable mineralized form, and potential to generate a saleable coproduct mineral carbonate stream. Acknowledged disadvantages of this CO2 storage paradigm include: relatively low CO_2 storage capacity of global caustic industrial byproduct streams as compared to global anthropogenic CO_2 emissions, need to handle large volumes of reactant and product materials, potential to mobilize heavy metals and other trace constituents of concern, and potential for mineralized CO_2 to react and release stored CO_2 under certain environmental conditions. Taken together, these suggest that CO_2 mineralization using caustic byproduct materials may be a viable approach to partially offset CO_2 emissions for select industries and at specific facilities, but may not make a significant contribution toward stabilizing atmospheric CO_2 concentrations.

Given this context, select proposed approaches for CO_2 mineralization with caustic materials will be considered. Discussion will include summary of experimental efforts to evaluate reactivity of CO_2 with coal utilization byproducts and bauxite residue materials, consideration of the applicability of geochemical equilibrium modeling to estimate CO_2 mineralization potential, and discussion of possible implications for viability of this CO_2 storage paradigm.

(290g) The NETL Aqueous Process: History and Recent Developments

William O'Connor, NETL, Office of Fossil Energy, US DOE, Albany, OR, G.E. Rush, Geology, NETL, Office of Fossil Energy, US DOE, Albany, OR and Circe Verba, NETL, Albany, OR

An aqueous mineral carbonation process was developed at the NETL with emphasis on the utilization of calcium and magnesium silicate ores as the primary reactants. The studies included cursory regional resource assessments, process development, and a process evaluation for the carbonation of flue gas-derived CO_2 at the industrial scale. While successful demonstration of the aqueous process was achieved, several opportunities for further process optimization were identified. For example: 1) inclusion of a magnetic separation step in the serpentine process flow sheet could produce an iron oxide by-product while limiting deleterious iron oxidation during heat treatment; 2) utilization of a pipeline reactor in place of the high-pressure stirred tank reactor was demonstrated at the bench-scale, and could dramatically reduce plant costs at scale; and 3) studies by other researchers indicate that ~75% of the heat required for thermal treatment of serpentine ore could be recovered and utilized for additional operations in the plant by efficient integration of the overall plant operations. While each of these process modifications was investigated with positive results, they are but a few examples of the potential for overall process integration and improvement.

While the recent NETL focus has been on geological sequestration, mineral carbonation research has continued. These efforts have included the investigation of alternative feedstock materials, such as coal fly ash, electric arc furnace (EAF) dust, and steel grinding swarf. Results indicate that the reactivity of the various materials is highly dependent on their phase composition, which can be used as a predictive tool when considering such materials as suitable feedstock. Finally, modification of the carrier solution, namely removal of the NaCl addition, was demonstrated successfully, which could reduce corrosion of the process equipment and eliminate the detrimental impact of chloride on the potential commercial applications for the final carbonate product.

(317a) Summary Results of Epri's Engineering and Economic Case Studies of Post Combustion Capture Retrofit Applied to Various North American Host Sites

Des Dillon, Advanced Generation, EPRI, Palo Alto, Ca, CA

Summary Results of EPRI's Engineering and Economic Case Studies of Post Combustion Capture Retrofit Applied to Various North American Host Sites.

Dr. Desmond Dillon[1], Robert Chu[2] and Dr. Gerald Choi[2]

[1]The Electric Power Research Institute, EPRI, Palo Alto, Ca, USA

[2] Nexant Inc, San Francisco, Ca, USA

Abstract

With more than 300 GW of installed coal-fired plants in the US, accounting for about half of all electricity being generated domestically, it is anticipated that the existing coal fleet would need to be retrofitted for CO_2 capture in order to significantly reduce greenhouse gas emissions to acceptable levels. Amongst the available and developing capture technologies of pre-combustion, oxy-combustion and post-combustion capture (PCC), the latter is viewed as a viable and commercially available technology for existing plant retrofit application.

However, retrofitting PCC to an existing coal fired plant presents significant challenges in itself. Key issues arise such as:

· Limited space for new plant equipment

· Limited heat available for process integration

· Limitations of the existing steam turbine

· Cooling water limitations

· Replacement power considerations

· Complicated pipe routings.

With an objective to better understand these factors and quantify their effect, the Electric Power Research Company (EPRI) of California USA, recently completed a series of detailed economic and engineering studies examining the feasibility of retrofitting post combustion capture (PCC) to existing pulverised coal (PC) and/or circulating fluidized-bed (CFB) power plants, for five different North American "host" sites.

To effectively undertake this study EPRI assembled a Retrofit Team with engineering consultants of Nexant Inc. and architect engineers of Bechtel Power Corporation of Frederick, Maryland. Nexant's expertise in process plant economic optimization was complimented with Bechtels practical knowledge and experience in power plant retrofit costing, design and build.

The sites selected for the study included:

· Edison Mission Groups 1500 MW Powerton Station, operated by Mid West Generation in Pekin, Illinoi, USA

· Intermountain Power Agencies 950 MW Intermountain Generating Station in Delta, Utah USA

· Nova Scotia Powers 320MW (2 x 160MW units) Lingan Generating Station in Nova Scotia, Canada

· Great River Energys 1100MW Coal Creek Station in Underwood, North Dakota, USA

· First Energys 176MW CFB Unit at Bayshore Generation Station in Oregon, Ohio, USA

Each of the host sites offered a unique combination of unit sizes and ages, planned emissions control systems, fuel types, steam conditions, boilers, turbines, and cooling systems. Collectively, they represented a critical sample of five different types of existing coal-fired plants within the North American fleet. The data from these "real world" sites provided a broad insight into generalized cost and performance impact estimates for a variety of power plant configurations.

Specifically the studies:

- Assessed the most practical CO_2 capture efficiency configuration based on the existing site constraints
- Determined the space required for the CO_2 capture technology and the interfaces with the existing systems
- Estimated performance and costs for the post combustion CO_2 capture and compression plants
- Assessed the features of each plant that materially affect the cost and feasibility of the retrofit

The paper will give a summary overview of the studies.

Electrochemical Membrane for Carbon Dioxide Separation and Power Generation

Stephen Jolly, Hossein Ghezel-Ayagh, Jennifer Hunt, Dilip Patel
FuelCell Energy, Inc., 3 Great Pasture Road, Danbury, CT 06813

William A Steen, Carl F Richardson
URS Corporation, 9400 Amberglen Blvd, Austin, TX 78729

Olga A Marina
Pacific Northwest National Laboratory, 902 Battelle Blvd., Richland, WA 99352

FuelCell Energy, Inc. (FCE) has developed a novel system concept for separation of carbon dioxide (CO_2) from greenhouse gas (GHG) emission sources using an electrochemical membrane (ECM). The salient feature of the ECM is its capability to produce electric power while capturing CO_2 from flue gas, such as the flue gas from an existing pulverized coal (PC) plant. Laboratory scale testing of the ECM has verified the feasibility of the technology for CO_2 separation from simulated flue gases of PC plants as well as combined cycle power plants and other industrial facilities. Recently, FCE was awarded a contract (DE-FE0007634) from the U.S. Department of Energy to evaluate the use of ECM to efficiently and cost effectively separate CO_2 from the emissions of existing coal fired power plants. The overarching objective of the project is to verify that the ECM can achieve at least 90% CO_2 capture from flue gas of an existing PC plant with no more than 35% increase in the cost of electricity (COE) produced by the plant. The specific objectives and related activities planned for the project include: 1) conduct bench scale tests of a planar membrane assembly consisting of ten or more cells of about 0.9 m^2 area each, 2) develop a detailed design for an ECM-based CO_2 capture system applied to an existing PC plant, and 3) evaluate the effects of impurities (pollutants such as SO_2, NO_x, Hg) present in the coal plant flue gas by conducting laboratory scale performance tests of the membrane. The results of this project are anticipated to demonstrate that the ECM is an advanced technology, fabricated from inexpensive materials, based on proven operational track records, modular, scalable to large sizes, and a viable candidate for >90% carbon capture from existing PC plants. In this paper, the fundamentals of ECM technology including: material of construction, principal mechanism of operation, carbon capture test results and the benefits of applications to PC plants will be presented.

SYSTEM CONCEPT

To address the concerns about climate change resulting from emission of CO_2 by coal-fueled power plants and other industrial plants, FCE has developed **C**ombined **E**lectric **P**ower **a**nd **C**arbon-dioxide **S**eparation (**CEPACS**) system concept, as a novel solution (US Patent 7,396,603 B2) for greenhouse gas emission reduction. The ECM-based CEPACS system utilizes the Company's well-established Direct FuelCell® (DFC®) technology. The system concept works as two devices in one: it separates the CO_2 from the exhaust of other plants such as an existing coal fired plant and simultaneously, using a supplementary fuel, produces clean and environmentally benign electric power at a very high efficiency.

A simplified diagram of the CEPACS system concept is shown in Figure 1. CO_2-containing flue gas from a coal-fired (combustion-based) power plant, such as the exhaust from a PC power plant or other industrial sources, is utilized as the oxidant for the ECM cathode. The key feature is that the ECM utilizes the CO_2 of the flue gas as a reactant for the electrochemical reaction to produce power, while synergistically transferring CO_2 from the flue gas to the anode exhaust stream. A supplementary fuel such as natural gas or biogas (from a digester) is internally reformed in the fuel cell to (convert methane as shown in Equation 1 and) provide the hydrogen needed to complete the electrochemical power generation cycle.

$$CH_4 + 2H_2O \rightarrow CO_2 + 4H_2 \qquad (1)$$

Hydrogen is used as a reactant at the ECM anode.

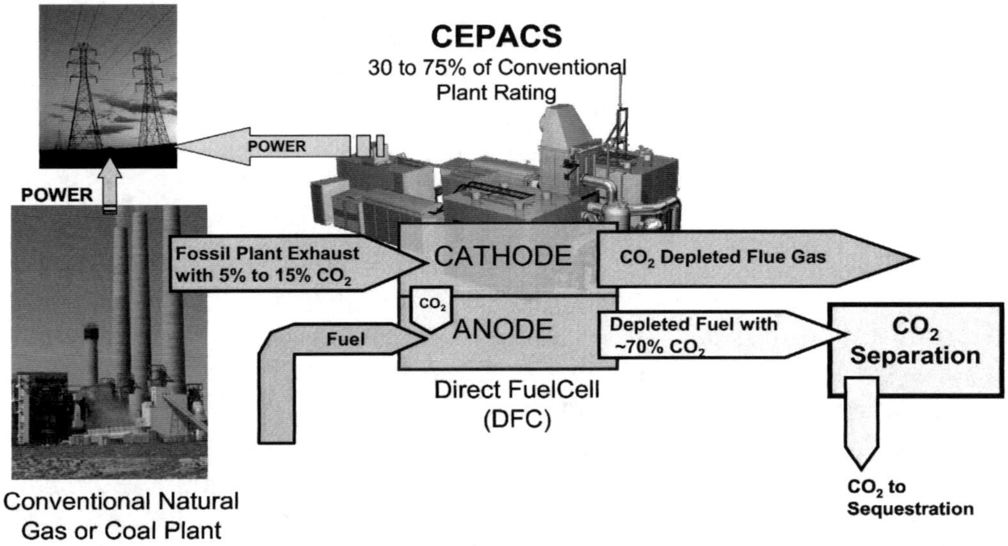

Figure 1. CEPACS CO_2 Separation and Power System Concept:
The system can be used with a variety of CO_2-containing greenhouse gases (GHG)

CEPACS separates the CO_2 from the flue gas because of the unique underlying mechanism of ECM operation. The operating principle of the ECM is shown in Figure 2, along with the electrochemical reactions involved. Carbon dioxide and oxygen present in the flue gas are used as reactants at the cathode. Carbonate ions produced at the cathode migrate to the anode side via the electrolyte in the cell. Carbon dioxide is liberated at the anode as carbonate ions react with H_2. In addition to CO_2, H_2O is produced at the anode as H_2 is consumed by the electrochemical reaction. The transport of carbonate ions internally in an ECM cell and the flow of electrons in the external circuit results in electric power generation as a byproduct of the electrochemical CO_2 separation process. Overall, the operating mechanism of the ECM cell results in the separation and transfer of CO_2 into the anode exhaust stream which has a much reduced volumetric flow rate (resulting in a CO_2-rich stream) compared to the original flue gas stream.

The CO_2-rich anode exhaust gas is processed (post-processing) in the ECM-based CEPACS system to further concentrate and compress the CO_2 for sequestration. The CO_2 rich anode exhaust contains water vapor which can be condensed out. It also contains a small amount of unused fuel (H_2) which can be easily separated. Water condensed during post-processing of the CO_2-rich anode exhaust stream is used to provide water (steam) needed for internal reforming of CH_4 in supplementary fuel, eliminating the need for external process water. H_2 separated during post-processing of CO_2-rich anode exhaust stream is recycled to provide additional preheat in the system and as part of the supplementary fuel (thereby reducing NG fuel needed). The captured CO_2 can then be processed for sequestration.

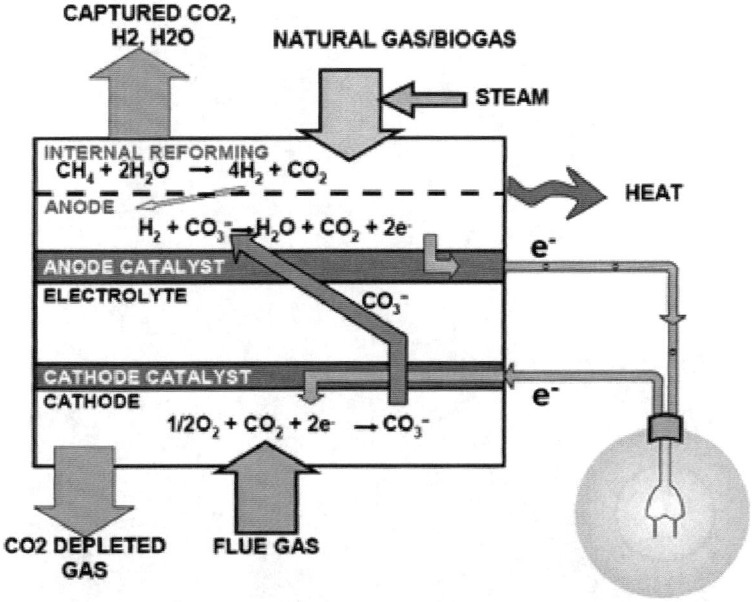

Figure 2. Separation of CO_2 in a Carbonate Fuel Cell:
Carbon dioxide is used at the cathode as an oxidant and transferred to the anode via the carbonate electrolyte

The CO_2-lean cathode exhaust (flue gas after CO_2 removal) is vented to atmosphere after recovering the heat for process use (e. g. preheating of feed streams).

The ECM cell operates at 550 - 650°C and atmospheric pressure. Unlike conventional membrane technologies that rely on pressure (partial pressure) differentials and permeability properties, ECM separates CO_2 at a rate dependent on the electrical current drawn. The flue gas does not need to be pressurized and vacuum operation is not required on the permeate side. Because of fast electrode kinetics, the ECM cell does not require high CO_2 concentration in the flue gas feed. Planar geometry and large gas flow channels of the ECM cell enable processing of large volumetric flow of flue gas feed without significant back pressure.

TECHNOLOGY STATUS

The fuel cells used in the CEPACS system are built from the same materials and by the same fabrication processes as those used in the Company's DFC-based power plant products. DFC (ECM) technology has reached commercial maturity and is readily available as a manufactured product used in building fuel cell stacks. Currently, the mass produced cells of about 9000 cm^2 area, suitable for large-scale applications, are commercially available. These cells have already been engineered for multi-layer stacking. The stacks of ECMs are incorporated in modules, which make up the MW scale building blocks for large power plants and CEPACS

Systems. Figure 3 shows a diagram of the ECM production and modularization process, which is performed at the Company's manufacturing plant in Torrington, Connecticut.

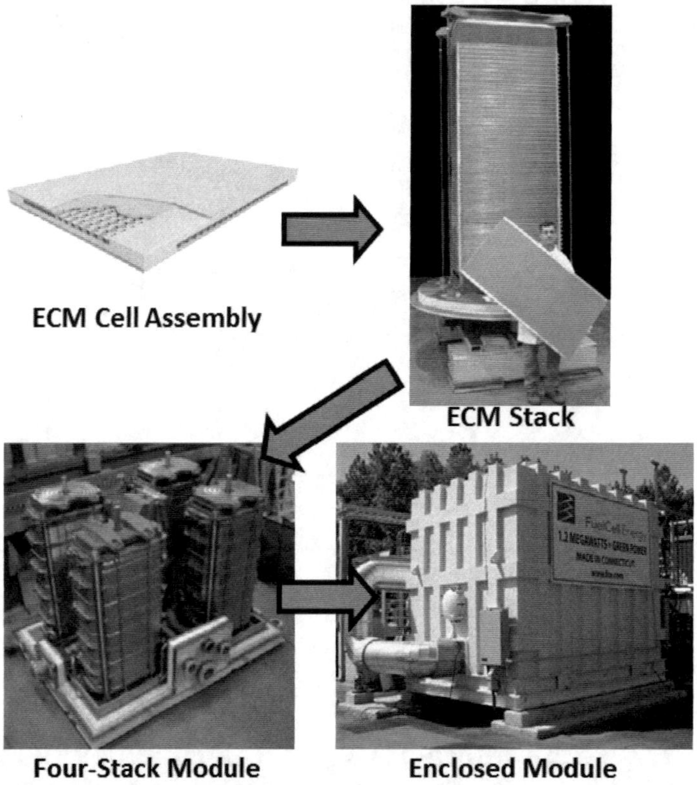

Figure 3. ECM Module Assembly Process: Cells fabricated at FCE's manufacturing plant in Torrington, CT are stacked and then modularized to form the building blocks usable in a CEPACS system

The CEPACS enabling technology has been deployed in MW-scale power plants for stationary power applications in three sizes: 300 kW, 1.4 MW, and 2.8 MW units. A 1 MW (nominal) unit (DFC1500) installed at a customer site is shown in Figure 4. FCE services over 50 DFC power plant sites around the globe.

Figure 4. 1MW DFC1500 Direct FuelCell Power Plant:
The DFC products are suitable for large multi-MW systems

CEPACS SYSTEM TECHNO-ECONOMIC EVALUATION

Monoethanolamine (MEA) based scrubbing technology is considered to be the current benchmark technology for separating CO_2. However, the energy and efficiency penalties of using amines for CO_2 capture in PC plants are substantial. About 22-30% of plant gross power is used up by the amine system, dropping the plant efficiency to <30%. Operation of the CEPACS system is set apart from other technologies by generation of electric power rather than consuming it, resulting in an increase in the power output of the retrofitted power plant.

A preliminary Technical and Economic Feasibility Study has been performed. The study included assessments of cost and performance of the ECM-based CO_2 capture and compression system for a 550 MW (net AC) PC plant. While the ECM-based CEPACS system captures 90% of CO_2 from the PC plant flue gas, it generates additional (net AC) power (~441.15 MW) after compensating for the auxiliary power requirements of CO_2 capture and compression. Figure 5 presents the plant performance results showing the impact on PC plant efficiency. The net electrical efficiency of the PC plant equipped with CEPACS-based CO_2 capture was estimated to be 39.8% (based on higher heating values - HHV of coal and natural gas fuels used by PC plant and CEPACS system, respectively). The efficiency of the PC plant with no CO_2 capture is estimated to be 36.8% (coal HHV). The later number is based on U.S. DOE/NETL system study[1]. The DOE study also reports efficiency of the PC plant with amine-based CO_2 capture to be 26.2% (coal HHV). For comparison, the DOE study assumes fixed (nominal) net plant output of 550 MW. To compensate for the auxiliary power requirements of CO_2 capture and compression, the PC plant with amine-based CO_2 capture is bigger (672.7 MW gross) than the PC plant with no CO_2 capture (582.6 MW gross). The CEPACS-based CO_2 capture also has other attractive features/benefits. The PC plant equipped with CEPACS system has lower emissions of NOx, SOx and Hg compared to a PC plant equipped with amine-based scrubber for CO_2 capture. The CEPACS system produces excess process water thereby reducing the total water usage of the PC plant equipped with CEPACS-based CO_2 capture.

[1] Cost and Performance Baseline for Fossil Energy Plants, Volume 1: Bituminous Coal and Natural Gas to Electricity, Revision 2, DOE/NETL-2010/1397, November 2010.

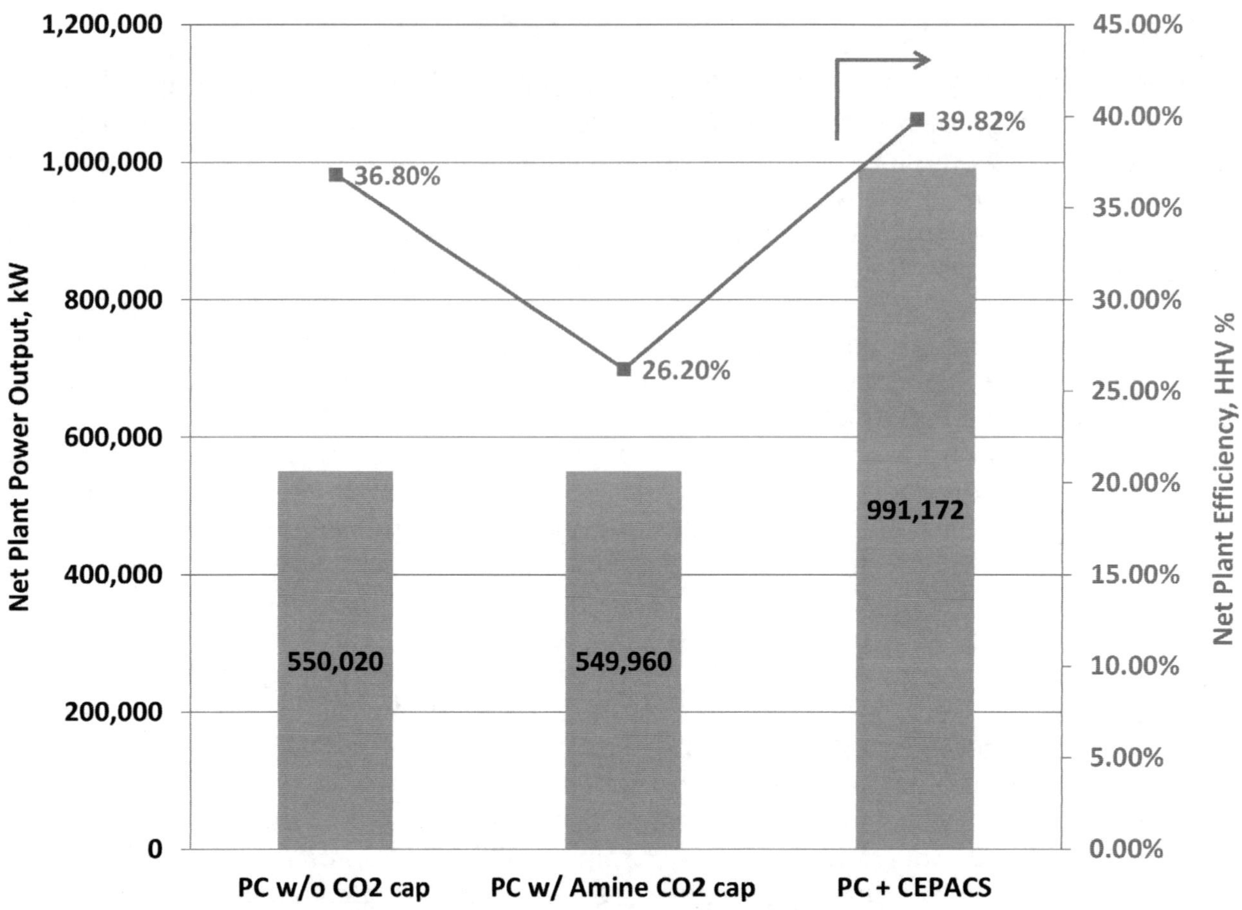

Figure 5. Impact of CO_2 capture on PC Plant Performance

The economic analysis included estimation of CEPACS plant cost and subsequently the incremental cost of electricity (COE) for ECM-based CO_2 capture. Figure 6 presents the results showing the impact on PC plant COE and provides a comparison with the competing amine-based CO_2 capture technology (PC w/o CO_2 capture is referred as Case 1). The fuel costs were based on the coal price of \$38.18/ton and the natural gas price of \$6.55/MMBtu. The cost estimation methodology outlines in DOE's guideline document was followed for the analysis[2]. The CEPACS technology has the potential to meet/exceed DOE's goal of <35% increase in COE for post-combustion CO_2 capture. The cost of CO_2 capture for a PC plant equipped with CEPACS system was estimated to be \$14/ton which is significantly lower than \$60/ton for a PC plant equipped with an amine-based scrubbing system.

[2] Quality Guidelines for Energy System Studies: Cost Estimation Methodology for NETL Assessment of Power Plant Performance, April 2011, DOE/NETL-2011/1455.

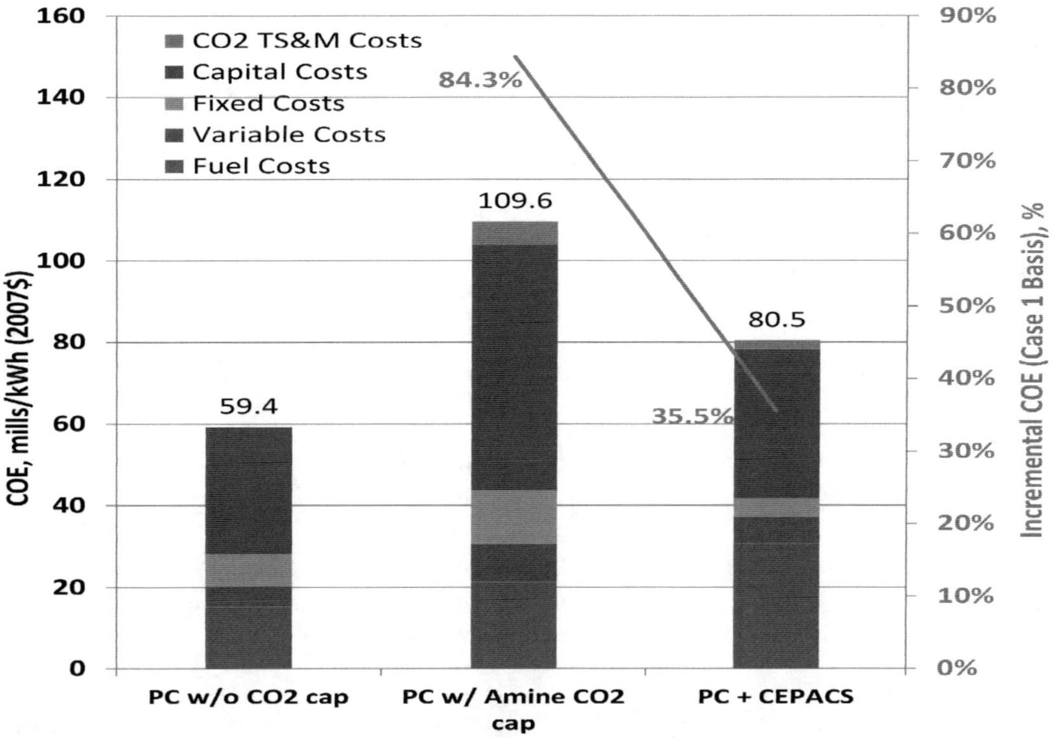

Figure 6. Impact of CO_2 capture on PC Plant Cost of Electricity

(317c) Modeling, Simulation and Optimization of Absorption, Membrane and Adsorption-Based CO2 Capture for Variable Feed CO2 Concentration and Feed Flow

M. M. Faruque Hasan, Richard C. Baliban, Josephine A. Elia and Christodoulos A. Floudas, Chemical and Biological Engineering, Princeton University, Princeton, NJ

Studies on four leading technologies for industrial CO_2 capture are performed. The technologies are MEA-based chemical absorption, multi-stage membrane, pressure swing adsorption (PSA) and vacuum swing adsorption (VSA). Each process is rigorously modeled, optimized, and energy integrated to capture and compress at least 90% of CO_2 from the feed for sequestration, over a range of feed CO_2 compositions and total feed flow rates. Emphasis is given on understanding and quantifying how the cost of CO_2 capture changes with CO_2 feed composition and flow rate of the gas mixtures coming from various chemical, power and heat generation plants. The novel features of this study include:

(i) variation in feed CO_2 composition (ranging from 1 to 90%) and total feed flow rate (ranging from 0.1 to 10 kmol/s);

(ii) CO_2 capture coupled with compression to 150 bar;

(iii) rate-based modeling of the MEA absorption-based CO_2 capture coupled with heat integration;

(iv) multi-stage superstructure model for CO_2 capture using membranes;

(v) detailed and grey-box modeling of PSA and VSA alternatives for the adsorption-based CO_2 capture;

(vi) cost-based comparison between the absorption, membrane, PSA and VSA-based CO_2 capture technologies for different emission scenarios; and

(vii) explicit expressions for the investment and operating costs as function of feed CO_2 composition and total feed flow rate.

In this talk, we will discuss our approach towards the modeling, simulation and optimization of each of the four technologies. A rigorous simulation-based optimization model is proposed to determine the minimum annualized cost of the MEA-absorption process. The MEA-absorption process is further heat integrated through heat exchanger network optimization. A novel mathematical model is developed for the optimization of multi-stage and multi-component separation of CO_2 using membranes, which can be also used for a range of membrane-based gas separation applications. A 4-step cyclic process with multiple adsorption columns, purging and product recovery at moderate to low vacuum is configured for the PSA and VSA processes. A PDE-based non-isothermal adsorption model is used, which are fully discretized and optimized via a Kriging model.

We will also present a cost-based technology comparison over a range of feed gas compositions and flow rates. This may assist the decision-makers to select the cost-appropriate technology for CO_2 mitigation by taking the diverse emission scenarios into consideration. Furthermore, we will discuss an explicit input-output model for the investment and operating costs for the first time. This would provide a basis towards scaling up a CO_2 capture process, and lay the foundation for (i) process synthesis studies, and (ii) CO_2 supply chain network studies in the carbon management domain.

(317d) Pressure Swing Adsoprtion: Design and Optimization for Pre-Combustion Carbon Capture

Alexander W. Dowling[1], Lorenz T. Biegler[1] and David C. Miller[2], (1)Chemical Engineering, Carnegie Mellon University, Pittsburgh, PA, (2)U.S. Department Of Energy, National Energy Technology Laboratory, Morgantown, WV

Pressure Swing Adsorption (PSA) is an efficient method for gas separation, and is a potential candidate for carbon dioxide (CO_2) capture from power plants. However, few PSA cycles have been designed for this purpose and the optimization of PSA systems for CO_2 capture remains a very challenging task. In this study, we present a systematic optimization-based formulation for the synthesis and design of novel PSA cycles for CO_2 capture in an IGCC flowsheet, which can simultaneously produce hydrogen (H_2) and CO_2 at a high purity. Here, we apply a superstructure-based approach to simultaneously determine optimal cycle configurations and design parameters for PSA units. This approach integrates automatic differentiation, efficient ODE solvers for the state and sensitivity equations of the PSA model and state of the art NLP solvers. Three optimization formulations are proposed and two PSA case studies are considered. The first case study considers a binary separation of H_2 and CO_2 at high purity, where specific energy is minimized, while the second considers a larger five component separation.

(317e) Assessment and Analysis of Post-Combustion Carbon Capture with Membrane Separation Technology

Xiangping Zhang, Norwegian University of Science and Technology, Trondheim, Norway, Xuezhong He, Norwegian University of Science and , Trondheim, Norway and Truls Gundersen, Energy and Process Engineering, Norwegian University of Science and Technology, Trondheim, Norway

Large scale CO_2 Capture and Storage has been considered to be one of the most promising ways of mitigating anthropogenic CO_2 emissions, but energy penalty, high economic costs and possible environmental impacts are important barriers for its deployment. One of the effective solutions is to develop lower cost and industrially feasible technologies, especially for the post-combustion capture technologies that have a major challenge due to the large gas volumes, low CO_2 concentration and low partial pressure. Among the different technologies, MEA-based systems have been the most likely candidate from an industrial perspective, but some inherent drawbacks, like solvent degradation and large energy consumption, limit its applications. Due to the rapid development of gas separation membrane technology and its inherent benefits for carbon capture[1, 2], it has attracted considerable attention. It is expected to have the potential to overcome the drawbacks of MEA-based systems, but most of the research is still done in lab level or pilot scale. From the viewpoints of industrial application and process design, understanding deeply the technical and economic benefits or drawbacks of the new technology along with fundamental experimental research is of significant importance for further development, and can also provide possible research directions to accelerate the industrial application of the technology.

However, there are only a few reports related to systematic analysis of membrane systems compared to the large amount of work focusing on material synthesis and experimental research[2, 3]. Bounaceur et al. performed relatively early a systematic analysis of the separation performance and energy cost of a single stage module and optimized the pressure ratio[4], but they did not perform any economic analysis. Zhao et al. described a detailed parameter study of mass and energy balances for a single membrane[5] and also the energetic and economic analyses of a multi-stage process[6]. Merkel et al. investigated the influences of membrane parameters and process configurations on the energy consumption and capture cost, and some important suggestions were proposed for future studies[7], however, more detailed economic models should be used. Hussain et al. investigated the feasibility of carbon capture from flue gas by facilitated transport membranes, and a real gas composition was considered[8]. However, some results were inconsistent and only a few of these studies focused on thermodynamic analysis for this energy intensive process. Britan et al. studied the feasibility of membrane technology of mixed gas separation with pure thermodynamic principles[9], but did not consider the real process. Thus, further systematic analysis and assessment is important and necessary.

In this work, a systems methodology integrated with process simulation, technological and economic analysis as well as exergy analysis is applied to assess a membrane capture technology and compare it with MEA-based systems. An Advanced Super Critical (ASC) Boiler and Turbine delivering 819 MWe (gross) being part of a coal based power plant is chosen as a base case for the study. The flue gas after ESP and FGD to remove ash and sulfur is fed into the membrane separation system. The aim of the work is to try to clarify the advantages and disadvantages with using membranes for post-combustion carbon, and to indicate what the main obstacles are and how far from commercialization the technology is.

At first, the current research on polymeric membranes for post-combustion CO_2capture is reviewed, including the type of membranes, application conditions, permeability and selectivity, and some emerging materials. Then the membrane separation process is simulated with a cross flow model in PRO/II. Under the different process configurations, membrane parameters and operational parameters, the mass and energy balances, the power consumption as well as the membrane area are estimated. Afterwards, the influence of key parameters on membrane area and energy consumption is studied. A cost model is established to estimate the capture cost, and used to optimize the key parameters. Both the energy consumption and the capture cost of the membrane technology are compared with MEA-based systems. Finally, the exergy analysis method is used to identify the bottlenecks of the energy consumption in a multi-stage membrane process.

The simulation results show that the energy consumption and membrane area of a membrane process are influenced greatly by CO_2 permeability and CO_2/N_2 selectivity, which are the two key parameters in carbon capture from flue gas. Unfortunately, there is frequently a trade-off between permeability and selectivity, which is also a big challenge in material development. Higher CO_2 permeance can reduce the specific energy consumption and membrane area considerably. For a given CO_2permeance, the specific energy consumption decreases with increasing selectivity, however, more membrane area is needed. But in general, continuously increasing selectivity will not influence the separation performance so much, which means that permeance could still be the dominating parameter.

Besides the membrane parameters, the configuration of the separation process also plays a key role for energy requirement and capture cost. Due to the limitation of the membrane performance for post-combustion capture at present, it is hard to reach both 90% capture ratio and 90% CO_2purify with 1 stage membrane separators, otherwise a very high pressure ratio across the membrane is required and tremendous energy is needed to compress the gas and create vacuum, which is unpractical. Multi-stage processes represent a good option to resolve this problem, but more than 2 stages will be very complex from an operational point of view. The simulation and optimization results in this work also show how the capture load in each stage influences the total energy consumption and capture cost.

Depending on the influence of membrane parameters and flowsheet structure, referring to the concept of Robeson Plot[10], a 3-dimensional profile which reflects the relationship between permeability, selectivity and capture cost is drawn, which provides a roadmap to develop the suitable membrane material which has the possibility to compete with an MEA-based system.

Different from the MEA-based system which is a typical chemisorption process, the membrane separation of $CO_2$from a gas mixture is a physical process and the driving force is the pressure differential of the gases between the feed and permeate side. The energy demand to produce the high pressure ratio between upstream and downstream is noticeable. Thus, understanding the energy consumption level is a key point for process design, and exergy analysis is a basic method for this purpose. It is helpful to indicate where the exergy losses occur, the magnitude of these losses, and what the bottleneck parts are. Taking a two-stage process as a case study, the total exergy loss due to irreversibilities, waste heat and waste mass in each unit is calculated and ranked. The results show that the largest exergy loss occurs in the 1st membrane stage, followed by the 2nd stage. The two separators contribute to about 80% of the total exergy loss due to irreversibilities.

Compared with an MEA-based system, the gas separation membrane technology has shown some advantages for post combustion carbon capture, such as less environmental footprint, ease for scaling up and operability, without auxiliary retrofitting measures or costs in the power plant. In this work, some important aspects of this new technology were investigated using an integrated assessment method and making a comparison with an MEA-based system. This study will be important both for membrane development and future industrial applications.

Acknowledgements

This publication has been produced with support from the BIGCCS Centre, performed under the Norwegian research program *Centers for Environment-friendly Energy Research (FME)*. The authors acknowledge the following partners for their contributions: Aker Solutions, ConocoPhilips, Det Norske Veritas, Gassco, Hydro, Shell, Statoil, TOTAL, GDF SUEZ and the Research Council of Norway (193816/S60). We are grateful to the anonymous referees for their constructive comments on an earlier version of this paper.

Reference:

1. E. Favre, *Chem Eng J*, 2011, **171**, 782-793.

2. C. E. Powell and G. G. Qiao, *J Membrane Sci*, 2006, **279**, 1-49.

3. C. A. Scholes, S. E. Kentish and G. W. Stevens, *Recent Patents on Chemical Engineering*, 2008, **1**, 52-66.

4. R. Bounaceur, N. Lape, D. Roizard, C. Vallieres and E. Favre, *Energy*, 2006, **31**, 2556-2570.

5. L. Zhao, E. Riensche, R. Menzer, L. Blum and D. Stolten, *J Membrane Sci*, 2008, **325**, 284-294.

6. L. Zhao, E. Riensche, L. Blum and D. Stolten, *J Membrane Sci*, 2010, **359**, 160-172.

7. T. C. Merkel, H. Q. Lin, X. T. Wei and R. Baker, *J Membrane Sci*, 2010, **359**, 126-139.

8. A. Hussain and M. B. Hagg, *J Membrane Sci*, 2010, **359**, 140-148.

9. I. M. Britan, I. L. Leites and T. N. Vasilkovskaya, *J Membrane Sci*, 1991, **55**, 349-352.

10. L. M. Robeson, *J Membrane Sci*, 1991, **62**, 165-185.

(317f) Dynamic Modelling and Analysis of Post-Combustion CO2 Capture Processes

Niall Mac Dowell, Chemical Engineering, Imperial College London, London, United Kingdom and Nilay Shah, Centre for Process Systems Engineering, Department of Chemical Engineering, Imperial College London, London, United Kingdom

There have been several design and optimisation studies in the context of amine-based post-combustion CO_2 capture (PC-CC) from coal-fired power stations, and it is well accepted that the principal contributor to the capital-cost (CAPEX) of amine-based PC-CC is the absorber, with the majority of operating cost (OPEX) being accounted for by solvent regeneration[1]. However, the majority of these studies consider only the maximal stable generation (MSG) point of the power station. In reality, many power-plants do not operate consistently at the MSG. Thus, the cost optimal design of a PC-CC process must account for the dynamic operation of many fossil-fuel power-plants. The economic operation of a decarbonised fossil-fired power-plant is a complex trade-off between the time varying cost-of-electricity and cost of fuel. These variables will in turn affect the cost-optimal degree of CO_2 capture, and thus the design and operation of the CO_2 capture plant.

In this work, we present a new approach for the multi-period design of an integrated coal-fired power-plant and an alkanolamine-based CO_2 capture process. A validated[2] rate-based model of an absorber/stripper system for the chemisorption of acid gas in aqueous solvent solutions is used to represent the CO_2 capture process. To account for the non-idealities that are typical of amines and water, the statistical associating fluid theory for potentials of variable range (SAFT-VR)[3],[4] is used. This is a molecular approach, specifically suited to associating fluids. The SAFT formalism is used to represent some of the equilibrium reactions characterising the system, thereby simplifying the description of the chemical reactions[5]. The dynamic operation of the power-plant is described using data obtained from a real power station describing the variation in fuel burn rates in response to changing power demand. All models are implemented in the gPROMS[6] software package.

---

[1] Mac Dowell, N. et al, Energy. Environ. Sci., 3, 2010, 1645-1669

[2] Mac Dowell, N., Galindo, A., Jackson, G and Adjiman, C. S., Comp. Aided Chem. Eng., 28, 2010, 1231-1236

[3] Chapman, W.G., Gubbins, K.E., Jackson, G. & Radosz, M., Ind. Eng. Chem. Res. 29, 1709-1721., 1990

[4] Gil-Villegas, A., Galindo, A., Whitehead, P. J., Mills, S. J. & Jackson, G., J. Chem. Phys. 106 (10), 1997

[5] Mac Dowell, N., Llovell, F., Adjiman, C. S., Jackson, G and Galindo, A., Ind. Eng. Chem. Res., 49(4), 1883-1899, 2010

[6] Process Systems Enterprise (PSE) Ltd. http://www.psenterprise.com/index.html

(317g) Comparisons of Solvents for Post-Combustion CO2 Capture Using a Multi-Objective Analysis Approach

Anita S. Lee and John R. Kitchin, Department of Chemical Engineering, Carnegie Mellon University, Pittsburgh, PA

Gas scrubbing using amine solvents is one of the most promising first generation technologies for post-combustion CO_2 capture. However, the high parasitic energy demand associated with the leading commercially available amine solvent, monoethanolamine, has motivated the development of new amine solvents. [1-2] As new solvents are developed, the need arises to be able to quantify and compare improvements in energy and utility demands of CO_2 capture at the process level. In this work we uses ASPEN Plus to model the CO_2 capture and compression process for three different amines, monoethanolamine (MEA), diethanolamine (DEA), and Amino-2-methyl-1-propanol (AMP). Utilizing ModeFrontier and the modeling framework of Eslik and Miller[3] we identify amine specific optimal operating and process conditions, considering energy demands, water consumption, and capital costs as design objectives. These solutions provide the basis of comparison of performance between amines. Additionally, we analyze trade-offs between design objectives for each amine. This work illustrates the complexities associated with the CO_2 capture process and identifies an approach to comparing the performance of different solvents.

(1) Singh, D.; Croiset, E.; Douglas, P. L.; Douglas, M. A., Techno-economic study of CO2 capture from an existing coal-fired power plant: MEA scrubbing vs. O-2/CO2 recycle combustion. *Energ Convers Manage* **2003,** *44* (19), 3073.

(2) NETL, Cost and Performance Baseline for Fossil Energy Plants. **2010**.

(3) Eslick, J. C.; Miller, D. C., A multi-objective analysis for the retrofit of a pulverized coal power plant with a CO(2) capture and compression process. *Computers & Chemical Engineering* **2011,** *35* (8), 1488.

(317h) Post-Combustion Gas Permeation Carbon Capture System Models

Juan E. Morinelly, National Energy Technology Laboratory, Morgantown, WV and David C. Miller, U.S. Department Of Energy, National Energy Technology Laboratory, Morgantown, WV

Post-Combustion Gas Permeation Carbon Capture System Models

Juan E. Morinelly, David C. Miller

National Energy Technology Laboratory (NETL)

Gas permeation membranes have been proposed as an alternative to amine absorption for post-combustion carbon capture [1,2]. Industrial-scale membrane systems have been successfully applied in natural gas sweetening, and their use in other separations is expected to grow in the near future [3]. Under the constraints imposed by the performance limits of state-of-the-art gas permeation membranes multi-stage processes are required to obtain a high purity CO_2 stream suitable for sequestration. An effective design for such a process entails balancing of numerous factors, including the pressure ratio across the membrane stages, which affects the parasitic power load, and the size of the membrane modules, which affects the capital cost. Merkel et al. (2010) introduced a multi-stage process that incorporates permeate recirculation, the use of the boiler's air feed as sweep, compression of the feed, and pulling vacuum on the permeate side. The main challenge from a modeling perspective is that, unlike conventional CO_2 capture processes, this is not a purely 'end of pipe' process. The proposed process is integrated with other parts of the power plant. A modeling strategy that captures undesired effects such as higher flows due to the recirculation of gases is required in order to accurately evaluate its performance.

Two steady-state process models were built in Aspen Custom Modeler® (ACM) based on the work by Merkel et al. (2010) and Toy et al. (2011). These models include modules for the boiler, air pre-heater, and FGD units that can be tuned to a given pulverized coal (PC) power plant system. A one-dimensional, multi-component, distributed model was developed for the membrane stages within the process models. This device-scale gas permeation model incorporates pressure drop effects and can be adjusted to varying dimensions for the module as well as membrane properties. The process models are linked to a simulation interface that allows the optimization of key process parameters with respect to system-level functions such as the cost of electricity (COE). In addition to a description of the process models and their features, this presentation will describe the results of varying levels of membrane properties. The base power plant in this analysis is a supercritical PC with a net output of 650 MWe before the capture and compression systems are integrated. Constraints on the capture level (>90%), the purity of the sequestration stream (>95%), and the increase in flow of the flue gas stream out of the boiler (<15%) were included in the optimization routine.

References

1. Merkel, TC, Lin, H, Wei, X, Baker, R. Power Plant Post-Combustion Carbon Dioxide Capture: An Opportunity for Membranes. *J. Membr. Sci.*, 2010;359:126-139

2. Favre, E. Carbon Dioxide Recovery from Post-Combustion Processes: Can Gas Permeation Membranes Compete with Absorption? *J. Membr. Sci.*, 2007;294:50-59

3. Bernardo, P, Drioli, E, Golemme, G. Membrane Gas Separation: A Review/State of the Art. *Ind. Eng. Chem. Res.*, 2009;48:4638-4663

4. Toy, L. CO2 Capture Membrane Process for Power Plant Flue Gas. Presented at the 2011 NETL Capture Technology Meeting. Pittsburgh, PA, August 22, 2011.

"Fuel Processing: Renaissance of the Second Decade"

Brant A. Peppley, Rajesh Parmar and Kunal Karan
Queen's-RMC Fuel Cell Research Centre
Queen's University
Kingston Canada

The early development of fuel processing technologies for fuel cell systems was, to a great extent, driven by fuel cell automobile programs. In the early 1990s, an on-board reformer was believed to be the best option for fuel cell automobile commercialisation. Regrettably, the belief that we simply needed to adapt design concepts from large-scale industrial steam reforming hampered progress at the time.

Figure 1: 100MM SCFD reformer and waste heat recovery section supplied by Selas-Linde Fluid in a hydrogen plant designed, supplied, and constructed by Linde. Source: http://www.selas-linde.com/range_of_services/modular_constructions.php

Very early attempts to modify reformer designs to address the need for rapid load following and better heat transfer lacked the fundamental understanding of reactor design and reaction kinetics.

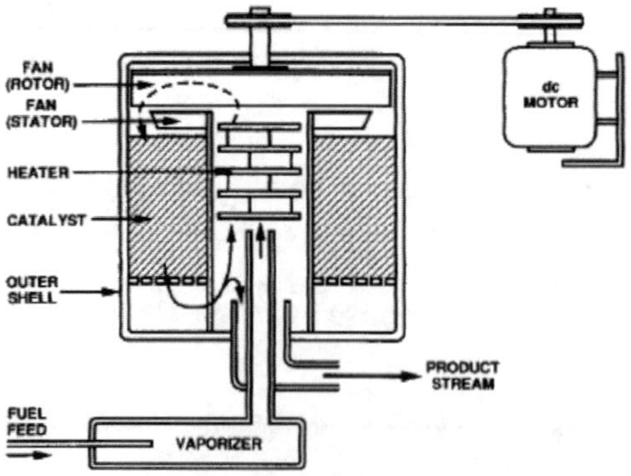

Figure 2: Schematic cross-section of a recirculating methanol steam reformer with enhanced heat transfer but very undesirable flow characteristics not to mention undesirable parasitic power requirements. (R.A. Lemons, J. Power Sources, 29 (1990) 251 – 264.

Autothermal (ATR) and Partial Oxidation (POX) reformers provided a partial solution to the heat transfer problems but resulted in a severe penalty in overall system efficiency. The idea of on-board reforming for automobiles was all but completely abandoned in the late-1990s and, the level of effort on fuel processing for fuel cell systems decreased drastically. A significant amount of progress, however, had been made prior to the reduction in support for fuel processor research and reformer designs based on structured catalysts and catalyst coated micro-structured heat transfer components had made significant progress.

Recent developments in stationary and portable fuel cell system products using hydrocarbons, alcohol and oxygenated fuels including diesel fuel, dimethyl ether, ethanol, methanol and propane as well as natural gas have created a renewed interest in developing improved fuel processors. Although these types of systems do not necessarily require fast start up and load following, significant improvements in overall efficiency, reductions in pressure drop through the bed and improved dynamic response can be and have been achieved. A better understanding of the coupling of heat transfer, catalytic kinetics and mass transfer limitations in conjunction with the discovery of new classes of catalysts that meet the demanding requirements of hydrocarbon fuel processing has led to some highly innovative new directions in fuel processor design.

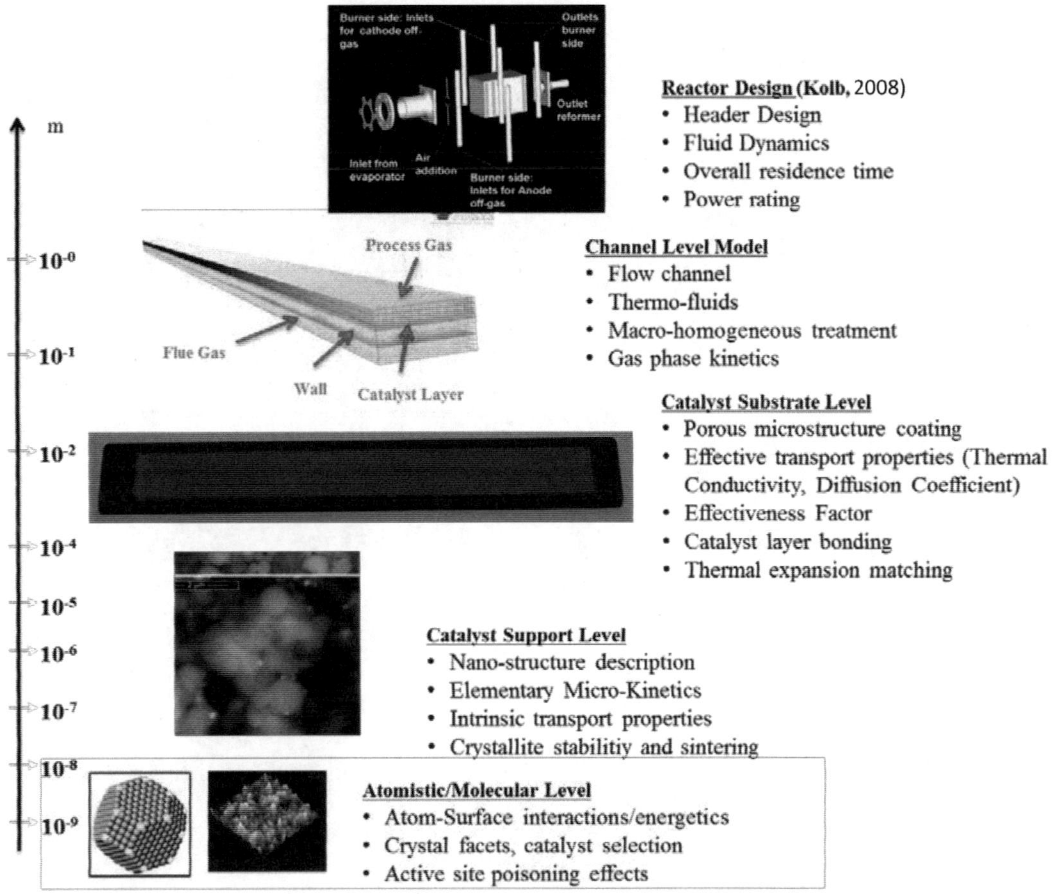

Figure 3: Understanding reactor design using a multi-scale approach where fundamental catalyst properties at the atomic level [3] are considered in the final design at the macro-level [4] offer the opportunity for significant improvements in fuel processor designs.

New directions in fuel processor design and the development of promising catalysts that are able to endure the challenges of high temperature reforming of heavier hydrocarbons has opened the door to many new opportunities for fuel cell product development.

References

[1] Selas-Linde website, Accessed 24 Sept 2012
http://www.selas-linde.com/range_of_services/modular_constructions.php

[2] Lemons, R. A., J Power Sources, 29 (1990) 251-264. Proc of First Grove Fuel Cell Symposium

[3] Haynes, D. J., D. A. Berry, D. Shekhawat and J. J. Spivey (2008). "Catalytic partial oxidation of n-tetradecane using pyrochlores: Effect of Rh and Sr substitution." Catalysis Today 136(3-4): 206-213.

[4] Kolb, G., T. Baier, J. Schürer, D. Tiemann, A. Ziogas, S. Specchia, C. Galletti, G. Germani and Y. Schuurman (2008). Chemical Engineering Journal 138(1-3): 474-489

(343b) Desulfurization of JP-8 Fuel Using Ag-Loaded Silica Supports Under Ambient Conditions

Dat T. Tran, Electrochemistry - SEDD, U.S. Army Research Laboratory, Adelphi, MD, Jessica St. John, Chemistry and Biochemistry, University of California, Santa Cruz, Santa Cruz, CA, Deryn Chu, Electrochemistry - SEDD, U.S. Army Research Laboratory, Adelphi and Scott R.J. Oliver, Chemistry and Biochemistry, University of California, Santa Cruz, Santa Cruz

2012 AIChE Annual Meeting

October 28 – November 2, 2012
Pittsburgh Convention Center

Pittsburgh, PA

Submit to **Symposium: Fuel Processing for Hydrogen Production**, as a **Talk**.

Title: Desulfurization of JP-8 fuel using Ag-loaded silica supports under ambient conditions

Dat T. Tran,[1] Jessica St. John,[2] Deryn Chu[1] and Scott R. J. Oliver[2]

[1]U.S. Army Research Laboratory, RDRL-SED-C, 2800 Powder Mill Road, Adelphi, Maryland 20783; dat.t.tran4.civ@mail.mil

[2]Department of Chemistry and Biochemistry, University of California, Santa Cruz, 1156 High Street, Santa Cruz, CA 95064; soliver@ucsc.edu

The U.S. Army demand for mobile, portable and stationary fuel cells based on hydrogen has increased rapidly. To meet these power generation demands, the U.S. Army is interested in using JP-8 fuel as a hydrogen carrier. The reformation of JP-8 to hydrogen offers an energy dense feedstock via on-demand fuel processing. Unfortunately, organic sulfur compounds intrinsic to JP-8 pose a significant challenge because they can poison the reformer and fuel cell catalysts.

In this work, mesoporous silica and aluminosilicate adsorbents loaded with silver (and other metal) ions were synthesized, characterized and examined for the adsorptive desulfurization of JP-8 fuel. The adsorbents were evaluated with JP-8 fuel containing ~700 ppmw sulfur under ambient conditions. This talk will describe the preparation and characterization of these adsorbents and their sulfur adsorption characteristics for the goal of fuel desulfurization.

(343c) Desulfurization of Commercial Fuels Using Mixed Oxide Supported Silver Adsorbents At Room Temperature for Fuel Cell Applications

A. H. M. Shahadat Hussain[1], Hongyun Yang[2] and Bruce J. Tatarchuk[1], (1)Chemical Engineering, Auburn University, Auburn, AL, (2)IntraMicron Inc., Auburn, AL

Sulfur removal from petroleum fuels is a major challenge globally due to ever growing demand of sulfur free feedstock for fuel cell applications. Adsorption of organosulfur compounds from hydrocarbon fuels at ambient conditions is a promising solution to the high energy and processing cost requirements for ultra clean fuel production. In this work, the performances of Ag adsorbents supported on titania dispersed alumina (TiO_2-Al_2O_3) for desulfurizing JP5, JP8 jet fuels and off-road diesel have been discussed. Sulfur adsorption capacities were obtained through observation of breakthrough of commercial fuels. Model fuels containing thiol and thiophene derivatives and aromatic compounds were used in breakthrough experiments to study the selectivity of the adsorbent toward sulfur species. N_2 physisorption, TPR, XRD, and *in situ* FTIR were employed to characterize the adsorbent. In the breakthrough experiments, 12% by weight Ag supported on TiO_2-Al_2O_3 adsorbent had respected saturation capacities of 11.83, 8.01, and 10.9 mg S/g sorbent for desulfurizing JP5, JP8, and off-road diesel (initial sulfur contents of 1172, 630, and 452 ppmw, respectively). Mixed oxide TiO_2-Al_2O_3 support provided seats for more dispersed and oxidized silver as observed from TPR and XRD. Breakthrough of model fuels revealed that the aromatic rings are essential for adsorption onto silver adsorbent. The sulfur capacity of the adsorbent was reduced in presence of other aromatic compounds. Infrared spectra of the samples treated with sulfur and aromatic compounds detected both physisorbed and chemisorbed sulfur species on Ag/TiO_2-Al_2O_3 adsorbent.

(343d) Multi-Physics Modeling of Auto-Thermal Diesel Surrogate Reforming

Rajesh D. Parmar[1], Amrit Jalan[2], Dushyant Shekhawat[3], William H. Green Jr.[2], Brant. A. Peppley[1] and Kunal Karan[1], (1)Queen's - RMC Fuel Cell Research Centre, Queen's University, Kingston, ON, Canada, (2)Department of Chemical Engineering, Massachusetts Institute of Technology, Cambridge, MA, (3)National Energy Technology Laboratory, Morgantown, WV

Abundant supply and existing infrastructure makes diesel a favorable candidate for portable energy supply. Diesel generators are widely used in locations where there is no reliable electric grid, but they are noisy, have emission issues, and often have poor fuel efficiency in practical operations. Because of these drawbacks, there is interest in developing Solid Oxide Fuel Cell (SOFC) systems which use diesel fuel, but which deliver electricity quietly and with lower emissions and higher efficiency.

The current study is part of a larger effort aimed at developing 1-5 kW diesel fed solid oxide fuel cells at SOFC-Canada and is mainly focused on understanding and deconvoluting the mechanism of diesel surrogate auto-thermal reforming in an experimental packed bed reactor. Gas phase kinetics play an important role during auto-thermal reforming and this study employs a detailed kinetic model developed using the automated Reaction Mechanism Generator (RMG) software. The generated model has about 9500 reactions and 450 species incorporating updated parameters from experiment and theory. The model has been validated against ignition delay data at different equivalence ratios and was found to perform reasonably well. The model has also been validated against the packed bed reactor steady state concentration data at different operating conditions. Coupling the fluid dynamics and heat transfer effects defined above with the large number of reactions and species was found to be very difficult using currently available commercial software such as Fluent and COMSOL. Hence an iterative approach was used in which simplified packed bed plug flow reactor model with heat transfer was solved using a finite element solver while the kinetics equations were solved using the CHEMKIN plug flow solver. The generated model shows the importance of entrance region effects for auto-thermal reformer design. Gas phase oxidation/pyrolysis consumes a large part of the hydrocarbon leading to lower molecular weight products that reach the catalyst surface and participate primarily in steam reforming reactions dominant on the surface of the catalyst.

(343e) Heat-Integrated Steam Reformer for Hydrogen Production

Saurabh A. Vilekar, Christian Junaedi, Kyle Hawley, Dennis Walsh and Subir Roychoudhury, Precision Combustion, Inc. (PCI), North Haven, CT

Precision Combustion, Inc (PCI) has been developing a novel heat-integrated steam reforming reactor for efficient syngas (i.e., H_2 and CO) generation based on its patented Microlith® technology, providing higher heat flux compared to other state-of-the-art reactors. The Microlith based reactor design results in a remarkably compact, lightweight, and efficient steam reforming reactor, which is otherwise widely known to be limited by the heat transfer resistance from the exothermic (heat source) to the endothermic reforming side. The burner component comprises of catalytic oxidizer instead of flame-stabilized combustion, resulting in improved temperature control and uniformity. In this paper, results from testing the reformer with a distillate fuel (containing 2-5 ppm_w sulfur) and natural gas will be presented, where near equilibrium product distribution was observed. The effect of steam-to-carbon ratio, operating pressure (i.e., up to 10 atm), and thermal inputs on the catalyst performance, fuel conversion and H_2 production will be discussed. The sulfur tolerance studies of the steam reforming catalyst and the burner catalyst will also be presented. Catalytic sulfur tolerance allows effective utilization of poor quality or waste fuels, thus significantly improving the overall process efficiency. Results from hundreds of hours of testing of the steam reformer unit showed minimal catalyst degradation, thus providing a measure of the reactor performance under realistic conditions and demonstrating the durability of the reformer. Finally, downstream hydrogen separation approaches are being examined and results will be reported. We will also discuss modeling analysis for system integration for high purity hydrogen generation.

(343f) Diesel Reforming Over Pyrochlore-Based Monolith

Daniel J. Haynes[1], David Berry[2], Dushyant Shekhawat[1], Mark W. Smith[3], Matthew M. Seabaugh[4], Richard Long[4] and James J. Spivey[5], (1)National Energy Technology Laboratory, Morgantown, WV, (2)NETL, Department of Energy, Morgantown, WV, (3)URS Corporation, Washington Division, Morgantown, WV, (4)NexTech Materials, Lewis Center, OH, (5)Chemical Engineering, Louisiana State University, Baton Rouge, LA

Diesel Reforming over Pyrochlore-Based Monolith

Daniel J. Haynes[1], David A. Berry[1], Dushyant Shekhawat[1], Mark W. Smith[2], Mathew Seabaugh[3], Rich Long[3], James J. Spivey [4]

[1] National Energy Technology Laboratory (NETL), Morgantown, WV 26505

[2] URS Washington Division, Morgantown, WV 26505

[3] NexTech Materials Ltd., Lewis Center, OH 43035

[4] Louisiana State University, Baton Rouge, LA 70803

NETL has been conducting research in developing a highly active and selective catalyst for the reforming of heavier hydrocarbon feed stocks (i.e. diesel) into a hydrogen-rich synthesis gas. This is a necessary enabling technology for the commercialization of fuel cells among other applications. Previously, NETL demonstrated 1000 hrs of continuous diesel reforming using a pyrochlore-based catalyst in powder form. However it is essential to develop a structured form, which can handle high gaseous throughput with minimal pressure drop. Key issues in the development of these structured forms include: homogeneity of coating layers, adhesion of catalyst layers (coating stability), thermal stability, and ensuring accessibility of the active sites on the catalytic material. In this study, a square channel alumina based monolith structure (400 cpi) is first coated with an oxygen-conducting support onto which an active pyrochlore phase was then deposited. These monoliths were tested for the partial oxidation (POX) and oxidative steam reforming (OSR) of pump diesel fuel. While no deactivation was observed over the duration of the tests (24h), the catalyst coating layer was not optimized and therefore the lack of homogeneity resulted in insufficient use of active phase. Imaging of the catalyst shows that there is less than optimum uniformity between the ZDC and the Pyrochlore, and current efforts are directed toward improving the contact between these two components of the catalyst.

(343g) Carbon Deposition Over Supported Rh-Ni Catalysts in Steam Reforming of Liquid Hydrocarbon Fuel for H2 Production: Influence of Support, Temperature, and Sulfur

Chao Xie[1], Yongsheng Chen[2] and Chunshan Song[1], (1)Energy and Mineral Engineering, Pennsylvania State University, University Park, PA, (2)Energy and Mineral Engineering, Penn State, State College, PA

Carbon Deposition over Supported Rh-Ni Catalysts in Steam Reforming of Liquid Hydrocarbon Fuel for H_2 production: Influence of Support, Temperature, and Sulfur

Chao Xie[a], Yongsheng Chen[a,b], Chunshan Song[a,b*]

[a] Clean Fuels and Catalysis Program, EMS Energy Institute, The Pennsylvania State University, 209 Academic Projects Building, University Park, PA, 16802, USA..

[b] Department of Energy and Mineral Engineering, The Pennsylvania State University, University Park, PA, 16802, USA.

*Corresponding author: E-mail: csong@psu.edu; Tel: 814-863-4466; Fax: 814-865-3573

This contribution aims at elucidating the effects of support (Al_2O_3, CeO_2-Al_2O_3, and CeO_2-SiO_2), reaction temperature (550 and 800 ºC) and the presence of sulfur (350 ppm by weight) on the carbon deposition over bimetallic Rh-Ni catalysts in steam reforming of liquid hydrocarbon fuel for H_2 production. The amount, morphology, and chemical nature of carbon deposits on used catalysts were studied by temperature-programmed oxidation (TPO), scanning electron microscopy (SEM), and X-ray absorption near edge structure (XANES) spectroscopy. It was revealed that (I) CeO_2 modification facilitates carbon removal due to its superior redox property, and (II) Al_2O_3 is much more effective than SiO_2 in suppressing the growth of filamental carbon. As revealed by X-ray Diffraction (XRD) and temperature-programmed reduction (TPR), the Ni particles in Rh-Ni/CeO_2-Al_2O_3 were remarkably smaller than those in Rh-Ni/CeO_2-SiO_2 very likely because of the stronger interaction of Ni-Al_2O_3 as compared with that of Ni-SiO_2. Therefore, the growth of filamental carbon on Rh-Ni/CeO_2-Al_2O_3 was greatly hindered as carbon deposition becomes less favorable on smaller Ni particles. Comparative study of the used Rh-Ni/CeO_2-Al_2O_3 catalysts after the reactions without and with sulfur at 550 and 800 ºC indicated that both reaction temperature and the presence of sulfur dramatically affected the catalyst behavior of carbon deposition: (I) in the absence of sulfur, the formation of filamental carbon was significant at 550 ºC, which was prohibited by increasing the temperature to 800 ºC, and (II) for the sulfur-containing reaction, sulfur was able to suppress the growth of filamental carbon at 550 ºC, while catalyst deactivation caused by sulfur poisoning gave rise to abundant carbon deposits at 800 ºC as a result of severe hydrocarbon pyrolysis at such a high-temperature condition.

Fuel Processing with Diesel Fuel and Kerosene for APU Applications

ID# 265665

Joachim Pasel[1], *Jan Meißner*[1], *Remzi Can Samsun*[1], *Ralf Peters*[1], *Detlef Stolten*[1,2]

[1]*Forschungszentrum Jülich GmbH, Institute of Energy and Climate Research, IEK-3: Electrochemical Process Engineering, 52425 Jülich, Germany*

[2]*Chair for Fuel Cells, RWTH Aachen University, Germany*

Introduction

The fuel processing unit of a fuel cell system with high temperature polymer electrolyte fuel cell (HT-PEFC) consists of three main components: the autothermal reformer (ATR), the water-gas-shift reactor (WGS) and the reactor for catalytic combustion (CAB). In combination with necessary balance-of-plant components such as pumps, blowers, heat exchangers and sensors as well as an HT-PEFC, a complete fuel cell system emerges, which can be used as an auxiliary power unit (APU) for onboard power supply in aircraft or trucks. The main benefits of fuel cell systems in comparison to conventional technologies are fuel savings and emission reduction. The following chemical reactions take place in the three main fuel processing reactors:

Autothermal reformer:

$C_nH_m + n\,H_2O$	$\longleftrightarrow$	$n\,CO + (m/2 + n)\,H_2$	endothermic
$C_nH_m + n/2\,O_2$	$\longleftrightarrow$	$n\,CO + m\,H_2$	exothermic
$CO + H_2O$	$\longleftrightarrow$	$CO_2 + H_2$	exothermic
$CO + 3\,H_2$	$\longleftrightarrow$	$CH_4 + H_2O$	exothermic

Water-gas-shift reactor:

$CO + H_2O$	$\longleftrightarrow$	$CO_2 + H_2$	exothermic

Catalytic burner:

$CH_4 + 2\,O_2$	$\longleftrightarrow$	$CO_2 + 2\,H_2O$	exothermic
$H_2 + 0.5\,O_2$	$\longleftrightarrow$	H_2O	exothermic
$CO + 0.5\,O_2$	$\longleftrightarrow$	CO_2	exothermic

Reactors for Fuel Processing

In this section, this extended abstract will focus on the methodology for the up-scaling of reactors for the water-gas-shift reaction. In recent years, three reactor generations were developed at Jülich (WGS 3, WGS 4 and WGS 5). Their progression was based on former investigations and on relevant data from the literature (1–3). Each reactor is equipped with a precious metal catalyst supported on ZrO_2-CeO_2 supplied by Clariant AG. Their design is characterized by two stages working adiabatically at different temperature levels. The high-temperature shift (HTS) stage covers a temperature range between 400 °C and 450 °C, the low-temperature shift (LTS) stage exhibits temperatures between 300 °C and 330 °C. Cooling between the two stages is achieved by quenching water into the gas stream.

The starting point of development was WGS 3 with a thermal power of 13 kW, a power density of 5.0 kW/l, and a specific power of 4.5 kW/kg. The value for the thermal power refers to the lower heating value of moles of hydrogen produced via autothermal reforming flowing through the water-gas-shift reactor. In the case of WGS 4, the thermal power was increased to 28 kW by linearly increasing the catalyst volume. The values for the power density and the specific power remained almost constant in comparison to those of WGS 3. WGS 5 with a thermal power of 140 kW was specifically developed for aircraft applications. In this case, the values for the power density and the specific power were higher and amounted to 8.5 kW/l and 8.8 kW/kg, respectively.

CFD simulations were performed on up-scaling WGS 4 to WGS 5 in order to design the area in which water is injected into the low-temperature shift stage. This area is called the evaporation chamber. Figure 1 (left) shows a sectional drawing of the first version of the WGS 5 evaporation chamber. In this version, water is injected into the evaporation chamber in liquid form at four equidistant points around the circumference. The reformate from the HTS stage flows into the evaporation chamber at a temperature of approx. 450 °C. Internal components for mixing the reformate flow with the evaporated water or for homogenizing the flow are not part of this design. However, the resulting temperature distribution in the evaporation chamber is unacceptable (see Figure 1, right). At the outlet of the evaporation chamber on the right-hand side, strong radial temperature differences between approx. 240 °C and 350 °C are clearly visible. If the catalyst in the low-temperature shift stage was fed a gas mixture with such a strong radial temperature gradient, a uniform reaction flow would be impossible.

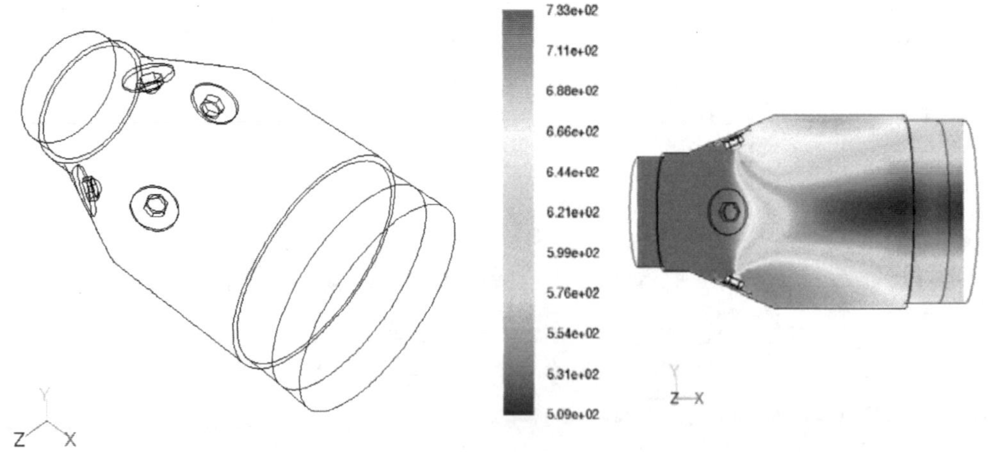

Figure 1. Sectional drawing of the first version of the evaporation chamber (left), temperature distribution in the evaporation chamber (right)

Two constructive measures were therefore implemented to prevent this temperature gradient forming. The first involved the integration of a rotating body in the flow direction above the water injection in order to ensure that the gas flow coming from the high-temperature shift stage rotates around its main flow axis. The second measure was the incorporation of a perforated plate in front of the gas flow outlet from the evaporation chamber into the low-temperature shift stage in order to homogenize the flow. Figure 2 (left) shows the second version of the evaporation chamber with the two internal components.

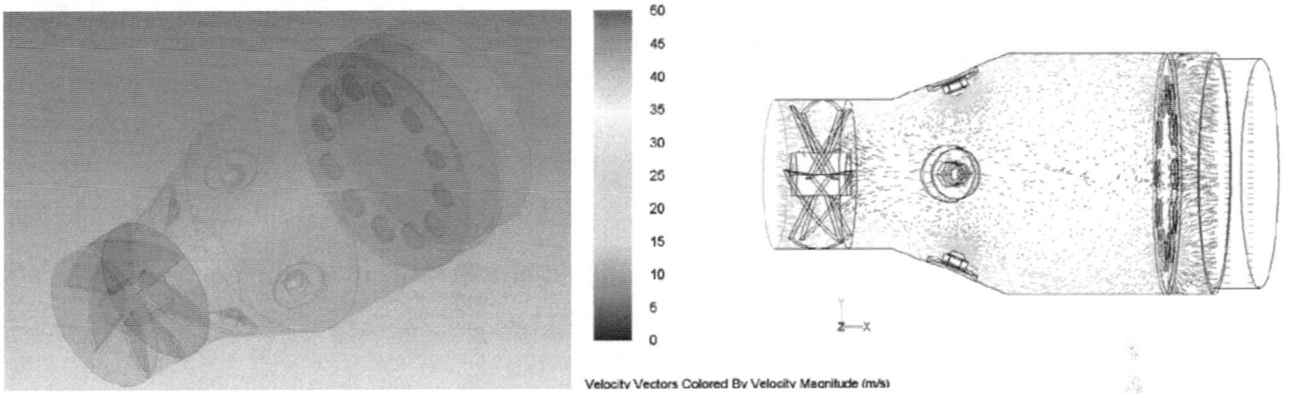

Figure 2. Sectional drawing of the second version of the evaporation chamber (left), flow profile in the evaporation chamber at 100 % load (right)

Figure 2 (right) shows the flow profile in the evaporation chamber at 100 % load produced with the aid of the components described above. The velocity vectors of the individual gaseous molecules are shown. The rotating body leads to the formation of a gas phase flow along the external wall of the reactor as a result of the centrifugal force exerted on the individual gaseous molecules. It also causes an area to develop in the center of the evaporation chamber with a backflow of some of the gaseous molecules. The perforated plate contributes to the formation of this area. The evaporation of the injected water droplets thus occurs in the counter flow: the water droplets collide with the returning gaseous molecules and evaporate. This prevents water droplets being transported to the catalyst, and a compact central evaporation zone is created.

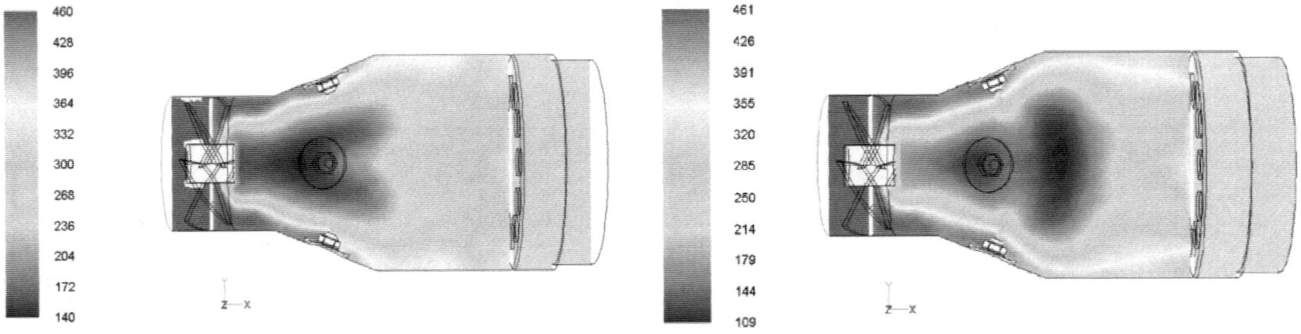

Figure 3. Temperature profile in the evaporation chamber at 100 % load (left) and at 25 % load (right)

Figure 3 shows the resulting temperature profiles in the evaporation chamber at 100 % and 25 % load. In both cases, the gas flow had a constant temperature of approx. 300 °C at the outlet of the evaporation chamber. Under these constant conditions, a water-gas-shift reaction is possible in the low-temperature shift stage with the necessary carbon monoxide conversion.

Experimental Results and Discussion

The experimental section of this abstract deals with a number of experiments performed with the autothermal reformer ATR 8 using different kinds of diesel fuel. The fundamental design of the autothermal reformers developed at Jülich is explained in detail in [4]. The chemical and physical characteristics of the different fuels are shown in Table 1. It becomes obvious that they differ from each other in terms of their boiling ranges, their mass proportions of aromatics, and their molecular formula. For these experiments, the autothermal reformer was run at a partial load of 60 % and educt water was not evaporated or preheated before entering the internal heat exchanger of the reactor.

Table 1. Different kinds of diesel fuel applied for autothermal reforming with ATR 8

	Supplier	90 % recovered [°C]	Aromatics [wt.%]	Chemical formula
EcoPar diesel	EcoPar Sweden	338	< 0.1	$C_{15}H_{32}$
NExBTL diesel	NESTE Oil Finland	295	< 0.1	$C_{17}H_{34}$
Winter grade diesel	Shell MDS Malaysia	307	< 0.1	$C_{16}H_{34}$
Summer grade diesel	Shell MDS Malaysia	350	< 0.1	$C_{18}H_{38}$
Ultimate diesel	ARAL Germany	350 – 360	approx. 15	$C_{19}H_{38}$

Figure 4 shows the concentrations of hydrogen, carbon monoxide and methane during operation of ATR 8 with the five different diesel fuels from Table 1 at a molar O_2/C ratio of 0.47 and a molar H_2O/C ratio of 1.9. Only small differences were observed in the gas concentrations in the reformate. The hydrogen concentrations were in the range of 36.5–37.5 vol.%, those of carbon monoxide varied between 9.2 vol.% and 11.2 vol.%, while the methane contents were stable showing values of 0.3 vol.%. However, the experiment with Ultimate diesel was an exception. A remarkable decrease in the hydrogen content to 33.7 vol.% in parallel to the highest value for the carbon monoxide concentration at 11.3 vol.% were observed. This finding can be explained by the high mass fraction of aromatics contained in Ultimate diesel. For stoichiometric reasons, the conversion of aromatics during autothermal reforming delivers less moles of hydrogen than that of aliphatic hydrocarbons.

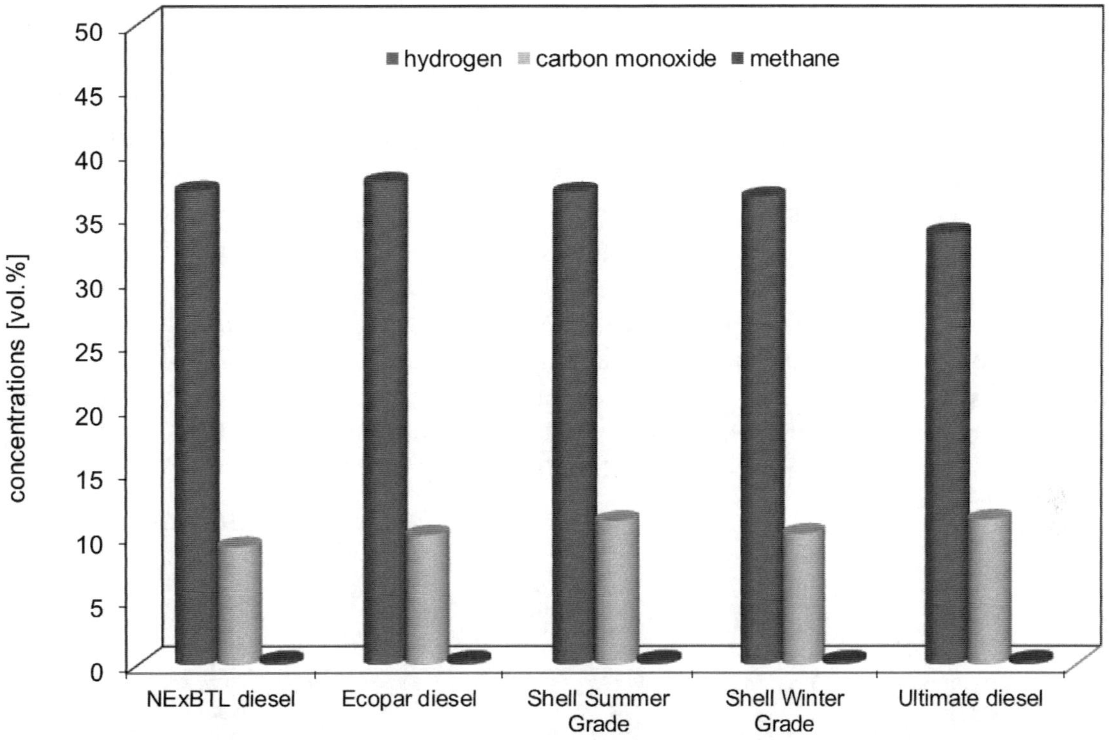

Figure 4. Concentrations of hydrogen, carbon monoxide and methane during the operation of ATR 8 with five different diesel fuels at a molar O_2/C ratio of 0.47 and a molar H_2O/C ratio of 1.9, 60 % reformer load, water injection at ambient temperature

In addition to the results shown in Figure 4 with respect to the main products obtained during operation of ATR 8 with five different diesel fuels, Figure 5 displays the respective concentrations of by-products in the gas phase and in the liquid phase at a molar O_2/C ratio of 0.47 and a molar H_2O/C ratio of 1.9. The abbreviation NMHC in Figure 5 stands for the concentration of non-methane hydrocarbons, i.e. the sum of the contents of several single by-products such as ethene, ethane, propene, propane, butenes, pentenes, hexenes and benzene. Among them, propene showed the highest values and is therefore exemplarily displayed in Figure 5. Since water is added to the reformer at large excesses (molar H_2O/C ratio = 1.9), samples of condensed, non-converted H_2O from the ATR can be collected directly behind the reactor. These samples were analyzed for the dissolved amounts of total organic carbon (TOC) and non-purgeable organic carbon (NPOC). These values are also shown in Figure 5. Acetic acid showed the highest contents among the substances in the condensed water and is also displayed. Other dissolved by-products found were acetaldehyde, propanal, acetone, butanal, 2-butanone, benzene, 2-pentanone, toluene and benzaldehyde.

Figure 5 shows very low values for the concentrations of the by-products when applying NExBTL diesel. The NMHC value was only 68 ppmv. According to Table 1, this diesel fuel is free of aromatics with the lowest temperature for 90 % recovery during fractionated distillation. When using Ecopar diesel, the concentrations for the by-products were slightly higher. The NMHC value reached

226 ppmv. A strong increase in the concentrations of all displayed by-products occurred when Shell summer grade diesel fuel was used. The NMHC value improved to 1474 ppmv, that of propene to 372 ppmv. The NPOC and TOC concentrations amounted to 134 and 170 ppmw, respectively. Acetic acid was formed at considerable amounts of 31 ppmw. Shell summer grade diesel fuel was also free of aromatics but had a remarkably higher temperature for 90 % recovery of 350 °C. Some of the single components in Shell summer grade diesel fuel demanded a higher enthalpy flow for their evaporation before entering the ATR catalyst. Therefore, the danger of liquid fuel droplets impinging on the catalyst, being submitted to pyrolysis on the catalyst, and thus damaging active sites of the catalyst is higher for Shell summer grade diesel fuel than for NExBTL diesel. Obviously, the catalyst inside ATR 8 was deactivated by the experiments with Shell summer grade diesel fuel since the concentration values for the by-products in Figure 5 when using Shell winter grade diesel fuel were much higher (NMHC = 931 ppmv) than those observed with NExBTL diesel and Ecopar diesel although their physical and chemical characteristics in Table 1 were comparable. Finally, Figure 5 shows the results obtained for Ultimate diesel, which has by far the highest mass fraction of aromatics among the fuels investigated (approx. 15 wt.%) and a comparably high temperature range for 90 % recovery during fractionated distillation of 350–360 °C. Accordingly, the concentrations for the by-products were found to be rather high. The NMHC value reached 1244 ppmv, that of propene 274 ppmv. The NPOC and TOC concentrations were 93 ppmw and 118 ppmw, respectively.

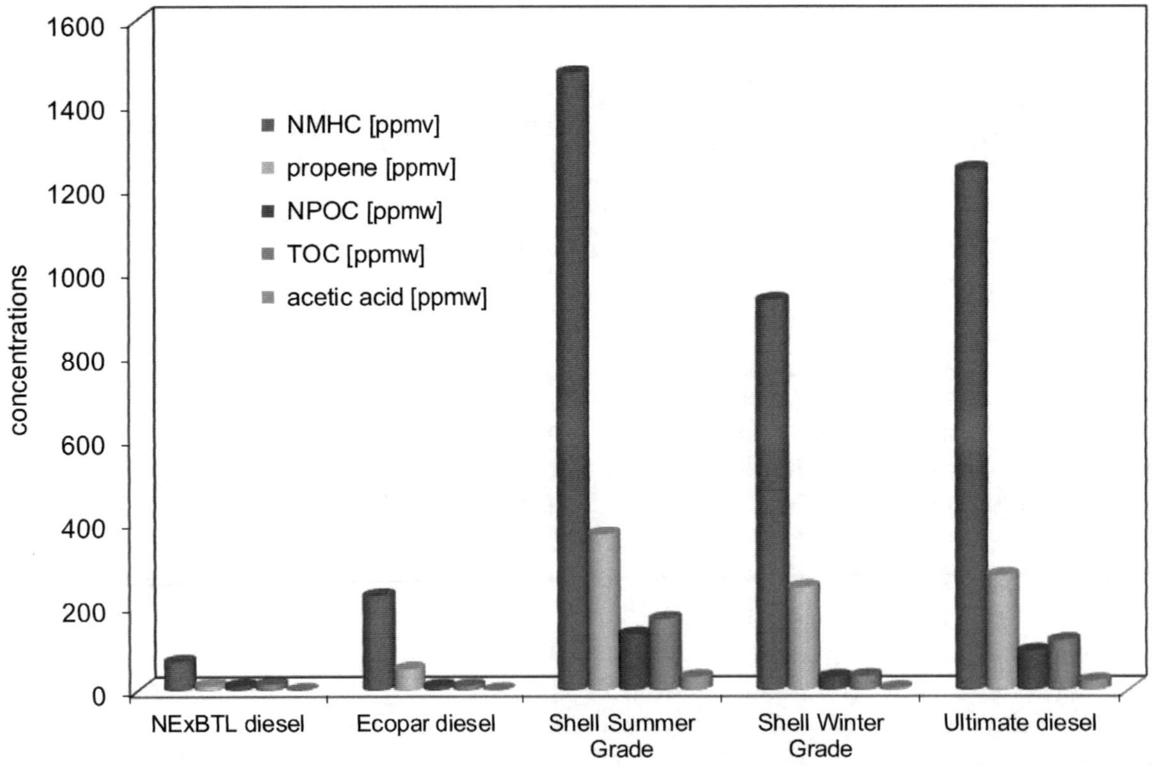

Figure 5. Concentrations of by-products in the gas phase and in the liquid phase during the operation of ATR 8 with five different diesel fuels at a molar O_2/C ratio of 0.47 and a molar H_2O/C ratio of 1.9, 60 % reformer load, water injection at ambient temperature

Outlook

More information about fuel processing with diesel fuel and kerosene for APU applications done at Jülich will be given during the respective presentation at the 2012 AIChE annual meeting and in the following paper for the special issue of Energy & Fuels.

References

1. Pasel, J.; Samsun, R.C.; Schmitt, D.; Peters, R.; Stolten, D. Journal of Power Sources 2005, 152, 189 - 195

2. Pasel, J.; Cremer, P.; Wegner, B.; Peters, R.; Stolten, D. Journal of Power Sources 2004, 126, 112 – 118

3. Koryabkina, N.A.; Phatak, A.A.; Ruettinger, W.; Farrauto, R.J.; Ribeiro, F.H. J. Catal. 2003, 217, 233

4. Pasel, J.; Meißner, J.; Pors, Z.; Samsun, R.C.; Tschauder, A.; Peters, R. Int. J. Hydrogen Energy 2007, 32, 4847 – 4858

(343i) Reactor Design for Diesel Reforming Using a Graded-Composition Monolith

Mark W. Smith[1,2,3], David A. Berry[2], Dushyant Shekhawat[2], Daniel J. Haynes[2], James J. Spivey[2,4] and John Zondlo[2,5], (1)Chemical Engineering, West Virginia University, Morgantown, WV, (2)National Energy Technology Laboratory, Morgantown, WV, (3)URS Corporation, Washington Division, Morgantown, WV, (4)Chemical Engineering, Louisiana State University, Baton Rouge, LA, (5)Chemical Engineering, West Virginia University, morgantown, WV

An NETL-developed Rh-substituted pyrochlore catalyst successfully reformed pump diesel into a H_2- and CO-rich synthesis gas for over 1000 hrs with no signs of deactivation. Even with an already low level of precious metal loading, the possibility of decreasing catalyst cost was later demonstrated using a graded catalyst bed approach. These demonstrations were performed using catalyst powders. In order to develop a commercially representative catalyst, this study examined the application of Rh-, and less expensive Ni- and Ru-substituted pyrochlore catalysts deposited onto alumina foam monoliths. The catalyst coated monoliths were placed into the reactor in several different configurations to examine the performance of a graded monolith under the catalytic partial oxidation (CPOX) of a diesel surrogate fuel. Catalytic testing was followed by carbon characterization of the individual coated foams using temperature programmed oxidation (TPO). This approach allowed for quantification of carbon formation on each catalyst in different sections of the reactor during CPOX experiments. The results of this study confirmed the trends observed in the previous powder studies and explored additional combinations of catalyst compositions in the different reaction zones of the reactor. Further it demonstrated the viability of applying the graded-composition approach to a commercially structured monolithic reactor.

(355a) Carbon Molecular Sieve Membranes for N_2/CH_4 Separation

Xue Ning, School of Chemical & Biomolecular Engineering, Georgia Institute of Technology, Atlanta, GA and William J. Koros, School of Chemical & Biomolecular Engineering, Georgia Institute of Technology, Atlanta, GA

Many natural gas reserves are contaminated by a high content of nitrogen which has to be removed to meet pipeline specifications. Currently available nitrogen rejection technologies include cryogenic distillation, pressure swing adsorption (PSA), and lean oil absorption. Polymer-based membrane separation for N_2/CH_4 has not been pursued widely because of the small size difference between N_2 and CH_4 and the competing effects of diffusion and sorption. These challenges lead to unattractive permselectivity using traditional polymeric materials due to the broad distribution of segmental motions of polymer chains. Carbon Molecular Sieve (CMS) membranes overcome this problem by providing a hyper-rigid structure and distinctive bimodal pore size distribution.

Different pyrolysis conditions for Matrimid® polyimide and resultant separation performance of CMS membranes were studied in this work. It was shown that N_2/CH_4 selectivity was highly improved with increased pyrolysis temperature. Sorption tests were performed to deconvolute permeability into diffusion coefficient and sorption coefficient. The temperature effects on permeability, sorption coefficient and diffusion coefficient were also studied.

(355b) High Performance Carbon Molecular Sieves (CMS) Hollow Fiber Membranes for Natural Gas Separation

Nitesh Bhuwania, School of Chemical and Biomolecular Engineering, Georgia Institute of Technology, Atlanta, GA and William J. Koros, School of Chemical & Biomolecular Engineering, Georgia Institute of Technology, Atlanta, GA

High Performance Carbon Molecular Sieves (CMS) Hollow Fiber Membranes for Natural Gas Separation

Nitesh Bhuwania, William J. Koros

Nitesh.Bhuwania@chbe.gatech.edu

bill.koros@chbe.gatech.edu

School of Chemical and Biomolecular Engineering

Georgia Institute of Technology, Atlanta, GA

CMS membranes show the enhanced separation capability for natural gas purification (processing of CO_2 from CH_4 stream). For industrial application, it is important to achieve this high performance in hollow fiber membrane configuration. Previous studies in CMS hollow fibers have shown the ability to achieve the good separation efficiency (i.e. selectivity) for CO_2/CH_4 but with a drop in transport flux in comparison to polymer precursor fibers. This is caused by the densification of porous morphology in CMS fibers, termed as"sub-structure collapse".

CMS membranes are formed by heating the polymer precursors at higher temperatures above its decomposition point. The phenomenon of sub-structure collapse occurs during the thermal transition of polymer, which begins at its glass transition temperature (T_g). Therefore in our study, we look at stabilizing the precursors before they undergo the thermal decomposition. Commercial available "Matrimid®" and custom-synthesized 6FDA-BPDA/DAM polyimides were used as the precursor platform for our research.

The impact of pre-pyrolysis stabilization on fiber morphology and separation performance for final CMS fiber membrane is significant. Our technique not only show attractive separation performance but also overcomes some key challenges during the processing of CMS membranes. We will go in detail of our method during the talk.

Evaluation of Facilitated Transport Membrane for Olefin/Paraffin Separation

Majid Keyvani, LyondellBasell

Ali Hamza, and Glenn Towe, Imtex Membranes Corporation

Abstract:

Currently, the separation of olefins (ethylene, propylene) from paraffins (ethane, propane) on a commercial scale is accomplished almost exclusively by cryogenic distillation in petrochemical industries and is considered highly energy intensive. Consequently, there is an enormous economic incentive to explore alternative separation technologies with lower energy consumption and an opportunity to achieve related reductions in environmental impact from air pollutant and greenhouse gas (GHG) emissions.

Several attempts have been made in the past to develop alternative separation technologies for this application, including significant work on facilitated transport membranes. One of the main unresolved challenges was membrane instability over time. Imtex Membranes Corporation has developed a membrane that has shown performance stability over thousands of hours of operation on both coupons and spiral wound test units. Imtex membranes are based on chitosan, a polysaccharide material, and silver nitrate as a facilitation agent.

LyondellBasell and Imtex Membrane Corporation have set up a collaborative effort to evaluate the performance of Imtex membrane in the olefin/paraffin separation, namely C2 and C3 splitter applications.

The performance results were very encouraging for C2 and C3 splitters as greater than 99.5% purity and 90% recovery were achieved in both cases. Based on the membrane performance results, we believe that the technology presents potential opportunities in upgrading refinery grade propylene to polymer grade propylene as the membrane was evaluated under operating conditions comparable to those of commercial plants.

There is also a promising potential for butene/butane separations where distillation separations require large numbers of trays and extremely high reflux ratios, or are difficult to achieve at all. Applying Imtex membranes in C4 separation can have a very positive impact on the economics of olefin metathesis, an increasingly important source of propylene.

Introduction and Background:

A recent Department of Energy (DOE) study [1] indicated distillation operations account for 95 percent of the total separation energy used in the refinery and chemical processing industries with 40,000 distillation columns operating in more than 200 different processes. Another study stated that olefin production is the single most energy consuming process in the chemical industry. The breakdown of the specific energy consumption shows that 30 percent of the total yearly energy consumption is associated with separation processes and 80 percent of that energy is utilized for olefin/paraffin separation [2]. This large energy requirement provides incentives to improve and enhance the energy efficiency in olefin/paraffin separation.

In light of the importance of olefin/paraffin separations, and because of the energy cost associated with distillation, alternative technologies continue to be the focus of industry studies. Some review literature [3, 4] describes the current status of research. It cites reactive absorption where silver or copper salts are used to extract the olefin from bulk gas in an absorber/stripper type process as promising new techniques. Union Carbide and BP/Stone and Webster took this type of process to the pilot plant stage. Other processes that were evaluated include extractive distillation and molecular sieves, such as UOP SORBUTENE technology. Despite all the reported studies, however, a viable alternative to conventional distillation has yet to be developed.

One of the most comprehensive research efforts in this subject was carried out by a DOE sponsored program led by Membrane Technology Research, Inc. (MTR) partner SRI International, ABB Lummus Global [5]. The study concluded that silver salt-based facilitated transport membranes are not stable even in the presence of ideal olefin/paraffin mixtures. Silver tetrafluoroborate (AgBF4) was used with two polymer matrices: polyether-polyamide block copolymer, poly ethylene oxide (PEO), poly vinyl pyrrolidone (PVP), and poly1-trimethylsilyl-1-propyne (PTMSP). However, regardless of the polymer matrix used, decline in membrane performance occurred. They attributed the issue with an unrecognized phenomenon labeled as " olefin conditioning" implying that the very species targeted for separation by the carrier membranes (olefins) are a source of membrane instability.

Imtex Membranes Corporation successfully developed a synthetic membrane with a viable alternative to distillation in separating olefins from their respective mixtures with paraffin. The patented [6] chitosan based Imtex membrane addresses the shortcomings of the aforementioned alternative approaches, including the avoidance of the "olefin conditioning" phenomenon experienced by MTR study.

Due to the presence of hydroxyl (-OH) and the amino ($-NH_2$) groups, Chitosan is a very hydrophilic and chelating polymer with excellent film-forming characteristics and has an excellent adsorption capacity for a number of metal ions such as the Silver cation. As presented here, excellent olefin selectivities and permeabilities were obtained when using chitosan in olefin/paraffin separation tests. The facilitation mechanism of the Imtex membrane is presumed to be through the olefin molecules hopping from one silver cation site to another as well as through the expected mobility of the olefin-silver ion complex within the aqueous phase of the membrane. The membrane structure can be balanced by the degree of deacetylation as well as by the silver loading level utilized. These aspects make chitosan an excellent base membrane for facilitated transport membranes

LyondellBasell and Imtex Membrane Corporation have set up a collaborative effort to evaluate the performance of Imtex membrane in C2 and C3 splitter applications.

In this paper we report performance data for propylene/propane and ethylene/ethane separation. Promising preliminary results for butenes/butanes are also presented.

Experimental:

The Imtex membrane uses a flat sheet composite structure based on Chitosan material and AgNO3 as a facilitation agent. Chitosan is a linear polysaccharide produced commercially by deacetylation of chitin, which is the structural element in the exoskeleton of crustaceans (such as crabs and shrimp) and other arthropods. The membrane units used for these studies included both small coupons as well as spiral wound elements. Each test coupon had a surface area of approximately 12.6 cm^2 while the spiral wound elements used in this study had each a nominal surface area of 600 cm^2. The chitosan layer thickness is estimated to be 7 micron. The tests were conducted with various ethylene/ethane, propylene/propane and butane/butane feed compositions. The feed pressure ranged from 40 psig to 360 psig.

Below is a picture of the front panel (flow lines and valves) of the Imtex test Apparatus used at LyondellBasell also showing two housings of spiral wound elements.

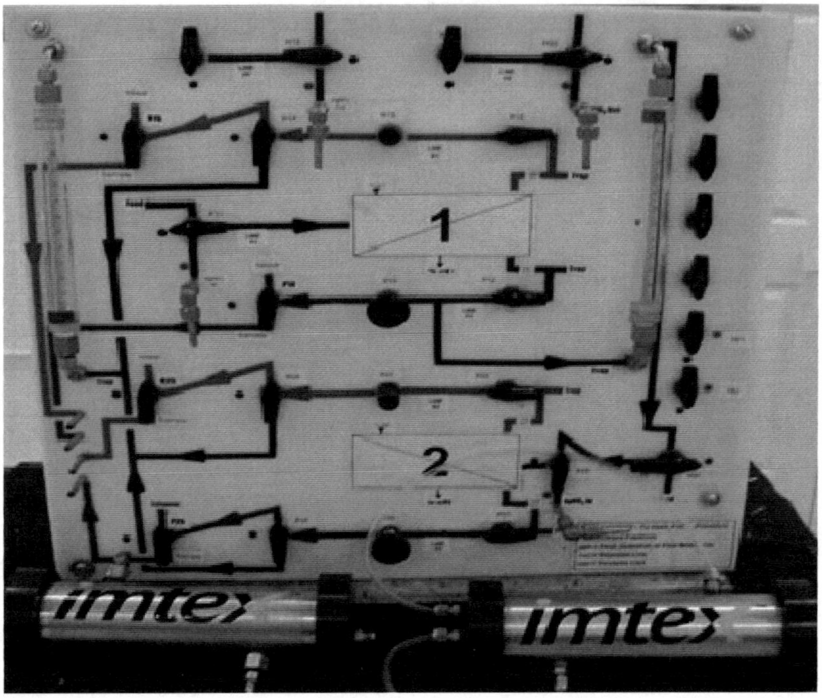

Results and Discussions:

Typical feed compositions to C2 and C3 splitters are 80/20 ethylene/ethane and propylene/propane. Using synthetic gas mixtures and operating at 60 psig feed pressure, selectivities greater than 100 were achieved using a spiral wound module made from first generation Imtex membrane LF1. This selectivity is higher than reported values in the literature for polymeric membranes (Figure 1) but not as good when compared with the reported data [7] for facilitated transport type membranes (Figure 2).

Figure 1. Membrane Performance Curve for Ethylene/Ethane

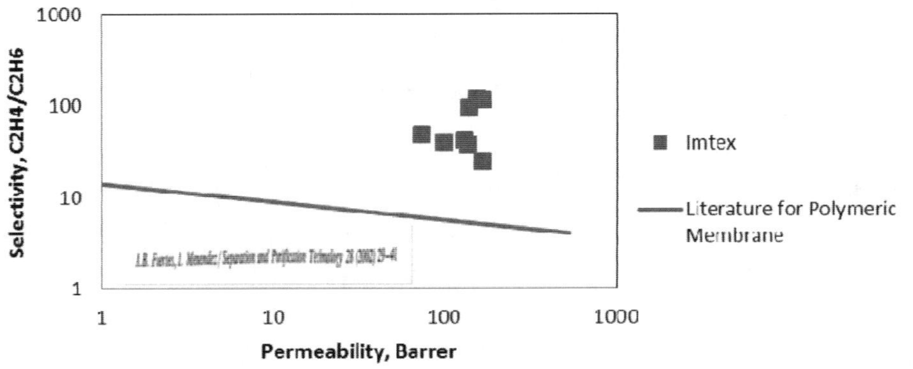

Figure 2-Imtex Membrane Performance against Literature (Facilitated Transport) for C2H4/C2H6

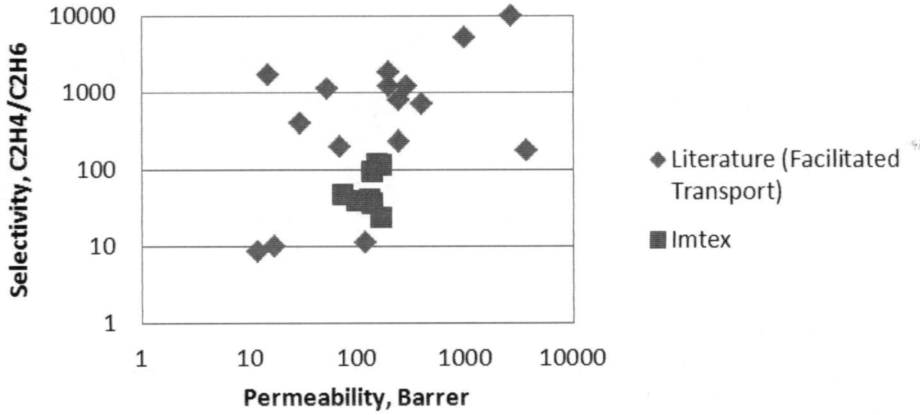

The ideal selectivity was calculated based on feed concentration instead of averaging feed and retentate concentrations:

Selectivity $\gamma = (x_{C2=} / x_{C2})_{perm} / (x_{C2=} / x_{C2})_{feed}$

Where x = mole fraction

The permeability was calculated based on the classic Fick's diffusion law. The STP (standard pressure, temperature) conditions defined as atmospheric and 68 °F and measured permeate flow rate was used as the flux (not total feed rate):

$$J = \frac{Perm(p_0 - p_l)}{l}$$

Where c= concentration, p = partial pressure, l= membrane thickness, J is flux, and Perm is permeability.

The excellent separation was seen over a wide range of feed composition. Figure 3 shows the performance of Imtex membrane LF1 as a function of ethylene content in the ethylene/ethane feed mixture.

Figure 3. Membrane Performance versus Feed Composition
Feed: Ethylene/Ethane Mixture at 66 psig and room temperature
Small spiral wound element with ~ 0.05 m^2 Imtex LF1 membrane

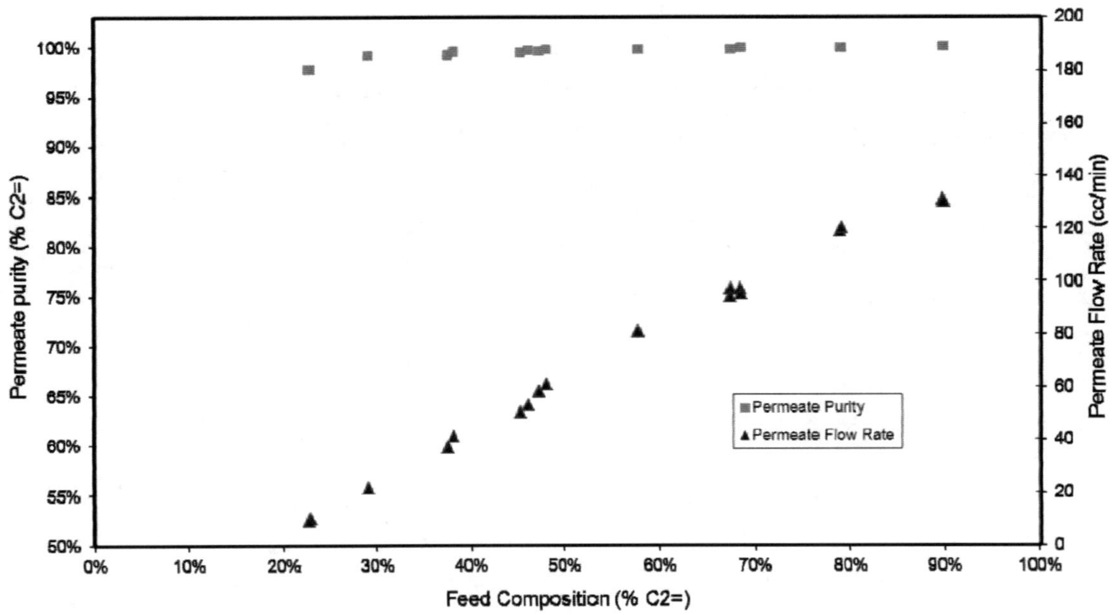

By increasing the feed pressure from 60 psig to 120 psig, the flux through the membrane almost doubled at the same product purity. Figure 4 shows performance (concentration of ethane in permeate stream) at varying pressures over time:

Figure 4- Ethylene/Ethane 80/20 Feed at Different Pressures:
Doubling Permeate Rate by increasing pressure while maintaining purity

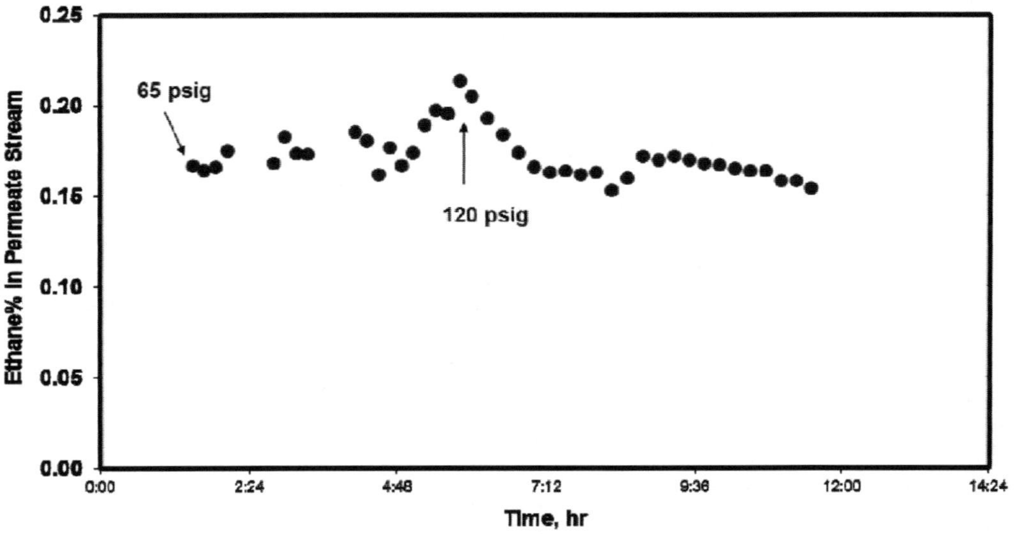

However, in spite of promising results for the C2 separation, we believe that the application for the C2 splitter faces many technical challenges, namely:

- C2 splitter is operated at cryogenic conditions and Imtex membrane has to be operated at ambient temperature (up to 50 ^{o} C).
- Permeate stream picks up water as it goes through the membrane matrix (has to be dried prior to returning back to column).
- There are acetylenic compounds present in the feed stream (e.g. acetylene, methyl acetylene (MA) and propadiene (PD), also known as MAPD). Ag complex strength in the membrane can be ranked strongest to weakest as followed PD>acetylene, methyl acetylene> ethylene> propylene. Thus any MAPD in the feed could end up with the olefin product on the permeate side.

On the other hand, the C3 splitter application does not have to deal with many of the above issues. Particularly, the membrane is best suited in upgrading the refinery grade propylene to a polymer grade.

The performance of Imtex membrane looks more favorable as compared against the published data on the application of facilitated membrane for propylene-propane separation. Figure 5 presents this comparison where a study was performed with 80/20 gas mixture of propylene/propane. Figure 6 shows the effect of feed pressure on the permeate stream purity and the flux:

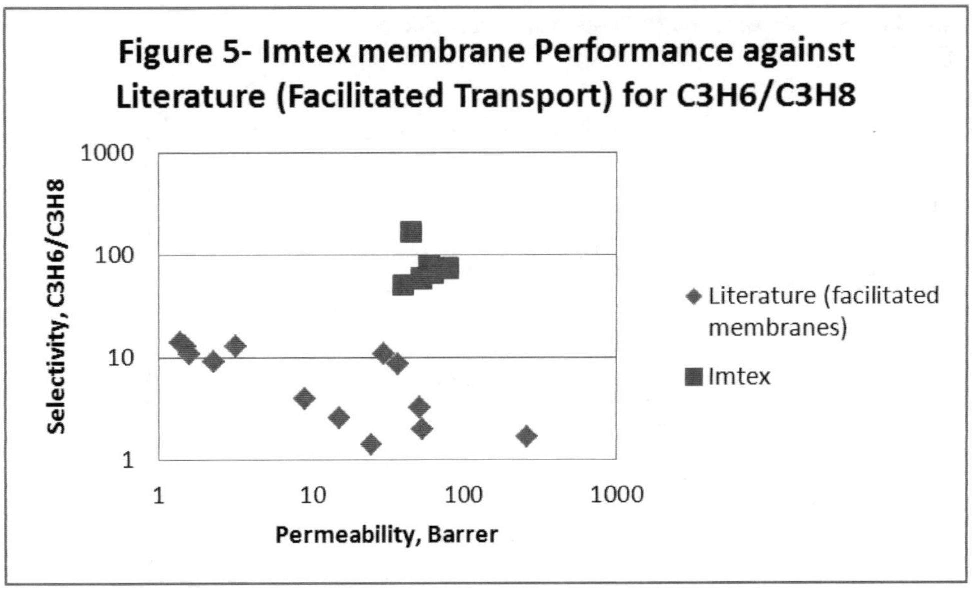

Figure 5- Imtex membrane Performance against Literature (Facilitated Transport) for C3H6/C3H8
Selectivity, C3H6/C3H8
1000
100
10
1
1
10
100
1000
Permeability, Barrer
Literature (facilitated membranes)
Imtex

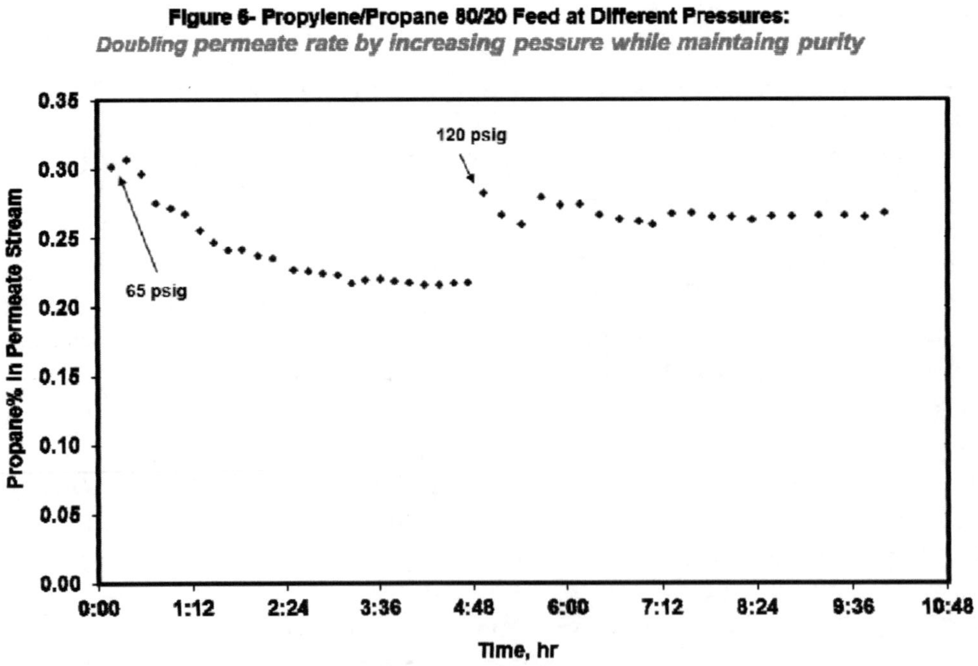

Figure 6- Propylene/Propane 80/20 Feed at Different Pressures:
Doubling permeate rate by increasing pessure while maintaing purity
Propane% in Permeate Stream
0.35
0.30
0.25
0.20
0.15
0.10
0.05
0.00
120 psig
65 psig
0:00
1:12
2:24
3:36
4:48
6:00
7:12
8:24
9:36
10:48
Time, hr

While Imtex membrane LF1 steadily showed high permeate purities (in both C2= and C3= cases) and higher permeabilities than those in the literature, for industrial applications, further improvement in flux was still needed.

Further development work led to making changes in the chemical and physical structure of the membrane. This resulted in the second generation of membranes referred to as HF3. This membrane still showed olefin product purities in the same range as LF1 but with an over three fold increase in the permeation rate.

Figure 7 below shows the performance of Imtex HF3 membrane for both C2s and C3s. The olefin mobility rate within the chitosan (along with a bonded Ag+ cation or as a hopping mechanism from one Ag+ site to another) seems to be similar for both ethylene and propylene. The molar or volumetric rates observed were at nearly the same level or at times slightly lower in the case of propylene. The difference in molecular weight between ethylene and propylene, however, reveals the more established mass flow rate difference as shown in Figure 7.

Figure 7. Imtex HF3 Membrane Performance
C2s and C3s Separation at 100 psig and room temperature

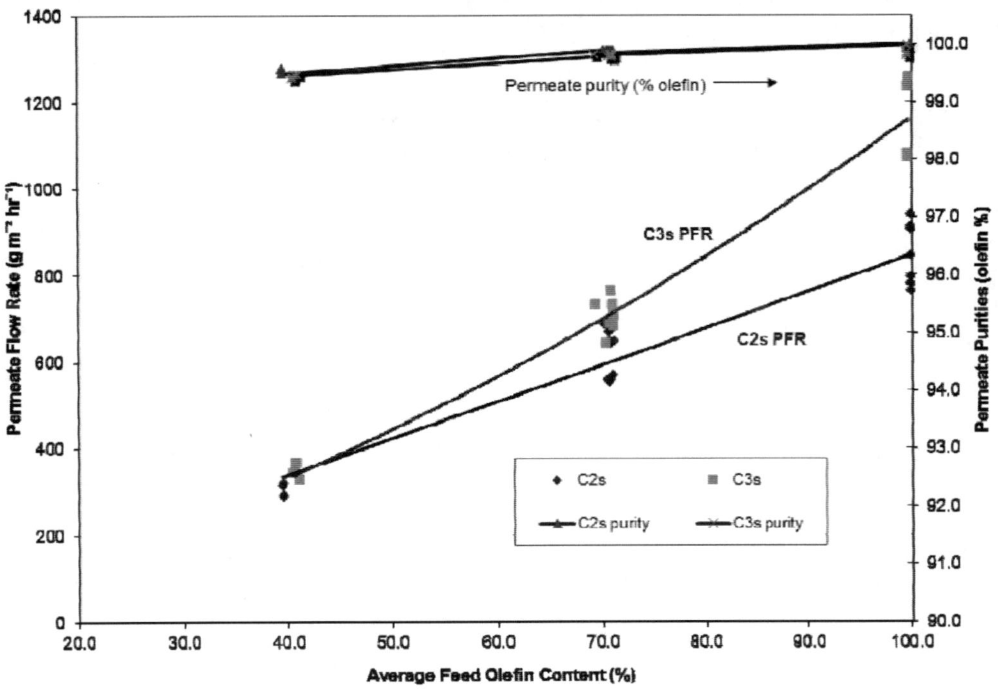

With HF3 being a promising candidate membrane for product scale-up and pilot demonstration purposes, R&D at Imtex is shifting more to product development and testing of spiral wound elements of various sizes. The picture below shows three sizes of rolled test elements.

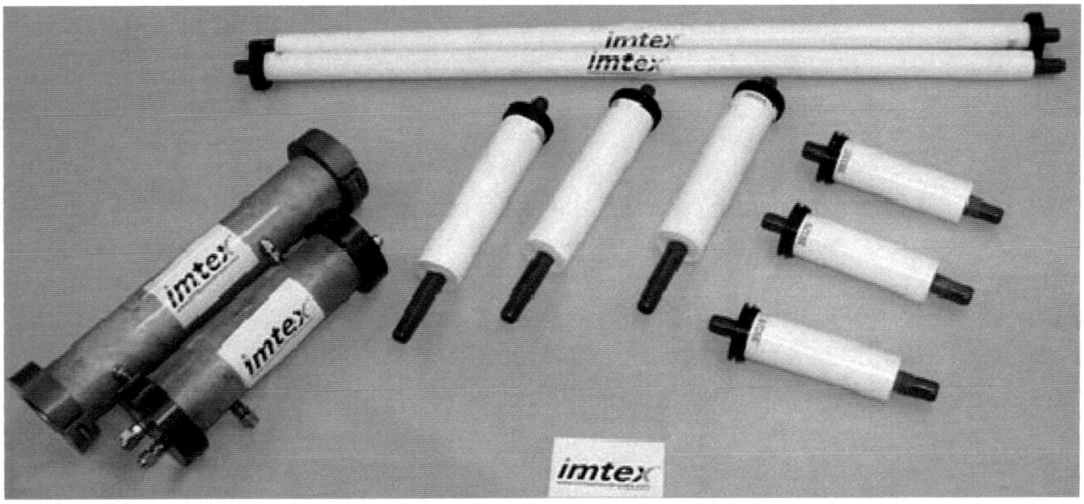

Using only one membrane module, we demonstrated reasonable propylene recovery producing polymer grade propylene from a 70/30 propylene/propane gas mixture as shown in Figure 8.

Figure 8- Effect of Stage Cut on Membrane Performance (60 Psig and room Temperature)

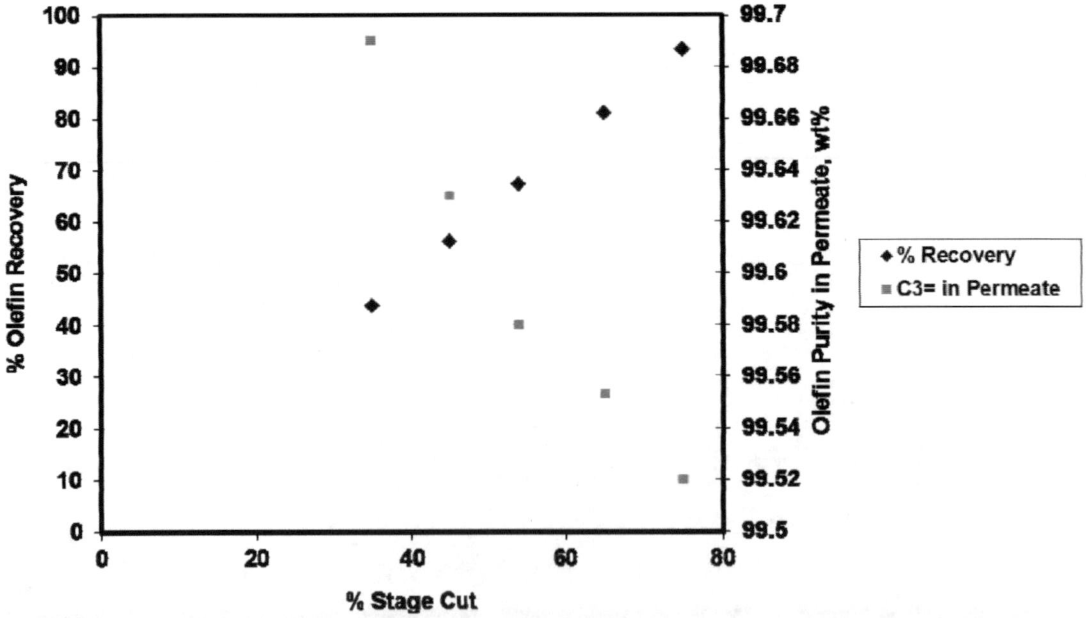

Multiple Membrane Modules in Series:

Currently the upgrading of the refinery grade to polymer grade is carried out in distillation columns; thus, the distillation bottom will produce 90%+ propane as second "product." Using the membrane technology to replace the distillation has the drawbacks of not being capable of producing propane in such high concentration. The implementation of a hybrid system to achieve debottlenecking may be an attractive option. A hybrid system will be technically easier and more economical than replacing the entire distillation system with a membrane unit. We presented various scenarios and performed an extensive simulation work for C2 splitter to quantify the energy saving through the retrofit of existing distillation columns with a membrane unit [8].

Another alternative approach is to use multiple membrane modules to get the propane to a reasonable concentration. We performed experiments with two (2) and four (4) modules in series.

The experimental results are presented below for 2 modules in series. As shown, while the polymer grade propylene is produced at the permeate side, the propane rich stream only was concentrated to 80 percent on the retentate side.

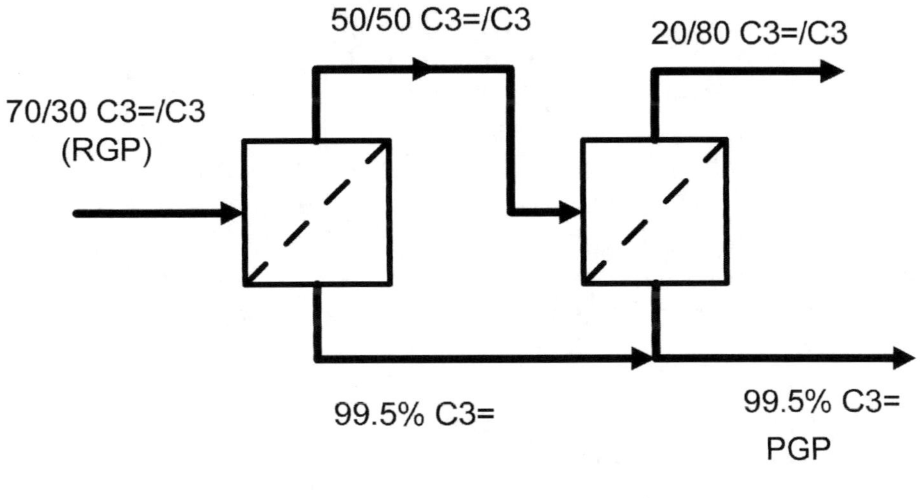

The experiments were performed at 66 psi. At this feed pressure and 20 percent olefin content, olefin feed side partial pressure would be approximately 16 psia. Then there is hardly any driving force against the nearly 14.7 psia on the permeate side. In theory, feed side pressure would be raised above 150 psig to get any permeation flow if olefin content in retentate is needed to be at 10 percent. Figure 9 shows the results.

Figure 9- Simulating RGP Feed to Memebrane: Propane/ Propylene 30/70 wt% @ 120 psig Feed Pressure

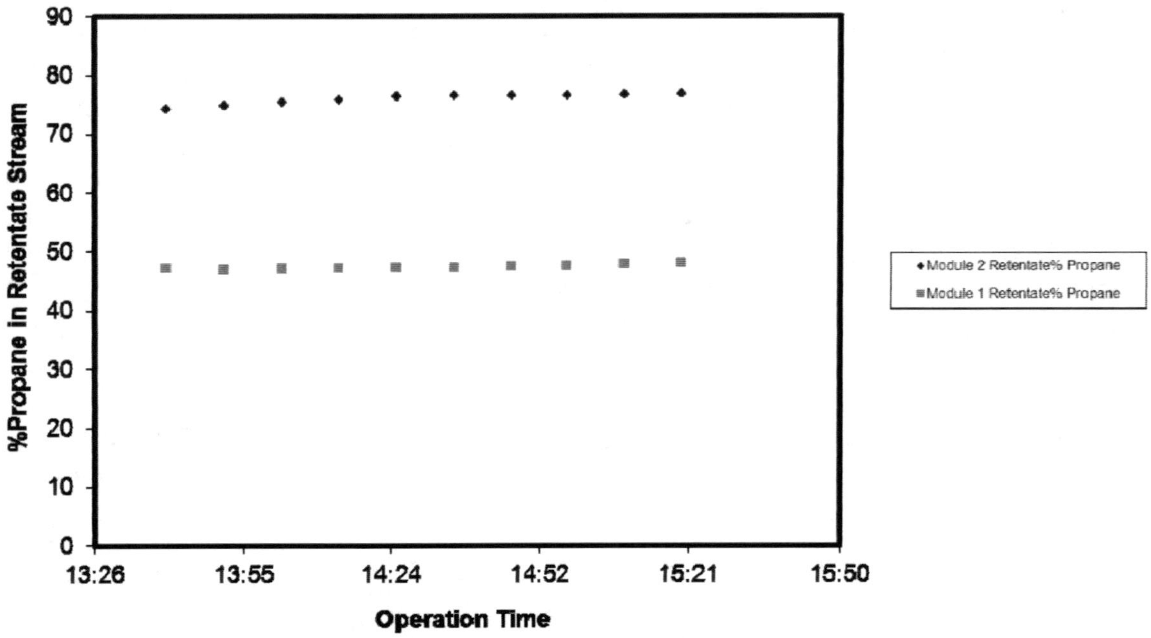

Potential Opportunities:

Energy savings of the membrane use in the olefin-paraffin separation is considered very attractive. Assuming that the membrane is in a hybrid distillation/membrane arrangement, energy savings were estimated to be on the order of $2 MM/yr for a commercial size C2 splitter. A retrofit of an existing splitter may include compressors for both permeate and retentate streams, and a dryer for the permeate stream in addition to the membrane skid. The $2MM energy savings alone does not justify the capital estimated for the retrofit. However, if the membrane were to be used for debottleneck purpose, and then the economics for C2 splitter are very favorable.

Another promising application requiring olefin-paraffin separation is in olefin metathesis where polymer-grade propylene is produced by combining the n-butene with ethylene. Butene/butane separation is desired for the operation and is not commercially attractive to use distillation. If successful, a membrane process would increase reactant utilization of metathesis plants and offers attractive energy savings by producing more propylene from the same volume of feedstock.

Imtex HF3 membranes were also tested in separating C4s. Testing was conducted with gas mixtures of 1-Butene/n-Butane at 40 psig and a temperature of 40 $^{\circ}$ C. As shown in Figure 10, n-butane content in the permeate stream was between 0.1 to 0.3%. The molar permeability of 1-butene was higher than either of ethylene and propylene. Accordingly, and in addition to the molar mass difference, the mass permeate rate for C4= would be significantly higher than for C2= or C3= if testing were performed at the same pressure.

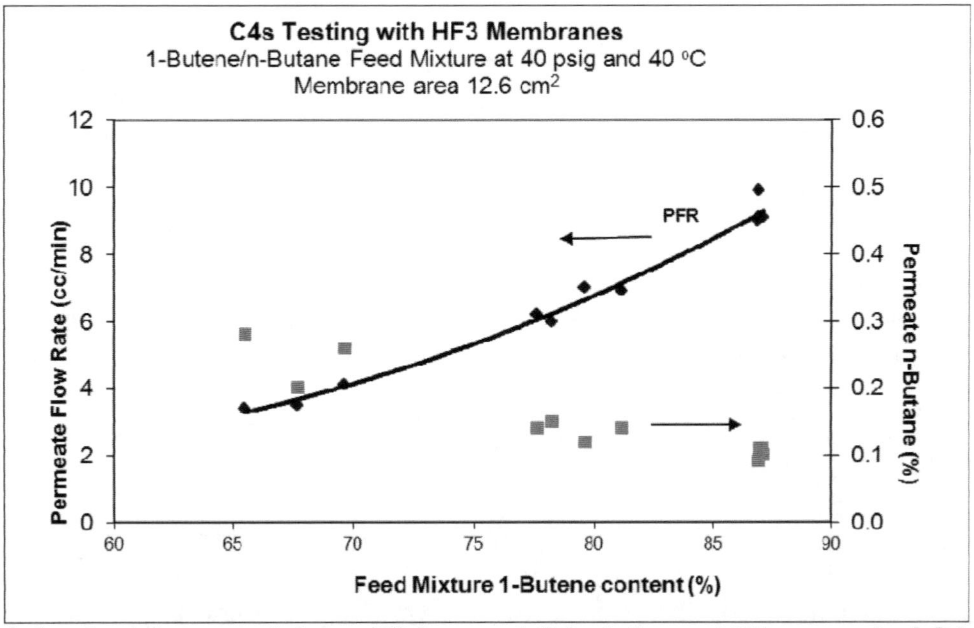

A feed mixture containing 73 percent C4 olefin and 27 percent of C4 paraffin was fed to the 500 cm^2 membrane unit. A total paraffin concentration of 0.57 weight percentage was measured in the permeate stream as shown in Figure 11.

Figure 11. Feed Mixture Containing 73% C4 Olefin and 27% C4 Paraffin fed to 600 Cm² Membrane Unit (65 psig pressure and room temperature)

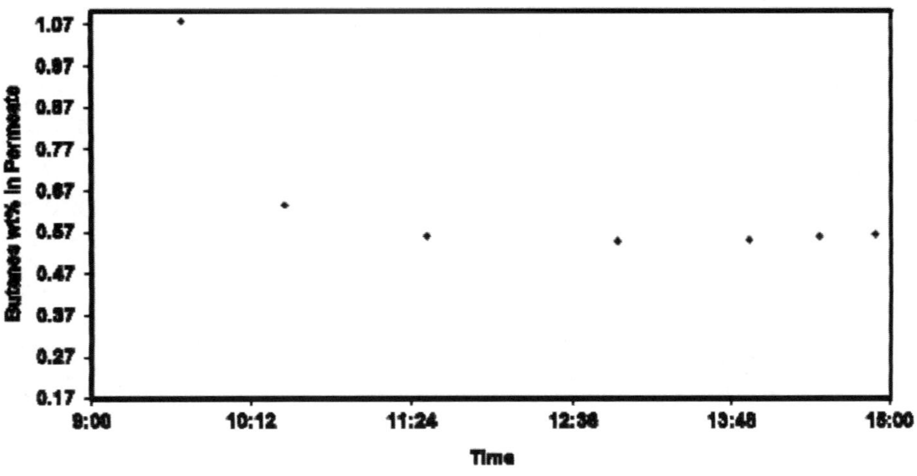

We believe that more investigations are needed to explore wider operating conditions (temperature and pressure) for C4s separations

Conclusions:

The preliminary work on various olefin-paraffin applications produced very encouraging results. In addition to the excellent selectivity and permeability, Imtex membranes demonstrated performance stability over time, which was one of the key shortcomings in prior attempts by others to use facilitated transport membranes in this application.

Currently Imtex is collaborating with The Shaw Group Inc. for the eventual commercial marketing and deployment of their technology. Work is underway at "proof of concept" stage. Prototype membrane elements have been made at commercial scale and a test program is underway at a Shaw pilot plant facility to test and refine the technology. Elements of this size will be the modular building blocks for commercial installations.

The technology has great potential to offer significant benefits in various applications beneficial to olefin manufacturing operations, such as debottlenecking C2 splitters and upgrading refinery grade propylene to chemical/polymer grade.

There is also a very promising potential for butene/butane separations where there are technical challenges facing distillation. A successful application of Imtex membranes can have a very positive impact on the economics of olefin metathesis which is used to produce polymer-grade propylene by combining n-butene and ethylene. Also, efficiencies can be realized in the production of n-butene co-monomers for linear low density polyethylene production.

References:

1- "Hybrid Separation/Distillation Technology: Research Opportunities for Energy and Emission Reduction", US DOE, Energy Efficiency and Renewable Energy April 2005.

2- "Energy Efficiency and Innovative Emerging Technologies for Olefin Production", Tao Ren, Martin Patel, and Kornelis Blok, presented in European Conference on Energy Efficiency, October 21-22, 2004.

3- R.B. Eldridge, "Olefin/Paraffin Separation Technology: A Review," Industrial & Engineering Chemistry Research, 32, 2208-12 (1993).

4- D.J. Safarik and R.B. Eldridge, "Olefin/Paraffin Separations by Reactive Absorption: A Review," Industrial & Engineering Chemistry Research, 37, 2571-81 (1998).

5- Separation of Olefin/Paraffin Mixtures with Carrier Facilitated Membranes; Final report by MTR for DOE; Contract # DE-FC36-04GO14151, March 2007.

6- P.S. Herrera, X. Feng, D. Payzant and J.H. Kim, "Process for the Separation of Olefins from Paraffins using Membranes," U.S. Patent 7,361,800 (April 22, 2008).

7- M. Azhin et al. "A Review on Olefin/Paraffin Separation Using Reversible Chemical Complexation Technology", Journal of Industrial and Engineering Chemistry, 14 (2008) 623-638.

8- Jose A. Caballero, et. al. "Design of Hybrid Distillation-Vapor Membrane Separation Systems"; Ind. Eng. Chem. Res. 2009, 48, 9151-9162.

(355d) Gas Transport of Paraffins and Olefins in Liquid Crystal Membranes

Feras Rabie, Chemical Engineering, Virginia Polytechnic Institute and State University, Blacksburg, VA and Stephen M. Martin, Chemical Engineering, Virginia Tech, Blacksburg, VA

Gas Transport of Paraffins and Olefins in Liquid Crystal Membranes

Abstract

A solubility induced phase separation (SIPS) method was used to form 8CB and MBBA liquid crystal domains in a polysulfone matrix. The membranes were characterized for their temperature dependent phase behavior and structural characteristics using DSC, POM, and FE-SEM. Paraffin and olefin separations were investigated at different temperatures and different liquid crystal phases using a mixed gas permeation apparatus. Propane and propylene were chosen to minimize size effects of diffusion and sorption.

We studied the effects of permeation and selectivity on intermolecular ordering of the small molecular liquid crystal. The liquid crystal phase plays an important role for diffusion and solubility on the solute. The crystalline state exhibited the lowest permeability with an increase in both propylene transport and propylene selectivity over propane as the crystalline domain sizes increased. Transport through the liquid crystals at their mesophase resulted in both an increase in permeability and selectivity. The contributing factor to this phenomenon was concluded to be solubility selectivity of propylene over propane due to enhanced intermolecular interactions between the ordered domains and the more polar propylene solute. A further increase in temperature resulted in a two fold magnitude increase in permeability while maintaining a selectivity of 4.

A cyanobiphenyl liquid crystal attached to a butadiene backbone with a flexible alkane spacer was synthesized for the investigation of its gas transport properties. Room temperature transport studies showed a helium selectivity of 60 over methane. This was primarily due to size selective diffusion across the membrane with very little solubility or sorption. The effect of propane and propylene transport on pressure and phase dependence was tested. Transport for temperatures under the LCPs glass transition was dominated by diffusivity. A transition to a smectic liquid crystal phase resulted in an increase in permeability and selectivity. The contributing factor was the increase of propylene solubility compared to that of propane. A feed pressure increase resulted in an increase in permeability selectivity, which was also due to an increase in solubility selectivity. Mixed gas transport experiments correlated well with single gas transport.

A series of methacrylate liquid crystal polymers with nitro azobenzene and cyano biphenyl mesogens were synthesized and characterized. These liquid crystal monomers were copolymerized with 2-ethylhexyl acrylate (2-EHA) using a free radical polymerization method. Liquid crystal phases were present for LC content of over 50 mol% for the cyanobiphenyl and 60 mol% for the nitro azobenzene. Mesophase stability increased with increasing LC content. A method to form acrylate networks for the fabrication of stable membranes was developed and partially characterized. We will test its transport properties in the future.

(355e) Synthesis and Characterization of ZIF-8 Membrane with Exceptionally High Propylene / Propane Separation Performance

Hyuk Taek Kwon, chemical engineering, texas A & M university, college station, TX, Sagar Lonkar, chemical engineering, Texas A & M university and Hae-Kwon Jeong, Chemical Engineering, Texas A & M University, College Station, TX

Synthesis and characterization of ZIF-8 Membrane with exceptionally high propylene / propane separation performance

Hyuk Taek Kwon[1], Sagar Lonkar[1], and Hae-Kwon Jeong*[1,2]

[1]Artie McFerrin Department of Chemical Engineering and [2]Materials Science and Engineering Program

Texas A&M University, College Station, TX 77843-3122

* Corresponding author: hjeong7@mail.che.tamu.edu

Metal-organic-frameworks (MOFs) have attracted intense research interest due to their potential application such as small-molecule separation[1], storage[2], catalysis[3], and others. Membrane-based gas separation is one of the most intriguing applications where MOFs can play a role due to their regular and sub-molecular size of pore structure, recognition and binding ability of specific molecules[4].

Several MOF membranes have been synthesized and tested for gas separation[5]. Despite their potentials, most of MOF membranes reported showed rather unimpressive separation performances mainly due to the flexible nature of their frameworks[6] as well as the likely poor microstructure. Recently Lai and co-worker have synthesized ZIF-8 membrane with exceptional hydrocarbon separation capability. The superior performance is likely due to improved microstructure[7], confirming the importance of controlling microstructure. Besides, there is also a strong need to develop a new synthesis method to prepare MOF membranes in a commercially-viable manner.

Here, we would like to present one-step in-situ route to prepare ZIF-8 membrane, which is highly reproducible. The formation of defect-free ZIF-8 membrane was completed within a couple of hours with exceptional mechanical stability. Most importantly, the membranes showed the exceptionally high separation performance of propylene / propane. We will also present synthesis of other MOF membranes such as HKUST-1 and ZIF-7.

References and Notes

1. J.-R. Li, R. J. Kuppler and H.-C. Zhou, Chemical Society Reviews **38** (5), 1477-1504 (2009).

2. L. J. Murray, M. Dinca and J. R. Long, Chemical Society Reviews **38** (5), 1294-1314 (2009).

3. J. Lee, O. K. Farha, J. Roberts, K. A. Scheidt, S. T. Nguyen and J. T. Hupp, Chemical Society Reviews **38** (5), 1450-1459 (2009).

4. M. Eddaoudi, J. Kim, N. Rosi, D. Vodak, J. Wachter, M. O'Keeffe and O. M. Yaghi, Science **295** (5554), 469-472 (2002).

5. M. Shah, M. C. McCarthy, S. Sachdeva, A. K. Lee and H.-K. Jeong, Industrial & Engineering Chemistry Research

51 (5), 2179-2199 (2011).

6. H. Bux, F. Liang, Y. Li, J. Cravillon, M. Wiebcke and J. r. Caro, Journal of the American Chemical Society **131** (44), 16000-16001 (2009).

7. Y. Pan and Z. Lai, Chemical Communications **47** (37), 10275-10277 (2011).

(355f) Design of ZIF/Polyimide Mixed Matrix Membrane for Olefin/Paraffin Separations

Chen Zhang, School of Chemical and Biomolecular Engineering, Georgia Institute of Technology, Atlanta, Georgia and William J. Koros, School of Chemical & Biomolecular Engineering, Georgia Institute of Technology, Atlanta, GA

<>C_2~C_4 olefins are consumed as feedstocks to produce a variety of significant chemicals. Membrane separations have been emerging as more energy-efficient alternatives to fractional distillation to enrich olefin from olefin/paraffin mixtures. The development of high-performance membrane materials has been the key to promote membrane separations as practical approaches for large-scale industrial gas separations.

Mixed matrix membranes, formed by dispersing molecular sieve particles/platelets in polymer matrices, are promising approaches for gas/vapor separations that combine the ease of processing polymers with the superior separation performance of molecular sieves. The selection of molecular sieving phase and polymer phase is crucial to fabrication of mixed matrix membranes with intact polymer-sieve adhesion and favorable gas transport properties. Zeolitic imidazolate frameworks (ZIFs) are microporous crystalline solids with zeolite or zeolite-like topologies. The favorable adsorption and molecular sieving properties of ZIFs enable the possibility of being used as selective adsorbents as well as membrane materials.

Our study suggests that with appropriate choice of ZIF and polymer, enhanced and industrially attractive permeability and permselectivity can be simultaneously obtained in the ZIF/polymer mixed matrix membrane. ZIF-8/6FDA-DAM mixed matrix dense films were fabricated with hydrophobic ZIF-8 as the molecular sieving phase and a high performance polyimide 6FDA-DAM as the polymer phase. Permeation properties of the neat 6FDA-DAM dense film and ZIF-8/6FDA-DAM mixed matrix dense films were studied with pure propylene (C_3H_6) and propane (C_3H_8) feeds as well as a 50/50 (vol %) C_3H_6/C_3H_8 mixture feed. Pure gas permeation showed significantly enhanced mixed matrix ZIF-8/6FDA-DAM membrane C_3H_6/C_3H_8 separation performance compared to the pure 6FDA-DAM membrane. A C_3H_6 permeability of 56 Barrers and C_3H_6/C_3H_8 ideal selectivity of 31 was found in ZIF-8/6FDA-DAM mixed matrix membrane with 48 wt % ZIF-8 loading. These results are 258 % and 150 % higher than the pure 6FDA-DAM membrane, respectively for permeability and selectivity. Mixed gas permeation also verified that selectivity enhancements were achievable in mixed gas environment by ZIF-8.

(355g) TR Polymers for Olefin/Paraffin Separation

Zachary P. Smith[1], David F. Sanders[1], Ruilan Guo[2], Benny Freeman[1], Donald R. Paul[1] and James E. McGrath[3], (1)Department of Chemical Engineering, The University of Texas at Austin, Austin, TX, (2)The University of Notre Dame, Notre Dame, IN, (3)Macromolecules and Interfaces Institute, Virginia Tech, Blacksburg, VA

Ethylene and propylene are currently the two largest, by volume, organic chemical feedstocks produced in the U.S [1]. These olefins must be purified for polyolefin synthesis by separating ethylene from ethane and propylene from propane, a thermally intensive process that requires approximately 0.12 Quads of energy per year and distillation columns containing nearly 200 trays [2]. This study focuses on determining the performance and plasticization resistance for a thermally rearranged (TR) polymer for olefin/paraffin separation.

Polyimides containing reactive functional groups *ortho*-position to the diamine can be chemically transformed into TR polymers at elevated temperatures [3]. For this study, a polyimide was prepared from 3,3'-dihydroxy-4,4'-diamino-biphenyl (HAB) and 2,2'-bis-(3,4-dicarboxyphenyl) hexafluoropropane dianhydride (6FDA) via chemical, thermal, and solid-state imidization [4, 5]. Chemical imidization produces polyimides with *ortho*-functional acetate groups, while thermal and solid-state imidization produce polyimides with *ortho*-functional hydroxyl groups. Furthermore, each imidization technique alters the non-equilibrium nature of the polymer film, thus producing polyimides with unique transport properties. All films were characterized by ^{1}HNMR, DSC, FT-IR, TGA-MS, and pure gas permeation.

For ethylene, ethane, propylene, and propane, increasing polyimide conversion dramatically increased gas permeation. For example, partial conversion of the chemically imidized HAB-6FDA polyimide to its corresponding TR polymer increased ethylene permeability by over two orders of magnitude. Furthermore, ethylene/ethane selectivity decreased by less than 10%. For ethylene/ethane and propylene/propane separation, HAB-6FDA TR polymers showed combinations of permeability and selectivity near the polymer upper bound [6, 7]. Propylene plasticization pressure curves were determined for TR polymers derived from all three synthesis routes, and the thermally imidized TR polymer had the highest plasticization pressure point of approximately 4 bar at 35°C.

1. Facts & Figures: Output Declines in U.S., Europe, in Chemical & Engineering News. 2010. 54-62.

2. Eldrige, R.B., Olefin/paraffin separation technology: A review. *Industrial & Engineering Chemistry Research*, **1993**. 32(10), 2208-2212.

3. Park, H.B., C.H. Jung, Y.M. Lee, A.J. Hill, S.J. Pas, S.T. Mudie, E. Van Wagner, B.D. Freeman, and D.J. Cookson, Polymers with cavities tuned for fast selective transport of small molecules and ions. *Science*, **2007**. 318(5848), 254-258.

4. Ghosh, M.K. and K.L. Mital, Polyimides: Fundamentals and applications. 1996, New York: Marcel.

5. Ohya, H., V.V. Kudryavtsev, and S.I. Semenova, Polyimide membranes: Applications, fabrications, and properties. 1996, Amsterdam: Gordon and Breach Publishers.

6. Burns, R.L. and W.J. Koros, Defining the challenges for C_3H_6/C_3H_8 separation using polymeric membranes. *Journal of Membrane Science*, **2003**. 211(2), 299-309.

7. Staudt-Bickel, C. and W.J. Koros, Olefin/paraffin gas separations with 6FDA-based polyimide membranes. *Journal of Membrane Science*, **2000**. 170(2), 205-214.

(355h) Asymmetric Hollow Fiber Membranes for Aggressive Sour Gas Purification

Brian E. Kraftschik and William J. Koros, School of Chemical and Biomolecular Engineering, Georgia Institute of Technology, Atlanta, GA

Carbon dioxide (CO_2) and hydrogen sulfide (H_2S) are present in high concentration in a large fraction of natural gas reserves. These species must be removed prior to transport through natural gas pipelines in order to prevent corrosion, increase the heating value of natural gas and because of the dangers associated with H_2S. A series of novel crosslinkable materials based on the glassy polyimide 6FDA-DAM:DABA, which has been used in the past for CO_2/CH_4 separations, has been developed for use in the simultaneous removal of H_2S and CO_2 from aggressive natural gas feeds. These materials exhibit significantly improved H_2S/CH_4 selectivity and plasticization resistance over the base polymer without sacrificing the outstanding CO_2/CH_4 permselectivity of 6FDA-DAM:DABA. Following dense film permeation and sorption studies, one of these crosslinkable 6FDA-DAM:DABA derivatives was identified as a promising candidate for asymmetric hollow fiber formation and testing. In dense film form, this material has an H_2S/CH_4 selectivity as high as 23.5 and a CO_2/CH_4 selectivity just below the intrinsic value when using a 20% H_2S/20% CO_2/60% CH_4 mixture at feed pressures up to 900 psi. The dense film characterization results as well as those for asymmetric hollow fiber spinning and preliminary permeation testing will be discussed.

(365a) Carbonate Ceramics and Concrete: A Disruptive Technology for CO2 Utilization and the Construction Business

Richard Riman, Materials Science and Engineering, Rutgers, The State University of New Jersey, Piscataway, NJ

Carbonate Ceramics and Concrete: A Disruptive Technology for CO_2 Utilization and the Construction Business

R. E. Riman,[1] S. Gupta,[1] Q. Li,[1] V. Atakan,[2] Ç. Vakifahmetoglu,[1] J. Azurdia,[2] J. Krishnan, M. A. Bitteto, L. Tang,[1] J. Czerepinski,[1] Larry E. McCandlish,[2] and Nick Decristofaro[2]

1. Department of Materials Science and Engineering, Rutgers, The State University of New Jersey, 607 Taylor Road, Piscataway, NJ 08854

2. Solidia Technologies Inc., 11 Colonial Drive, Piscataway, NJ 08854

Alternatives to Carbon Capture and Storage (CCS) are needed that can use captured CO_2 directly or convert it to valuable products, such as fuels, chemicals, plastics, or building materials. This paper will describe a breakthrough process invented and patented by Rutgers called hydrothermal liquid phase sintering. When CO_2 is incorporated into the hydrothermal liquid phase to react with ultramafic mineral silicates at temperatures below 100¢ªC, chemically bonded ceramics can be produced with outstanding properties. Our work has demonstrated that both building and infrastructure materials can be manufactured that meet or exceed property specifications for a range of applications to justify commercialization. At the same this process is green in terms of its carbon footprint and energy consumption. CO2 mass balance computations suggest that use of mineral-based CaSiO3 can yield a truly carbon-negative material that consumes a fraction of the energy required for cement or concrete processing. Solidia Technologies was founded as a Rutgers start-up to build construction materials businesses that yield materials with exceptional performance and cost while maintaining its green profile. Progress towards commercialization and its potential impact on global CO_2 emissions and energy will be discussed. This presentation will also provide a perspective with respect to the sustainability of this process relative to stone, ceramic and concrete processing.

(365b) Synthesis of CaSiO3 for Use As a Cementitious Material for Carbonate Concrete

Nick DeCristofaro, Solidia Technologies, Piscataway, NJ

Synthesis of $CaSiO_3$ for use as a Cementitious Material for Carbonate Concrete

V. Atakan[1], C. Vakifahmetoglu[2], L. McCandlish[1], J. Krishnan[1], L. Tang[2], Q. Li[2], R. E. Riman[2] and N. DeCristofaro[1]

1. Solidia Technologies Inc., Piscataway, NJ 08854

2. Rutgers, Department of Materials Science and Engineering, The State University of New Jersey, Piscataway, NJ 08854

Solidia Technologies is using mineral-based Wollastonite ($CaSiO_3$) as a cementitious material for a novel carbonate concrete slated for commercial production. $CaSiO_3$ is reacted with CO_2 to form SiO_2 and $CaCO_3$ via a patented process called hydrothermal liquid phase sintering. The SiO_2 and $CaCO_3$ products occupy ~65% more volume than the $CaSiO_3$ reactant and act as cementing phases to bond sand and aggregate in the carbonate concrete. Unfortunately, the availability of mineral-based Wollastonite does not ensure its wide-scale deployment as a cementitious material. Approximately 10^5 T/yr of Wollastonite are mined in North America. The vast majority of the mineral is used in the ceramics and metallurgical industries, either for its white color, its acicular particle shape or its fluxing capability. While current production quantities adequately supply these applications, the known reserves of Wollastonite are insufficient to consider this mineral as a substitute for Portland cement in the production of concrete. By comparison, over 10^8 T/yr of Portland cement are manufactured in North America. Additionally, Portland cement can be synthesized virtually anywhere from locally available raw materials such as limestone and clay. Thus, a more logical approach is to develop an infrastructure of $CaSiO_3$ manufacturing using the same precursor raw materials as those used for Portland cement. This presentation will describe efforts to use raw materials, reactors and unit operations common to Portland cement to synthesize $CaSiO_3$ and similar silicate compounds. The production of $CaSiO_3$ will be compared to Portland cement production in terms of $CaO\text{-}SiO_2\text{-}Al_2O_3$ phase equilibria, the energy requirements to drive the solid-state reaction, and the emissions of toxic Hg vapor and CO_2 greenhouse gases.

CO_2 UTILIZATION IN CONCRETE BLOCK PRODUCTION

Dr. Yixin Shao, McGill University, Montreal, Quebec

Dr. Liang Hu, 3H Company, Lexington, KY

Hilal El-Hansson, McGill University, Montreal, Quebec

Concrete blocks have been widely utilized as load-bearing and non-load- bearing partition walls in construction. The US market for concrete blocks and bricks is projected to increase to 4.3 billion units per year in 2014. In comparison to cast-in-place concrete, concrete block structures not only exhibit superior performance due to precast quality, but also represent a low environmental impact construction system. Concrete blocks are made of calcium silicate cement, porous in nature and mass produced. The concrete block plants are also demographically distributed and strategically located close to CO_2 sources such as power plants and cement plants. Concrete blocks are reactive with carbon dioxide. It is the calcium silicate cement in concrete blocks that captures carbon dioxide and converts it into calcium carbonate, a thermodynamically stable solid that contributes to strength gain and durability improvement of the final products. Currently concrete blocks are produced by steam curing. If this process can be replaced by CO_2 curing, concrete blocks can serve as CO_2 sink for carbon storage.

Carbonation of fresh dicalcium silicate (C_2S) and triclacium silicate (C_3S) pastes is an accelerated curing process, with the governing reactions shown by Eqs (1-2). The carbonation reaction products are a hybrid of calcium-silicate-hydrate (C-S-H) and calcium carbonate ($CaCO_3$). High early strength can be obtained within a few minutes to a few hours [1, 3].

$$C_3S+(3-x)CO_2+yH_2O\rightarrow C_xSH_y+(3-x)CaCO_3 \quad (1)$$

$$C_2S+(2-x)CO_2+yH_2O\rightarrow C_xSH_y+(2-x)CaCO_3 \quad (2)$$

The carbonation reaction could also occur in hardened concrete (Eqs.3-4). It is the reaction between carbon dioxide and hydration products, such as calcium hydroxide (CH) and calcium silicate hydrate (C-S-H), forming $CaCO_3$ and silica gel.

$$Ca(OH)_2 + CO_2 \rightarrow CaCO_3 + H_2O \quad (3)$$

$$C\text{-}S\text{-}H + 2CO_2 \rightarrow SiO_2 + 2CaCO_3 + H_2O \quad (4)$$

This paper presents a study on the feasibility of carrying out the carbonation curing between fresh concrete and hardened concrete such that that is a process window that allows the removal free water and the growth hydration products for more carbonation reaction. The effect of initial curing on CO_2 curing of lightweight concrete blocks was examined with both slab samples representing web sections and full size blocks. This early process window includes initial air curing of 0, 4, 6, 8, and 18 hours. They are denoted as 0a, 4a, 6a, 8a, and 18a respectively in Table 1. Without air curing, reaction efficiency was significantly reduced. The corresponding carbon uptake was only about 9%. Initial air curing from 4h to 18h removed certain amount of free water, making more space for gas to permeate and carbonates to precipitate. The early age open air curing yielded a carbon uptake of 21-24% based on dry cement. Carbon content was measured by mass gain method, mass curve method and thermal analysis. The results are in good agreement (Table 1). For a carbonation of 96 hours (18a+96c), the carbon uptake reached 35%, which was corresponding to 70% reaction efficiency. The CO_2 curing was carried out in a chamber under a constant pressure of 0.1 MPa (14.5 psi). The early-age and 28-day compressive strength of steam cured, carbonated, and hydrated samples were compared. Due to loss of water caused by initial curing, the carbonated concretes exhibited lower 28-day strength in comparison to hydrated and steam-cured samples. A water spray mechanism was devised to restore the lost water, and ultimately, the late compressive strength was comparable among samples.

Table 1—Carbon uptake in carbonated concrete slabs (3"x5"x1.5")

	CO_2 Uptake (% Cement Mass)		
Condition	**Mass Gain**	**Mass Curve**	**Thermal analysis**
0a	-	-	0.0
0a + 4c	7.5 ± 1.8	9	8.9
4a + 4c	21.3 ± 1.4	22.1	21.6
6a + 4c	22.8 ± 1.8	23.6	22.7
8a + 4c	23.5 ± 1.6	24.0	23.2
18a + 4c	24.2 ± 0.35	24.4	24.0
18a + 2c	20.4 ± 0.4	21.7	21.5
18a + 96c	34.9 ± 1.6	35.2	35.0

Based on the results from small concrete slab shown in Table 1, carbonation of standard 8" concrete blocks were carried out. The fresh blocks underwent an initial curing of 18 hours at 50% RH and 25°C (77°F). Water loss due to initial curing was about 50%, very close to that observed in slab tests. The blocks were then carbonated at 0.1 MPa (14.5 psi) gas pressure for 4 hours. Water loss due to carbonation was about 5%. Carbon uptake by three block samples, at the age of 1 day, is presented in Table 2. The three methods were used in carbon estimation. In comparison, the three blocks demonstrated an uptake in the range of 23-25%. These results agreed with that of the slab samples in Table 1, whose uptake was approximately 24%.

Table 2—Carbon uptake in full-size concrete blocks (8"x8"x16")

	Carbon Uptake (%)		
CMU #	**Mass Gain**	**Mass Curve**	**Thermal analysis**
1	23.8	24.0	23.3
2	23.1	23.6	23.9
3	23.6	23.8	23.5

Initial curing is beneficial for carbon uptake in an early CO_2 curing process for an accelerated hydration and carbon dioxide storage. However, initial curing could be detrimental to late strength development because of the water loss due to the process. With water compensation immediately after carbonation, high early strength, equivalent late strength, and superior carbon uptake can be achieved.

The process parameters obtained from slab samples are successfully applied to full size block production in laboratory. Both carbon uptake and strength gain in blocks are of close value to slab tests. Therefore, laboratory work can be scaled up to commercial production. XRD patterns illustrate that the carbonation products are stable calcite with the consumption of C_2S, C_3S, CH and possibly CSH.

Reaction of cement with carbon dioxide at early age is a CO_2 sequestration process. If one 20cm (8") full sized block with a mass of 15 kg (33 lb) contains 13% cement could have a CO_2 uptake of 24% based on cement, one block has the capacity to store 0.47 kg (1 lb) of CO_2 in a thermodynamically stable calcium carbonate form. Assuming every block or brick has the same carbon storage capacity, the projected annual production of 4.3 billion units in US market can thus consume 2 million tons of CO_2 per year. The capacity for carbon capture and storage (CCS) in geologic formation is approximately 1 million ton per year per site. CO_2 utilization in concrete blocks and bricks production is equivalent to carbon sequestration in two geologic formation sites. The cement annual production in the United State is about 100 million tons with CO_2 emission of 80 million tons. If all block and brick plants adopt carbonation curing with the same carbon uptake rate, CO_2 utilization in their production lines alone could reduce carbon emission by 2.5% for cement industry.

CO_2 curing process can replace steam to reduce embodied energy in concrete blocks, utilize sufficient amount of carbon dioxide in the vicinity of CO_2 sources, and produce the final products with equivalent performance.

References:

[1] J. F. Young, R. L. Berger, J. Breese, Accelerated Curing of Compacted Calcium Silicate Mortars on Exposure to CO_2, Journal of the American Ceramic Society, 57 (1974) 394-397.
[2] R. L. Berger, J. F. Young, K. Leung, Acceleration of Hydration of Calcium Silicates by Carbon-Dioxide Treatment, Nature-Phys Sci, 240 (1972) 16-&.
[3] C. J. Goodbrake, J. F. Young, R. L. Berger, Reaction of Beta-Dicalcium Silicate and Tricalcium Silicate with Carbon Dioxide and Water Vapor, Journal of the American Ceramic Society, 62 (1979) 168-171.

(365d) Catalyzed Activation of CO2 by Lewis Base Site in Hybrid Metal Organic Frameworks

Liang Chen, Institute of Materials Technology and Engineering, Chinese Academy of Sciences, Ningbo, China

Activation of carbon dioxide (CO_2) is a crucial and challenging step in recycling CO_2 as a source of carbon to produce useful chemicals, because CO_2 is an inert and thermodynamically stable compound. In this work, a metal-organic framework (MOF)-based catalyst W-Cu-BTC is designed by hybridizing highly active W atoms into the Cu-BTC frameworks based on density functional (DFT) calculations. We show that the hybrid W-Cu node plays a pivotal role in activating CO_2 according to the frontier molecular orbital theory. In contrast to the Lewis acid nature of open metal sites in most MOFs, the exposed W ion in W-Cu-BTC is identified as a Lewis base site, evidenced by the substantial electron donation from W ion to CO_2. Kinetically, the linear CO_2 molecule can be readily bent by forming a CO_2-W complex after overcoming a negligible activation barrier of 0.09 eV. In addition, we present calculated infrared spectra (IR) and X-Ray spectra (XPS) for reference in future experimental studies.

UTILIZATION TECHNOLOGIES FOR CO_2 DERIVED FROM NORTH DAKOTA LIGNITE

Jason D. Laumb, Melanie D. Jensen, Robert M. Cowan, Alexander Azenkeng, Loreal V. Heebink, Laura J. Raymond, and Peter A. Letvin, University of North Dakota Energy & Environmental Research Center, Grand Forks, ND

Introduction

Because of the abundant supply of coal resources such as lignite in North Dakota and some of the midwestern states, the United States will rely on the use of fossil fuels for its energy needs for many years to come, thus sustaining or increasing the level of CO_2 emissions. Since lignite produces more CO_2 per unit of energy compared to the other ranks of coal, it will be the fuel most impacted by any move to force CO_2 emission reductions from power plants. Given that the primary use for North Dakota lignite is minemouth electrical power plants, any CO_2 regulations will potentially have more impact on the North Dakota power industry and economy. To lessen the impact on the state, it is important to begin exploring ways of addressing the CO_2 emission issue in a sustainable manner. To this end, approaches that focus on CO_2 capture and reuse are attractive, particularly in areas where geologic storage may not be an optimal solution. Also, finding a potential use for the captured CO_2 might generate additional revenue to help offset some of the costs associated with CO_2 capture and, hence, act as an incentive for capturing CO_2 from lignite-fired power plants.

The primary purpose of this Energy & Environmental Research Center (EERC) study was to assess existing and emerging technologies for the beneficial use of CO_2 captured from North Dakota lignite-fired power plants. In the context of this study, the best technology is defined as one that not only utilizes CO_2 efficiently and cost-effectively but also results in or produces the most marketable product. By specific recommendation of the North Dakota Industrial Commission's (NDIC's) Lignite Research Council, technologies related to enhanced oil recovery (EOR), enhanced coalbed methane (ECBM), and enhanced gas recovery (EGR) were not considered in the investigation. All other areas that show both near- and long-term promise were considered in the study.

Methodology

While carbon capture, utilization, and storage (CCUS) technologies are developing rapidly worldwide and possess the potential to offer a significant contribution to CO_2 mitigation, there is interest in also exploring the possibility of CO_2 utilization in other end uses, such as in building material production, fuels, and chemicals. Several broad technology areas are being developed to facilitate the beneficial use of captured CO_2 from large point sources, such as coal power plants. These technologies encompass various applications, including EOR, ECBM recovery, CO_2 conversion to chemical feedstocks and fuels, biological conversion (photosynthesis) processes, and CO_2 mineralization for the production of materials. EOR is one of the largest uses of CO_2 captured from anthropogenic point sources, especially in the last few decades when there has been a greater desire to rely more on locally produced oil as opposed to oil from foreign countries. During times of high natural gas prices in the United States, some of the CO_2 has been directed to ECBM recovery to boost production. In the last several years, more attention is being directed to other technologies such as algae growth, electrofuels, and chemicals as a means of providing a beneficial use for the captured CO_2, which could offset some of the high costs associated with CCUS technologies. An overview of these technologies is presented in Figure 1.

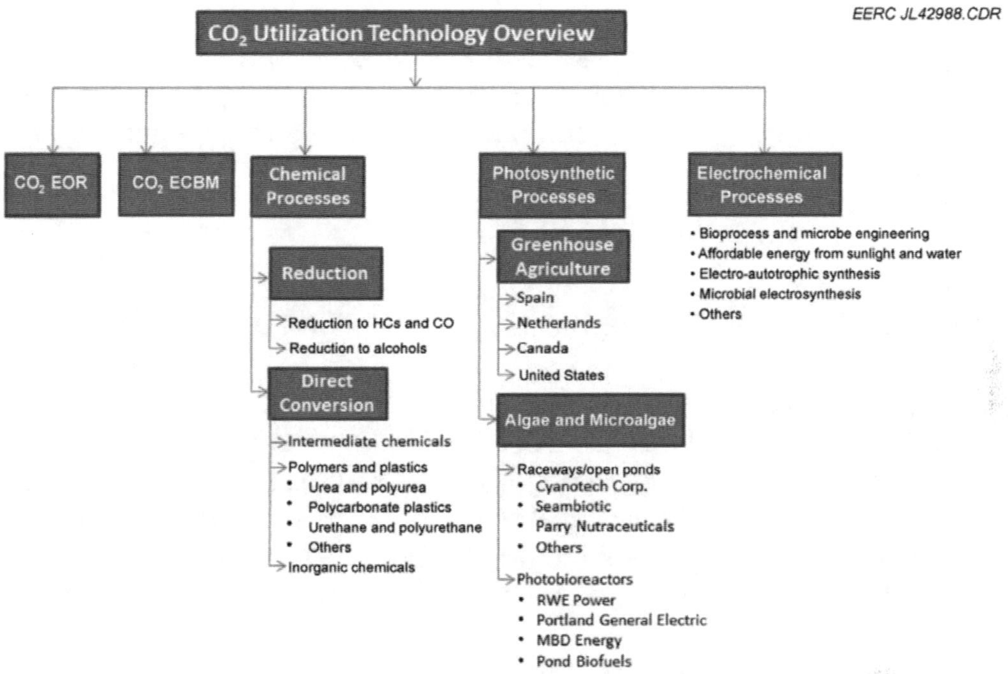

Figure 1. An Overview of CO_2 Utilization Technologies

Currently, the major areas of CO_2 utilization technologies include CO_2 EOR, CO_2 ECBM, CO_2 EGR, mineralization, chemical conversion processes, biological (photosynthetic) processes, and electrochemical processes. However, the specific goals of this project did not include further research into EOR, ECBM, or EGR. Thus, the remaining areas that were the focus of this study are mineralization, chemical conversion processes, biological processes, and electrochemical processes. The technologies in these areas that are under development were evaluated for their suitability to use CO_2 captured from North Dakota lignite-fired power plants. The criteria used for evaluating these technologies include the following:

- Technological feasibility
- Lignite-specific challenges/CO_2 purity requirements
- Regional and climatic challenges
- Source of CO_2
- Amount of CO_2 used
- Potential market for products
- Initial process economics

- Integration with the North Dakota fleet of plants
- Maturity and commercialization prospects

Since there are several options within each main technology area mentioned above, only the top few options, particularly those that have advanced to at least the pilot-scale demonstration stage, are evaluated in detail based on the criteria listed above. These are grouped under the main technology area. Because there is no one technology option that clearly meets all of the criteria set above, it is not possible to declare any option as the best available technology option for enabling the beneficial use of CO_2 captured from North Dakota lignite-fired power plants. Every option has its merits and demerits depending on its stage of development and/or the principal concept of the technology. The technologies identified as the most promising based on these criteria then underwent a more comprehensive market analysis and economic evaluation.

Results

CO_2 utilization technologies were divided into the following six categories:

- The direct use of CO_2, such as in carbonated beverages, as a dry cleaning solvent, or for energy recovery processes like EOR or ECBM production.
- The mineralization of CO_2 by reacting it with metal oxides or metal hydroxides to form metal carbonates or metal bicarbonates that may be used in construction materials.
- Use as a feedstock in the manufacture of chemicals, including chemical products or precursor chemicals that require chemical reduction of the carbon to a less oxidized form.
- Use as a feedstock in the manufacture of chemicals, including chemical products of precursor chemicals like urea or bicarbonate that do not require chemical reduction of the carbon.
- Photosynthesis-based technologies that reduce the carbon in CO_2 to organic carbon for use as food, fuel, or a chemical feedstock.
- Novel technologies based on the direct use of engineered microorganisms, electricity, and/or the direct use of sunlight for the production of fuels and/or chemical precursors.

Direct-use technologies include the use of CO_2 for EOR, ECBM production, and as a solvent, refrigerant, or in foods and beverages. These technologies are well known, have been extensively documented elsewhere, and were not a focus area of this project. However, it should be noted that the supply of CO_2 for EOR and ECBM projects could represent a good near-term opportunity for North Dakota lignite users.

Mineralization

Mineralization to form products from CO_2 is a relatively new concept. It is the formation of a carbonate or bicarbonate solid from CO_2; thus, the CO_2 becomes a part of the solid product. CO_2 captured from any source can be used as a feedstock for mineralization reactions. The process also requires a source for the alkalinity required by the reaction; lignite fly ash could potentially provide this alkalinity. The most advanced of the mineralization technologies is still only at a pilot scale of development, and the products that will be generated by the various technologies, should they become commercial, are more likely to fill niche markets than be widely employed. Nationwide, 19 mineralization technology developers were identified. Most of the companies working in the area of CO_2 mineralization have provided lists of potential products but have not provided a clear path to making and marketing those products. The market will dictate the type and quantity of products that are made. The entry-level product for most mineralization companies will likely be aggregate that can be used for roads and/or as a component of concrete. Although the concept shows promise, it does not appear to offer an economically viable opportunity for the lignite industry in the near term because 1) aggregate made from mineralization of CO_2 is estimated to cost roughly double the current rate for gravel aggregate because of the value of the materials required to supply the metal cations and alkalinity for the mineralization and 2) lignite fly ash is more valuable as a raw material used for solidification of waste pits in the western North Dakota oil fields than as a source of alkalinity for mineralization reactions.

Chemical Production (Reductive and Nonreductive)

CO_2 can be used in the production of chemicals and fuels. Many approaches are being developed to utilize CO_2 captured from various sources to produce useful fuels and chemical feedstocks and in the direct conversion of CO_2 to chemical products such as polycarbonate plastics or urea. The potential for these technologies to use CO_2 from coal-fired power plants is limited because 1) substantial energy input is needed to convert the carbon in CO_2 from its fully oxidized state into a reduced state where it can serve as a fuel and 2) the industries using CO_2 as a feedstock in chemical production also perform upstream processes that produce CO_2 either directly (typically at high temperature and pressure) or when they consume fuel in order to provide energy for the overall process.

The status of CO_2 reduction to fuels is currently limited to research and development studies, mostly in academic laboratories. When a fuel is made from CO_2, energy is used to reduce the carbon from a fully oxidized state to a more reduced state. The amount of energy required for this reduction process is greater than the amount of energy that can be obtained either from the process or from use of the newly produced fuel. A CO_2-to-fuel process only makes sense where the product formed is of very high value, the fuel is used as a storage product made from an intermittent energy supply source (e.g., wind, solar), and/or the fuel produced is useful in ways that the original source fuel was not (e.g., production of a transportation fuel from coal-derived CO_2).

When CO_2 is directly converted into chemicals, it is reacted with another feedstock that had to be produced in an upstream process. The quantity of CO_2 produced in this upstream process, along with the CO_2 produced from energy generation associated with this upstream process, will exceed the CO_2 demand of the step that uses CO_2. Therefore, most companies performing these processes will not use externally supplied CO_2. Additionally, most of the CO_2 use processes, or the upstream processes used to generate the reactive intermediates, require reaction conditions such as high pressure and/or high temperature, with fossil fuel combustion typically used to provide the heat and power necessary to meet these needs. In fact, more CO_2 is produced during polycarbonate plastic or urea production than is used to make the products. Some of the largest postcombustion CO_2 facilities operating in the world are located at urea plants where CO_2 is captured from natural gas combustion flue gas in order to supply some of the CO_2 used for converting ammonia to urea.

Photosynthesis-Based Technologies

Photosynthesis-based processes using externally sourced CO_2 include algae production and greenhouse agriculture. In order for these technologies to provide a favorable CO_2 demand, the energy input for CO_2 reduction to organic carbon needs to be primarily from sunlight rather than from electric lights unless the electricity is derived from a zero-carbon source (i.e., wind, solar, nuclear). Greenhouse agriculture and algae systems can use low-concentration CO_2 streams. In algae production, CO_2 must be supplied both as a source of carbon for growth of the algae and to control the pH of the growth media. In greenhouse agriculture, CO_2 serves as the carbon source for plant growth and can increase plant growth rates. CO_2 supply to greenhouses is particularly important in colder climates where increasing air exchange to supply CO_2 from the outside air would result in excessive heating costs.

The microalgae production industry is a small and well-developed industry that has a proven ability to make money. The industry purchases externally sourced CO_2, but the size of its markets is small relative to the amount of CO_2 that is potentially available from power plants. Less than 20,000 tons/yr (18,150 tonnes/yr) of algae is produced worldwide, primarily for use as nutritional supplements (1). There has been a recent explosion of algae start-up companies (some estimate more than 200 since 2005) that are trying to break into potential algae product markets that promise to be much larger than the nutritional supplement market but require much less expensive algae. These larger markets include the production of biofuels, animal feeds, and fish meal replacements. Since these products have a relatively low value, production costs must be substantially reduced from current commercial production costs. Production of these lower-value products cannot be performed in an economically viable manner even under the most favorable conditions.

Algae and microalgae technologies are not economically feasible for North Dakota. The successful algae-producing companies are located in environments that favor the manufacture of their products (i.e., moderate temperatures and sunlight are available without extra cost). Their high-value nutrient supplement products are dry, shelf-stable and, therefore, relatively inexpensive to transport, making them readily available to the local population even without local producers. Irrespective of location, algae and microalgae products that could utilize a substantial amount of CO_2 (e.g., fuels and feed) are currently more expensive to produce than their potential market value can fetch.

Greenhouse agriculture, or controlled-environment agriculture, involves growing plants in a greenhouse. High-technology greenhouses are supplied with CO_2, heat and humidity control, and supplemental light as required to ensure high productivity. The common products from this type of agriculture include flowers, specialty fruits, and vegetables. North Dakota power plants may potentially benefit from the development of greenhouse agriculture operations in the state. These facilities can use both CO_2 and low-grade heat from the power plants and will also be customers for electricity used in supplemental lighting. The total demand for CO_2 is unlikely to be high, but the market and economic indicators investigated appear to indicate that a profitable venture could be developed based on this CO_2 use technology.

Novel Technologies

Twelve novel CO_2 utilization technologies were evaluated. These are primarily conceptual and laboratory-scale proof-of-concept processes of the type being supported by the U.S. Department of Energy's ARPA-E (Advanced Research Projects Agency-Energy) Program. They include processes that involve the electrochemical conversion of CO_2 to fuels and/or other chemicals, bioelectrochemical systems such as reverse microbial fuel cells that combine microbial processes and electrochemistry to produce chemicals, the use of microorganisms that convert H_2 and CO_2 to desirable chemicals, and other processes that use sunlight to power chemical synthesis reactions. All of the novel CO_2 utilization technologies are at a very early stage of development and are not close to moving out of the laboratory. In addition, these processes require the input of energy to convert CO_2 into a useful product. The hope is that some of these concepts will, at the very least, contribute to the development of useful technologies that can be commercially relevant sometime in the future, perhaps within the next 25 years. A great deal of work and the investment of substantial time and money will be required if that is to happen.

Market Analysis

The CO_2 mineralization technologies do not yet have well-defined products. The market will dictate the type and quantity of products that are made, but the entry-level product for most mineralization companies will likely be aggregate that can be used for roads and/or as a component of concrete. There is a substantial need for aggregate in North Dakota, particularly in the Devils Lake Basin and in the Bakken–Three Forks shale oil development area. The cost of gravel is roughly one-half of the developer-estimated cost of aggregate formed by mineralization, indicating that product improvements are needed for this technology to compete economically. Another use for mineralization products might be as solidifying agents for drilling waste pits formed during oil field operations. The fly ash is currently more valuable for this use than as an alkalinity source for a mineralization process.

Greenhouse agriculture appears to be the most promising technology for which products are obvious and can be assessed for potential markets. Based on the market analysis conducted as part of this study, it is known that greenhouse agriculture can produce much higher yields of some vegetable crops than traditional agriculture. The market price for greenhouse tomatoes is quite high, and there is also a high demand in the United States, which imports nearly 1.3 million tons (1.2 million tonnes) of tomatoes annually (2). The capital and operating expenses for such an enterprise could be large and the amount of daylight during the winter in North Dakota could pose a seasonal production issue. However, people and companies are interested in buying locally sourced vegetables such as tomatoes, peppers, and cucumbers because they are fresher and transportation costs are lower (it should be noted that the added expense associated with greenhouse agriculture probably eliminates the cost advantage to the individual consumer). Large grocery chains and food service companies that are headquartered in or near the region have expressed interest in the availability of produce from this type of regional source. Finally, there are few greenhouse agriculture enterprises in the region around western North Dakota, so the competition would be minimal.

Economics – Greenhouse Agriculture

Product value that can be derived from the greenhouse depends on how productive the operation is and the price of the product. Table 1 shows average productivities for two of the major greenhouse agriculture areas: Almeria, Spain, and the Netherlands. As seen in Table 1, the greenhouses in the Netherlands are much more productive. This is a function of the intensity and sophistication of the greenhouse operations. Figure 2 shows the U.S. producer price for tomatoes from 1991 through 2007 reported for all farm operations. For a producer price of \$8/kg for tomatoes and a productivity of 42 kg/m^2/year, the potential production cost would be \$336/m^2/year. If this production is performed using liquid CO_2 priced at \$110/tonne and the supply of CO_2 costs \$2.41/m^2/year, then the CO_2 cost represents only 0.7% of the producer cost, providing a good price for the CO_2 supply.

In light of the above analysis, North Dakota power plants will potentially benefit from any greenhouse agriculture operations in the state. In most cases, these greenhouses have been subsidiary companies owned by the power plants themselves so as to facilitate integration with the current plants and to derive additional synergistic benefits such as a supply of low-grade heat for maintaining the temperature in the greenhouses in the winter or a supply of power to cool the greenhouses in the hot summer months. Also, collocation removes extra transportation requirements and associated costs, which makes the operation even more economically feasible. Based on 2003 estimates, the United States imports a total of about 280,000 tonnes of greenhouse-grown tomatoes annually (3). In 2009, the U.S. imported a total of 1.3 million tons (1.2 million tonnes) of tomatoes (not all of which were greenhouse-grown) (2). This high demand, coupled with very good prices, could mean a significant revenue source for North Dakota power plants.

Table 1. Vegetable Yield in Greenhouses, Annual Productivity, kg/m^2 (4)

Crop	Almeria, Spain	The Netherlands
Tomatoes	10–12	42
Peppers	6–7	26
Cucumbers	8–9	58
Snap Beans	5	32

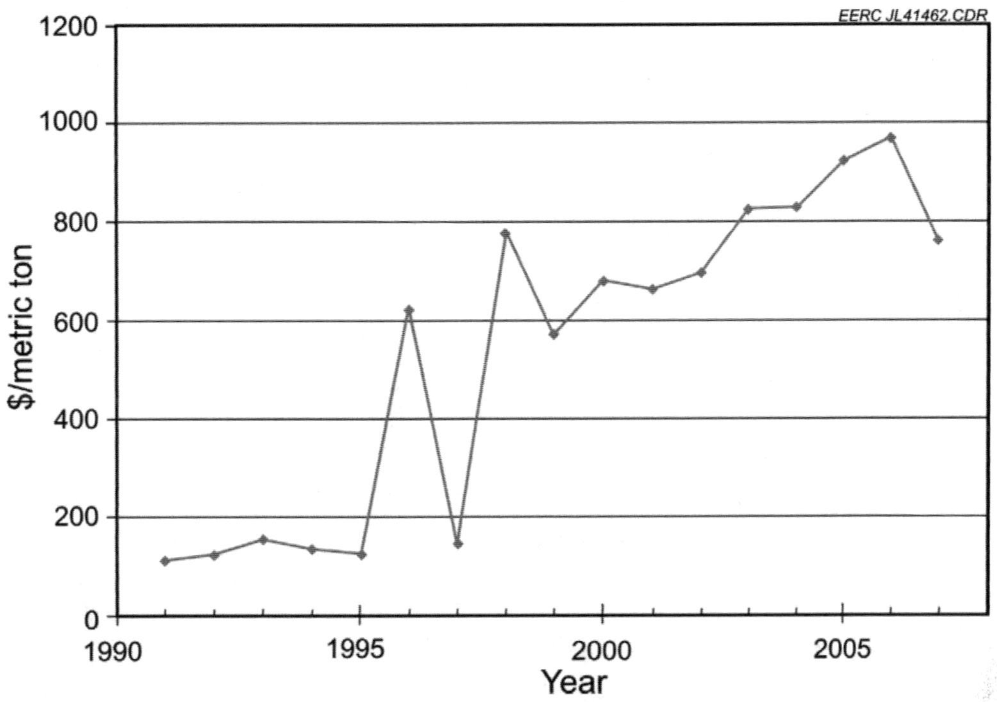

Figure 2. U.S. producer price for tomatoes (5).

Conclusions

Based on information obtained about the technical aspects of the technologies, the following key challenges are expected to be faced by any potential CO_2 utilization technologies intended for North Dakota lignite power utilities:

- The weather will be a serious challenge to algae and microalgae technologies in North Dakota. Raceway or open-pond technologies will be particularly impacted because of freezing of the water in the ponds during winter months. Photobioreactors (PBRs) will also be affected, although to a lesser extent. The only remedy is to use heated systems, but additional heating will drive up the cost even more.
- The alkaline content of lignite fly ash is lower than the desired levels for mineralization processes; external sources of alkalinity may be needed to make this a feasible option.
- Flue gas contaminants are a potential threat to the highly sensitive catalyst systems currently being developed for use in most chemical conversion processes.
- Land space limitations in the immediate vicinity of existing power plants may limit the application of some technology options such as greenhouse agriculture or algae cultivation. PBR algae systems may not have excessive space demands, but the technology is currently too expensive to be a practical option.

Currently, no CO_2 utilization option is ready for implementation or integration with North Dakota power plants. Mineralization technologies suffer from the lack of a well-defined product and insufficient alkalinity in lignite fly ash to make any significant reduction in the produced

CO_2. Algae and microalgae technologies are not economically feasible and will have serious weather-related challenges as well. Algae oil productivity is also currently very limited; in order to sustain the demand of any algae-derived biofuel products such as biodiesel, much R&D is still warranted. However, the most promising option appears to be greenhouse agriculture, which has a readily available market and is carried out in closed environments where temperatures can be easily controlled relatively cheaply by using waste heat from power plants.

Based on the market analysis conducted as part of this study, it is known that greenhouse agriculture can produce larger yields of some vegetable crops than traditional agriculture. The market price for greenhouse tomatoes is quite high, and there is also a high demand in the United States, which imports nearly 1.2 million tonnes of tomatoes annually (2). The capital and operating expenses for such an enterprise could be large, and the amount of daylight during the winter in North Dakota could pose a seasonal production issue. However, people and companies are interested in buying locally sourced vegetables such as tomatoes, peppers, and cucumbers because they are fresher and transportation costs are lower (although the added expense associated with greenhouse agriculture probably eliminates the cost advantage to the individual consumer). Large grocery chains and food service companies that are headquartered in or near the region have expressed interest in the availability of produce from this type of regional source. Finally, there are few greenhouse agriculture enterprises in the region around western North Dakota, so the competition would be minimal.

A preliminary economic feasibility analysis indicates that none of the currently available technologies would be feasible for North Dakota power plants. However, greenhouse agriculture appears to be a promising option and there seems to be a demand for the products both locally and across the United States. A more detailed economic study based on an actual greenhouse in operation would be needed to determine the exact profitability of greenhouse farming in North Dakota. Algae technologies, although seemingly very appealing, are still expensive to operate; more importantly, the product yield is still too low to meet demand.

Acknowledgments

This material is based upon work supported under contract with the NDIC and the U.S. Department of Energy (DOE) National Energy Technology Laboratory under Award Number DE-FC26-08NT43291. The efforts of the EERC's office services, editing, and graphics teams are greatly appreciated. Finally, the authors gratefully acknowledge Angela Morgan for her support and assistance in the preparation of this document.

DOE Disclaimer

References

1. Benemann, John R. (2011), "Microalgae Biofuels and Animal Feeds: An Introduction," johnbenemann@microbioengineering.com (obtained Nov 2011).
2. U.S. Department of Agriculture (2010) "U.S. Tomato Statistics," Table 80, Jul 2010 data set.usda.mannlib.cornell.edu/MannUsda/viewDocumentInfo.do?documentID=1210 (accessed Jan 2012).
3. Cook, Roberta and Linda Calvin (2005), "Greenhouse Tomatoes Change the Dynamics of the North American Fresh Tomato Industry," U.S. Department of Agriculture, Economic Research Report No. ERR2, (http://151.121.68.30/publications/err2/ (accessed Dec 2011).
4. Cantliffe, Daniel J. and John J. Vansickle (2003), "Competitiveness of the Spanish and Dutch Greenhouse Industries with the Florida Fresh Vegetable Industry," Document HS918, Horticultural Sciences Department, Florida Cooperative Extension Service, Institute of Food and Agricultural Sciences, University of Florida, http://edis.ifas.ufl.edu/cv284 (accessed Sept 2011).
5. Food and Agriculture Organization of the United Nations (2011), http://faostat.fao.org/site/570/DesktopDefault.aspx?PageID=570#ancor (accessed Sept 2011).

(365f) Harnessing Applied Potential for Sunlight Driven CO2 Utilization

Benjamin R. Buckley, Department of Chemistry, Ashby Road, Loughborough University, Loughborough, United Kingdom

Escalating CO_2 emissions due to increasing fossil fuel consumption is generally acknowledged as one of the major contributors to global warming. Therefore, the development of methods for reducing the amount of CO_2 released is a major challenge with significant societal consequences. Pressures on fossil fuel resources in the coming decades also mean that raw materials for the chemical industry must be found that can reduce our reliance on fossil fuel-based feedstocks. One promising approach towards reducing our reliance on fossil fuel based raw materials for chemical synthesis is to use CO_2 from the atmosphere as a starting material. We herein report an electrosynthetic process for converting atmospheric CO_2 into organic molecules (cyclic carbonates (5, 6 and 7 membered rings), acyclic carbonates, carbamates and alcohols) that is powered by sunlight. This is one of only a handful of approaches that activates CO_2 at atmospheric pressure and ambient temperature, and provides a step change in the area.

See for example our initial communication in the area: Electrosynthesis of Cyclic Carbonates from Epoxides and Atmospheric Pressure Carbon Dioxide, Benjamin R. Buckley, Anish P. Patel and K. G. Upul Wijayantha. Chem. Commun. 2011, 47, 11888-11890.

Accelerating Progress Toward Operational Excellence of Fossil Energy Plants with CO_2 Capture

Stephen E. Zitney[*], Eric. A. Liese[*], Priyadarshi Mahapatra[*], Richard Turton[*,†], and Debangsu Bhattacharyya[*,†]

Abstract

To address challenges in attaining operational excellence for clean energy plants, the National Energy Technology Laboratory has launched a world-class facility for *Advanced Virtual Energy Simulation Training And Research (AVESTAR^TM^)*. The AVESTAR Center brings together state-of-the-art, real-time, high-fidelity dynamic simulators with operator training systems and 3D virtual immersive training systems into an integrated energy plant and control room environment. This paper will highlight the AVESTAR Center simulators, facilities, and comprehensive training, education, and research programs focused on the operation and control of an integrated gasification combined cycle power plant (IGCC) with carbon dioxide capture.

Keywords: IGCC, CO_2 capture, Dynamic simulator, Operator training system, Plant operations, Control, Immersive training system, Virtual plant

1 Introduction

The National Energy Technology Laboratory's (NETL) world-class *Advanced Virtual Energy Simulation Training And Research (AVESTAR[TM]) Center* is dedicated to accelerating progress toward achieving operational excellence for commercial-scale, electricity-producing power plants capable of carbon dioxide (CO_2) capture (Zitney, 2012a; Liese et al., 2012; Zitney and Turton, 2011). The AVESTAR Center is accomplishing this vital mission by bringing together advanced dynamic simulation and control technologies, state-of-the-art simulator facilities, and leading energy experts to address R&D challenges facing industry on the efficient, safe, and reliable operation of clean energy plants. The center is also satisfying NETL technology transfer and outreach objectives by providing comprehensive, hands-on, simulation-based workforce training to engineers and operators in the energy industry. In addition, AVESTAR's education program offers experiential simulation-based learning to engineering students and researchers at major universities and trainees in power plant certificate programs at community and technical colleges.

[*] AVESTAR Center, National Energy Technology Laboratory, Morgantown, WV 26507, USA

[†] Department of Chemical Engineering, West Virginia University, Morgantown, WV 26506, USA

NETL launched the AVESTAR Center in March 2011 with the deployment of a high-fidelity, real-time dynamic simulator for an integrated gasification combined cycle (IGCC) power plant with CO_2 capture (DOE, 2011; Zitney, 2011; Zitney et al., 2011). Developed in collaboration with Invensys Operations Management (Invensys), the IGCC dynamic simulator combines, for the first time, a "gasification with CO_2 capture" process simulator with a "combined-cycle" power simulator into a single high-performance dynamic simulation framework. Based on the DYNSIM® dynamic process simulation software from Invensys, the rigorous plant-wide IGCC model combines heat and material balances, thermodynamics and complex gasification reaction chemistry to provide an accurate and continuous view of the highly-integrated energy system in action by calculating its transient behavior over time.

The IGCC dynamic simulator at NETL's AVESTAR Center is designed as a full-scope operating training system (OTS) to provide extensive hands-on experience with plant operations and control. Invensys's InTouch® software, a human-machine interface (HMI), enables trainees to interact with the real-time simulator and recreates the look and feel of the operator station in the IGCC power plant control room (Figure 1).

The OTS enables engineers and operators to analyze IGCC performance over a wide range of operating scenarios, including normal full-load operation, plant startup, shutdown, power demand load following, and variable CO_2 capture rates. The IGCC dynamic simulator also lets users analyze the plant's response to disturbances (e.g., fluctuating coal feed flow rate and composition) and malfunctions (e.g., gas turbine trip). In addition, the dynamic simulator allows for switching and co-firing of different types of feedstocks, including coals, petroleum coke, and biomasses (e.g., switchgrass, yellow pine).

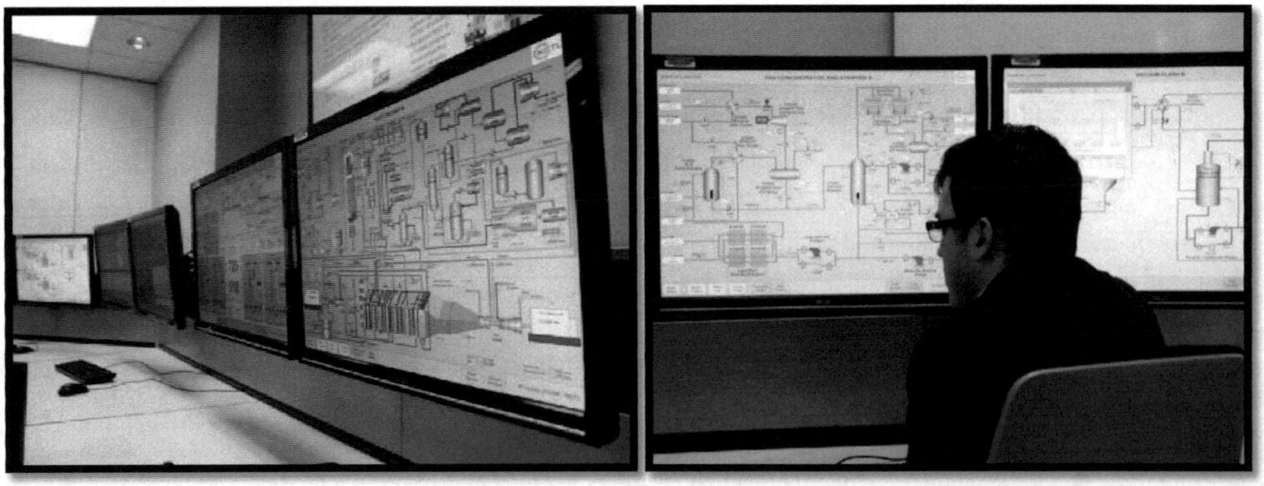

Figure 1. IGCC dynamic simulator and operator training system at the NETL AVESTAR Center

NETL's AVESTAR Center also offers an immersive training system (ITS) for the IGCC power plant with CO_2 capture (Zitney, 2012b). Built using the EYESIM® virtual reality software technology from Invensys, the ITS adds another dimension of realism to the real-time dynamic OTS by providing a 3D plant walk-through environment. Wearing 3D glasses or a stereoscopic headset, ITS users can interact with plant equipment (e.g., gasifier), activate transparent views (e.g., liquid level in a tank), display pop-up trends (e.g., gas turbine combustor temperature over time) and web pages, and experience equipment sound effects (e.g., pump engines), malfunctions (e.g., leaks, fires), and visual training scenarios (e.g., CO_2 absorber column operation). Figure 2 shows the internals of the gasifier where the animations of the slag flow and the gasification process are linked to the slag viscosity and gasification efficiency in the dynamic simulator.

Figure 2. Transparent view of gasifier operation from IGCC 3D virtual ITS

Using the ITS, IGCC field operators can coordinate activities with control room operators. Immersed in the virtual environment, field operators can move and interact as if they were in the real plant. The environment is fully interactive with the dynamic simulation models, so actions taken by a field operator will have an impact on the process and actions performed in the control room will change the information visible to the field operator. As a result, field and control room operators will be trained to coordinate their activities and perform collaboratively as a team. Additional benefits include training for safety-critical tasks, rare abnormal situations, and emergency shutdowns.

The AVESTAR training program offers a variety of courses, ranging from IGCC orientation to an advanced operators' course, all developed for NETL by Fossil Consulting Services (FCS) (Provost, 2011). The IGCC courses merge classroom learning, simulator-based OTS learning in a control-room operations environment (Figure 3), and immersive learning in the interactive 3D virtual plant environment or ITS. All of the courses introduce trainees to base-load plant operation, control, startups, and shutdowns. Advanced courses require participants to become familiar with coordinated control, fuel switching, power-demand load shedding, and load following, as well as to problem solve equipment and process malfunctions.

Designed to ensure work force development, training is offered for control room and plant field operators, as well as engineers and managers. The importance of teamwork and communication is reinforced. Such comprehensive simulator-based instruction allows for realistic training without compromising worker, equipment, and environmental safety. It also better prepares operators and engineers to manage the plant closer to economic constraints while minimizing or avoiding the impact of any potentially harmful, wasteful, or inefficient events.

The training platform can also provide dynamic simulations that assist in the design, construction and performance improvement of IGCC power plants with CO_2 capture. Designers can run and validate an entire control strategy before the plant is even built. And once the plant is up and running, the training solution can be integrated with the control system for continuous operation and refinement based on real-world data.

Figure 3. Trainees use IGCC OTS in AVESTAR's control room environment

An AVESTAR facility similar to the one at NETL is also in place at the West Virginia University's (WVU) National Research Center for Coal and Energy (Dillon, 2011). It also features a DYNSIM-based OTS and EYESIM-based ITS for an IGCC power plant with CO_2 capture. With support from the NETL Regional University Alliance (RUA), the AVESTAR Center at WVU is used to augment graduate and undergraduate engineering education in the areas of process simulation, dynamics, control, and safety (Bhattacharyya et al., 2011a). Students and researchers gain hands-on simulator-based training experience and learn how the commercial-scale power plants respond dynamically to changes in manipulated inputs, such as coal feed flow rate and power demand. Students also analyze how the regulatory control system impacts power plant performance and stability. In addition, students practice start-up, shutdown, and malfunction scenarios. The 3D virtual ITSs are used for plant familiarization, walk-through, equipment animations, and safety scenarios.

Furthermore, the IGCC dynamic simulator and ITS are advanced computational tools that are leveraged to conduct internationally-recognized R&D aimed at accelerating clean fossil energy technology (Zitney, 2012c). The dynamic simulator allows for testing of different types of fossil fuels, including coals, petroleum coke, and biomasses, as well as the co-firing of these feedstocks. The simulator also enables valuable analysis of key equipment response times and interactions between major plant sections, including power generation and CO_2 capture. In addition, it facilitates evaluation of alternative IGCC control strategies without the expense and the unexpected hazards of plant experimentation. Additional AVESTAR R&D focus areas include high-fidelity dynamic modeling of gasification and CO_2 capture processes, model predictive control, sensor placement, plant-wide optimization, and virtual plant simulation.

In the following sections, we highlight the AVESTAR Center simulators, facilities, and comprehensive training, education, and research programs focused on the operation and control of gasification-based power plants with carbon dioxide capture.

2 AVESTAR's IGCC Dynamic Simulator/OTS Development and Deployment

NETL used a multi-year, multi-phase approach to establish the world-class AVESTAR Center with the development and deployment of a full-scope, high-fidelity, real-time dynamic simulator/OTS for a generic IGCC power plant with CO_2 capture. The overall effort consisted of six overlapping phases ranging from project scoping and planning to simulator development, testing, and deployment to the establishment and on-going support of the AVESTAR Center:

- Phase 1 – Scoping Study for IGCC Dynamic Simulator/OTS
- Phase 2 – Planning and Function Design Specification for IGCC Dynamic Simulator/OTS
- Phase 3 – Development and Model Validation for IGCC Dynamic Simulator/OTS
- Phase 4 – Factory Acceptance Testing for IGCC Dynamic Simulator/OTS
- Phase 5 – Deployment and Site Acceptance Testing for IGCC Dynamic Simulator/OTS
- Phase 6 – Establishment and Operation of the AVESTAR Center

2.1 Phase 1 - Scoping Study for IGCC Dynamic Simulator/OTS

NETL conducted a detailed nine-month scoping study to determine the IGCC dynamic simulator requirements, capabilities, and features, as well as potential software and hardware suppliers, R&D technology collaborators, training partners, and industry advisors to guide simulator development and establishment of the AVESTAR Center (Erbes and Zitney, 2006a). The scoping study also identified major project milestones and deliverables, as well as constraints and risks. In addition, it provided estimates for cost, staffing, and schedule.

A market research study was done to solicit information and feedback from prospective vendors and other potential project bidders. The information obtained from this market research study was used to improve the request for proposal and detailed simulator specification documents. A plan for establishing a group of R&D partners and an industry experts group was developed and discussed with potential members from other research organizations, end users, engineering firms and potential software vendors. Results of the scoping study were published as an NETL technical report by Erbes and Zitney (2006b).

2.2 Phase 2- Planning and Function Design Specification for IGCC Dynamic Simulator/OTS

During the one-year planning phase, NETL collaborated with its project partners, WVU, FCS, Enginomix, and the Electric Power Research Institute (EPRI) to develop the functional design specification (FDS) containing the detailed technical requirements to supply a generic, high-fidelity, real-time dynamic simulator and operator training system for a generic IGCC power plant with CO_2 capture. The FDS scope included the design, development, fabrication, performance testing, training, documentation, preparation for shipment, shipment, installation, commissioning, fine-tuning and technical support for the IGCC dynamic simulator/OTS for use by NETL and its partners at the AVESTAR Center.

The FDS also provided a conceptual design of the IGCC dynamic simulator reference plant including all major plant areas (e.g., coal slurry preparation, air separation, gasification, acid gas removal including sulfur and CO_2 capture, gas turbines, heat recovery steam generator, steam turbine), cooling and water systems, major equipment items, and relevant control loops. The basis of the simulator conceptual design is the steady-state engineering model of the IGCC power plant published as Case 2 in the systems analysis study by DOE/NETL (2007). The detailed process descriptions specific to the AVESTAR IGCC dynamic simulator reference plant can be found in the NETL technical report by Zitney et al. (2008).

The AVESTAR dynamic simulator is designed to model a generic IGCC power plant with CO_2 capture that produces 746 MW gross (556 MW net) electric power (Figure 4). The gasification plant consists of dual-train gasifiers producing shifted, high-hydrogen syngas. CO_2 capture and desulfurization of the sour syngas are accomplished by a two-stage selective acid gas removal (AGR) process. Sulfur is recovered using a Claus unit with tail gas recycle to the AGR. A treatment unit for sour water strips off the acid gases for utilization in the Claus unit. In the combined cycle, the high-hydrogen syngas is combusted in two (2) F-class combustion turbines designed to operate with syngas as their primary fuel. When firing with syngas under standard conditions, each gas turbine is capable of producing power at its maximum shaft torque limit of 232 MW, for a combined output of 464 MW. The gas turbines are partially integrated with an elevated-pressure air separation unit (ASU). The exhaust of each gas turbine is sent through a heat recovery steam generator (HRSG) and produces steam that is supplied to a steam turbine having a net output of 275 MW. A syngas expander supplies an additional 6.3 MW of net output.

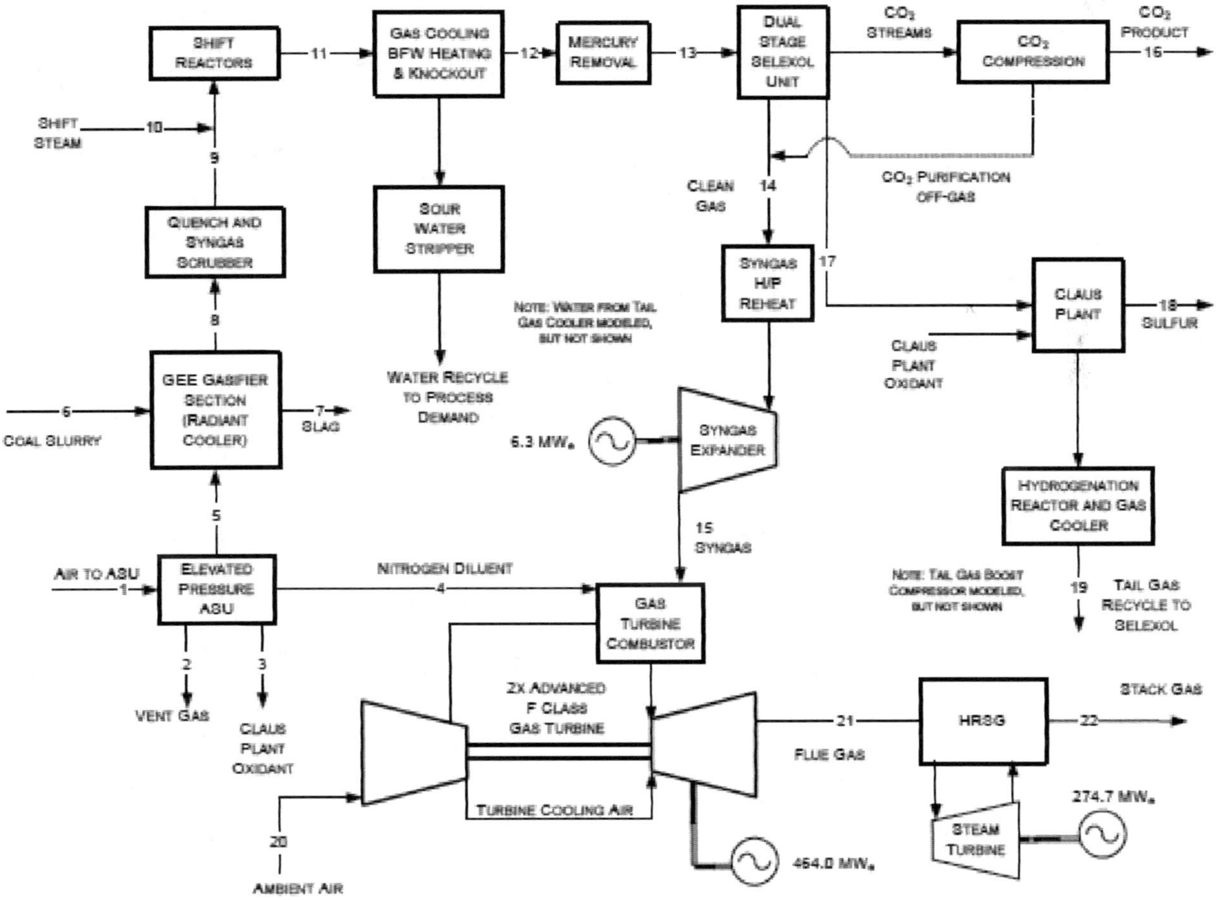

Figure 4. Simplified block diagram for IGCC with CO_2 capture reference plant (DOE/NETL-2007/1281)

2.3 Phase 3 - Development and Model Validation for IGCC Dynamic Simulator/OTS

In October 2008, NETL, Invensys, and its other project partners initiated development of the dynamic simulator/OTS for the generic IGCC power plant with carbon capture. The development phase consisted of the following major tasks:

Task 1. Create engineering design package (EDP)

For all 20 major IGCC plant areas plus electrical system (Table 2), the EDP contains:

(Figure 5)

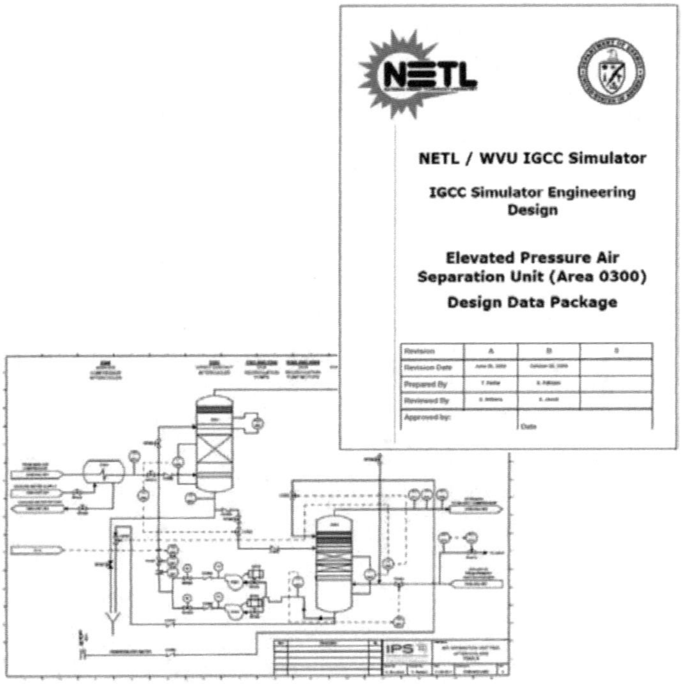

Figure 5. Sample P&ID

Table 2. IGCC Plant Areas

- **Area 100: Plant Common System and Controls**
- **Area 200: Slurry Preparation**
- **Area 300: Elevated-Pressure Air Separation Unit (ASU)**
- **Area 400: Gasifier**
- **Area 500: Syngas Scrubber System**
- **Area 600: Shift Reactors System**
- **Area 700: Gas Cooling System**
- **Area 800: Sour Water Stripper**
- **Area 900: Mercury Removal System**
- **Area 1000: Acid Gas Removal (AGR)**
- **Area 1100: CO_2 Compression**
- **Area 1200: Syngas Reheat and Expansion**
- **Area 1300: Sulfur Recovery Unit**
- **Area 1400: Hydrogenation**
- **Area 1500: Combustion Turbine**
- **Area 1600: Cooling Water System**
- **Area 1700: Boiler Water System**
- **Area 1800: Circulating Water System**
- **Area 1900: HRSG Steam System**
- **Area 2000: Steam Turbine**
- **Area 2100: Selective Catalytic Reduction (SCR)**
- **Area 2200: Electrical System**

Note - Gasification areas in black; Combined Cycle areas in blue; Common areas in green

Task 2. Develop and validate steady-state models

Given the EDP and P&IDs from Task 1, a detailed plant-wide IGCC with CO_2 capture steady-state model was developed using the PRO/II® software from Invensys. This steady-state model provided additional intermediate stream information and a more detailed heat and material balance. Invensys' ModThermo® physical properties package was used for all thermodynamic calculations which were based on public physical property data. WVU researchers provided modeling and validation support for the elevated pressure ASU (Jones et al., 2011), multizonal kinetics-based gasifier model, water gas shift reactors system, dual-stage AGR for H_2S and CO_2 capture (Bhattacharyya et al., 2011b), and Claus plant for sulfur recovery (Jones et al., 2012). The PRO/II steady-state IGCC results were validated against the steady-state Case 2 IGCC with CO_2 capture plant design from DOE/NETL (2007).

Task 3. Design and develop operator trainee station HMI displays

Invensys Operation Management's Wonderware® InTouch® human-machine interface (HMI) software platform was used to develop well over 100 operator training displays for the IGCC OTS including for all 20 plant areas, as well as electrical system and turbine controls. The simulator was provided with a fully emulated DCS (Distributed Control System) HMI that is generic but typical of an advanced full-scope control system that is found in IGCC plants. The emulated DCS was designed so that it can be used for realistic operator training. The P&IDs were used to build the HMI displays for the various plant areas. Figure 6 shows the P&ID for gasifier feed and cooling for train A and the corresponding operator HMI.

Task 4. Generate operating procedures, malfunctions, and remote functions

<u>Operating Procedures</u>

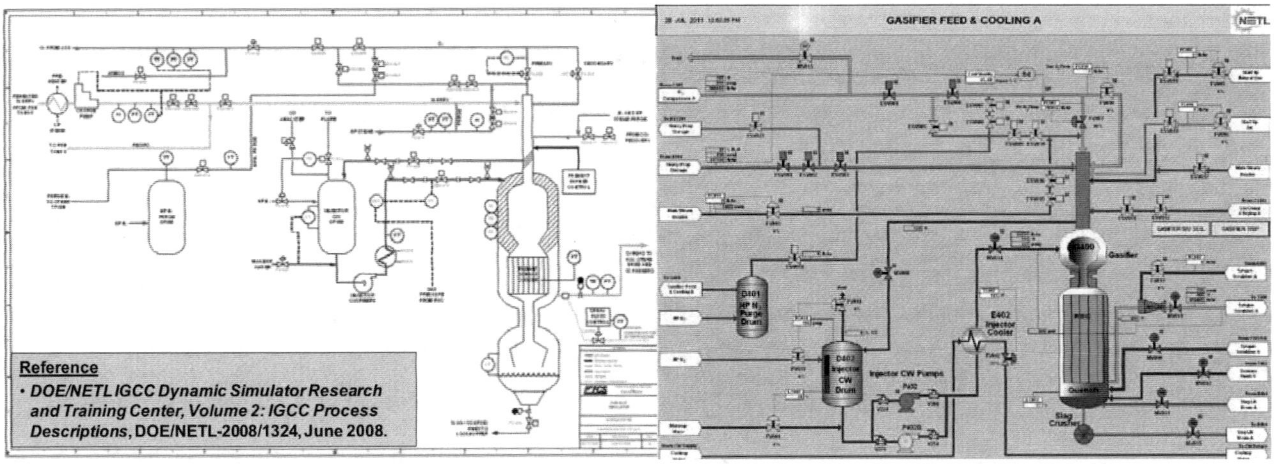

Figure 6. P&ID and HMI display for Gasifier Feed & Cooling for IGCC Train A

In this task, detailed step-by-step operating procedures were generated for IGCC plant cold startup and shutdown, as well as for using the four plant-wide control modes. For example, Table 3 shows the operating procedure for transferring to the coordinated control mode whereby the gas turbine controls the power produced by the plant and the gasifier controls the gasifier pressure.

Table 3. Operating procedure for transferring IGCC to coordinated control mode

Step	Action	DCS Screen	Notes
1.	CLOSE boot spray valve, MV008.	Condensate	
2.	PLACE FC002 in AUTO and verify that the controller is in CASCADE. If FC002 is not in CASCADE PLACE the controller in CASCADE.	Syngas Reheat & Expansion A	
3.	MONITOR FT001 (Syngas from E1201) and verify that the syngas flowrate through FT001 is roughly half of the nitrogen flow seen as the process variable for FC002.	Syngas Reheat & Expansion A	After the unit is steady proceed to the following step
4.	VERIFY that GT MW on the Train A Gas Turbine Master controller is tracking actual Gas Turbine A MW.	Unit Master Control Screen	
5.	Place the GT A in Remote by selecting "REMOTE" in the Load Select pane. Select "Yes" when prompted.	GT Control and Startup A	Unit Master control will track and hold unit at same MW
6.	Enter following values on the Unit Master Control screen: <u>GT</u> MW Min = Max (Operator Min, 0) MW Max = Min (Operator Max, 200) when the masters are in manual. GT Syngas Pressure Setpoint = Set Syngas Pressure to 434 psig. <u>ST</u> Operator STG Min=0 Operator STG Max=275	Unit Master Control Screen Syngas Reheat and Expansion	For Train A the values are: A side to side Operator Bias is also available to the operator.
7.	Set PC003 flare set point to 440 psig.	Syngas Reheat Expansion	Causes PV003 to close and the flare to stop flaring.
8.	SET the Turbine Master MC002 to Auto. This will transfer the controls to GT Follow.	Unit Master Control Screen	The PV for the master will transition to syngas pressure as the GT is now controlling pressure.
9.	Select "TRAIN A RELEASED TO PLANT CONTROL"	Unit Master Control Screen	Button should stay highlighted
10.	SET Slurry Pump Speed Controllers SC204 and SC205 to AUTO CASCADE.	Slurry Preparation Storage	
11.	PLACE Slurry Train A and Slurry Train B controllers FC204 and FC205 to AUTO CASCADE.	Slurry Preparation Storage	
12.	Verify that the G400 Primary O_2 Flow Controller FC003 is in AUTO CASCADE	Gasifier Feed and Cooling	
13.	Match plant MW demand to Plant Gross MW	Unit Master Control Screen	
14.	When ready place the Gasifier Master in Auto	Unit Master Control Screen	This will place the Unit Master controls in GT Follow Gasifier Lead.
15.	Train A MW can now be increased by entering a new setpoint in "PLANT MW DEMAND" box	Unit Master Control Screen	

Malfunctions

The IGCC dynamic simulator is capable of simulating abnormal and emergency events in real-time, including malfunctions and equipment failures to demonstrate inherent plant response and automatic plant control functions. Approximately 200 malfunctions (including redundant components) are pre-programmed on the instructor station. A subset of those is shown in Table 4. In addition, it is possible to create new malfunctions and editing existing ones.

Table 4. Sample of pre-programmed IGCC power plant malfunctions

SYSTEM	MALFUNCTIONS
Air Separation Unit	1. Oxygen leakage 2. MAC motor driver trip 3. N_2 diluent compressor motor driver trip 4. Filter plugging (MAC, DNC) 5. Heat exchanger tube leakage 6. Gas piping / Component leakage 7. CO_2 break through mol-sieve 8. Steam valve failure in regeneration sequence causing low temperature during regeneration. 9. Valve failures in mol sieve sequence. 10. Failure of direct contact cooler causing a higher H_2O load to the mol-sieves 11. Malfunction of compander, causing less cooling capacity 12. Reduced compressor capacity as a result of recycle valves of compressors opening 13. Rising level in low pressure column 14. Reboiler valve fails closed or not allowing sufficient flow 15. Drifted analyzer(s) causing fault interpretation of presented value
Slurry Preparation	1. Weighfeeder Malfunctions (improper scaling, speed, broken belt, etc.) 2. Plugged Silo Chutes 3. Loss of Mill Lube Oil System 4. High mill reject rates, low mill KW or combustion changes that indicate rod must be added to mills 5. Failure of Slurry Feed Pump and Mill Discharge Tank Pump
Gasifiers	1. Ignition Failure 2. Ignition indication false positive 3. Gasifier feed control valve failures 4. Improper Coal / Air Ratio 5. Gasifier Emergency Shutdown 6. Slurry Charge Pump inlet or outlet valve 7. Hi/Lo RSC Drum Level problems including trips (Emergency Shutdown System) 8. Loss or low RSC steam drum circulation flow trip (loss of circulating water pumps) 9. Injector cooling coil leak and or trip 10. RSC sump level trip 11. Lock hopper recycle flow reduced by fouling 12. Forming of slag lumps due to wrong gasifier conditions 13. Bad coal conversion due to wrong gasifier conditions 14. Blockage or overloading of slag crusher.

Remote Functions

Using the IGCC dynamic simulator, the instructor is capable of operating equipment included in the simulation for which no controls are included on operator workstation HMI. The instructor controls for this equipment are commonly called remote functions (Table 5). These remote functions, which include such items as remote manual valves, skid mounted pumps, and motors with local control only, are implemented to realistically simulate field operation of the component. Any item required for startup, shutdown, or malfunctions whose control is local but is observable from the control room is typically a remote function. The remote functions are provided on the instructor station in a graphical from a process diagram or in tabular form where the instructor simply selects from the list of available remote functions.

Table 5. Sample remote functions for the IGCC dynamic simulator

Plant Areas	Remote Functions
Air Separation Unit	1. Prepurifier Cross Connect Valves 2. Prepurifier Vent Valves 3. Prepurifier Isolation Valves 4. Cold Nitrogen Recirculation Valve 5. Seal Gas Isolation to LCT Expander 6. Aftercooler Outlet Valves to the Primary Heat Exchanger
Slurry Preparation	1. Caustic Supply to Slag and Fines Water
Gasifiers	1. Gasifier Isolation Valves 2. Flare Operations 3. Flash Gas Scrubber Warm-up Isolation 4. Caustic Supply to the Vacuum Tank 5. Process Condensate Isolation on Repressurization Line
Quench and Syngas Scrubber	1. Scrubber Isolations 2. Quench Water Strainer Vents 3. Air Ejector Isolations 4. Utility / Grey Water Isolation to Nozzle Scrubbers

Task 5. Design plant control system

The IGCC dynamic simulator was designed with a low-level regulatory control system based on proportional–integral–derivative (PID) controllers. PID is a generic control loop feedback mechanism (controller) widely used in industrial control systems. A PID controller calculates an error value as the difference between a measured process variable and a desired setpoint. The controller attempts to minimize the error by adjusting the process control inputs.

All models of plant areas and equipment included controls, interlocks, trips, and alarms that are associated with them. This included modulating PID controls as well as permissives, and overrides that are provided.

In addition, the IGCC dynamic simulator offers four plant-wide control modes: Gasifier Lead, Gas Turbine Lead, Coordinated, and Manual. All controls are emulated in the DYNSIM® software in a separate controls engine. Process control logic diagrams were generated for IGCC plant areas.

Task 6. Develop and validate real-time dynamic models

The IGCC dynamic model was designed to have adequate fidelity to verify operating procedures and general plant transients within the limits of the simulator development data that are based upon the NETL IGCC Case 2 model (DOE/NETL, 2007) and good process systems engineering practices. The gasifier model is the centerpiece of the IGCC simulator and was developed as a multizonal kinetics-based reactor model. With this level of fidelity, the simulated gasifier responds to changes in fuel characteristics due to switching, blending, or changing the properties or quality of the fuel or slurry. The overall IGCC dynamic model reflects problems with air emissions resulting from trouble in operation of the gasification/combustion systems, fuel and feeds, slag removal equipment, and H_2S and CO_2 removal equipment as specified. The simulator was designed to have sufficient fidelity such that an experienced IGCC operator would find simulator operations realistic making them feel as if they were operating a real IGCC plant.

In June 2010, NETL and its project partners completed two weeks of dynamic model validation for NETL's IGCC dynamic simulator and operator training system. To validate the gasification side of the IGCC plant, the coal feed rate to the gasifier trains was slowly ramped down until emergency shutdown conditions were triggered, at which time all equipment was isolated and shut down. From these warm shutdown conditions, the gasification trains were restarted, followed by startup of the syngas scrubbers, shift reactors, gas cooling system, sour water stripper, dual-stage acid gas removal process for capture of hydrogen sulfide and CO_2, Claus plant, and CO_2 compression. As an additional dynamic test, the coal feed to the IGCC plant was transitioned from Illinois #6 coal to a mixture of Illinois #6 subbituminous and West Virginia bituminous coal. The combined cycle side of the IGCC plant was validated with a startup from cold conditions, followed by a transition to syngas and full load. Both gas turbines were then ramped down, transitioned back to natural gas, and put through a controlled shutdown.

2.4 Phase 4 - Factory Acceptance Testing for IGCC Dynamic Simulator/OTS

In October 2010, NETL and its collaboration partners, WVU, FCS, EPRI, and Invensys, completed two separate two-week factory acceptance tests (FATs) on the "gasification with CO_2 capture" dynamic simulator and the "combined cycle" dynamic simulator, which were combined into the plant-wide IGCC OTS. The objective of the FAT was to demonstrate that the plant-wide dynamic model functions properly, the simulator package operates in accordance with the associated equipment models, and the software configuration and operator interface have been implemented correctly. During the gasification FAT, the team completed the cold startup of both gasifier trains, including equipment filling and warm-up, coal slurry preparation, gasifier warm-up and ramp to 60% design rate, CO_2 capture and compression startup, and fuel feed rate ramp to

100% load. Gasifier train A was then shut down, while train B continued to operate at design conditions. The team also tested switching fuel from Illinois #6 coal to Powder River basin coal, as well as co-firing of coal and biomass. In addition, several process control scenarios were tested for the gasifier and dual-stage acid gas removal unit for CO_2 capture.

For the combined cycle FAT, the team started both gas turbines on natural gas as a function of exhaust temperature control. The rated temperature was achieved at about 60% gas turbine output. After starting the HRSG and steam turbine, the system was ramped up to full load and switched over to synthesis gas. The combined cycle was then ramped down, transitioned back to natural gas, and put through a controlled shutdown. The team also tested all preprogrammed process malfunctions and alarms, as well as the effects of external parameters such as ambient temperature and pressure, humidity, external wind speed, and cooling water inlet temperature.

In February 2011, NETL completed the final two-week FAT for its plant-wide IGCC dynamic simulator and operator training system. With collaborators from WVU, FCS, EPRI, and Invensys, the team completed plant-wide IGCC startup and shutdown, as well as dynamic load and coordinated control testing. The team also tested all pre-programmed process malfunctions, alarms, and equipment trips.

2.5 Phase 5 - Deployment and Site Acceptance Testing for IGCC Dynamic Simulator/OTS

In March 2011, NETL and its partners conducted rigorous site acceptance tests for the two plant-wide IGCC dynamic simulators deployed by Invensys Operations Management at the AVESTAR Center facilities at NETL and WVU, both in Morgantown, WV. Two separate teams of engineers from NETL, WVU, FCS, and Invensys successfully completed acceptance test procedures for the simulator hardware and software installed at each facility. To fully test the simulator software installations, the teams performed a variety of plant startup, shut-down, load-following, and coordinated control tests.

2.6 Phase 6 - Establishment and Operation of the AVESTAR Center

To address challenges in attaining operational excellence for clean energy plants, NETL launched in March 2011 the world-class facility for Advanced Virtual Energy Simulation Training and Research (DOE, 2011; Zitney, 2011; Zitney et al., 2011). The AVESTAR Center brings together state-of-the-art, real-time, high-fidelity dynamic simulators with operator training systems (OTSs) into an integrated control room environment. In collaboration with its R&D partners and energy experts, NETL's AVESTAR team developed, tested, and deployed a dynamic simulator with full-scope OTS for an IGCC power plant with CO_2 capture. The AVESTAR team is now leveraging the IGCC dynamic simulator in ongoing advanced R&D, workforce training, and engineering education programs. NETL and its partners will continue building the AVESTAR portfolio of dynamic simulators and advanced research capabilities to satisfy industry's growing need for training and experience with the operation and control of clean energy plants.

3 AVESTAR's IGCC Dynamic Simulator Accelerates Clean Coal Technology

Relative to power efficiency and environmental conservation, gasification technology represents the future for coal-derived electricity. Simply defined, gasification is a process that uses heat, pressure, and steam to convert coal into a synthesis gas (also known as syngas), which is mainly hydrogen and carbon monoxide. The syngas is cleaned to remove impurities and sent to a gas turbine where it will undergo combustion to produce electricity. The hot exhaust gas from the gas turbine is used to generate steam, which is then fed to a steam turbine to produce additional electricity. This process is known as IGCC because a coal-fired gasification process is integrated with a combined cycle system that produces electricity from both a gas turbine and a steam turbine driven by the gas turbine's exhaust (Figure 7).

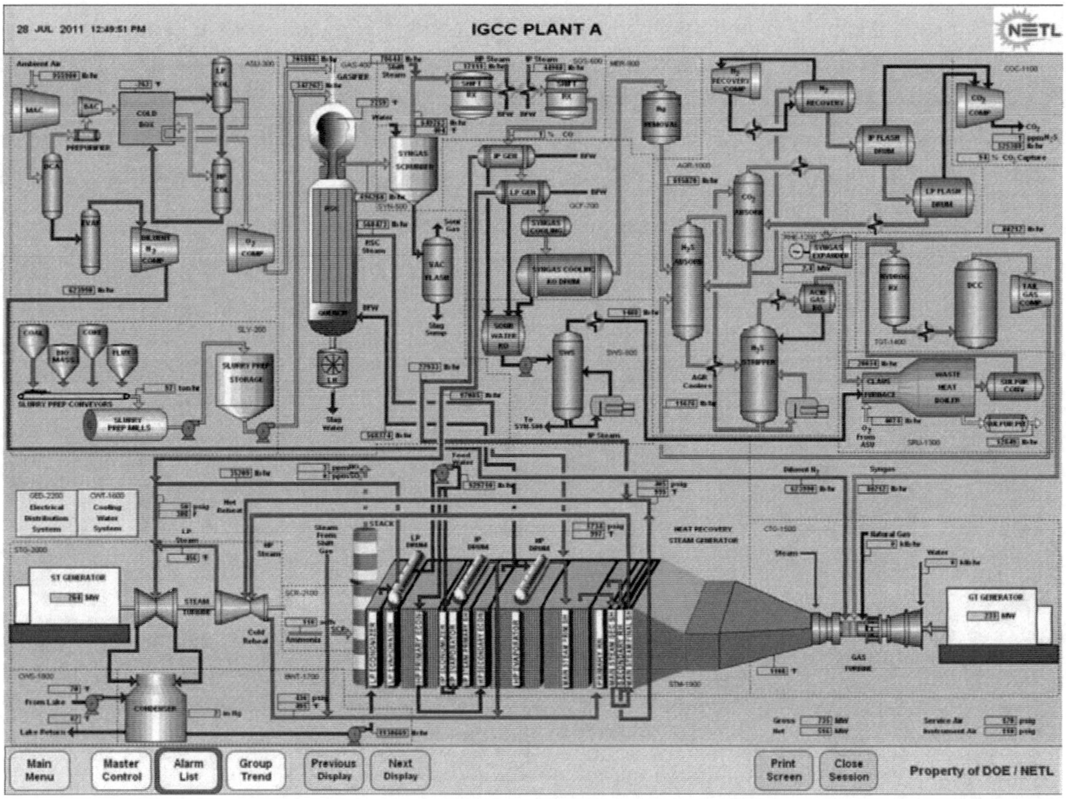

Figure 7. Overview of IGCC power plant flowsheet for Train A from operator training station at NETL's AVESTAR Center

The AVESTAR dynamic simulator is designed to model a generic IGCC power plant with dual-train gasifiers producing shifted, high-hydrogen synthesis gas (syngas). CO_2 capture and desulfurization of the sour syngas are accomplished by a two-stage selective acid gas removal process. In the combined cycle, the high-hydrogen syngas is combusted in two gas turbines (GTs) which are partially integrated with an air separation unit (ASU) which supplies nitrogen for syngas dilution to reduce NOx emissions. The ASU also produces high-purity oxygen for use in the gasifiers. The exhaust of each GT is sent through a heat recovery steam generator and produces steam that is supplied to a steam turbine for additional power generation.

Compared to traditional combustion-based pulverized coal power plants, IGCC offers many advantages, including increased power plant efficiency, resulting in lower-cost electricity. Because gasification occurs at higher temperature and pressure, with limited oxygen, environmental contaminants are easier to remove from the flue gas stream. The system also makes it possible to concentrate CO_2, providing for more efficient removal. Aside from coal, a wide range of carbon-containing materials, such as petroleum coke and biomass, can be used to produce syngas. In addition to being used for electricity, syngas can be converted to other end products, such as specialty chemicals, clean hydrogen, and transportation fuels. Coal gasification electric power plants are now commercially operational in the United States and other nations, and energy experts predict that this process represents the future of clean coal technology.

AVESTAR's IGCC dynamic simulator enables engineers and operators to analyze gasification and plant-wide performance over a wide range of operating scenarios, including plant startup (cold, warm, hot), shutdown, fuel switchovers, on-load cycling, high-load operations of 90-120% of rated capacity, and high frequency megawatt changes for automatic generation control. The IGCC dynamic simulator also lets users to analyze the plant's response to disturbances and malfunctions (Figure 8).

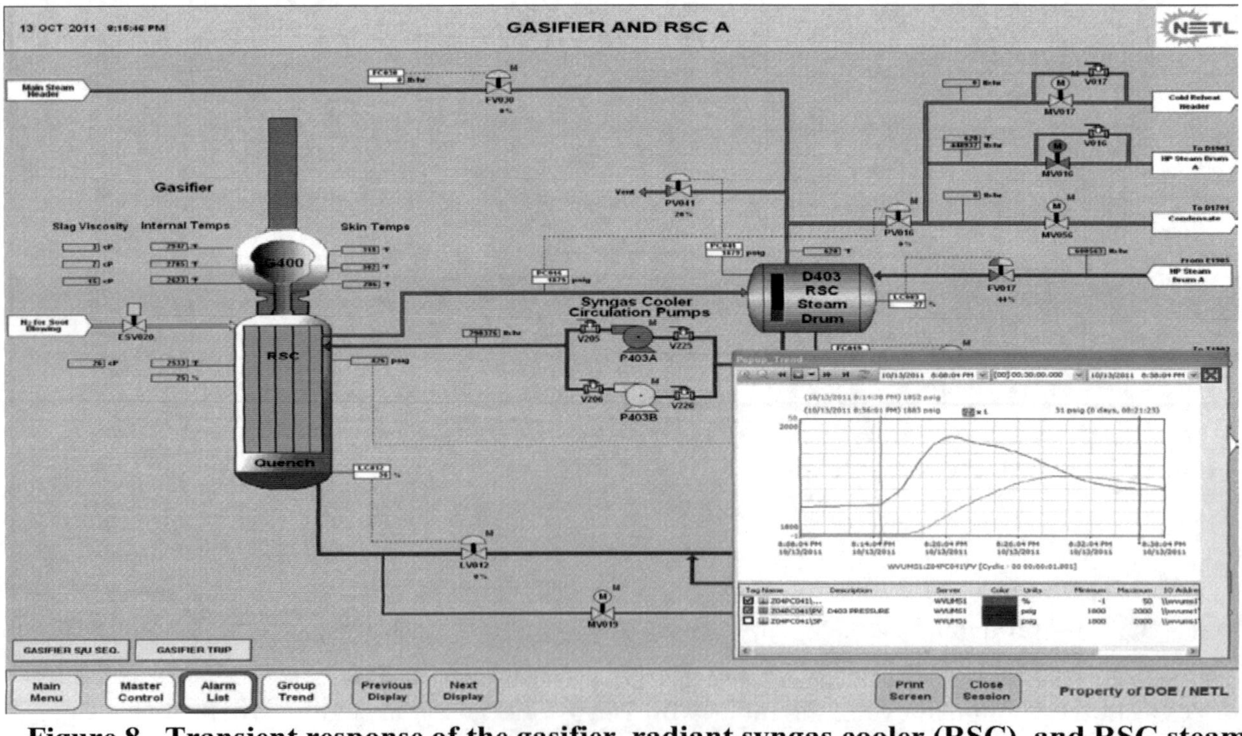

Figure 8. Transient response of the gasifier, radiant syngas cooler (RSC), and RSC steam drum to disturbance in coal feed flow rate

The AVESTAR team is also using the IGCC dynamic simulator to develop effective strategies for the operation and control of pre-combustion capture technology capable of removing at least 90% of the CO_2 emissions (Figure 9) (Bhattacharyya, 2012c,d). Achieving operational excellence can have significant impact on the extent and the rate at which commercial-scale capture processes will be scaled-up, deployed, and used in the years to come. If deployment of new CO_2 capture technologies is to be accelerated, power generators must be confident in ensuring efficient, flexible, reliable, environmentally-friendly, and profitable plant operations.

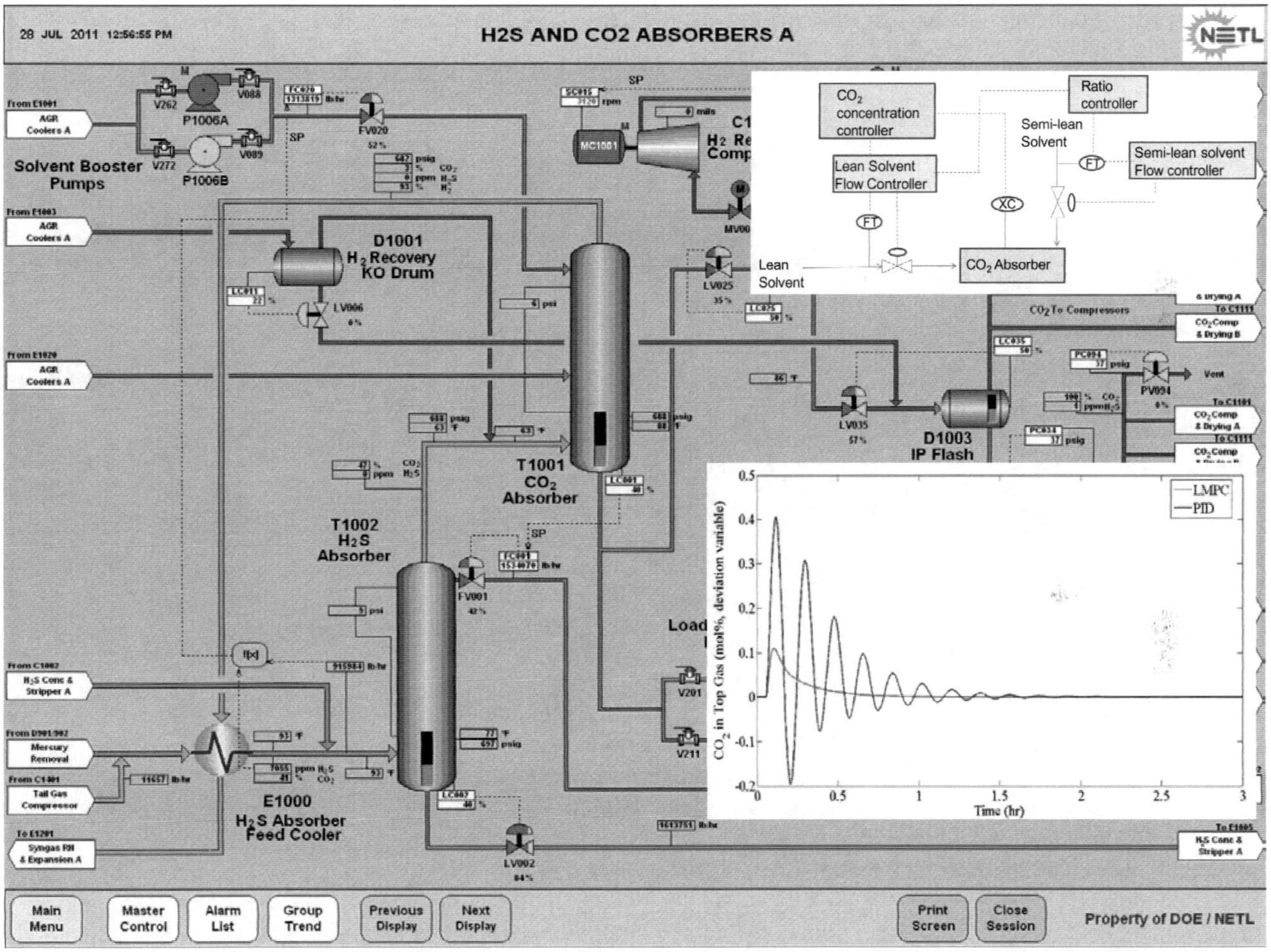

Figure 9. Controlling CO_2 emissions while load-following with IGCC power plant

AVESTAR's IGCC dynamic simulator enables users to analyze capture process performance over a wide range of transient operations. Under carbon-constrained scenarios, IGCC power plants with CO_2 capture will need to exploit variable operating strategies for capture facilities. AVESTAR's IGCC simulator enables researchers to examine how adjusting CO_2 emissions per unit of power generated impacts generation efficiency, cost, and net power output.

AVESTAR's IGCC dynamic simulator also lets researchers ensure that the CO_2 capture facilities do not impose serious restrictions on power plant controllability, flexibility, and emissions of other pollutants. CO_2 capture processes must be able to recover smoothly from disturbances and move to new controller set points in a measured and timely manner (Figure 10). Potential disturbances include variable fuel input flow rates and compositions and fluctuating ambient conditions. Moreover, dynamic plant interactions, particularly heat integration issues, between CO_2 capture and power generation processes must be considered. For improved profitability, effective control strategies are required to maintain plant operation close to economic constraints under conditions different than nominal design conditions, including varying rates of CO_2 capture. In addition, the emission levels of other pollutants must be regulated in concert with the desired CO_2 removal.

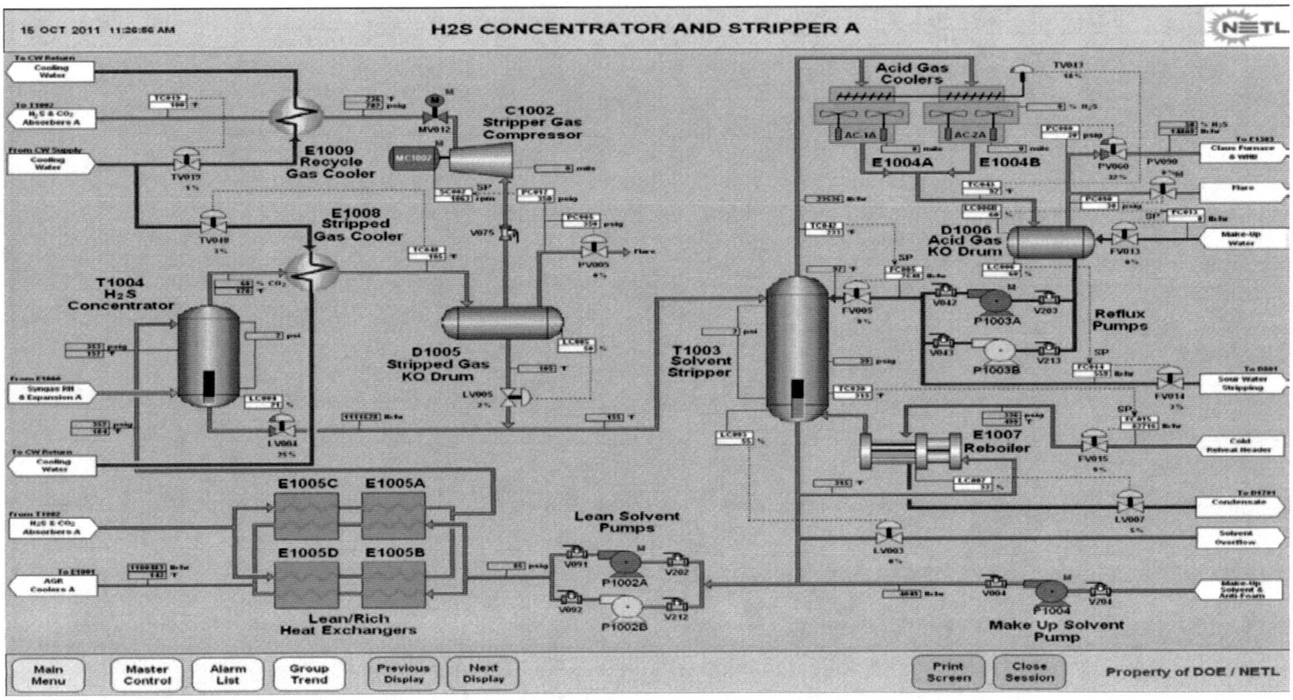

Figure 10. Cascade control for disturbance rejection in solvent stripper for H_2S/CO_2 capture

Comprehensive strategies are required for handling emergency shutdowns and rare abnormal situations, including equipment malfunctions and failures. AVESTAR's IGCC dynamic simulator facilitates the development and analysis of effective operations and control procedures without comprising worker, equipment, and environmental safety.

In the future smart grid era with variable power generators from renewable energy sources (e.g., wind, solar), IGCC power plants may be required to respond to grid dispatching quickly with fast ramp rates in order to ensure the security of power grid operation. AVESTAR's IGCC dynamic simulator will be instrumental in providing the software platform required to improve the load-following capabilities of next-generation gasification-based power plants with carbon capture.

4 IGCC Dynamic Simulator Enhances Power Plant Workforce Training and Experience

Dynamic simulator training has proven to be the most effective way to train control room operators who are responsible for the safe and efficient operation of large, distributed control system (DCS)-based power plants. The IGCC dynamic simulator at NETL's AVESTAR Center improves workforce training by providing operators and engineers with realistic hands-on experience to manage the power plant closer to economic and environmental constraints while minimizing or avoiding the impact of any potentially harmful, wasteful, or inefficient events (Figure 11). The IGCC simulator can:

- Improve understanding of general IGCC power plant and CO_2 capture theory and concepts
- Increase knowledge of plant systems and their function and interaction with other systems
- Enhance understanding of IGCC plant control theory and operation
- Gain operating experience, confidence, and accuracy in normal and abnormal IGCC plant operations
- Provide practice following specific IGCC operating procedures
- Demonstrate effective recovery from various plant upsets and equipment malfunctions

Figure 11. IGCC dynamic simulator provide hands-on workforce training and experience

The AVESTAR workforce training program offers a variety of IGCC courses, ranging from introductory to advanced, that merge classroom learning, simulator-based OTS learning in a control-room operations environment (Provost et al., 2010). Developed in collaboration with training partner, FCS, the portfolio of courses includes:

- **Integrated Gasification Combustion Cycle (IGCC) Orientation for Engineers and Managers:** This four-day familiarization course is designed to increase understanding of IGCC with CO_2 capture processes to personnel assigned as engineers and managers in the power industry.

- **Introduction to IGCC Operations:** This comprehensive ten-day course is designed to introduce engineers, managers, and operators to IGCC process operations including gasification, CO_2 capture, and combined cycle operations.

- **Gasification Process Operations:** This five-day familiarization course is designed for engineers, managers, and operators to increase understanding of gasification and CO_2 capture operations.

- **Combined Cycle Process Operations:** This five-day familiarization course is designed for engineers, managers, and operators to increase understanding of combined cycle process operations for power generation.

- **IGCC Operations for Advanced Operators:** This comprehensive ten-day course is designed increase understanding of IGCC plant operations and control, including gasification, CO_2 capture, and combined cycle operation for personnel assigned as IGCC control room operators.

The IGCC simulator training program has a combination of classroom instruction on background and theory that is complemented with hands-on simulator sessions in which the theory and background is put into practice. The classroom sessions are designed to introduce the basic knowledge required to effectively operate the power plant. The hands-on training is developed to maximize the effective use of the simulator and includes lesson plans that provide the instructor with detailed, day-by-day guidance that helps ensure consistent and comprehensive training. A simulator training manual complements the lesson plans by providing the technical information and procedures for the operators to learn and operate the simulated plant.

All of the IGCC courses introduce participants to base-load plant operation, control, startups, and shutdowns. Advanced courses require trainees to become familiar with coordinated control, fuel switching, power-demand load shedding, and load following, as well as to problem solve equipment and process malfunctions. Control room protocol is also covered including communications, teamwork, equipment monitoring, alarm management, troubleshooting and diagnosing equipment and process problems.

Designed to ensure workforce development, training is offered for control room and plant field operators, as well as engineers and managers from electric utilities, engineering & construction firms, and equipment vendors. The target audience for each of the IGCC courses offered at the AVESTAR Center is shown in Table 6.

Table 6. Target audience for IGCC courses offered at the AVESTAR™ Center

Course	Target Audience
Integrated Gasification Combined Cycle (IGCC)	
IGCC Orientation for Engineers and Managers	Engineers, managers, and trainers from electric utilities, E&C firms, equipment vendors, and research, consulting, and training entities. Researchers and students from academia and national laboratories.
Introduction to IGCC Operations	Engineers and operations personnel from electric utilities, E&C firms, and equipment vendors with an interest in IGCC operations including air separation, gasification, syngas cleanup, carbon capture and compression, and combined cycle.
IGCC Operations for Advanced Operators	Operations personnel from electric utilities, engineering & construction firms, and equipment vendors with an interest in IGCC operations including air separation, gasification, syngas cleanup, carbon capture and compression, and combined cycle.
Gasification	
Gasification Process Operations	Engineers, operations personnel, and managers from electric utilities, E&C firms, and equipment vendors with interest in gasification-based plants, air separation units, syngas cleanup, and carbon capture and compression.
Combined Cycle	
Combined Cycle Operations	Engineers, operations personnel, and managers from electric utilities, E&C firms, and equipment vendors with interest in combined cycle plants.

Expert power plant training professionals from FCS provide instruction for all of the IGCC courses offered at the AVESTAR Center (Figure 12). Continuing Education Units (CEUs) are available from West Virginia University. The AVESTAR team also works directly with clients to analyze their needs and goals to design customized training programs. Instructor training is also available to help train prospective instructors on how to get the most out of their IGCC simulator.

Figure 12. Instructor provides IGCC dynamic simulator-based training to engineers and researchers at NETL's AVESTAR Center

Based on early interest, the AVESTAR Center's IGCC training program is getting off to a good start and generating significant attention. In several inaugural demonstrations in spring 2011, more than 50 engineers and researchers from industry, government, and academia received hands-on experience with the gasification section of the IGCC simulator. In summer 2011, the Center offered its first IGCC training course for participants from the power industry. The overall feedback was extremely positive with the trainees clearly impressed with the simulator, course content, and their "learn-by-operating" experience. In fall 2011, the AVESTAR team provided training on the carbon capture section of the IGCC dynamic simulator to the industry advisory board for the DOE's Carbon Capture Simulation Initiative aimed at accelerating the widespread deployment of CO_2 capture technology to hundreds of power plants.

In spring/summer 2012, the combined cycle portion of the IGCC dynamic simulator started to see considerable interest (Figure 13). NETL's AVESTAR Center partner, FCS and Invensys, provided comprehensive combined cycle training to an integrated energy company. Using AVESTAR's combined cycle dynamic simulator, experiential learning scenarios covered the operation and control of all major combined cycle systems, including gas and steam turbines; condensate, feed water, and circulating water systems; heat recovery steam generator; and sulfur recovery unit. Base-load plant operations were emphasized, along with startup, shutdown, and abnormal situation handling. The customer is planning additional hands-on dynamic simulator courses throughout 2012.

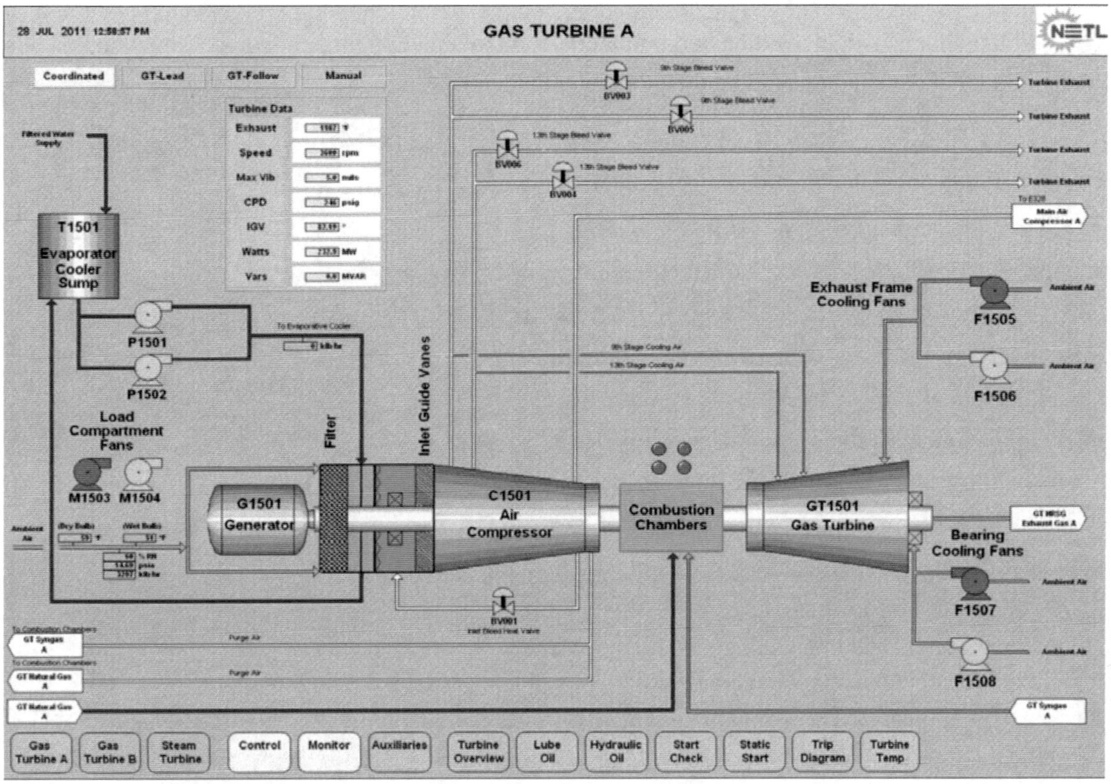

Figure 13. Operator training interface for Gas Turbine in Combined Cycle (IGCC Train A)

In June 2012, NETL's AVESTAR team provided hands-on IGCC dynamic simulator-based instruction to a dozen power plant training students from Pierpont Community & Technical College (PCTC) in Fairmont, West Virginia. After an overview of AVESTAR's IGCC dynamic simulator and operator training system, the two training sessions focused on combined cycle startup, including starting the gas and steam turbines, synchronizing them to the grid, raising power load, and monitoring drum levels in the heat recovery steam generator. Operated in partnership with FirstEnergy, PCTC's one-year workforce development program is aimed at training skilled power plant control room and field operators; successful students receive a Certificate Degree in Power Plant Technology. Due in part to an aging employee population, power generation stations in the region have a need for skilled operators, and excellent employment opportunities exist in West Virginia and throughout the U.S. for students who complete the PCTC program. NETL and PCTC, along with West Virginia University (WVU), have recently completed of a memorandum of understanding which will enable PCTC to leverage the AVESTAR facilities and simulators at WVU for PCTC instructor-led dynamic simulator-based training on power plant operations.

To more rapidly accelerate technology transfer to the energy industry, the AVESTAR team is providing run-time executable versions of the IGCC dynamic simulator to EPRI and several industry members of its CoalFleet Program. These energy industry partners will use the dynamic simulator for in-house training, education, and demonstration on IGCC operations and control.

5 IGCC Dynamic Simulator/OTS Advances Chemical Engineering Education

With support from the NETL Regional University Alliance (NETL-RUA), the AVESTAR Center team is using the IGCC dynamic simulator and operator training system to augment graduate and undergraduate engineering education (Figure 14). Initial activities are focused on using the IGCC simulator located at AVESTAR's West Virginia University (WVU) facility in chemical engineering courses at WVU (Bhattacharyya et al., 2012a). Particular areas of emphasis include process modeling, simulation, dynamics, operations, control, and safety.

Figure 14. AVESTAR's IGCC dynamic simulator augments engineering education

At WVU, Professor Richard Turton (Figure 15) is teaching a first-of-a-kind semester-long course for chemical engineering juniors and seniors devoted to the theory of steady-state and dynamic process simulation, with a strong emphasis on operator training simulators. During the spring 2012 semester, the students spent a total of 3-4 weeks on the IGCC dynamic simulator to learn key OTS features and capabilities and practice start-up, shutdown, load shedding, fuel switching, and malfunction scenarios. Students also learn how the commercial-scale power plants respond dynamically to changes in manipulated inputs, such as coal feed flow rate and power demand.

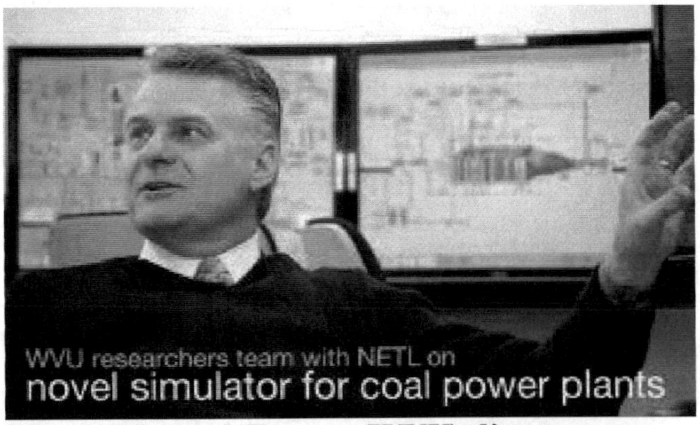

Figure 15. Prof. Turton (WVU) discusses use of the IGCC dynamic simulator for chemical engineering education

Professors Turton and Debangsu Bhattacharyya at WVU are also exploiting the IGCC dynamic simulator in process design and control courses for chemical engineers. Students and researchers spend 4-6 hours during the semester gaining hands-on, interactive, simulator-based training experience. In the control course, students analyze how the regulatory control system impacts power plant performance, stability, robustness, and disturbance rejection characteristics.

The students also investigate abnormal situations in real-time, including equipment malfunctions and failures. In addition, the students observe the real-life transient response of the processes and face the challenge of controlling these systems with regulatory control. They study the step and ramp response of the simple linear, lower order processes along with more complicated higher-order and nonlinear processes. Various open-loop unstable processes are studied and the role of feedback mechanism in making the system closed-loop stable is explained by demonstrating a number of IGCC operation scenarios (Figure 16). In addition to conventional control structures, the dynamic simulator is utilized for studying more complex control structures such as cascade control, feed forward control, ratio control, override control, and split-range control. The IGCC dynamic simulation system is also capable of monitoring, evaluating, and reporting the performance of the students through a trainee performance monitoring module. The students are challenged to control the plant within a set of operating limits in the face of the disturbances and simulated process faults.

The AVESTAR team is working to extend use of the IGCC dynamic simulator to other NETL-RUA universities, including Carnegie Mellon University, the Pennsylvania State University, the University of Pittsburgh, and Virginia Tech University. In addition, the team is pursuing wider university collaboration through the not-for-profit CACHE organization which promotes cooperation among universities, industry and government in the development and distribution of computer-related educational aids for the chemical engineering profession.

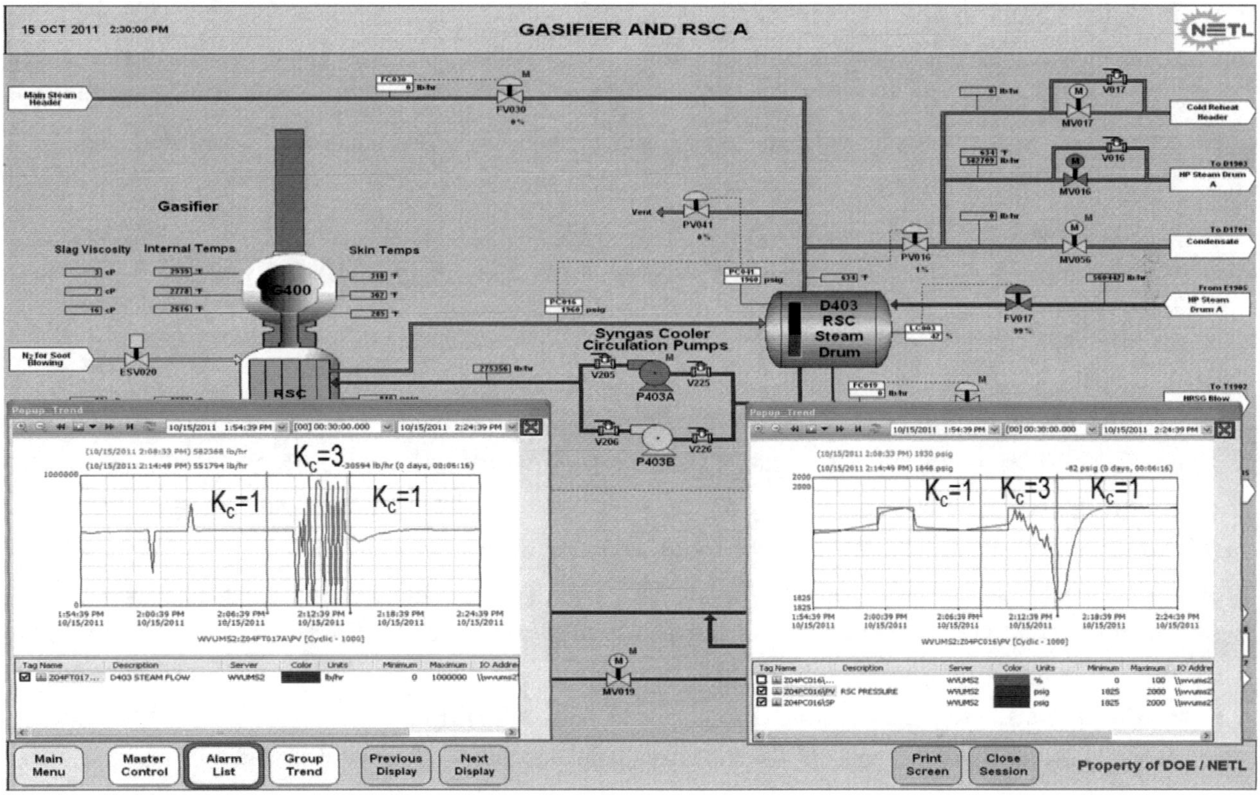

Figure 16. Students use IGCC dynamic simulator to carry out closed-loop stability analysis to determine the limiting proportional gain

6 IGCC Dynamic Simulator/OTS Offers Opportunities for Collaborative R&D

To further leverage the IGCC dynamic simulator, the AVESTAR team is pursuing a collaborative, innovative, internationally-recognized R&D program to achieve operational excellence for next-generation IGCC power plants with carbon capture. Attaining operational excellence requires maximizing the efficiency and profitability from IGCC operations through excellent process control and driving maximum business value from all plant assets, all while reducing negative environmental impact and improving safety. To help address these challenges, the AVESTAR Center brings together dynamic simulation-based technologies, state-of-the-art facilities, and leading energy researchers. Current R&D efforts are focused on the areas of high-fidelity dynamic process modeling, advanced process control, and optimal sensor placement.

6.1 High-Fidelity Dynamic Modeling and Simulation

The AVESTAR team is pursuing research on the dynamic modeling and simulation of IGCC power plants with carbon capture. Coal-fired gasifiers are the centerpiece of IGCC plants, and generate synthesis gas that is cleaned and sent to the gas turbine for power production. AVESTAR's IGCC dynamic simulator was originally developed using a multizonal kinetics-based reactor model to describe the physical and chemical processes taking place inside the single-stage, downward-firing, oxygen-blown, slurry-fed, entrained-flow gasifier. Higher fidelity gasifier models range from simple one-dimensional (1D) equilibrium models to sophisticated dynamic 3D computational fluid dynamics models that describe coupled solid–gas hydrodynamics, heat and mass transfer, and reaction kinetics over the complex gasifier geometry. At the AVESTAR Center, researchers are developing a 1D dynamic model of the entrained-flow gasifier to increase the fidelity of the multizonal model currently used in the IGCC dynamic simulator (Kasule et al., 2012a,b; Kasule et al., 2011).

In the distributed-parameter gasifier dynamic model, 1D partial differential equations (PDEs) representing mass, momentum and energy balances for the solid (char) and gas phases are written. The gasifier model also includes a number of heterogeneous and homogeneous chemical reactions along with devolatilization and drying of the slurry feed. The unreacted-shrinking-core model is used for calculating the rates of the heterogeneous reactions. A detailed model for the radiant heat transfer has been developed considering interactions between all internal surfaces (side wall, top, and bottom surfaces) of the gasifier and the solids along with the interactions between the surfaces themselves. In addition, the model includes the heat loss from the gasifier wall to the environment. In slurry-fed gasifiers, recirculation near the inlet of the gasifier is promoted for rapid mixing of the slurry feed with a portion of the hot reaction products. This vigorous mixing results in a significant rise in temperature at the inlet of the gasifier that helps in evaporating the water and devolatilizing the coal. The recirculation may be achieved by appropriately designing the feed burner, and feeding the oxygen through a swirling annular injector. In the new gasifier model, a heuristic recirculation model has been developed and the conservation equations have been appropriately modified. As shown in Figure 17, the higher-fidelity dynamic gasifier model provide insight into gasifier performance as the inlet and operating conditions change. The 1D PDE-based reactor model also provides profiles along the length of the gasifier for species concentration and gas, solid, and wall temperature (Figure 18).

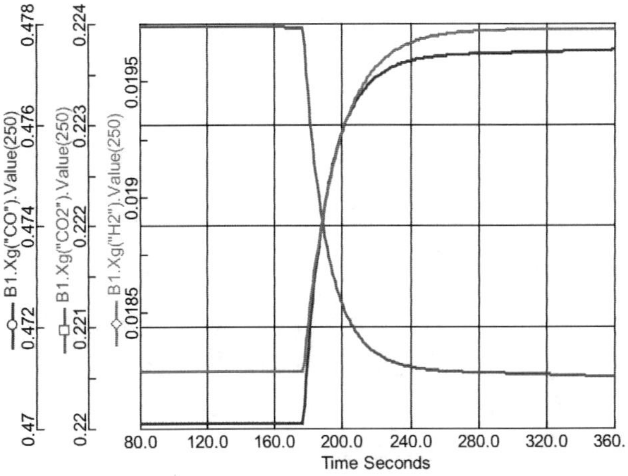

Figure 17. Dynamic response of syngas species concentrations at gasifier exit to a step-change in IGCC oxygen-coal ratio

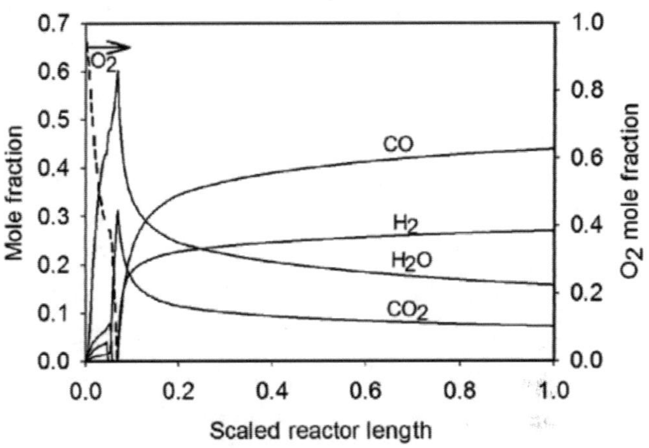

Figure 18. Major gas species concentration profiles along the dimensionless gasifier length

Distributed-parameter dynamic models such as the gasifier model described above are typically too computationally expensive for use in plant-wide dynamic applications, especially if real-time simulation is to be achieved. The use of process control with these high-fidelity PDE-based dynamic models is also beyond current technology. As a result, AVESTAR researchers are developing and applying innovative methods for reducing the complexity of dynamic PDE-based models, while preserving input-output behavior. The dynamic reduced-order models (D-ROMs) approximate the high-order, high-fidelity, PDE-based models, thereby offering a tradeoff between accuracy, range of applicability, and computational cost. Typically the goal is to achieve a D-ROM as accurate as possible at a reasonable cost. Properly implemented model reductions and nonlinear transformations allow the generation of D-ROMs without compromising validity or scope. The PDE model can be viewed as an input/output system, where inlet boundary/stream conditions and operating conditions/parameters represent the inputs and outlet streams and calculated parameters represent the outputs. A D-ROM can be developed which reproduces the output behavior of the PDE model over a limited range of input conditions. The range of validity of the D-ROM is determined by the specifics of the model order reduction technique. Work is underway at the AVESTAR Center on the development of D-ROMs for gasifiers and CO_2 capture devices.

6.2 Advanced Process Control

The AVESTAR team is conducting research on advanced regulatory control (ARC) and advanced process control (APC) strategies for IGCC systems. ARC and APC systems are applied to adapt, predict and adjust to dynamic changes in complex, multivariable processes. Standard regulatory control using PID (proportional, integral, derivative) loops is typically sufficient for relatively simple processes. However, PID control requires repeated manual tuning and has trouble, however, dealing with more complicated processes which may be nonlinear (fluctuating); involve long time delays; are subject to frequent dynamic changes due to plant upsets or load changes; involve complex relationships between process variables; or require control of multivariables which cannot be handled by single-loop PID controllers.

Advanced process control (APC) generates and controls supervisory set points to optimize plant performance. AVESTAR researchers are developing APC strategies based on model predictive control (MPC) for application to complex energy processes (Bhattacharyya et al., 2011d; Mahapatra et al., 2011a). At the core of MPC technology is a mathematical model of the process that is used to predict future process behavior. Using this predictive model the controller is able to calculate an optimum set of process control moves that minimize the error between actual and desired process behavior subject to process constraints, thereby reducing process variability and driving the process closer to its optimum performance.

In recent years, MPC solutions have been applied to advanced energy plants including combined cycle systems, IGCC plants, and hybrid fuel cell/ gas turbine systems. Considerable research challenges and opportunities exist in the development and application of advanced MPC strategies for advanced energy systems with carbon capture. For example, MPC strategies are required for driving power production to satisfy load demands while meeting energy plant integration, performance, and environmental objectives, including CO_2 capture.

Future IGCC plants with CO_2 capture may have to adjust their power output as demand for electricity from the grid fluctuates over time. Such load-following requirements will become far more challenging as power produced by renewable energy is connected to the grid and where seasonal and diurnal change in the load is expected. In view of this, AVESTAR researchers are leveraging the IGCC dynamic simulator to develop novel model predictive control strategies to improve ramp rates and load-following operation of IGCC power plants while satisfying CO_2 emission constraints. Transient studies show that the air separation unit poses a bottleneck in the entire IGCC plant operation by limiting the ramp-rate during load-following operation. As a result, a multiple MPC (MMPC) strategy has been developed to provide better composition control at high ramp-rates compared to the MPC approach (Figure 19). This in-turn allows for higher ramp-rates of ASU, without violating the oxygen-purity constraints. The proposed MMPC control strategy provides a better response to the plant-wide IGCC load-following problem compared to conventional PID and MPC approaches, circumventing the need for larger liquid oxygen/air storage requirements (Mahapatra et al., 2011b).

The AVESTAR team is pursuing a wide variety of collaborative R&D opportunities on advanced

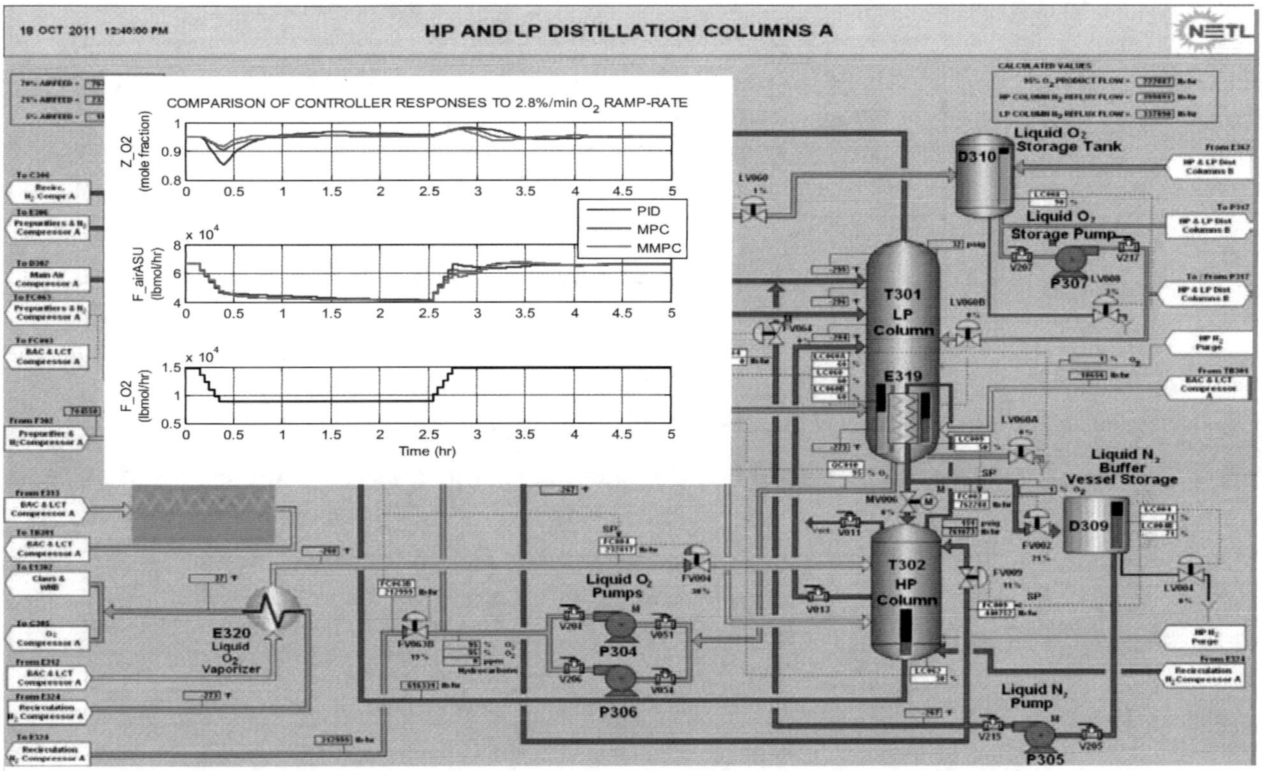

Figure 19. Multiple model predictive control of air separation unit as part of IGCC power plant during rapid load changes

process control for IGCC plants with carbon capture with the ultimate goal of implementing a plant-wide MPC layer around the AVESTAR IGCC dynamic simulator. Research is also underway on optimal sensor placement strategies and algorithms for IGCC systems (Paul et al., 2011).

7 IGCC Dynamic Simulator Operation, Capabilities, Features, and Architecture

The IGCC dynamic simulator is capable of simulating, in real time, plant operations similar to a typical IGCC power plant with CO_2 capture. In addition to normal base load operation, the simulator is capable of simulating a number of operating conditions including: complete cold, warm and hot start-ups; shutdowns; load changes; abnormal and emergency operating conditions; performance tests; and fuel switching and co-firing. As shown in Table 7, the IGCC dynamic simulator handles a number of built-in coal types, petroleum coke, and biomass feedstocks (switchgrass and yellow pine). New feedstocks can be added by providing the fuel analysis information.

Table 7. Coal types handled by IGCC dynamic simulator

Coal Type /Constituent	Illinois #6	Lignite	Powder River Basin	W. VA Bituminous	Pet Coke
HHV	10,300	7,225	8790	15,600	14,100-15,700
Moisture %	17.6	37	31	1	0
Volatile % Matter	44.2	26	31	17	6-12
Sulfur %	4.3	0.4	0.55	0.74 (4 Max)	1.5-10
Ash %	10.8	11.1	5.2	9.1	0.5-0.7

Using operator action, the IGCC dynamic simulator is capable of performing:

- Integrated IGCC startup from cold conditions to 100% load (combustion turbine not in service to both gas turbines, HRSG and steam turbine, and gas expander at full load).
- Integrated IGCC start-up with combustion turbine and HRSG off and the plant hot (pressures, temperatures and other conditions as would be expected within one hour of a trip of the plant from full load) to 100% load.
- Integrated IGCC start-up with one gasifier and one combustion turbine generator in operation with the HRSG and steam turbine generator at partial load, and the shutdown gasifier and combustion turbine generator in cold condition, to 100% load.
- Combustion turbine generator start-up and generator synchronization.
- Loss of one and/or both combustion turbines at full load followed by recovery to rated load.
- Loss of gas expander and gas expander turbine at load followed by recovery to rated load.
- Integrated IGCC shutdown from 100% load to hot standby and cool-down to cold shutdown
- Plant maneuvers (equipment changeover, fuel switching, load changes, adding/removing a gasifier, combustion turbine generator and/or control by the load dispatch [instructor]).
- Integrated IGCC operations with CO_2 compressor unavailable for operation
- Operations to fine tune (optimize) unit heat rate, including but not limited to accurate control of:
    - Coal/Coke (fuel) flow
    - Coal/Coke blend
    - HRSG steam temperatures
    - Condenser pressure
- Operations to reduce emissions.

7.1 Process/Equipment Controls

The IGCC power plant is operated using a basic regulatory control system consisting of conventional PID controllers. Controller faceplates are accessed by clicking on the equipment item of interest as shown in Figure 20. The figure shows a typical modulating controller faceplate with the following controls and indications: Setpoint, Process Variable (PV), Controller Output (Out), Auto (Local and Cascade), Manual/Auto, and Alarms from Low-Low to High-High.

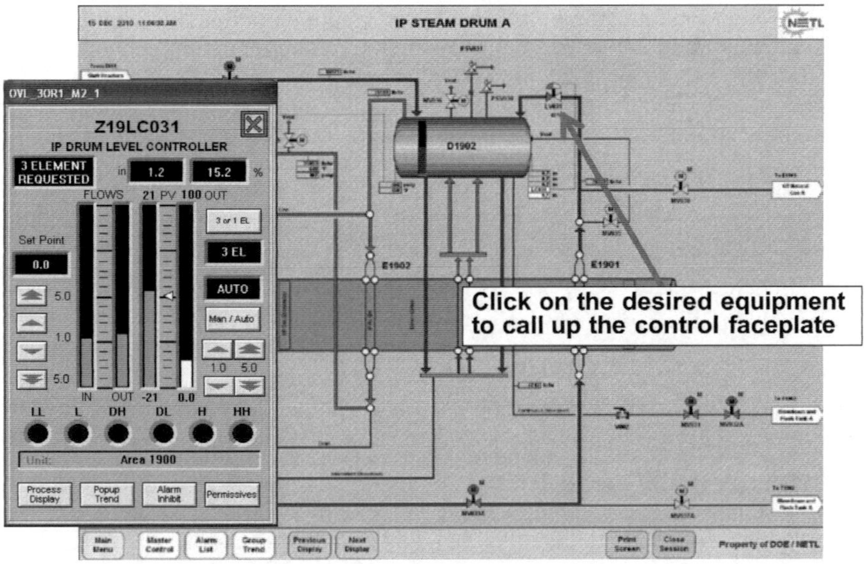

Figure 20. Controller faceplate with controls and indications

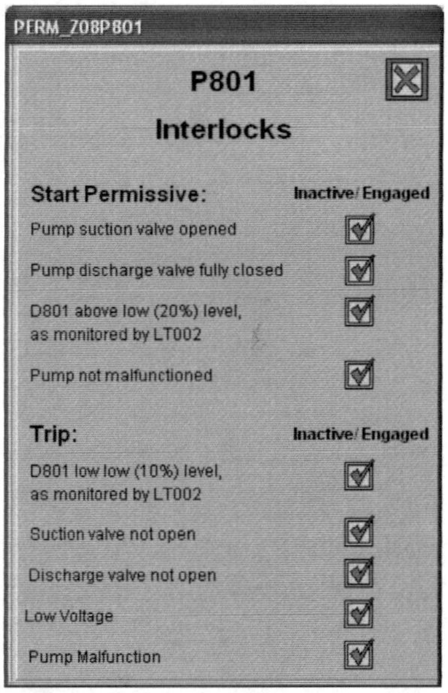

Figure 21. Permissive/Trips

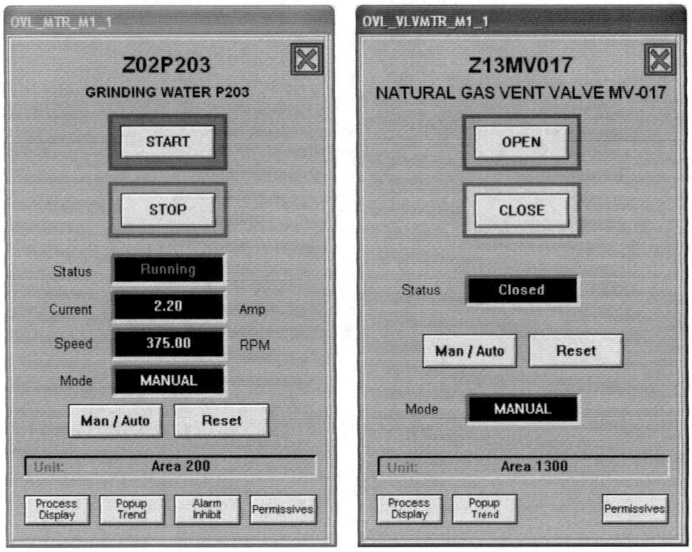

Figure 22. Pump motor and motor operated valve faceplates with controls and indication

Pop-up permissive and trip faceplates can be accessed from the controller faceplates to view logical or operational permissive conditions (Figure 21). Figure 22 shows a typical pump motor faceplate with the following controls and indications: Start, Stop, Status, Current, Speed, Mode, Manual/Auto, and Reset. The figure also show a typical motor operated valve faceplate with the following controls and indications: Open, Close, Status, Manual/Auto, Reset, and Mode. Figure 23 illustrates a Control Group.

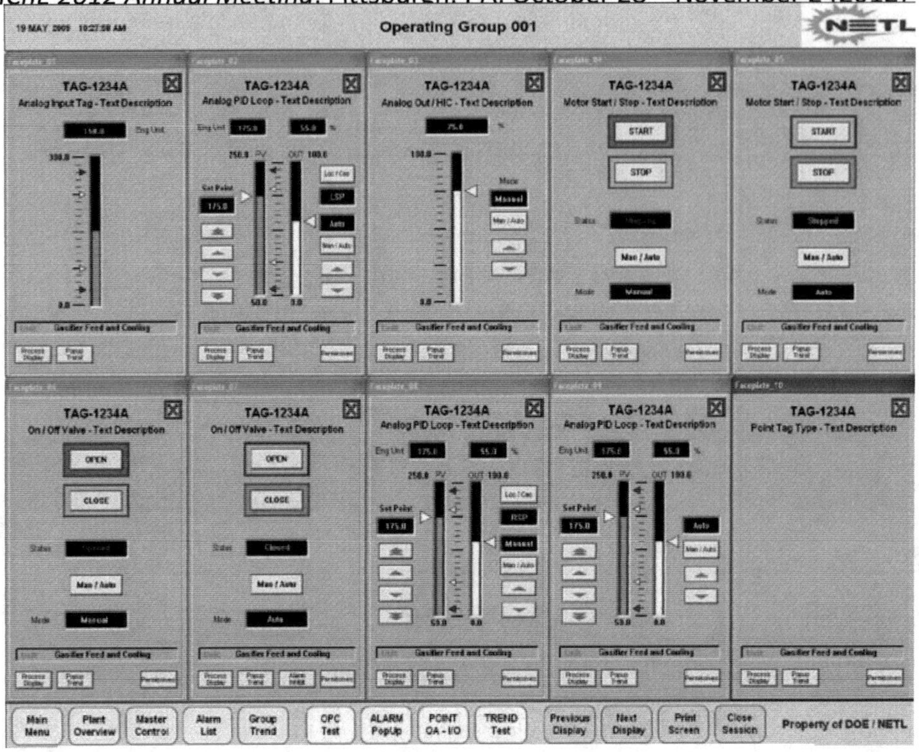

Figure 23. Sample control group

As shown in Figure 24, trends can be displayed for any IGCC dynamic process variable by clicking on a control faceplate using the "Popup Trend" button.

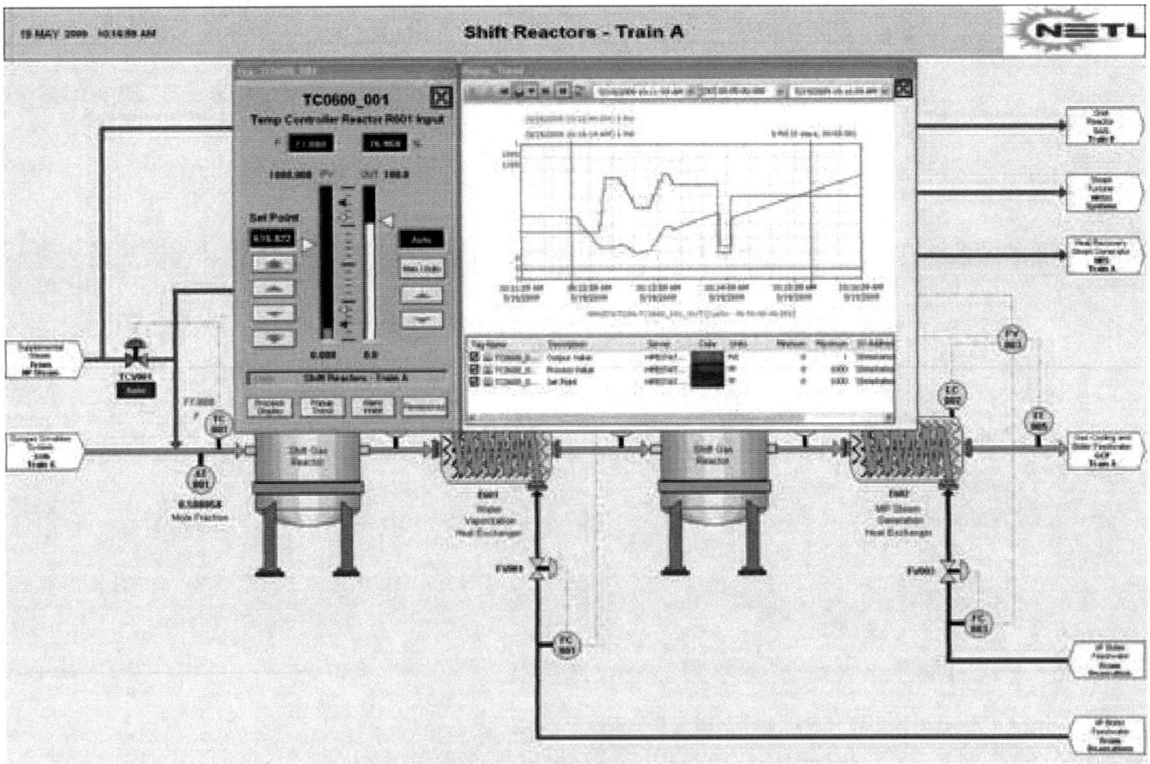

Figure 24. Popup trend showing process variable value over time

7.2 IGCC Coordinated Control

In addition to the regulatory PID controls, there are four plant-wide control modes: Gasifier Lead, Gas Turbine (GT) Lead, Coordinated, and Manual (Table 8). To complete the controls definition, the concept of 'Controller Follow' is necessary. The equipment that controls gasifier pressure is said to be in follow mode and the equipment that controls the MW is defined to be the lead controller. In all four of the required control modes, both gasifier and GT Masters are in AUTO. This means each of the four modes implies a follow mode as well.

Mode selection is achieved by determining the order in which master controllers are set to AUTO. The first Master to be set to AUTO will be in gasifier pressure control mode and the second master to be set in AUTO will be in MW control mode. To enter Coordinated Control mode, the operator has to first select GT lead and then press an additional button to transition to Coordinated mode.

Table 8. IGCC plant-wide control modes

First Controller to be Automated	Second Controller to be Automated	Coordinate Button	Control Mode Description
None	None		Manual
GT Master	None		GT Follow
GT master	Gasifier Master		Gasifier Lead, GT Follow
Gasifier Master	None		Gasifier Follow
Gasifier Master	GT Master		GT Lead, Gasifier Follow
Gasifier Master	GT Master	Enabled	Coordinate Control

<u>Gasifier Lead Control Mode</u>

In this mode, the Gasifier controls the MW produced by the plant and the Gas Turbine controls the Gasifier pressure. The MW demand from plant-wide controls is split to each train and the MW setpoint passes to the Gasifier Master. This MW setpoint is compared to the actual MW produced and the output of the Gasifier Master drives the slurry flow and the O_2 flow to the Gasifier. In this mode, the GT master controls the Gasifier pressure. The pressure controller output is passed through the GT master to the gas turbine remote setpoint.

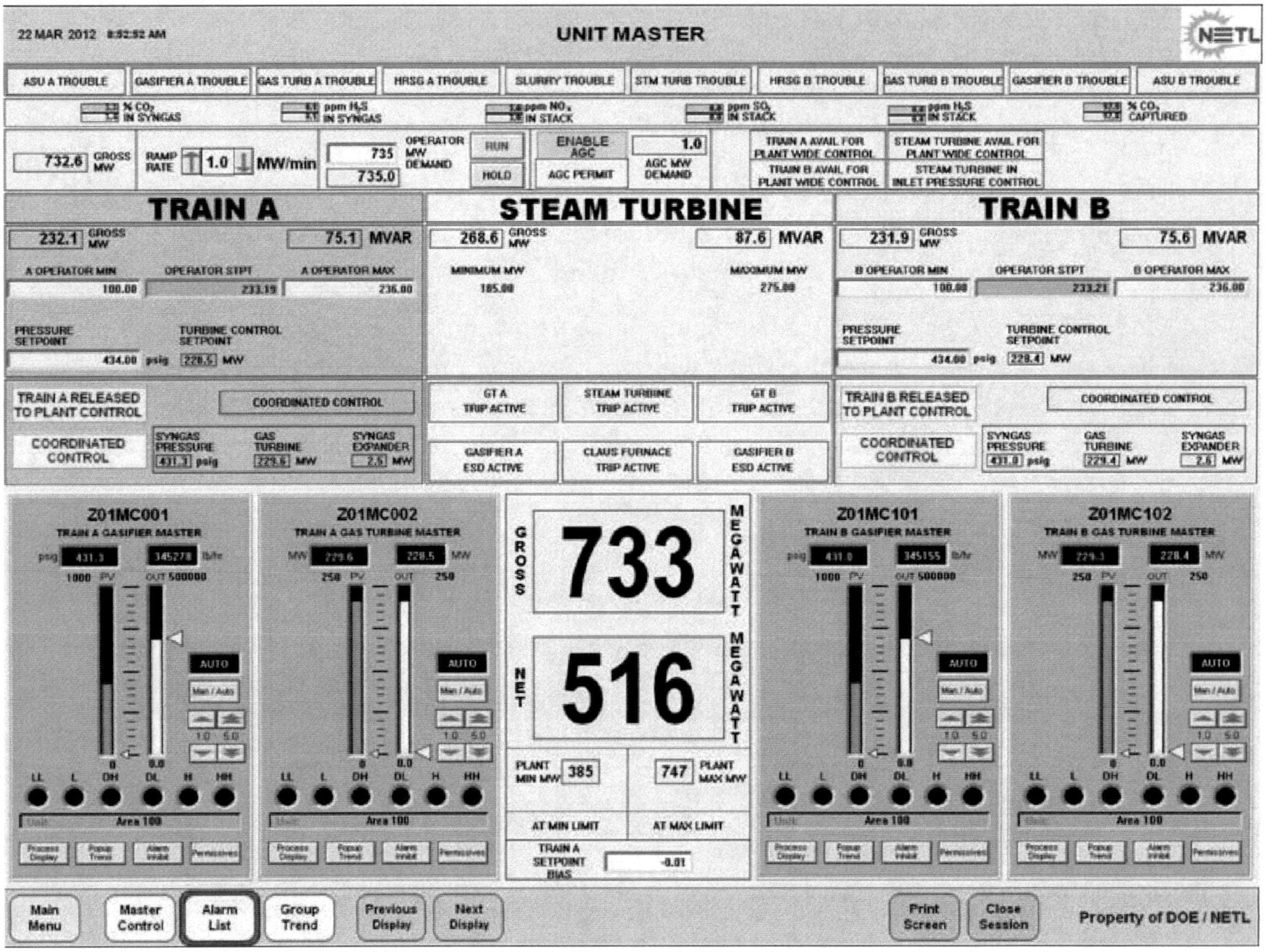

Figure 25. IGCC Unit Master HMI showing Trains A and B in Coordinated Control

<u>GT Lead Control Mode</u>

In this mode, the Gas Turbine controls the MW produced by the plant and the Gasifier controls the Gasifier pressure. The MW demand from plant-wide controls is split to each train and the MW setpoint passes to the GT Master. This MW setpoint is passed to the Gas Turbine controls as a remote setpoint. The Gas Turbine controls have to be set to Remote mode for the GT Master to come out of tracking.

In this mode, the Gasifier Master controls the Gasifier pressure. The pressure controller output is passed thru the Gasifier Master to the Slurry flow controller and O_2 Flow controller.

<u>Coordinated Control Mode</u>

In this mode (Figure 25), the Gas Turbine controls the MW produced by the plant and the Gasifier controls the Gasifier pressure. In addition, the actual MW's produced are used as a feedforward signal to the Gasifier Master and deviation in the Gasifier pressure is used to limit the ramp rate of the GT Master. The MW demand from plant-wide controls is split to each train and the MW setpoint passes to the GT master. This MW setpoint is passed to the Gas Turbine controls as a remote setpoint. This signal passes through a ramp block. The ramp rate is a function of Gasifier pressure deviation from setpoint. The Gas Turbine controls have to be set to Remote mode for the GT Master to come out of tracking. In this mode, the Gasifier Master controls the Gasifier pressure. The pressure controller output is passed through the Gasifier Master to the Slurry Flow controller and O_2 Flow controller. A feed forward signal is passed to the Gasifier Master that is a function of the plant's gross MW. This allows for a faster response to MW change.

Manual Control Mode

When the Gasifier Master and Gas Turbine Master are both in Manual, the Unit Master control is considered to be in Manual. It is possible to have one train in Manual and the other train in one of the automatic control modes. In Manual control mode, the Gas Turbine is controlled separately from the Gasifier. It is incumbent on the operator to maintain Gasifier operating pressure. If the Gas Turbine is in remote mode, the MW setpoint is set from the Gas Turbine Master. If the GT is not in remote, the MW setpoint is set from turbine controls and the GT Master will track the MW setpoint from turbine controls.

7.3 Process/Equipment Malfunctions

The IGCC dynamic simulator has powerful malfunction and scenario capabilities. The instructor, through the malfunction feature, has the ability to fail every pump, fan, motor operated or control valve, heat exchanger, transmitter, and control element on the simulator (Figure 26). In addition to the common and standard malfunctions, special malfunctions can also be included. Each malfunction can be triggered by the instructor or through scenarios. The introduction of a malfunction does not alert the operator to the impending malfunction. Where operator actions are a function of the degree of severity of the malfunction (e.g., gasifier shutdown, HRSG tube leaks, gasifier fail to lightoff, and heat exchanger leaks, etc.) the simulator has adjustable rates for the malfunction to represent the range of plant malfunction conditions.

The malfunctions include both discrete and variable malfunctions which may be applied to IGCC equipment, controls, and processes. Discrete failures include the failure of the IGCC equipment to specific states. Variable malfunctions include the failure of equipment, controls, or processes to a variable state. On variable malfunctions, the severity can be set instantaneously or may be ramped over time. An artificially imposed variation in heat exchanger fouling or an imposed drift in the signal from the sensing element are typical of variable malfunctions.

The IGCC dynamic simulator is capable of simulating a minimum of 10 non-conflicting simultaneous malfunctions. Malfunctions have the ability to occur simultaneously if they can occur by virtue of actual plant design or simulator configuration, or if they have occurred in actual operational experience of existing plants of similar configuration.

It is possible to introduce a malfunction by setting a finite time delay from time of entry. It is also possible to insert a malfunction based on a trigger that will activate or de-activate the malfunction when the triggers are satisfied. The trigger shall be any parameter in the simulator reaching a preset limit. An example of such a parameter is combustion turbine load reaching 150 MW.

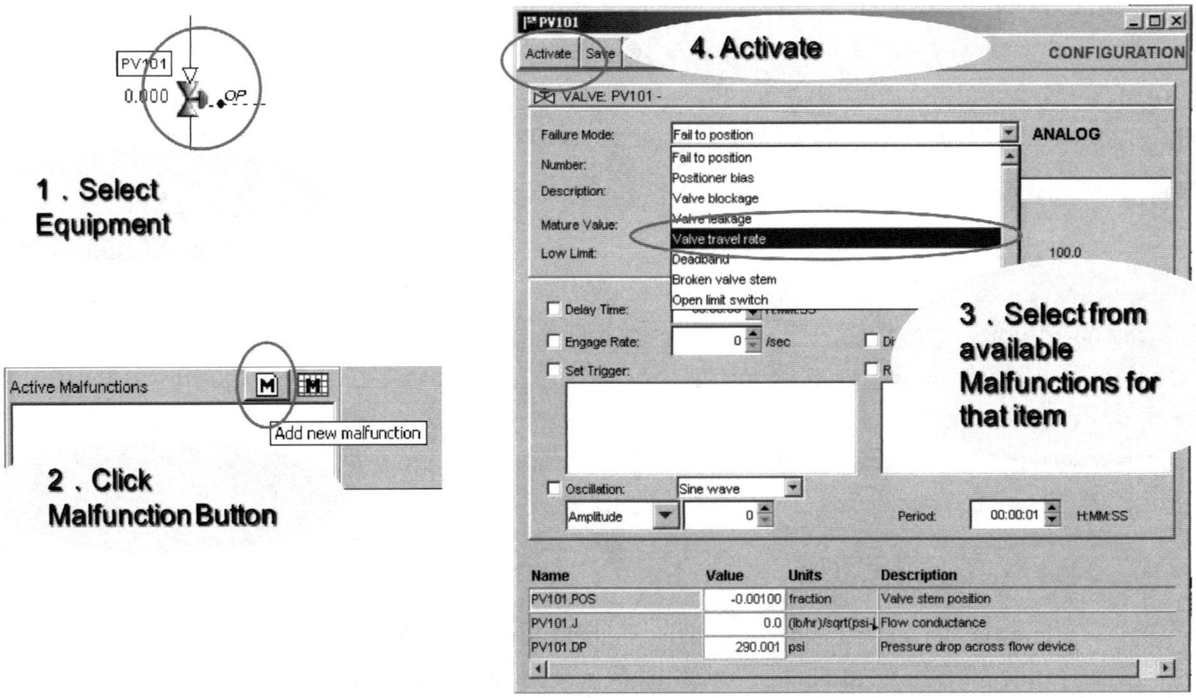

Figure 26. Activation an equipment model malfunction from instructor station

7.4 Alarms

A dedicated IGCC alarm screen is included as part of the operator HMI (Figure 27). The alarm display screen presents alarm messages related to digital inputs, state change, absolute analog, deviation, and other alarm conditions that are programmed into the model. The alarm display has the look of a modern DCS control system alarm display with the following features:

- Dynamically updated alarm status and value information
- Color-coded priority and status indicators that allow quick identification of critical alarms
- Ability to configure alarm presentation by chronological sequence, priority and/or acknowledged status
- Visual and audible alarm notification at any operator console
- Simulator-wide alarm acknowledgement that allows any console to acknowledge alarm(s)
- An historical list of alarms (relevant to the current simulation session)

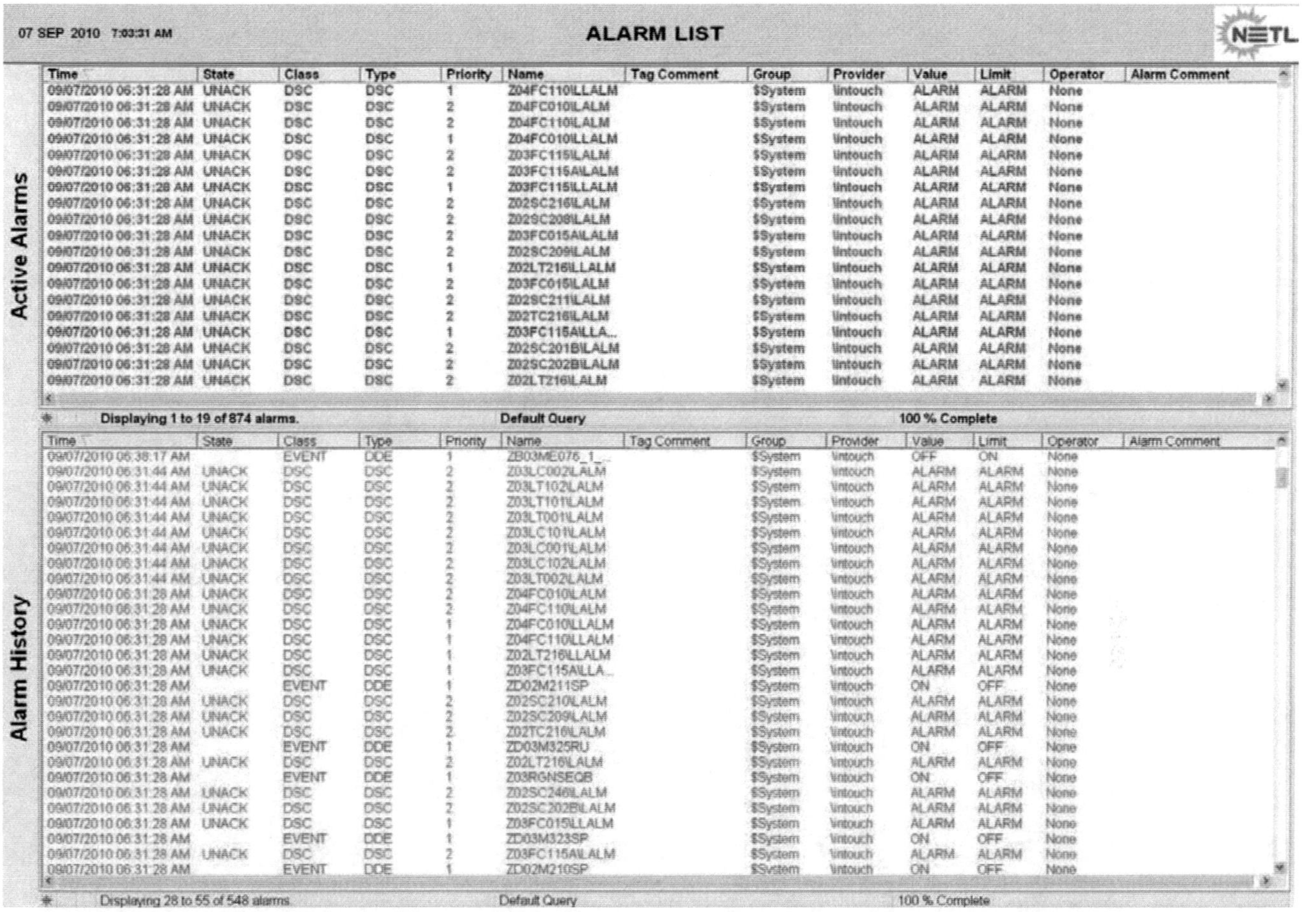

Figure 27. IGCC operator trainer alarm list with active alarms and alarm history

7.5 Operator Training HMI Displays

The human-machine interface (HMI) for the IGCC dynamic simulator is similar to a distributed control system operator interface. The operator training interface is menu-driven allowing the user to navigate to the HMI screens for all processes associated with the plant. Figure 28 shows the main menu used to navigate to all of the individual plant area displays. The HMI screen graphics contain a schematic representation of the process or equipment in addition to pop-up control stations, permissive pop-ups, trending, and an alarm suite as discussed above. Other examples of HMI displays are shown in Figures 7-10, 13, 16, and 19-27.

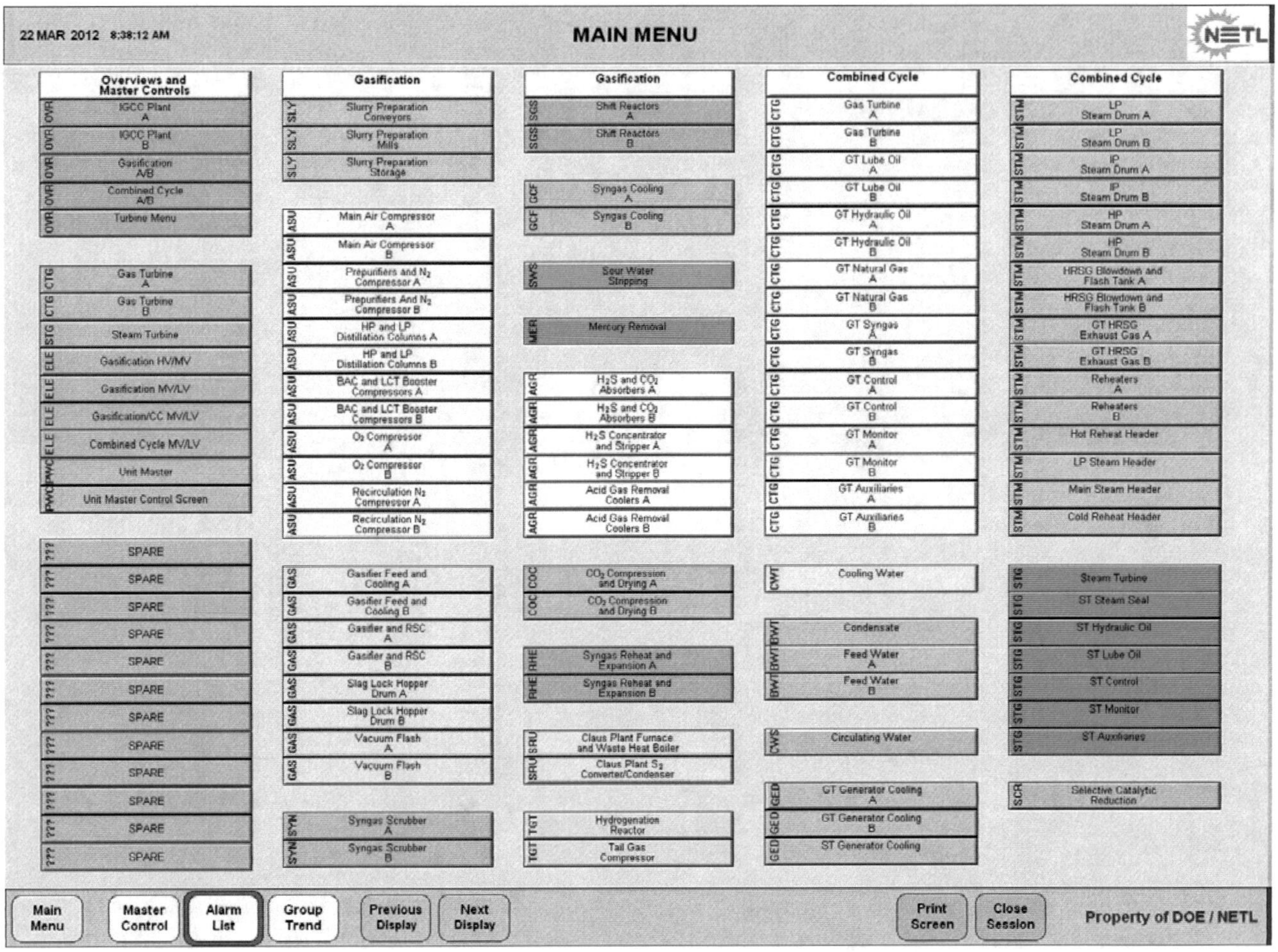

Figure 28. Main menu HMI display for navigation to all IGCC plant area displays

7.6 Instructor Station

The instructor's control station is configured to give the trainer complete control of the IGCC dynamic simulation (Figures 29 and 30). The console is designed using good human factors practices for ease and convenience of use by the instructor, ease of maintenance, and to reduce errors. The following functions are implemented on the instructor's console:

- Run time controls including run, freeze, snapshot, and backtrack
- Save and load initial conditions
- Define and initiate plant malfunctions
- Operate remote functions such as hand valves and pump control switchers
- Record and initiate automated scenarios of scripted commands
- Manipulate external parameters which affect simulated IGCC plant such as ambient temperature, barometric pressure, wind speed, cooling water inlet temperature, electrical grid voltage
- Run trainee performance monitor

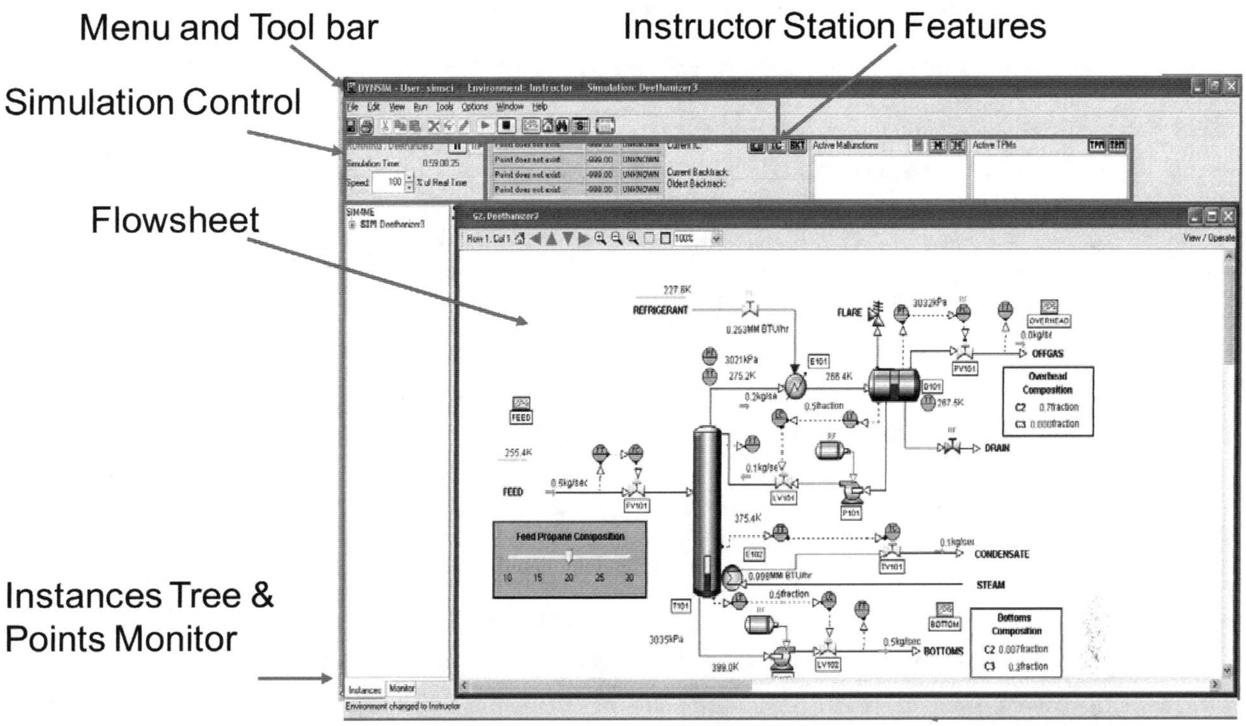

Figure 29. Graphical user interface for instructor station for IGCC dynamic simulator

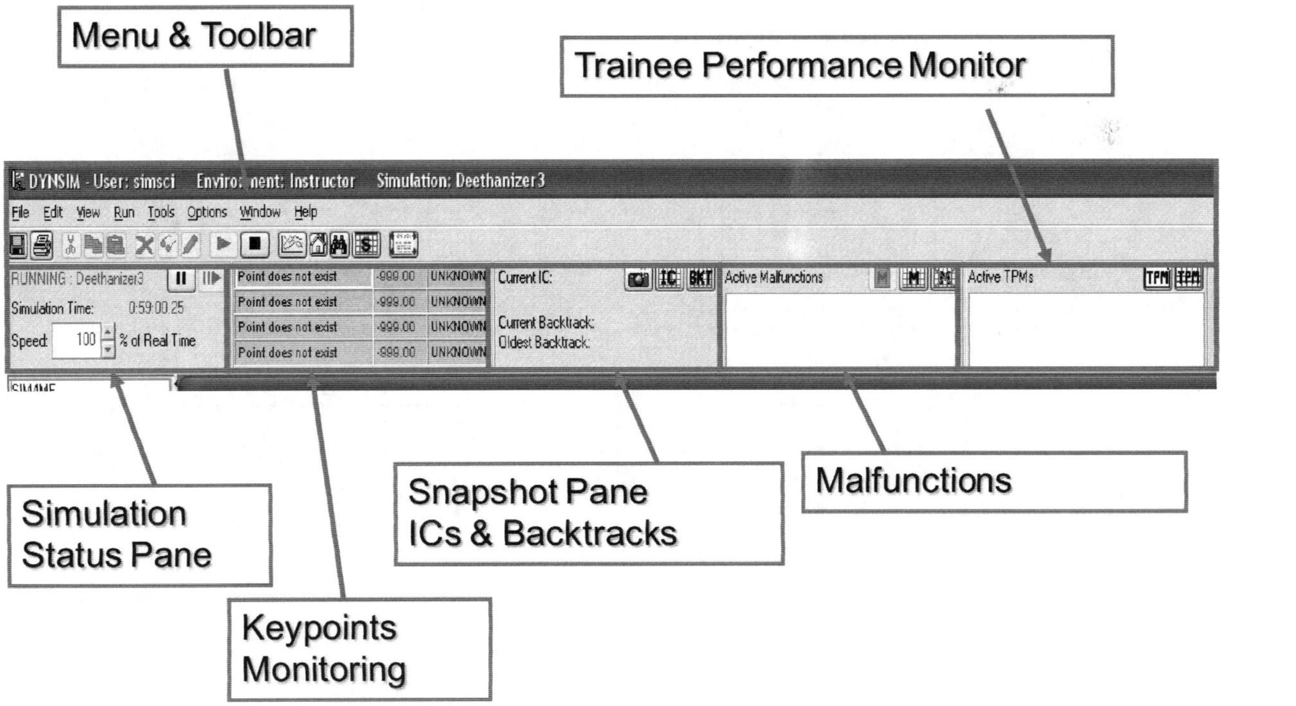

Figure 30. Instructor station run-time controls and functionality for IGCC dynamic simulator

7.7 IGCC Dynamic Simulator Hardware Architecture and Configuration

The IGCC dynamic simulator consists of two major plant sections or cycles:

- Gasification with CO_2 capture plant
- Gas-turbine-based, combined-cycle power generation plant

The dynamic simulator is designed and operated both as a single, integrated model of an IGCC plant and as multiple, independent models consisting of, at a minimum, separate models for the gasification and combined-cycle plant sections. The plant section models are designed such that they can be operated in a separated mode, to allow training on the gasification and combined cycle models independently at the same time, or they can be operated together in an integrated mode, to model a unified IGCC power plant with CO_2 capture. When running multiple instances of the simulator, users can control and monitor each instance from a select group of operator workstation HMI screens.

As shown in Figures 31 and 32, the IGCC dynamic simulator hardware configuration consists of eight operator training computers, two instructor station computers, two model server computers, and two engineering workstations that allows two different instructors, 16 students, and two engineering developers to work independently at the same time. The instructor stations provide the interface for the instructor(s) to control the operational and educational features of the simulation software. The engineering workstations provide for engineering development while the simulator is being used for training or demonstrations, and the operator workstations are used by the trainees for controlling and monitoring the IGCC processes.

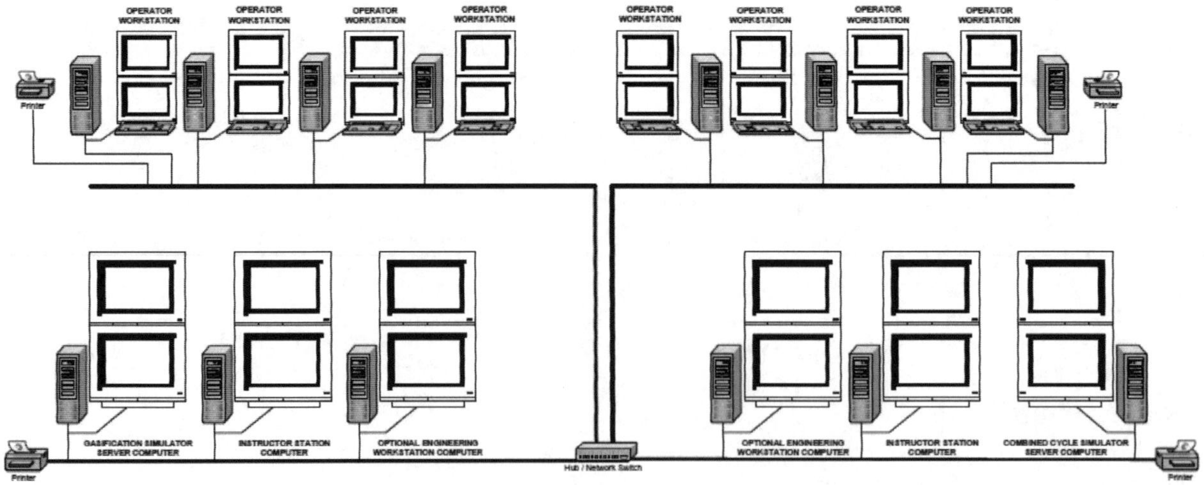

Figure 31. Hardware architecture for IGCC dynamic simulator

Figure 32. Hardware configuration for IGCC dynamic simulator at NETL AVESTAR Center

7.8 IGCC Dynamic Simulator Software

The IGCC dynamic simulator/OTS is built on the following software standards, technologies, and components (Figure 33):

- **DYNSIM®** dynamic process simulation software for plant-wide IGCC dynamic model
- **SIM4ME®** real-time simulation executive software for integration, interoperability, and distribution of dynamic simulation engines
- **InTouch®** human machine interface software for display of operator station displays
- **OPC** software standard for communication of real-time data

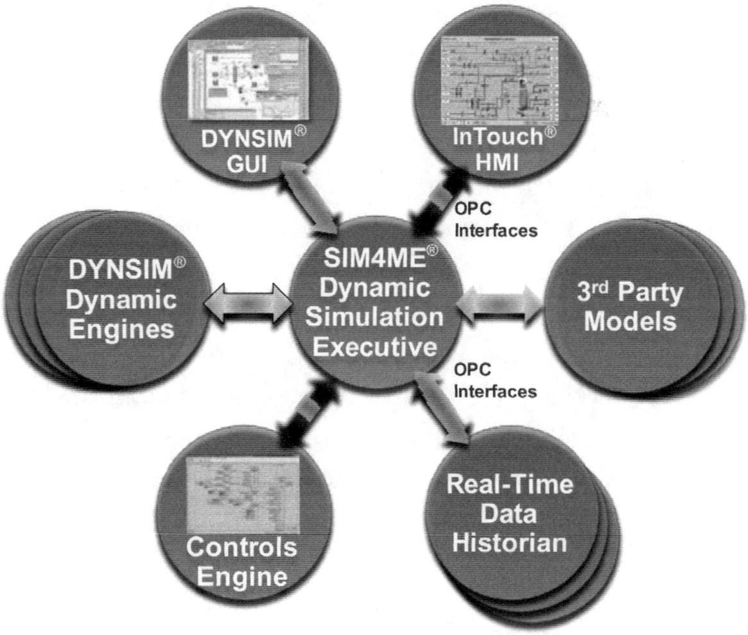

Figure 33. Real-Time Dynamic Simulation Software Architecture for IGCC Operator Training System

DYNSIM® Dynamic Simulation Software

Dynamic simulation is a highly sophisticated and powerful computational approach that can be used to help achieve aggressive operation and control objectives for clean energy plants. Dynamic simulation tools provide a continuous view of a process in action by calculating its transient behavior over time. Typical applications include normal and faulted operations, as well as process startup, upsets, and shutdown. Dynamic simulation can also be used to determine key equipment response times and to investigate interactions between major plant sections, such as power generation and CO_2 capture and compression. In addition, dynamic simulation can be used to evaluate alternative control strategies without the expense and the unexpected hazards of plant experimentation. Advanced process control strategies can also be applied to drive power production to satisfy load demands while meeting energy plant integration, performance, and environmental objectives, including CO_2 capture.

DYNSIM® from Invensys Operations Management's SimSci-Esscor® is a state-of-the-art, field-proven dynamic process simulation program that enables users to design and operate a modern process or energy plant safely and profitably. DYNSIM is an advanced, unified process plant-modeling environment for use throughout a power plant's lifecycle:

- Rigorous dynamic process simulation for engineering design
- High-fidelity modeling used for control system checkout
- Rigorous process modeling for Operator Training for first time startup
- Plant performance and operation improvement tool for operating plants

DYNSIM is unique in its ability to seamlessly scale from engineering design applications to control checkout and operator training, all within the same graphical user environment. Its modular architecture and open interfaces ensure that DYNSIM meets simulation requirements throughout the power plant lifecycle. DYNSIM's first principles models, including rigorous thermodynamic and fluid flow calculations, bring a superior level of robustness and accuracy to dynamic process simulation. DYNSIM rigorously handles even the most complex plant layouts and systems, using a robust and high performance solution algorithm. Furthermore, DYNSIM robustly models cold startup from purged conditions to normal operation and then back to shut-down. DYNSIM also takes full advantage of modern, open software standards (e.g., OPC – see discussion below) to interface with other important plant applications, such as distributed control systems.

SIM4ME® Real-Time Simulation Executive Software

SIM4ME® is a next generation real-time simulation executive user environment from Invensys Operations management. As shown in Figure 34, SIM4ME provides all the architecture and functionality for DYNSIM-based operator training simulators, such as Run, Freeze, Save/Load Initial Conditions, Scenarios, Malfunctions, Trainee Performance Monitor (TPM), etc. All this functionality is provided within a single GUI which is tightly integrated with the process modeling software, DYNSIM. The SIM4ME architecture is written using the Common Object Request Broker Architecture (CORBA) object model as its basis, which provides greater flexibility and speed in linking to other software than more traditional approaches. Therefore, the SIM4ME architecture gives OTS flexibility, stability, robustness, and an open and scalable structure. These features expedite the comprehensive engineering workflow: design, operational analysis, dynamic simulation, 3rd party packages checkout, operator training, performance monitoring and real time optimization.

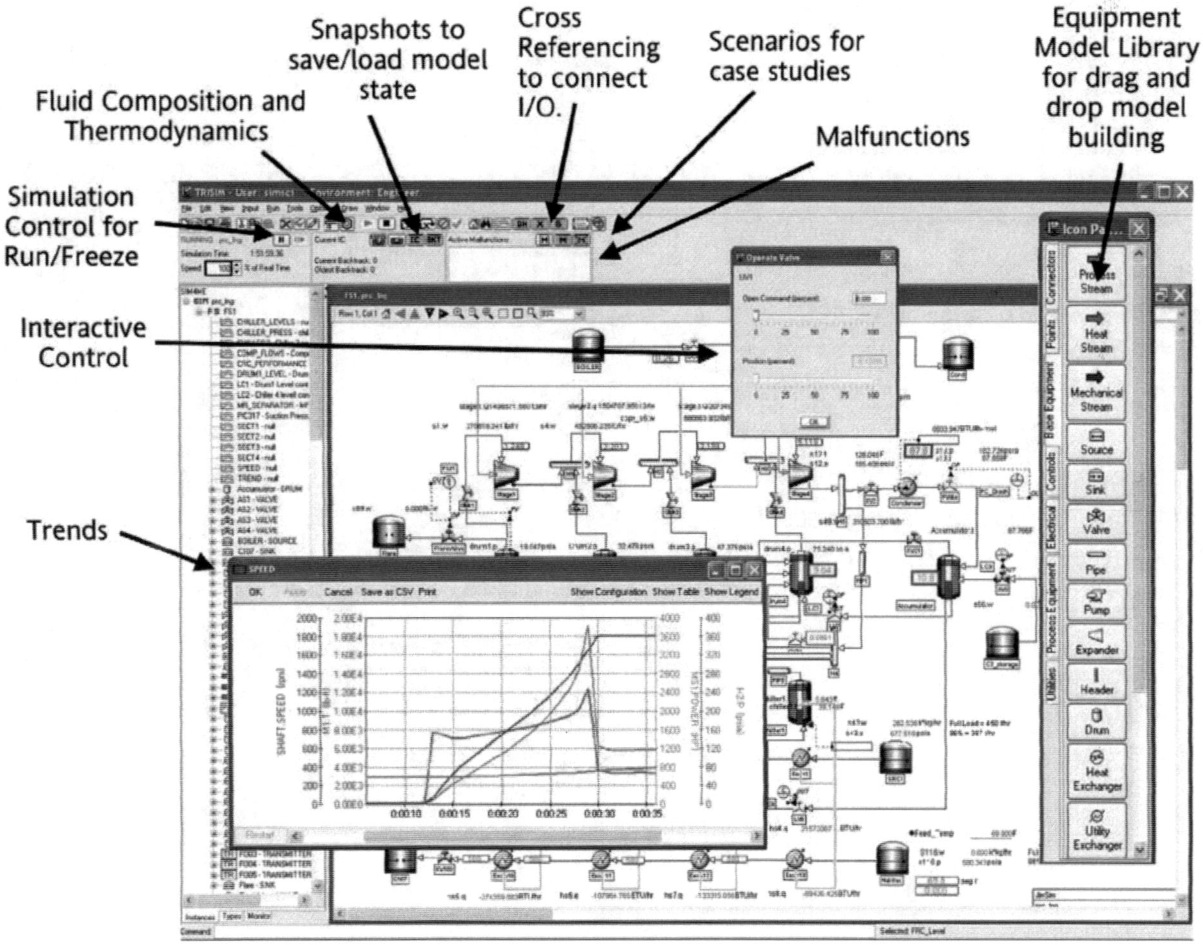

Figure 34. SIM4ME real-time simulation executive user environment

InTouch® Human Machine Interface Software

Used in over one-third of the world's industrial facilities, Invensys Operation Management's Wonderware® InTouch® is the world's leading human machine interface (HMI) software, offering market leading innovation, brilliant graphics for operator training displays, legendary ease of use, and unsurpassed connectivity. InTouch software is an open and extensible supervisory HMI and supervisory control and data acquisition (SCADA) solution that enables users to quickly create standardized, reusable visualization applications and then deploy them across the entire enterprise. Key capabilities of InTouch include:

- Resolution independent graphics and intelligent symbols that visually brings a power plant control room facility to life right on the computer screen
- Sophisticated scripting to extend and customize applications for specific needs
- Real-time distributed alarming with historical views for analysis
- Built-in, real-time and historical dynamic process variable trending
- Microsoft ActiveX controls and .NET controls integration
- Extensible library of over 500 pre-designed intelligent and customizable graphic and object symbols

Figure 35 shows an InTouch-based plant overview HMI used by the IGCC dynamic simulator operators. As shown in Figure 34 above, each of InTouch-based HMIs interoperates with the SIM4ME dynamic simulation executive via the software standard OPC interfaces. Dynamic process variable tags and values are passed in real-time from the DYNSIM engines to the InTouch-based displays.

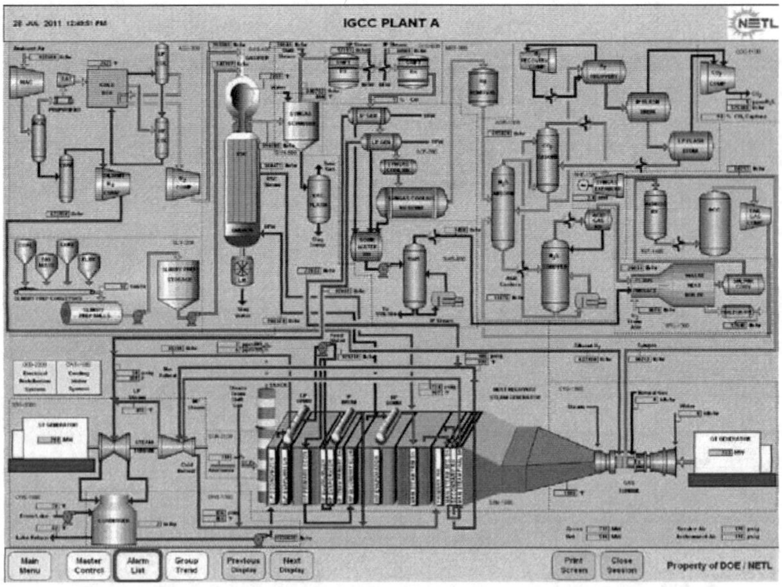

Figure 35. InTouch-based Overview HMI for Train A of IGCC Power Plant

OPC Software Standard

OLE for Process Control (OPC), which stands for Object Linking and Embedding (OLE) for Process Control, is an interoperability standards specification for the communication of real-time data between dynamic simulation and control devices from different vendors. Maintained by the OPC Foundation, the specification was originally based on the OLE, Component Object Model (COM), and Distributed Component Object Model (DCOM) technologies developed by Microsoft for the Microsoft Windows operating system family. The specification defined a standard set of objects, interfaces, and methods for use in process control and manufacturing automation applications to facilitate interoperability. The most common OPC specification was OPC Data Access (DA), which is used to read and write real-time data. DYNSIM v4.5.3 and earlier versions are compliant with the OPC-DA interoperability standard.

The OPC Unified Architecture (UA) is the next generation OPC standard that provides a cohesive, secure and reliable cross platform framework for access to real time and historical data and events (Figure 36). OPC-UA is based on cross-platform capable Web Services and SOA (Service Oriented Architecture). SOA involves creating services that perform a very specific function and are available to any remote software application that has both the authority and the need for the service. Whereas the first-generation OPC leveraged Microsoft standards, OPC-UA is purposely built for the needs of the automation industry. This makes it very effective in terms of both performance and security. OPC-UA uses a compressed and binary data transfer, while managing secure access through security certificates. Finally, OPC-UA builds on earlier concepts of the OPC data specifications and extends them with new complex data and the ability to maintain context and data relevance (by allowing clients to access structured information) from the plant floor through to enterprise-level intelligence applications. Starting with Version 4.5.5, DYNSIM is compliant with the OPC-UA interoperability standard.

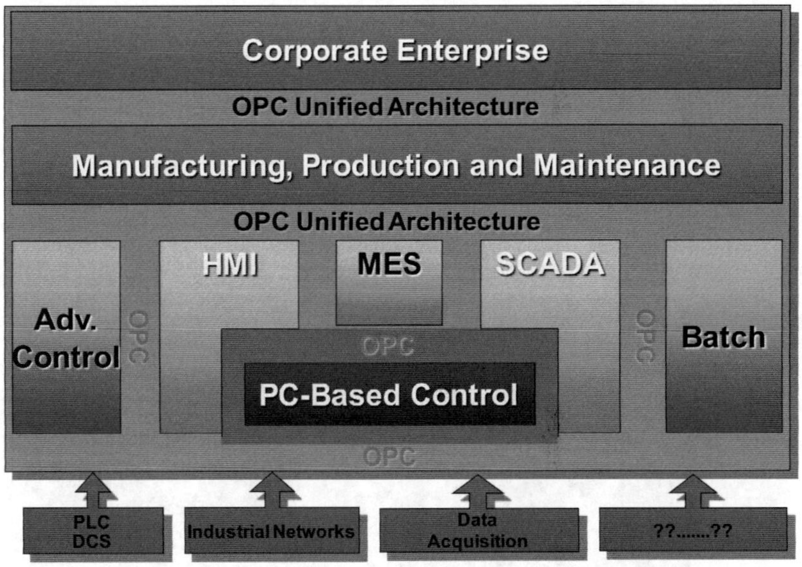

Figure 36. Industry-standard OPC-UA Architecture for Interoperability, Productivity, and Collaboration

8 AVESTAR's IGCC Immersive Training System

In the fossil energy industry, electric power utilities are continually striving to better train their workforces to operate power generation plants in a safe, reliable, environmentally friendly, and profitable manner. Real-time dynamic simulators with immersive training system (ITS) capabilities are one of the best ways to prepare engineers for the operation and control of advanced next-generation power plants with carbon capture.

NETL's AVESTAR team and its partners from Invensys, FCS, and WVU have developed an immersive training system for an IGCC plant with carbon capture (Zitney, 2012b). Using the 3D virtual reality software, EYESIM®, from Invensys, the ITS adds another dimension of realism to the real-time IGCC dynamic simulator deployed at the AVESTAR Center in July 2012 (Figure 37). Using gamepads for navigation and 3D visors with head tracking, ITS users can interact with plant equipment items (e.g., gasifier), activate transparent views (e.g., liquid level in a tank), display pop-up trends (e.g., gas turbine combustor temperature over time), and experience equipment sound effects (e.g., pump engines), malfunctions (e.g., leaks, fires), and visual training scenarios (e.g., CO_2 absorber column operation).

Figure 37. Immersive training system for IGCC plant with CO_2 capture

The ITS extends the training scope to both control room and field operators, allowing them to work as a team. The training covers base load operation, normal shutdown, load shedding and following, fuel switching, and complete cold, warm, and hot startups. The combined OTS and ITS solution also handles emergency shutdowns and rare abnormal situations in real-time (including equipment malfunctions and failures), together with changes initiated through actions from the instructor station or from field operators. Such comprehensive simulator-based instruction allows for realistic training without compromising worker, equipment, and environmental safety. It also better prepares operators and engineers to manage the plant closer to economic constraints while minimizing or avoiding the impact of any potentially harmful, wasteful, or inefficient events. Additional benefits include improved communication and collaboration among work crews, off-line evaluations of procedures, and training for safety-critical tasks.

8.1 ITS Software and 3D Immersive Virtual Reality Model

The IGCC immersive training system is implemented as a virtual reality (VR) software application that provides a real-time, 3D immersive, plant walk-through environment for training field operators and engineers. The VR environment lets the user move and interact freely within the plant, without being bound to prefixed paths or animations. Any action that the field operator carries out in the 3D environment is immediately reflected in the IGCC dynamic simulator and, conversely, any value that is updated by the simulation interface is also updated in the VR platform. A 3D visor with head tracking is provided for training a single field operator. Stereoscopic projectors are provided for training multiple operators using shutter glasses. Other peripherals include gamepads for navigation and a backup and storage system.

A photorealistic 3D VR model has been generated to provide the context in which the ITS user feels immersed in an actual IGCC power plant. First, a 3D computer-aided design (CAD) model was built on a real-world scale (1:1) from the P&IDs for all major IGCC plant sections, including air separation, entrained-flow gasification, synthesis gas cleanup, water gas shift, dual-stage acid gas removal process for hydrogen sulfide and CO_2 capture, Claus plant for sulfur capture, CO_2 compression process, and combined cycle for power generation. Next, the 3D CAD model was used to create an interactive and immersive VR plant environment composed of geometric component models with appropriate material and lighting information. The four main categories of component models include the active equipment items (e.g., pumps, heat exchangers), active instrumentation (e.g., valves, control panels), non-interactive structures (e.g., piping, stairs, floors), and the background scenery (roads, distant plant sections, sky, landscape).

To increase the visual realism, all 3D component models make use of textures and shading, derived mainly from pictures taken of real plants. In this project, the 3D VR model has been enhanced using high-resolution photographs from commercial-scale plants, including IGCC systems. Stereoscopic 3D pictures were taken during the photography sessions at the visited plants. In addition, videos recorded in high definition were used to create a 3D spatial point grid for correct positioning of all plant components. The final photorealistic 3D VR model of the entire IGCC plant has been implemented as an immersive and interactive training application.

8.2 Virtual Interaction

For all processes in the IGCC system, the ITS is capable of providing virtual plant interactions for those equipment items that are operated from the control room and/or from the field and necessary to support all training objectives. The user can navigate an avatar (field operator) through the 3D virtual plant using a gamepad, a gesture-sensitive handheld device used as the primary means of input for most modern video games. During plant navigation, the user can select and interact with active instrumentation and equipment components that are linked to the operator training system. The IGCC dynamic simulator realistically responds in real-time to the field operator action where actions such as opening or closing a manual valve are required within the scope of the simulation.

The ITS user can select equipment components and activate a transparent view. Figure 2 shows the internals of the gasifier where the animations of the slag flow and the gasification process are linked to the slag viscosity and gasification efficiency in the dynamic simulator. In Figure 38, the shell of a reboiler is transparent, visualizing the tube sheet with corresponding boiling liquid and generated vapor flow. Figure 39 shows a transparent view for a distillation column in which the animation of the flow rates of the vapor and liquid up and down the column, respectively, are tied to real-time process variables in the dynamic simulator. For all transparent equipment items, the visual representations will be interactive and depend on the process conditions as generated by the dynamic simulator linked in real time.

Figure 38. Reboiler in transparent mode with fill level visible

Figure 39. Distillation in transparent mode with animation of liquid and vapor flows

In addition to interactive actions and reactions with transparent equipment objects, collision geometry and pop-up variable trends over time are some of the additional functionalities that are provided. The ITS also offers several weather and lighting combinations, including a clear sunny day, a cloudy day, and a cloudy day with rain or snow. The VR environment also provides sound rendering effects that increase the sense of immersion for the user. These include constant low-intensity background sounds (e.g., distant equipment), random environment sounds (e.g., flock of seagulls), and equipment-specific sounds (e.g., pump engines, sirens).

The virtual plant can be used to practice training routines for malfunctions and emergencies that could not be attempted in the real plant (e.g., leaks, fires). Figure 40 shows a gas piping leaking in the Claus plant area. In another malfunction example, rupture of the pure oxygen piping at the outlet of the oxygen compressor in the air separation unit progresses to light smoke and then heavy smoke followed by a large fire that requires isolating the oxygen compressor by closing the appropriate valve, shutting down the compressor, and switching over to the oxygen supply from the oxygen storage tank. Gasifier operation will be affected as appropriate, based upon the severity of the rupture and reduction in oxygen supply to the gasifier.

Figure 40. Gas leak scenario displayed in the ITS

A number of visual training scenarios are also designed to promote a better understanding of gasification and CO_2 capture technologies. For the gasifier, a visual representation is displayed as a semi-transparent view showing operational changes from initial ignition to full combustion and subsequent gasification at normal operation. For the CO_2 capture process, a detailed transparent view of the liquid and vapor flow rates in the CO_2 absorber column displays a noticeable change as the absorber responds to load shifts and switching from one acid gas removal train to a two-train operation and back. This will promote better understanding of absorber column operation as well as enhance the simulator as a showpiece for carbon capture.

Looking to the control room of the future, AVESTAR researchers have the vision that immersive 3-D virtual reality technology can become the new operator interface for next-generation energy plants.

9. Summary

The AVESTAR Center's R&D, training, and education programs are focused on transforming how companies in the energy industry achieve operational excellence for clean energy plants. Attaining operational excellence requires that energy companies maximize the efficiency and profitability from their operations through excellent control and drive maximum business value from all their industrial assets, all while reducing negative environmental impact and improving safety. Another important requirement to addressing all of these goals lies in empowering the energy industry's most valuable asset – its people. With the deployment of the IGCC dynamic simulator/OTS, NETL has established the world-class AVESTAR Center to bring together simulation-based technology, state-of-the-art facilities, and leading energy experts to help address the critical R&D, workforce training, and education challenges facing industry on its drive toward achieving operational excellence for next-generation energy plants with carbon capture.

The high-fidelity IGCC dynamic simulator combines–for the first time–a "gasification with CO_2 capture and compression" process simulator with a "combined-cycle" power simulator together in a single real-time dynamic simulation framework. At the AVESTAR Center, this powerful and easy-to-use software tool facilitates comprehensive, hands-on, simulation-based workforce training, thereby serving as an effective vehicle to transfer IGCC and CO_2 capture technology knowledge and expertise to engineers, operators, and managers in the energy industry around the world. The IGCC dynamic simulator also offers experiential learning to engineering students and researchers at major universities and trainees in power plant certificate programs at community and technical colleges. In addition, the dynamic simulator is helping to accelerate the world-wide deployment and commercialization of clean coal technology by addressing R&D challenges on the safe, reliable, and efficient operation and control of next-generation IGCC systems with CO_2 capture.

Glossary

3D	Three dimensional
AGR	Acid Gas Removal
APC	Advanced Process Control
ARC	Advanced Regulatory Control
ASU	Air Separation Unit
AVESTAR™	Advanced Virtual Energy Simulation Training And Research
CACHE	Computer Aids for Chemical Engineering
CEU	Continuing Education Unit
CO_2	Carbon dioxide
COM	Component Object Model
CORBA	Common Object Request Broker Architecture
DCOM	Distributed Component Object Model
DCS	Distributed Control System

DNC	Diluent Nitrogen Compressor
DOE	Department of Energy
D-ROM	Dynamic Reduced Order Model
EPRI	Electric Power Research Institute
FAT	Factory Acceptance Test
FCS	Fossil Consulting Services
FDS	Functional Design Specification
GT	Gas Turbine
GUI	Graphical User Interface
H_2S	Hydrogen Sulfide
HMI	Human Machine Interface
HRSG	Heat Recovery Steam Generator
IGCC	Integrated Gasification Combined Cycle
ITS	Immersive Training System
K	Kelvin, units of temperature
MAC	Main Air Compressor
MMPC	Multiple Model Predictive Control
MPC	Model Predictive Control
MW	Megawatts
N_2	Nitrogen
NETL	National Energy Technology Laboratory
NETL-RUA	NETL Regional University Alliance
NOx	Nitrogen Oxides
NRCCE	National Research Center for Coal & Energy
O_2	Oxygen
OLE	Object Linking and Embedding
OPC	OLE for Process Control
OPC-DA	OPC Data Access
OPC-UA	OPC Unified Architecture
OTS	Operator Training System
NOx	Nitrogen Oxides
PCTC	Pierpont Community & Technical College
PDE	Partial Differential Equation
P&ID	Process & Instrumentation Diagram
PID	Proportional, Integral, Derivative
R&D	Research and Development
ROM	Reduced Order Model
RSC	Radiant Syngas Cooler
SAT	Site Acceptance Test
SCADA	Supervisory Control and Data Acquisition
SCR	Selective Catalytic Reduction
SOA	Service Oriented Architecture
SOx	Sulfur Oxides
ST	Steam Turbine
TPM	Trainee Performance Monitor
VR	Virtual Reality
WVU	West Virginia University

References

Bhattacharyya, D., R. Turton, and S.E. Zitney, "Students Learn Fundamentals of Process Operations and Control using Dynamic Simulator of an Integrated Gasification Combined Cycle (IGCC) Plant with CO_2 Capture," Presented at the *AIChE 2011 Annual Meeting*, Minneapolis, MN, October 16-21 (2011a).

Bhattacharyya, D., R. Turton, and S.E. Zitney, "Steady-State Simulation and Optimization of an Integrated Gasification Combined Cycle Power Plant with CO_2 Capture," *Industrial & Engineering Chemistry Research*, 50(3), 1674-1690 (2011b).

Bhattacharyya, D., R. Turton, and S.E. Zitney, "Model Predictive Control for Load-Following of an Integrated Gasification Combined Cycle (IGCC) Plant with CO_2 Capture," Presented at the *AIChE 2011 Annual Meeting*, Minneapolis, MN, October 16-21 (2011c).

Bhattacharyya, D., R. Turton, and S.E. Zitney, "Transient Studies of an Integrated Gasification Combined Cycle (IGCC) Plant with CO_2 Capture," Presented at the *AIChE 2011 Annual Meeting*, Minneapolis, MN, October 16-21 (2011d).

Department of Energy, "DOE Launches World-Class Virtual Energy Simulation Training and Research Center," Fossil Energy Techline, August 22 (2011). http://www.fossil.energy.gov/news/techlines/2011/11046-DOE_Launches_AVESTAR_Center.html

Department of Energy, National Energy Technology Laboratory, "Cost and Performance Baseline for Fossil Energy Plants, Volume 1: Bituminous Coal and Natural Gas to Electricity Final Report," DOE/NETL-2007/1281 (2007).

Dillon, M. "WVU Researcher Teams with NETL to Advance Clean Coal Technologies," *EngineeringWV*, WVU College of Engineering and Mineral Resources, 7(2), pp. 2-3, Fall (2011). http://www.statler.wvu.edu/news/magazine/EngineeringWV-Fall11.pdf

Erbes, M.R. and S.E. Zitney, "Dynamic Simulation and Training for IGCC Power Plants," *Proc. of the 23rd Annual International Pittsburgh Coal Conference*, Pittsburgh, PA, September 25-28 (2006a).

Erbes, M.R. and S.E. Zitney, DOE/NETL IGCC Dynamic Simulator Research and Training Center, Volume 1: Scoping Study, DOE/NETL-2008/1321, May (2006b).

Mahapatra, P., W. Bequette, and S.E. Zitney, "Application of Model Predictive Control for Rapid Combined-Cycle Load Following Using Custom Dynamic Models In Process-Simulation Software Environment," Presented at the *AIChE 2011 Annual Meeting*, Minneapolis, MN, October 16-21 (2011a).

Mahapatra, P., W. Bequette, and, "Multiple Model Predictive Control of Air Separation Unit As Part of IGCC Power Plant During Rapid Load Changes," Presented at the *AIChE 2011 Annual Meeting*, Minneapolis, MN, October 16-21 (2011b).

Jones, D., D. Bhattacharyya, R. Turton, and S.E. Zitney, "Optimal Design and Integration of an Air Separation Unit (ASU) for an Integrated Gasification Combined Cycle (IGCC) Power Plant with CO_2 Capture," *Fuel Processing Technology*, 92(9), 1685-1695 (2011).

Jones, D., D. Bhattacharyya, R. Turton, and S.E. Zitney, "Rigorous Kinetic Modeling, and Optimization, and Operability Study of a Modified Claus Unit for an Integrated Gasification Combined Cycle (IGCC) Power Plant with CO_2 Capture," *Industrial & Engineering Chemistry Research*, 51(5), 2362-2375 (2012).

Kasule, J., R. Turton, D. Bhattacharyya, and S.E. Zitney, "Rigorous Modeling of An Entrained Flow Gasifier Considering a Heuristic Recirculation Model and a Detailed Energy Balance," Presented at the *AIChE 2011 Annual Meeting*, Minneapolis, MN, October 16-21 (2011).

Kasule, J., R. Turton, D. Bhattacharyya, and S.E. Zitney, "Dynamic Modeling of a Single-Stage, Downward-Firing, Entrained-Flow Gasifier," *Proc. of the 29th Annual International Pittsburgh Coal Conference*, Pittsburgh, PA, October 15-18 (2012a).

Kasule, J., R. Turton, D. Bhattacharyya, and S.E. Zitney, "Mathematical Modeling of a Single-Stage, Downward-Firing, Entrained-Flow Gasifier," *Industrial & Engineering Chemistry Research*, 51(18), 6429-6440 (2012b).

Liese, E.A., S.E. Zitney, P. Mahapatra, R. Turton, D. Bhattacharyya, and G. Provost, "AVESTAR Center for Operational Excellence of Clean Energy Plants," Presented at *2012 Power Plant Simulation Conference (PowerPlantSim'12)*, San Diego, CA, February 12-17 (2012).

Paul, P., D. Bhattacharyya, R. Turton, and S.E. Zitney, "Constrained Nonlinear State Estimation of an Acid Gas Removal Process as part of an Integrated Gasification Combined Cycle (IGCC) Power Plant with CO_2 Capture," Presented at the *AIChE 2011 Annual Meeting*, Minneapolis, MN, October 16-21 (2011).

Provost, G., "NETL Establishes Dynamic Virtual Reality Simulation Research and Training Center for Demonstration of IGCC Technology with CO_2 Capture," Presented at ELECTRIC POWER Conference, May 10-12, Rosemont, IL (2011).

Provost, G.T., S.E. Zitney, R. Turton, "NETL Virtual Reality Dynamic Simulation Research and Training Center Promotes IGCC Technology with CO_2 Capture," Presented at ASME 2010 Power Conference, Chicago, IL, July 13-15, Paper no. POWER2010-27249, pp. 577-587 (2010).

Zitney, S.E., "Advanced Virtual Energy Simulation Training And Research (AVESTAR™) Center," CACHE Trustees Meeting, Nashville, TN, June 2-3, 2011, CACHE Summer Newsletter (2011). http://cache.org/site/news_stand/summer11/sum11%20AVE.pdf

Zitney, S.E., "AVESTAR Center for Clean Energy Plant Operators of the Future," Presented at the *Sixteenth Annual ARC World Industry Forum: Transforming Industry through New Processes and Technologies*, Orlando, FL, February 6-9 (2012a).

Zitney, S.E., "Virtually Simulating the Next Generation of Clean Energy Technologies," *Control Magazine*, January 5 (2012b).

Zitney, S.E., "Coal Can Fuel a Cleaner Tomorrow," *Times West Virginian*, 112(086-B), 1B, March 26 (2012c).

Zitney, S.E., E.A. Liese, P. Mahapatra, R. Turton, D. Bhattacharyya, and G. Provost, "Advanced Virtual Energy Simulation, Training, and Research: IGCC with CO_2 Capture Power Plant," *Proc. of the 28th Annual International Pittsburgh Coal Conference*, Pittsburgh, PA, September 12-15 (2011).

Zitney, S.E. and R. Turton, "AVESTAR Center for Operational Excellence of Clean Energy Plants," Presented at *Invensys OpsManage'11*, Nashville, TN, November 8-10 (2011).

Zitney, S.E., R. Turton, M.R. Erbes, H.P. Stone, G.T. Provost, M. McClintock, J.N. Phillips, M. Quintrell, and J. Marasigan, DOE/NETL IGCC Dynamic Simulator Research and Training Center, Volume 2: IGCC Process Descriptions, DOE/NETL-2008/1324, June 30 (2008).

(389b) Characterization and CO_2 Sorption Properties of Materials for CCS Applications

Jiajun He, Erik C. Rupp and Jennifer Wilcox, Department of Energy Resources Engineering, Stanford University, Stanford, CA

Carbon capture and sequestration (CCS) has become an important option for mitigating CO_2 emission and global climate change by capturing CO_2 from large point sources and sequestering it underground. However, the feasibility of CCS is mainly limited by the high cost of carbon capture, and to be more specific, the significant energy requirements during the regeneration of aqueous amine solutions. To overcome this limitation, micro and mesoporous sorbents have become promising candidates. A considerable variety of sorbent materials have been investigated for the application of CO_2 capture, including metal organic frameworks, zeolites, carbon nanotubes and mesoporous silicas, including SBA-15 and MCM-41. However, the CO_2 sorption capacity and kinetics under realistic flue gas conditions remain unclear. In this study, a variety of potential sorbents are investigated by several characterization techniques to gain insights into the relationship between sorbent properties and adsorption performance under flue gas conditions.

Breakthrough experiments are performed in a temperature-controlled packed-bed reactor with small amounts of sorbent (<250 mg). Experiments are performed with a resolution of 0.1 mmol CO_2 per gram of sorbent using an Extrel MAX300-LG quadrupole mass spectrometer downstream of the packed bed. The inlet gas is either a mixed gas (CO N_2 and H_2O) for testing under ideal conditions or a simulated flue gas created by burning methane in air, with potential contaminants such as SO_2, HCl and NO_x doped in at part per million levels post-combustion. The breakthrough experiments provide an understanding of CO_2 uptake, sorbent repeatability and impact of flue gas contaminants on capture. In addition, a Quantachrome Autosorb iQ_2 automated gas sorption analyzeris used to determine surface area, pore volume and pore size distribution of the testedCO_2 sorbents.

The sequestration of captured CO_2 is the other key aspect of CCS. Enhanced natural gas production in gas shale formations using captured CO_2
(primarily clay minerals with small amounts of kerogen) has non-negligible effects on the interaction of a particular shale and injected CO_2, which may affect the extent of sequestration and enhanced gas recovery. In this study, we investigate the compositions of various shale formations, including the Barnett, Eagle Ford, Haynesville and Fort St. John formations. X-ray photoelectron microscopy is used to analyze the atomic composition of the samples while Fourier transform infrared spectroscopy is utilized for the characterization of functional compounds in the samples. Knowledge of the physical and chemical properties of these materials will provide insight into the surface-CO_2 interactions that take place in these systems.

Reducing Uncertainties in Model Predictions via History Matching of CO_2 Plume Migration at the Sleipner Project, Norwegian North Sea

Peng Lu[1], Chen Zhu[1], and Per Aaggard[2]

[1]Department of Geological Sciences, Indiana University, Bloomington, IN 47405, USA;
[2]Department of Geosciences, University of Oslo, PO Box 1047 Blindern, 0316 Oslo, Norway

The Sleipner Project in Norway is the world's first industrial-scale geological carbon sequestration project. Starting in 1996, CO_2 separated from natural gas has been injected into the Utsira Sand at the rate of approximately 1 million metric ton of CO_2 per year. To date, more than ~15 Mt of CO_2 has been injected. Seismic surveys of the site were conducted prior to injection in 1994, and then repeated in 1999, 2001, 2004, 2006, and 2008; 2010 survey data have not yet been released by StatOil. These surveys have produced high fidelity 4D seismic data that delineated the CO_2 plume migration history. Therefore, the Sleipner Project provides a somewhat unique opportunity to simulate the dynamics of CO_2 plume migration in a real geological system, which in turn gives insights into the uncertainties in CO_2 plume migration prediction. The results will complement the numerous "concept models" proliferated in the literature, which use ideal geometry and homogenous geological properties.

Through a project sponsored by the U.S. Department of Energy/National Energy Technology Laboratory's Carbon Storage Program, we have developed a multi-phase reactive flow reservoir model of Layer 9 of the Utsira Sand for the Sleipner project using the Computer Modeling Group's reservoir simulator GEM®. We calibrated the model against the time-lapsed seismic monitoring data. Our simulation results match with the extents of CO_2 plume migration history from 1999 to 2008. The successful match with historic plume development was aided by introducing permeability anistropy and a second feeder to Layer 9. Predicted gas saturation, thickness of the CO_2 accumulation and CO_2 solubility in brine are also comparable with interpretations of the seismic data in the literature. The model calculated that ~9% of total injected CO_2 is dissolved in brine, which is comparable to estimates (5-10%) based on seismic data interpretation. Our reservoir model was based on StatOil's geological model of the Utsira formation and grid mesh. Our simulation results illustrate that the actual behaviors of the injected CO_2 plume conform to the modeled behaviors.

The Sleipner project is on-going. The good match of plume history provides a calibrated model for making predictions of CO_2 plume migration into the future. By comparing the model prediction and monitoring data that are not used in the calibration, we will have well-grounded assessment of modeling uncertainties.

Although the work was partly sponsored by an agency of the United States Government, the views and opinions of authors expressed herein do not necessarily state or reflect those of the United States Government or any agency thereof.

(389d) Gas Hydrates Modeling: Spanning Multiple Scales

Nagasree Garapati[1], Srinath Velaga[2] and Brian Anderson[2], (1)Chemical Engineering, West Virginia University, morgantown, WV, (2)Chemical Engineering, National Energy Technology Laboratory/West Virginia University, Morgantown, WV

Gas hydrates are nonstoichiometric crystalline inclusion compounds formed by the physically-stable interactions between water and relatively small guest molecules. They are stable at relatively high pressures and low temperatures and are found naturally below permafrost on shore or under the seafloor, within ocean sediments or even in deep lake sediments. The gas hydrate stability zone (GHSZ) is the range of depths of subsea or subsurface at which natural gas hydrates can form and remain stable at *in situ* temperature, pressure and local gas composition. The stable regimes are governed by the hydrate thermodynamic phase equilibria**.** The phase equilibria of hydrates containing single or multiple gas components can be calculated using our cell potential method. This method is based on an analytical solution to the Lennard-Jones Devonshire approximation to the van der Waals-Platteeuw statistical mechanics model for hydrate equilibrium, along with variable reference parameters to account for lattice distortion. Bazant and Trout proposed temperature-dependent Langmuir constants that define the volumetric interaction of the potential energy between the guest and host molecules. The Langmuir constants of various guest molecules like CH_4 and CO_2 have been calculated by integrating the full, six-dimensional configurational integral for the guest gases in a hydrate lattice using our calculated *ab initio* potentials. Three-dimensional (*P-T-y*) phase equilibria surfaces and hydrate phase compositions of various mixed hydrates including CH_4-N_2, CH_4-CO_2 and CH_4-N_2-CO_2 obtained using the cell potential method have been incorporated into macro scale reservoir simulation tools such as HydrateResSim for the prediction of production of CH_4 by injecting either pure CO_2 or N_2+CO_2 gas mixtures. The CH_4 hydrate reservoir behavior is studied during injection of pure CO_2 and N_2 and found that in both cases there is a possibility of blockage near the injection well due to secondary hydrate and ice formation respectively. A solution may be to inject a mixture of N_2 +CO_2 therefore HydrateResSim has been modified for the ternary gas system CH_4-N_2-CO_2.

(389e) High-Resolution Numerical Simulation of CO2 Migration in Saline Aquifers

Ruslan Iskhakov, Denis Voskov and Hamdi Tchelepi, Energy Resources Engineering, Stanford University, Stanford, CA

We study the migration of CO2 plumes in deep saline aquifers during the post-injection period using high-resolution numerical simulation. The interactions between buoyancy driven flow, dissolution of CO2 into the brine, capillary mixing, and diffusion/dispersion effects lead to complex behaviors in space and time. Here, a two-component, two-phase model is used, and the phase behavior is represented using an equation-of-state. In order to resolve the nonlinear coupling, the Fully Implicit Method (FIM) is used for time discretization. We find that the numerical solutions are quite sensitive to both grid resolution and time truncation errors. These challenges are compounded by the presence of miscible convective instability, which can dominate the overall evolution of the plume.

A lock-exchange configuration is set up, in which the total plume mass is emplaced near the bottom (inlet) of the aquifer. Following an initial equilibration period, the pressure gradient and the streamlines become nearly constant. In order to improve the overall computational efficiency of the simulations, a moving-outer-boundary (MOB) strategy was implemented as follows. The domain is split into two sub-domains: one that stretches from the inlet to just beyond the leading edge of the CO2 plume, followed by a region that extends all the way to the actual outer (updip) boundary of the aquifer. In the second sub-domain, the pressure and saturation distributions remain unchanged. Information at the end of a time step about the pressure, velocity, and saturation distributions is used to delineate the boundary between the two sub-domains. This is facilitated by the observation – made by many investigators including our group - that the leading-edge speed is nearly constant during the long post-injection period. Every time when the boundary between the sub-domains need to by moved, a few small time steps are performed on the global domain. This ensures that the evolution of the capillary transition zone and other mixing phenomenon is fully resolved before operating only on the "active" sub-domain.

This method shares similar idea with adaptive implicit method (AIM) when only the part of the domain where flow occurs is treated fully-implicitly. However, AIM approach has significant stability issues for this problem and can improve the overall performance by only a small margin. All simulations have been performed with the Automatic-Differentiation General Purpose Research Simulator (ADGPRS). The results demonstrate that compared with the standard approach of solving a global problem all the time, the MOB method improves the total computational time by 40-50% with 0.1% error (measured in terms of the updip speed). Simulation results are then shown for extremely large aquifer models using our MOB approach.

(389f) Pore Scale CO2 Displacement in Sandstone with Comparison to Core Scale Dynamics

Dustin Crandall[1], Robert P. Warzinski[1], William O'Connor[2], Aslam M. Kabir[3] and Grant Bromhal[4], (1)National Energy Technology Laboratory, U.S. Department of Energy, Pittsburgh, PA, (2)NETL, Office of Fossil Energy, US DOE, Albany, NY, (3)U.S. Department of Energy, National Energy Technology Laboratory, Pittsburgh, PA, (4)National Energy Technology Laboratory, Morgantown, WV

Combining micro-computed tomography scanning and computational fluid dynamics this study examined how CO_2 displaces brine in a small volume of pore space. The pore volume (4.41 (10^{-11}) m^3) was obtained from an Xradia Micro XCT-400 scanner with a voxel resolution of 27.1 microns from a sub-core of Mt Simon sandstone, recovered from a depth of ≈1770m. The complex micro-structure was preserved via careful three-dimensional reconstruction of the scanned data into a fully meshed volume, with over 3.5 million hexahedral cells. A series of simulations was performed with the commercial CFD code FLUENT to evaluate the immiscible flow behavior of CO_2 and brine within this volume, under a range of conditions that might be expected at a CO_2 storage site. This was done using the volume-of-fluid method to handle multiphase flow, and solving the full Navier-Stokes equations. Initial results show that an increase in the CO_2 viscosity, which could be obtained by adding a surfactant or by increasing depth of the injection, results in an increase in the residual volume of CO_2 within the pore space.

MODELING OF PRESSURE AND DISPLACEMENT SIGNATURES FOR DETECTING A FRACTURE IN THE CAPROCK DURING GEOLOGIC SEQUESTRATION *

Hema Siriwardane [(1)], Raj K. Gondle [(1)] and Grant Bromhal [(2)]

Topic: CO_2 Storage in Geologic Formations

Extended Abstract

Finite difference and finite element methods were used to determine changes in fluid pressure and overburden deformations caused by the injection of carbon dioxide (CO_2) in a typical geologic formation such as a saline/brine aquifer. The objective was to evaluate the long-term storage potential by investigating the overburden geologic response and flow behavior due to the presence of fractures and faults in the caprock that could act as leakage pathways. The presence of geologic uncertainties such as fractures/faults in a caprock may lead to possible communication of injection fluid with overlying or underlying geologic media, which over a period of time can cause changes in fluid pressure and overburden deformations.

In this study, single-phase and multi-phase fluid flow modeling was performed to simulate water and CO_2 injection into a geologic formation. Geomechanical effects were incorporated in the analysis. Coupled flow-deformation finite element analyses were used to model single phase fluid injection. Multi-phase fluid flow modeling based on the finite difference method was also performed. Three different fracture locations were selected to investigate the influence of fracture, fracture location, and fracture permeability on the pressure and overburden response of the system. A comparison of numerical results based on single-phase and multi-phase fluid flow modeling is also presented in the paper. Results of the study show that the pressure and displacements signatures can be used to detect the presence of an existing fracture or the activation of a new fracture in the caprock layer. Results from this study can be used in the development of smart monitoring technologies with limited field instrumentation coupled with numerical modeling work.

Keywords: carbon sequestration, numerical modeling, fracture/fault, single-phase, multi-phase

(1) Permanent Address: NETL-RUA and Department of Civil and Environmental Engineering, West Virginia University, Morgantown, WV 26506-6103

(2) National Energy Technology Laboratory (NETL), U.S. Department of Energy, P.O. Box 880, Morgantown, WV 26507-0880

* Extended Abstract submitted to the AIChE Annual Meeting, Pittsburgh, PA, October 28 – November 2, 2012

The presence of geologic uncertainties such as caprock fractures/faults in an overlying seal layer or in an unlikely event of fracture activation during CO_2 injection may lead to possible communication of injected fluid with surrounding formations, which over a period of time may cause changes in fluid pressure and ground deformations. In the present paper, numerical modeling was used to evaluate overburden geologic response and flow behavior due to the presence of a fracture/fault in the caprock that could act as a leakage pathway. Results from flow modeling coupled with geomechanical effects, and combined with limited field instrumentation can be used in the development of geophysical monitoring techniques to detect a fracture in the caprock during large-scale CO_2 sequestration operations. Field monitoring technologies such as surface tiltmeters has been used extensively in the past to monitor ground deformations during fluid injection. Some studies have also shown that such ground monitoring techniques can also be used to provide constant updates on fluid movements and notify warning of any fluid intrusion into inadvertent regions. Moreover, ground displacements and pressure signals in the monitoring layer may vary depending on the number of fractures and their respective locations in the caprock formation.

In this study, finite difference and finite element methods were used to determine changes in fluid pressure and overburden deformations caused by the CO_2 injection at a hypothetical injection site. Single-phase and multi-phase fluid flow modeling coupled with geomechanical effects was performed to simulate water and CO_2 injection. A finite element based approach was to model single-phase fluid injection, and a finite difference approach was used to model multi-phase fluid injection. Since the objective of the present study is to investigate the influence of a caprock fracture on the overburden geologic response and pressure response due to changes in fluid flow, an overburden monitoring layer was considered right above the caprock layer. The overburden monitoring layer was assumed to be a high-permeability region similar to the target reservoir. A hypothetical fracture was simulated at 100m, 500m and 1,000m away from injection source, and only one fracture was assumed to be active at any given period of time.

Injecting CO_2 causes the reservoir pressure to increase with time, and the pressure increase in the monitoring layer can be significant when a caprock fracture is present. The pressure variation in the monitoring layer above a caprock fracture helps to identify the fracture location. The injected CO_2 leaks through the caprock and escapes to overlying monitoring layer when a caprock fracture is present. The simulated fractures were located at 100m, 500m, and 1,000m away from the injection source. The significant variation of pressure signatures in an overlying monitoring zone may more likely be an indicator to detect any presence of fractures in the caprock that could act as leakage pathways for migration of injected fluid. The pressure distribution in the monitoring layer can also be used as a monitoring aid in identifying the potential location of a caprock fracture.

Ground displacements are symmentric when there is no caprock fracture. When a caprock fracture is present, surface displacements are not symetric and are larger than when there is no fracture. The increase in surface displacements may be because of the pressure increase in the monitoring layer and the overburden layers. These ground displacement patterns could be helpful in identifying the location of a caprock fracture. Results show larger magnitudes of ground deformations when a caprock fracture is located closer to the injection source. The variation in displacement signals due to the presence of fractures at different locations could also be helpful in identifying the presence of a fracture/fault. Similar results were found for the overburden ground response and pressure response by using single-phase fluid injection (water injection).

Single-phase fluid flow and multi-phase fluid flow analyses were performed to study the pressure and displacement signatures when a fracture/fault gets activated due to activation of an existing dormant fracture during CO_2 injection. A distinct change in the pressure signature was observed. Similar results were found with single-phase fluid flow and by using finite element analyses. Results of this study can certainly provide some input to the development of geophysical monitoring technologies to evaluate the long-term CO_2 storage potential in deep brine/saline aquifers.

(389h) Effect of Calcite Dissolution On the Leakage Through Abandoned Wells in Geological CO2 Storage

Ezzeddine Laabidi and Rachida Bouhlila, LMHE, National Engineering School of Tunis-Tunis EL Manar University, Tunis, Tunisia

Deep saline aquifer is used for the CO_2 and radioactive waste disposal geological storage. Several works have been published in the last few decades that describe the CO_2 storage in deep saline aquifer (Holger C. et al., 2009; Andreas K. et al., 2010; George S.H. Pau, 2010;…). During the last years the behavior of CO_2 and other substances in the groundwater has become of great interest, since techniques are developed and investigated to inject CO_2 in the subsurface in order to counteract the greenhouse effect. A very undesirable effect would be the leakage from the storage. When the CO_2 injection is made into the brine layer in oil or gas field production, The greatest risk for escape of CO_2 may come from other wells, typically those for oil and gas production, which penetrate the storage formation. Injection of CO_2 in the subsurface leads to a relatively rapid response (decrease) on the pH. This decrease will then be buffered by mineral reactions, such as calcite dissolution. Calcite dissolution can induce an increase in the porosity and can enhance further leakage and provide more pathways for the CO_2 (K. Caldeira, G.H. Rau, 2000). The majority of the studies quantify these hydrogeological systems using traditional multi-phase flow model and do not consider the effects of the density dependent flow and the chemical reactions. Interdependence of density dependent flow and chemical reactions and their effects on the porosity and permeability are the most important keys towards a better knowledge of these coupled processes for the design of safer installations. The density dependent multi-phase flow and reactive multispecies model GEODENS (Bouhlila and Laabidi, 2008) is presented here and used to simulate the effects of the calcite dissolution on the flow patterns around a CO2 injection well and on the leakage through abandoned wells in a deep carbonate layer. The simulated domain is 1000 × 160 × 1 m and concerns a vertical slice confined aquifer separated by a thin aquitard, a leaky well is located at the centre of the domain and an injection well is 100 m away. A constant head boundary is applied in the two lateral boundaries (Nordbotten, J.M., Celia, M.A., 2006). The target of the first simulation is to quantify the leakage of CO2 from the aquifer through the leaky well, and the second simulation consists on the evaluation of the effect of the geochemical reactions (calcite dissolution and precipitation) on this leakage ratio.

Hydrogen Production from Glycerol Reforming over Ni Based Catalysts

Su Hyun Cho[a,c], Seung Hoon Kim[a,c], Seung-hwan Lee[d], Bang Hee Kim[d], Sang Woo Kim[a], Arunabha Datta[a], Kwan Young Lee[c] and Dong Ju Moon[a,b]*

a: Clean Energy Research Center, KIST, Seoul, South Korea
b: Clean Energy and Chemical Engineering, UST, Daejeon, South Korea
c: Dep. of Chemical and Bio Engineering, Korea University, Seoul, South Korea
d: Technolgy& Research Center, J& K heaters, Seoul, South Korea

Summary: Aqueous phase reforming of glycerol over Ni-based catalysts was carried out at 225℃, 23 bar and LHSV=4 h^{-1}. The Ni-based catalysts were prepared by an incipient wetness impregnation method. The catalysts were characterized by N_2 physisorption and XRD. It was found that Ni (20 wt%)-Co(3 wt%)/γ-Al_2O_3 catalyst showed higher glycerol conversion with H_2/CO = 4 than the other catalysts.The results suggest that the Ni-Co/γ-Al_2O_3 catalyst can be applied to the synthesis gas production system using APR of glycerol for application in synfuel production.

Keywords : Ni-Co/ γ-Al_2O_3 , Glycerol, Synthesis gas , Aqueous Phase Reforming

Introduction

Biomass has been considered as one of the most efficient sources to meet future energy needs. Recently, increased the production of biodiesel resulted in overproduction of glycerol [1-2]. Glycerol is produced in about 10% as a main byproduct and an important industrial feedstock for applications in food, cosmetics, pharmaceutical and other industries. Among possible uses for glycerol, synthesis gas production from glycerol has been considered as an attractive process [3-6].

An aqueous-phase reforming (APR) of glycerol consists of the two reactions, reforming reaction (1) by decomposition of glycerol, and water gas shift reaction (WGS) (2). The overall equation is (3).

$$C_3H_8O_3 \rightarrow 4H_2 + 3CO \quad (1)$$

$$3CO + 3H_2O \rightarrow 3CO_2 + 3H_2 \quad (2)$$

$$C_3H_8O_3 + 3H_2O \rightarrow 7H_2 + 3CO_2 \quad (3)$$

It was known that unlike steam reforming for producing synthesis gas from glycerol, the generation of synthesis gas by APR decreases underasible decomposition reactions encount - ered when heated above 300℃.

In this work, the APR of glycerol over Ni/MgO and Ni/ γ-Al_2O_3 catalysts were investigated and catalytic performance was compared with that of Ni-Co/γ-Al_2O_3 catalyst.

Experimental

The Ni/MgO, Ni/γ-Al_2O_3 and NiCo/γAl_2O_3 catalysts (see Table 1) were prepared by an incipient wetness impregnation method. γ-Al_2O_3 (Sigma Aldrich.co, BET S.A.=155 m^2/g , Pore Vol.= 0.25 cc/g) was used after pretreatment at 400℃ for 10 h. The samples were dried for overnight in an oven at 60℃ and then calcined at 400-500℃ for 5 h. Before the reaction, the catalyst was reduced by H_2 flow at 400℃ for 10 h.

Aqueous phase reforming of glycerol was conducted over Ni-based catalysts temperature at 225℃, pressure at 23 bar and LHSV=4 h^{-1}. The catalysts were characterized by N_2 physisorption and XRD.

Results & Discussion

Table 1 shows the metal loading, BET surface area, total pore volume, Ni particle size of the catalysts along with H_2/CO ratio.These characterization results indicate that properties of the catalysts were changed by the support and addition of promoter to the Ni catalyst. The surface area of the prepared catalyst increases in order of Ni-Co/γ-Al_2O_3 < Ni/MgO < Ni/γ-Al_2O_3.

*** Corrresponding author**
E-mail : djmoon@kist.re.kr
Phone/fax : +82-2-958-5867/+82-2-958-5809

Table1. Physico-chemical properties of Ni catalysts

Catalyst	Ni loading (wt%)	Co loading (wt%)	BET S.A. (m^2/g)	Total Pore Vol. (cc/g)	Ni Particle Size (nm)	H_2/CO ratio
Ni/MgO	20.0	-	150	4.0	5.2	4.2
Ni-Co/γ-Al_2O_3	20.0	3	109	2.0	7.1	4.0
Ni/γ-Al_2O_3	20.0	-	170	3.6	5.3	4.0

.

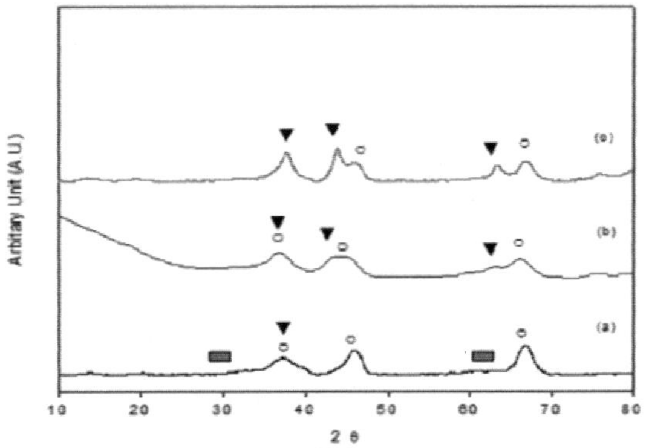

Fig. 1. The XRD patterns of prepared catalysts. (a) Ni/MgO, (b) Ni/Al_2O_3 (c) Co-Ni/ Al_2O_3 (▼NiO, ○ Al_2O_3, ▬ MgO).

From the XRD results of Ni/MgO NiO, and MgO phases were the predominate. In the case of the Ni/ γ-Al_2O_3 and Ni -Co γ-Al_2O_3 NiO and Al_2O_3 phases were predominate. No diffractions peaks observed for cobalt oxide phase.

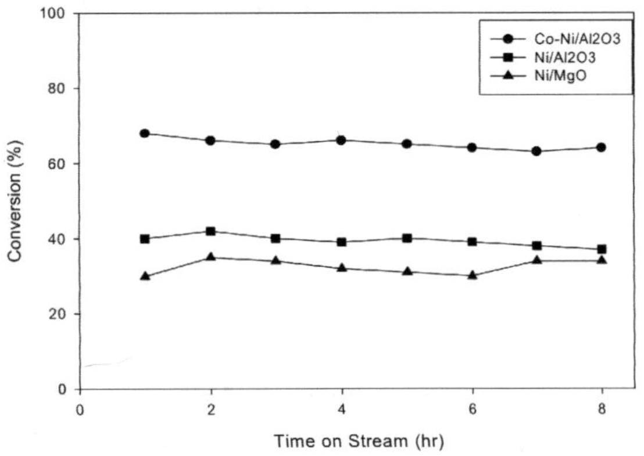

Fig. 2. The glycerol conversion in the APR of 10 wt% glycerol over Ni based catalyst. (Reaction conditions: Temperature=225 ° C, Pressure=23 bar and LHSV=4h^{-1}).

Figure 2 presents the catalytic performance of Ni based catalysts in the APR of glycerol. The conversion of glycerol ranges from 30% to 70% to the Ni/MgO, Ni/γ-Al_2O_3, Ni-Co/γ-Al_2O_3 respectively, under the tested conditions. Among the all Ni-Co/γ-Al_2O_3 catalysts shows higher activity compared to other catalysts.

Synthesis gases can be converted to methanol, DME and GTL products as following equations [6].

The synthesis gas, a mixture of CO, CO_2 and H_2 can be adjusting by separation of H_2.

$$2H_2 + CO \rightarrow CH_3OH \quad (1)$$
$$2CO+3H_2 \rightarrow CH_3OCH_3 \quad (2)$$
$$CO+ 2H_2 \rightarrow -(CH2)_{n-} \quad (3)$$

Synthesis gas manufactured by reforming of glycerol can be used for the production of synfuels as shown in Fig. 3 [6].

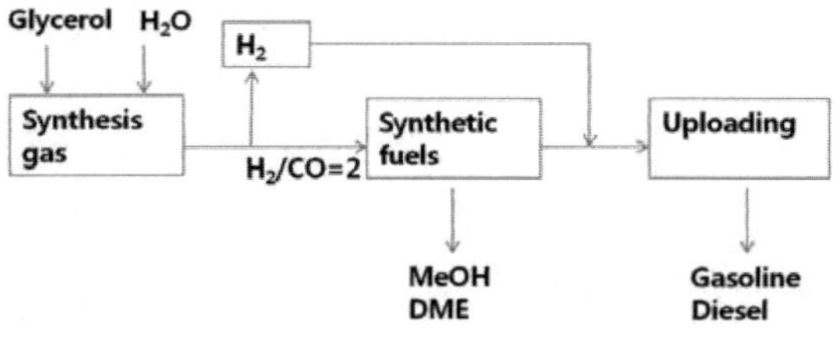

Fig.3. Schematic diagram of application technology for the production of synfuels by using synthesis gas prepared from APR of glycerol.

Conclusions

Aqueous phase reforming of glycerol over Ni-based catalysts was investigated in a fixed-bed reactor system. It was found that Ni-Co/γ-Al_2O_3 catalyst showed higher glycerol conversion than the other catalyst. The results suggest that the Ni-Co/γ-Al_2O_3 catalyst can be applied to the production of synthesis gas for in Mehtanol, DME, and GTL-FPSO processes applications.

References

[1] S.H. Cho and D.J. Moon, J. of Nanoscience & Nanotechnology , 11 (2011) 4202.

[2] D.J. Moon, S. H. Cho, S. H. Lee, S.H. Kim, J. S. Jung, E. H. Yang, S. B. Lee, J.H.Lee, B. H. Kim, G. J. Han, M.S. Chae, in Application to Korea Patent (2012).

[3] X.Wen, Y.Xu, H.Ma and Z.Xu, Int. J. of Hydrogen Energy. 33 (2008) 6657.

[4] K.Lehnert and P. Claus, Catal. Comm., 9 (2011) 2546.

[5] T. Furusawa and A. Tsutsumi, Applied Catalysis A: General 11 (2010) 2200.

[6] D.J. Moon, Y.J. Lee, J.S. Kang, M.J. Park, S.J. Kim, S.D Lee, H.J. Lee, Korea Patent Application, 10-2009-0116000 (2009) and U.S. Patent Application, 2011/0130474A (2011).

(389j) Computer Simulation of Water and Ammonia Electrolysis

Ali Estejab, Damilola A. Daramola and Gerardine G. Botte, Chemical and Biomolecular Engineering Department, Center for Electrochemical Engineering Research, Ohio University, Athens, OH

Hydrogen is considered as new fuel generation coupled with high efficiency fuel cell, but its storage and transportation is a matter of concern. Recent research has been done on substances that contain hydrogen but easier to carry. Water is abundant and contains hydrogen and always considered as one of substances that can be electrolyzed to produce hydrogen. Ammonia is also one of the richest containers of hydrogen and can be found in waste water. Thus one of the most interesting and environmentally considered ways of production of hydrogen is through electrolysis of ammonia. Ammonia electrolysis [1, 2] is done by oxidizing the ammonia into benign nitrogen at the anode:

$$2NH_3 + 6OH^- \leftrightarrows N_2 + 6H_2O + 6e^- \quad (1)$$

while producing hydrogen, a premium fuel, at the cathode:

$$2H_2O + 2e^- \leftrightarrows H_2 + 2OH^- \quad (2)$$

Therefore, pure nitrogen and hydrogen are obtained in the process with an overall reaction:

$$2NH_3 \leftrightarrows N_2 + 3H_2 \quad (3)$$

The theoretical voltage for the reaction is 0.058 V vs. SHE [1] which represents an energy consumption of 1.55 W-h per gram of H_2 produced. This energy does not account for the energy that can be recovered from the hydrogen. Despite of that the energy used for the removal of ammonia is much lower than any other current solution for ammonia removal including biological and chemical treatments.

Most of the research performed on improving ammonia electrolysis has been based on experimental observation. On the other hand, development of a mathematical model could result in enhancing these experimental efforts. For a system of any complexity it is invariably more cost-effective to perform a series of computer experiments than to perform the same experiments in the field. Therefore, in this work a mathematical model of ammonia electrolysis using ASPEN PLUS is presented.

In this model the behavior of the cell will be predicted (using a V-I curve) by considering the overall cell potential which is based on the reversible potential with the activation, concentration and ohmic overpotentials:

$$V = V_0 + \eta_{act} + \eta_{ohm} + \eta_{con} \quad (4)$$

- The activation overpotential or surface polarization is due to chemical reactions that takes place at surface of the electrodes. The parameter that is dependent on the exchange current density which structural factors such as the age of the electrocatalyst, the type of electrocatalyst, electrode morphology, pressure, temperature etc.
- The ohmic overpotential is due to ohmic resistance of the cell including the electronic losses due to the resistance of membrane, electrode, current collector etc.
- The concentration overpotential or mass transfer polarization is due to mass transport limitation and is more important at high current densities.

Based on the expression in (4), Garcia-Valverde, Espinosa and Urbina [3] proposed a model for water electrolysis which predicted the cell performance and the amount of hydrogen produced. In this present study, the aforementioned model will be reproduced and contrasted with a model for ammonia electrolysis to calculate the amount of hydrogen produced. To validate this computer simulation, an experimental setup will be built and measurements will be taken. The accuracy and predictions of the model will be presented at the meeting.

References

[1] F. Vitse, M. Cooper, and G. G. Botte, "On the use of ammonia electrolysis for hydrogen production," *Journal of Power Sources*, vol. 142, pp. 18-26, 2005.
[2] G. G. Botte, F. Viste, and M. Cooper, "Electrocatalysts for the oxidation of ammonia and their application to hydrogen production, Fuel cells, Sensors, and Purification Processes," vol. No. 7485211 USA, 2003.
[3] R. Garcia-Valverde, N. Espinosa, and A. Urbina, "Simple PEM water electrolyser model and experimental validation," *International Journal of Hydrogen Energy*, vol. 37, pp. 1927-1938.

(389k) Desulfurization of JP-8 Fuel Using Metal-Loaded Meso/Macroporous Aluminosilicates

Jessica St. John, Univeristy of California Santa Cruz, Santa Cruz, CA, Dat T. Tran, Electrochemistry - SEDD, U.S. Army Research Laboratory, Adelphi, MD, Deryn Chu, Electrochemistry - SEDD, U.S. Army Research Laboratory, Adelphi and Scott R.J. Oliver, Chemistry and Biochemistry, University of California, Santa Cruz, Santa Cruz

Title: Desulfurization of JP-8 fuel using metal-loaded meso/macroporous aluminosilicates

Jessica St. John,[1] Dat T. Tran,[2] Deryn Chu[2], and Scott R. J. Oliver[1]

[1]Department of Chemistry and Biochemistry, University of California, Santa Cruz, 1156 High Street, Santa Cruz, CA 95064; soliver@ucsc.edu

[2]U.S. Army Research Laboratory, RDRL-SED-C, 2800 Powder Mill Road, Adelphi, Maryland 20783; dat.t.tran4.civ@mail.mil

The U.S. Army is interested in using JP-8 for fuel cells, but the fuel must first be processed to remove organic sulfur compounds to avoid catalyst poisoning. These organic sulfur compounds are more difficult to remove than those found in lighter fuels. Sorbents capable of removing these challenging impurities under ambient conditions would allow for on-site fuel processing.

In this work, mesoporous and macroporous aluminosilicate materials were explored for sulfur removal from JP-8 fuel at room temperature. The desulfurization capacities of cerium and silver loaded aluminosilicate frameworks are compared. The metals were loaded both as extraframework species and metal-substituted within the framework. Varying the silicon to aluminum ratio in the frameworks as well as pore size are reported for maximization of desulfurization and breakthrough threshold.

A NOVEL KMgAl SORBENT FOR CO_2 CAPTURE IN A FLUIDIZED BED

Lei Li, Xia Wen, Feng Wang, Ning Zhao, Fukui Xiao, Wei Wei**
State Key Laboratory of Coal Conversion, Institute of Coal Chemistry, Chinese Academy of Sciences, Taiyuan, 030001, P.R. China
**Email: xiaofk@sxicc.ac.cn; weiwei@sxicc.ac.cn*

Abstract

The novel The KMgAl sorbents for CO_2 capture was investigated in this study. The sorbent showed a large CO_2 capture capacity with MgO loading of 10 wt % and 20 wt % at low temperature in simulated flue gas. The presence of an appropriate amount of water vapor pretreatment had a positive effect on the CO_2 capture capacity. The MgO played an important role in the forming of hydrotalcite-like compounds on the surface and in the pore of Al_2O_3 after water vapor pretreatment which leads to the high CO_2 absorption capacity of KMgAl sorbents. Furthermore, the performance for CO_2 sorption of KMgAl sorbent (KMgAlI3010) was investigated in the fluidized bed.

Introduction

CO_2 is a major greenhouse gas that is released into the air due to the use of fossil fuels [1-5]. At present, carbon capture and storage (CCS) is recognized as a promising option for CO_2 emission reduction. The post-combustion process is more suitable for CO_2 capture for the traditional pulverized coal power plants [6-8]. CO_2 chemical adsorption using dry regenerable solid sorbents had attracted much attention due to the low cost of raw materials and non-corrosion property [9-13]. In previous study, we had found that MgO/Al_2O_3 sorbent with 10 wt% MgO loading showed a better performance for CO_2 capturing at low temperature in the fixed-bed [11]. Nevertheless, the absorption amount of CO_2 was not high and heat control is difficult since the adsorption process is highly exothermic. Therefore, the novle sorbents, MgO/Al_2O_3 supported K_2CO_3, were prepared and investigated for CO_2 capture in the fluidized bed [12]. However, the effect of factors on CO_2 capture and adsorption mechanisms still need to be further studied.

Experimental

Table 1. The detailed properties of sorbents

Sorbents	BET (m^2/g)	Pore volume (cm^3/g)	Crushing strength (N/cm^2)	CO_2 capacity* (mmol/g)
KMgAlI3000	75.24	0.37	963.0	1.52
KMgAlI3005	57.48	0.30	965.0	1.54
KMgAlI3010	51.89	0.27	998.0	2.45
KMgAlI3020	37.70	0.19	988.0	2.43
KMgAlI3025	26.74	0.16	991.0	2.10

*Sorbent pretreated with 12 vol% H_2O 40 min (carrier gas: Ar); pressure, 1 bar; adsorption temperature, 60 °C; flow rate, 80 mL/min; feed gas, 10 vol % CO_2, 12 vol % H_2O, and balance gas, N_2; amount of KMgAl sorbent, 7.2 g.

Materials Preparation. The new sorbents with different MgO loadings were prepared by impregnating K_2CO_3 on MgO/Al_2O_3 using a wet impregnating method. Reagent-grade $KHCO_3$ was supplied by Tianjin Fuchen Chemical Reagents Factory. The preparation procedure of the sorbent consisted of three steps: mixing and impregnation, drying at 110 °C for 12 h, and then calcining the solid product in a furnace under N_2 flow (20 mL/min) at 500 °C for 4 h. The sorbent was denoted as KMgAlIXXYY. K represents K_2CO_3, I represents the impregnation method. XX and YY represent the loading of K_2CO_3 and MgO, respectively.

Materials Characterization. The structural change of sorbents before/after adsorption was examined by X-ray diffraction. The XRD analysis was performed on a Bruker D8 Advance X-ray diffractometer with a Cu Kα source at 40 kV and 40 mA. The surface area of the sorbent was determined by nitrogen adsorption with the BET method on a Micromeritics ASAP-2020 instrument. The pore volume was obtained with the Bopp-Jancso-Heinzinger method from the adsorption isotherm.

The crushing strength of the sorbent was measured by a digital crushing strength tester (Jiangyan Global Instrument Factory KC-3C), which was a dynamometer measuring the force progressively applied to the solid object during the advancement of a piston. The applied force increases until the solid breaks and collapses into small pieces and eventually powder. The corresponding value of the collapsing force is defined as crush strength [11]. The detailed properties of sorbents are shown in Table 1.

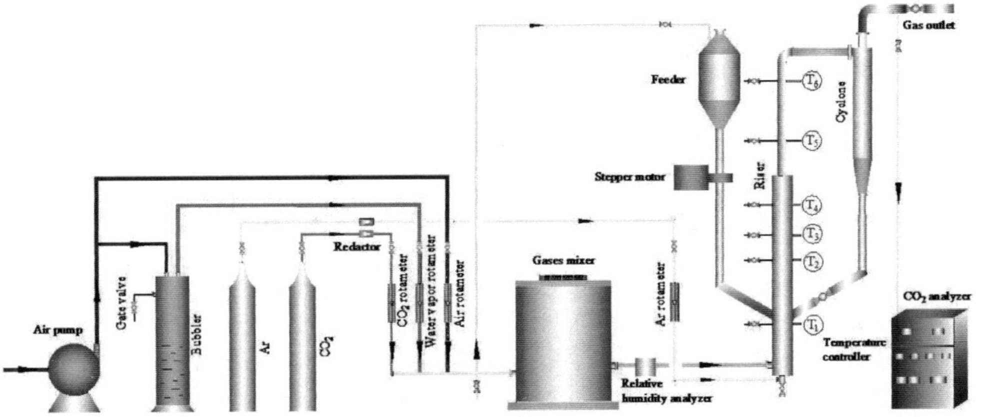

Figure 1. The flow chart of the fluidized bed experimental apparatus [12].

Apparatus and Operation Description. The comparison of the sorption capacity of CO_2 over various KMgAl sorbents was investigated in the fixed bed. To magnify experiments, the optimized sorbent was used for CO_2 capture in the fluidized bed. Figure 1 gives the flow chart of the fluidized bed experimental apparatus. It mainly consists of three parts including a gas injection part, fluidized bed absorption system, and CO_2 analysis system. A simulated flue gases composition of 10 vol % CO_2, 12 vol % H_2O, and 78 vol % air were used in this study [12].

Results and Discussion

The results of the comparison of the sorption capacity of CO_2 over various KMgAl sorbents in the fixed bed are shown in Table 1. It was found that upon increasing of MgO loadings, the amount of adsorbed CO_2 became larger at 10 wt % and 20 wt % MgO loadings, clearly, the total CO_2 absorption capacity was 2.45 and 2.43 mmol/g sorbent, respectively. It might originate from the balance of physical adsorption (high BET surface area) and chemical absorption (high MgO and K_2CO_3 loading) of the sorbent. Thus, the sorbent KMgAlI3010 and KMgAlI3020 were selected for the further investigation.

Figure 2 shows the XRD pattern of KMgAlI3010 and KMgAlI3020 with different H_2O steam pretreatment time and after adsorption CO_2. When the KMgAlI3010 sorbent was pretreated water vapor, the K_2CO_3 (PDF 27-1348) mainly transformed into $K_4H_2(CO_3)_3 \cdot 1.5H_2O$ (PDF 20-0886) (see Figure 2 b). After CO_2 adsorption, the main product was $KHCO_3$ (PDF 12-0292). Simultaneity, $K_2Mg(CO_3)_2 \cdot 4H_2O$ (PDF 29-1017) was produced by the reaction of K_2CO_3 and MgO (see Figure 2 c). However, the KMgAlI3020 sorbent was pretreated with 20 min, the hydrolyzate of Mg $(OH)_2$ from part of the MgO dissociates into Mg^{2+} and OH^-. In a basic medium, such as in HCO_3^- or CO_3^{2-} solution, the dissociation of $Mg(OH)_2$ is somewhat inhibited but an increase of $Al(OH)_4^-$ [14]. Thus, the deposition of $Al(OH)_4^-$ into the $Mg(OH)_2$ lattice to form $Mg_4Al_2(OH)_{12}CO_3 \cdot 3H_2O$. Simultaneity, a small number of $K_4H_2(CO_3)_3 \cdot 1.5H_2O$ appeared (see Figure 2 e). After CO_2 adsorption, the CO_3^{2-} replaced the OH^- of the $Mg_4Al_2(OH)_{12}CO_3 \cdot 3H_2O$ (PDF51-1525). Then it transformed into $Mg_2Al_2(CO_3)_4(OH)_2 \cdot 15H_2O$ and $MgCO_3$. Furthermore, the $MgCO_3$ and $KHCO_3$ generated new phase of $KHCO_3 \cdot MgCO_3 \cdot 4H_2O$ (PDF 24-0876), while a small amount of $K_2Mg(CO_3)_2 \cdot 4H_2O$ was produced (see Figure 2 f).

The effect of the adsorption temperature on the capacity of CO_2 adsorption was also investigated in the fluidized bed. Both total and breakthrough adsorption capacities of CO_2 are listed in Figure 3. Clearly,. the adsorption capacity of CO_2 decreased from 2.49 mmol/g sorbent at 60 °C to 1.49 mmol/g sorbent at 140 °C because both physical and chemical adsorption (usually) were exothermic [15, 16]. Even though lower temperature is favorable for the carbonation reaction, the process should be operated above the dew point of H_2O to prevent water condensation [12].

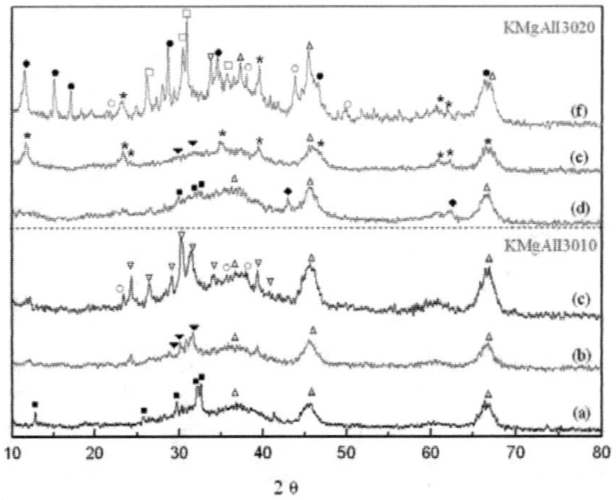

Figure 2. The XRD pattern of KMgAlI3010 and KMgAlI3020 with different H_2O steam pretreatment time and after adsorption CO_2. (a) pretreatment with water vapor 0 min; (b) pretreatment with water vapor 20 min; (c) pretreatment with water vapor 265 min; (d) pretreatment with water vapor 20 min after CO_2 sorption; (e) pretreatment with water vapor 20 min after CO_2 sorption; ■,K_2CO_3;Δ,Al_2O_3;▼,$K_4H_2(CO_3)_3 \cdot 1.5H_2O$; ★ ,$Mg_4Al_2(OH)_{12}CO_3 \cdot 3H_2O$; ▽ , $KHCO_3$; ●,$Mg_2Al_2(CO_3)_4(OH)_2 \cdot 15H_2O$; □, $KHCO_3 \cdot MgCO_3 \cdot 4H_2O$; ○, $K_2Mg(CO_3)_2 \cdot 4H_2O$.

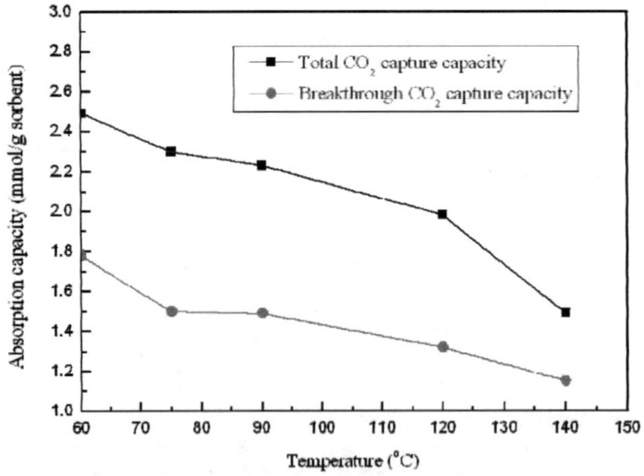

Figure 3. The effect of temperature on CO_2 sorption of KMgAlI3010 sorbent in the fluidized bed. Adsorption temperature, 60 °C; flow rate, 12 Nm^3/h; feed gas, 10 vol % CO_2, 12 vol % H_2O, and 78 vol % air; mass of the sorbent, 1000 g.

Conclusions

Experiments on CO_2 capture over the novel KMgAlI3010 and KMgAlI3020 sorbents showed high CO_2 uptake of 2.45 and 2.43 mmol CO_2/g sorbent at 60 °C under simulated flue gas conditions, respectively. XRD results revealed that there were different mechanisms of KMgAlI3010 and KMgAlI3020 sorbents pretreated with water vapor and after CO_2 sorption. In addition, experiments on CO_2 capture in the fluidized bed showed that the KMgAlI3010 sorbent had a high CO_2 uptake of 2.49 mmol/g sorbent at the flow rate of 12 Nm^3/h at 60 °C under simulated flue gas conditions.

Acknowledgements

This work was financially supported by Lu'an Group, Innovation Key Program of the Chinese Academy of Sciences (KGCX2-YW-215), the Ministry of Science and Technology of the People's Republic of China (2009BWZ003), ''Strategic Priority Research Program-Climate Change: Carbon Budget and Related Issues'' of the Chinese Academy of Sciences (XDA05010109, 05010110), the Natural Science Foundation of China (50976116) and the Instrument Developing Project of the Chinese Academy of Sciences (YZ201139).

References

1. Abu-Khader, M.M., *Recent progress in CO_2 capture/sequestration: A review.* Energy Sources Part a-Recovery Utilization and Environmental Effects, 2006. 28(14): pp. 1261-1279.
2. Murai, S. and Y. Fujioka, *Challenges to the Carbon Dioxide Capture and Storage (CCS) Technology.* IEEJ Transactions on Electrical and Electronic Engineering, 2008. 3(1): pp. 37-42.
3. Samanta, A., et al., *Post-Combustion CO_2 Capture Using Solid Sorbents: A Review.* Industrial & Engineering Chemistry Research, 2012. 51 (4): pp. 1438-1463.
4. Yang, H.Q., et al., *Progress in carbon dioxide separation and capture: A review.* Journal of Environmental Sciences-China, 2008. 20(1): pp. 14-27.
5. Li, L., et al., *A review of research progress on CO_2 capture, storage, and utilization in Chinese Academy of Sciences*, in *Fuel* 2011. Doi: 10.1016/j.fuel.2011.08.022.
6. Fan, L.-S., et al., *Chemical looping processes for CO_2 capture and carbonaceous fuel conversion - prospect and opportunity.* Energy & Environmental Science, 2012. 5(6): pp. 7254-7280.
7. Liang, Y., et al., *Carbon dioxide capture using dry sodium-based sorbents.* Energy & Fuels, 2004. 18(2): pp. 569-575.
8. Qi, G.G., et al., *High efficiency nanocomposite sorbents for CO_2 capture based on amine-functionalized mesoporous capsules.* Energy & Environmental Science, 2011. 4(2): p. 444-452.
9. Lee, S.C., et al., *CO_2 absorption and regeneration of alkali metal-based solid sorbents.* Catalysis Today, 2006. 111(3-4): pp. 385-390.
10. Zhao, C.W., X.P. Chen, and C.S. Zhao, *Multiple-Cycles Behavior of K_2CO_3/Al_2O_3 for CO_2 Capture in a Fluidized-Bed Reactor.* Energy & Fuels, 2010. 24: pp. 1009-1012.
11. Li, L., et al., *MgO/Al_2O_3 sorbent for CO_2 capture.* Energy and Fuels, 2010. 24: pp. 5773-5780.
12. Li, L., et al., *CO_2 Capture over $K_2CO_3/MgO/Al_2O_3$Dry Sorbent in a Fluidized Bed.* Energy & Fuels, 2011. 25(8): pp. 3835-3842.
13. Park, Y.C., et al., *Long-term operation of carbon dioxide capture system from a real coal-fired flue gas using dry regenerable potassium-based sorbents.* Energy Procedia, 2009. 1(1): pp. 1235-1239.
14. Xu, Z.P. and G.Q. Lu, *Hydrothermal Synthesis of Layered Double Hydroxides (LDHs) from Mixed MgO and Al_2O_3: LDH Formation Mechanism.* Chemistry of Materials, 2005. 17(5): pp. 1055-1062.
15. Ruthven, D.M., *Principles of Adsorption and Adsorption Processes.* John Wiley and Sons, Inc.: New York, 1984.
16. Atkins, P.W., *Physical Chemistry.* Oxford University Press: Oxford, U.K., 1990.

(389m) Microscopic Behavior of Chemical Demulsifiers At the Water-in-Brazilian Asphaltene Model Oil Emulsions: Langmuir Monolayers and Their Corresponding Langmuir-Blodgett Films

Juliana S. A. Carneiro[1,2], Montserrat Fortuny[1], Alexandre F. Santos[1], Cláudio Dariva[1], Elton Franceschi[1], Bianca M. S. Ferreira[3] and Sandro R. P. da Rocha[1,2], (1)Núcleo de Estudos em Sistemas Coloidas – NUESC, Universidade Tiradentes, Aracaju, Brazil, (2)Dept. of Chemical Engineering and Materials Science, Wayne State University, Detroit, MI, (3)CENPES, PETROBRAS, Rio de Janeiro, Brazil

Heavy oil fields represent a challenge for the oil industry because of its undesirable tendency to form stable water-in-oil emulsions during the crude oil exploration, causing high transportation costs and damage to processing equipment. Despite the recent new findings of light medium crude oil reserves in Brazil, the heavier crude oil fields still account for most of the overall oil production. The presence of asphaltenes in heavy oil is known to enhance emulsion stability owing to the formation of a viscoelastic film at the water-crude oil interface. To break-up these stable emulsions to accelerate the process of dehydration of the crude oil, chemical agents are generally employed. It is known that the demulsifiers act by reducing the interfacial tension between oil and water and also by impacting the viscoelastic properties of the interfacial film. Nevertheless, the disruption mechanism of the continuous interfacial film is still not well understood, and highly dependent on the chemistry of the asphaltene film, which in turn is highly depended upon the type of crude.

In our previous studies, a class of chemical additives was found to be effective in demulsifying water-in- Brazilian crude oil emulsions. In this work, we investigate the impact of chemical additives on the asphalene films from a Brazilian crude oil formed at the water-heptol (heptane/toluene) interface *in situ*, by determining the surface pressure vs. area isotherms in absence and presence of a range of chemical demulsifiers. We also probe the microstructure of the corresponding films as they are transferred onto solid substrates using atomic force microscopy (AFM). These results combined help us gain a deeper / more microscopic understanding of the effect of the chemistry of the demulsifiers on their ability to break up water-in-Brazilian crude oil emulsions.

(415a) New Reservoir Opportunities for Ccus: The Bio-Geochemistry of Mobile Water Intervals with Residual Oil: A Permian Basin Case History

David Vance, ARCADIS-US, Midland, TX, Steve Melzer, Melzer Consulting, Midland, TX and Robert Trentham, Geology, University of Texas Permian Basin, Odessa, TX

New Reservoir Opportunities for CCUS:

The Bio-Geochemistry of Mobile Water Intervals with Residual Oil:

A Permian Basin Case History

L. Stephen Melzer[i], Robert Trentham[ii] and David Vance[iii]

ASChE Annual Meeting

Session on CO_2 Utilization/EOR

Pittsburgh, PA

October 28 – November 2, 2012

Abstract

An entirely new set of CO_2 EOR targets has exploded on the scene in the Permian Basin region of the SW U.S. The commercial success of this group of reservoirs called residual oil zones (ROZs) has prompted a systematic reexamination of the science and distribution of their origins. Three types of ROZs have been identified. Much of the work to date has focused on carbonate reservoirs and the lateral sweep of a paleo-oil entrapment of dramatically larger thickness and lateral extent that would be suggested by looking only at the "conventional" main pay zone (MPZ) fields with a history of oil production. The work is demonstrating that the ROZ rock properties display a slightly higher porosity and permeability when likened to the nearby or overlying MPZs. A characteristic "bow" shape on the resistivity and porosity logs is generally present in the ROZs with decreasing resistivity and increasing porosity below the oil/water contacts. The long held explanation of a downward transition from oil saturations of 75-85% in the MPZs linearly grading to zero oil saturation fails to explain both the shapes of many ROZ saturation profiles and the rock property changes observed in the ROZs.

Evidence will be shown to support a late stage diagenetic event(s) creating more pervasive dolomitization in the ROZs. This overprint of dolomitization is accompanied by an otherwise uncharacteristic lack of disseminated anhydrite; evidenced not only in the observed cores but also in both the photoelectric cross section and spectral gamma ray logs. The implicated process is sulfur extraction from the anhydrites by anaerobic (sulfate reducing) microbes (with consequential effects on porosity and permeability). Occasional observation of free sulfur in vugs, sour oils and gas, and in sulfur-rich formation waters adds additional credence to the microbial process at work.

Finally, the combination of understanding the lateral sweep origins, successful modeling of the hydrodynamics, the rock property changes, and the success of CO_2 EOR ROZ projects is leading to a belief that the opportunities for CCUS may be much larger than previously believed including greenfields that are not associated with primary production. The large reservoir targets for CO_2 storage and oil production identified by NETL and DOE in previous studies may be understated when the ROZ targets are more fully defined and characterized.

[i] Consulting Engineer, Midland, Tx

[ii] Senior Lecturer, Geology, Univ. Tx of the Permian Basin, Odessa, Tx

[iii] Consulting Bio- Geochemist, Arcadis-US, Midland, Tx

(415b) Evolution of Carbonate Dissolution Features Produced Under Variable pCO2 Conditions Relevant to CO2-EOR and Geologic CO2 Storage

Susan Carroll, Megan Smith and Yue Hao, Lawrence Livermore National Laboratory, Livermore, CA

The role of mineral and pore-scale heterogeneity was explored in this study by observing the development of dissolution features in relatively heterogeneous and homogeneous carbonate units from the Weyburn-Midale hydrocarbon reservoir (Saskatchewan, Canada). The Weyburn-Midale reservoirs are dominated by carbonate (calcite and dolomite ± anhydrite, quartz) mineralogies that will partially dissolve as the acidic CO_2-rich waters used in EOR operations react with formation rock, altering porosity and permeability. Our experimental study aimed to quantify the relationship between fluid flow, heterogeneity, and reaction specific to carbon storage at the Weyburn-Midale field by integrating characterization, pressure/permeability data, and solution chemistry. We chose to study the porosity evolution using fluids equilibrated at pCO_2 levels representative of those calculated from site monitoring data and carbonate units reflecting different initial mineral abundances, porosity distributions, and permeabilities .We combine X-ray computed microtomography (XCMT) data with higher resolution scanning electron microscopy (SEM) images and statistical analysis to capture submicron pores to centimeter-scale or larger vugs. This methodology was chosen in an effort to use void space and mineral characterization of the carbonate cores before and after reaction to identify the type and mechanism of dissolution fronts in reactive systems.

Stable and unstable dissolution fronts in Marly dolostone and Vuggy limestone cores developed due to coupled flow and reaction processes, which were primarily affected and controlled by pore-scale heterogeneity. Here we discuss the role of heterogeneity on the development of distinct reaction fronts relevant to enhanced oil recovery and carbon storage using this rich experimental dataset. We utilize the data to rigorously constrain parameter input values such as kinetic rate constants and effective reactive surface areas to quantify relationships between dissolution and porosity production and evolving permeability in reactive-transport simulations of CO_2-carbonate interactions.

(415c) Opportunities for Utilizing Anthropogenic CO2 for Enhanced Oil Recovery (EOR)

Michael L Godec, Advanced Resources International, Inc., Arlington, VA

The purpose of this presentation is to address two common myths associated with the potential for carbon dioxide (CO_2) storage in associations with enhanced oil recovery. The first is that CO_2-EOR represents only a small fraction of the potential for storage the second is that CO_2 is not stored during EOR. The paper sets out to demonstrate that CO_2-EOR offers large CO_2 storage capacity potential, and can accommodate a major portion of the CO_2 captured from industrial facilities for the next 30 years. It will show that in fact CO_2 is stored with when deployed in association with CO_2-EOR, and that the amount stored depends on the priority placed on maximizing/optimizing storage. It goes on to demonstrate that carbon capture and storage (CCS) benefits from CO_2-EOR, in that the revenues (or cost reduction) from sale of CO_2 to EOR helps CCS economics, overcomes significant deployment barriers, while producing oil with a lower CO_2 emissions "footprint." Finally, the paper shows that CO_2-EOR needs CCS; large-scale implementation of CO_2-EOR is dependent on CO_2 supplies from industrial sources.

(415d) Potential for CO2 Ehanced Oil Recovery in the Appalachian Basin

Timothy Carr, Geology and Geography, West Virginia University, Morgantown, WV

The Appalachian basin oil and gas producing region has a large amount of oil which will be left in the ground, or "stranded", following the use of today's oil recovery practices. A major portion of this "stranded oil" is in mature reservoirs that based on this study appear to be technically amenable to enhanced oil recovery (EOR) using carbon dioxide (CO2) injection. Beyond its potential to augment U.S. oil production, CO2 EOR has the potential for permanently storing CO2, while adding value by maximizing oil recovery. CO2 EOR effectively reduces the cost of capturing and storing CO2 by earning revenues for the CO2 emitter from sales of CO2 to oil producers.

CO2 EOR, a tertiary recovery method in which supercritical CO_2 is injected into depleted oil reservoirs and has been in use by the oil industry for over 45 years. Utilizing CO2 EOR has a distinct advantage over waterflood-only recovery in that above a certain pressure the CO_2 becomes miscible with the in-situ oil (minimum miscibility pressure – MMP). This causes the oil to swell and become less viscous and more mobile allowing for production from the residual zones. The implementation of CO_2 EOR methods are limited to the reservoirs in which MMP can be achieved and have adequate sweep, but there are major economic limiting factors, especially if the injected CO_2 is procured from high-cost anthropogenic sources.

A focused study of the Jacksonburg-Stringtown field in north-central West Virginia demonstrates the potential in the region. The field has been under waterflood for a number of years. A minimum miscibility pressure analysis determined that the MMP of the oil within the Jacksonburg-Stringtown field was 941 psi at a reservoir temperature of 80 °F (25 °C). Using an average hydrostatic gradient of 0.465 psi/ft, the target reservoir must be at a minimum depth of 2,500 feet (800 m) in order for the injected CO_2 to reach MMP. The Lower Gordon Stray reservoir at the Jacksonburg-Stringtown field is at a depth of approximately 2800 to 3000 feet. In addition crude oils with API gravity greater than 22° are best suited for CO_2 EOR. The oil in the Jacksonburg-Stringtown field has an API value of 46.5° at reservoir conditions which satisfies this general requirement.

(415e) Carbon Dioxide Pipeline Infrastructure Dependence On Reservoir Risk

Jeffrey Bielicki[1], Richard Middleton[2], Melisa Pollak[1], Elizabeth Wilson[3], Jeffrey Fitts[4] and Catherine Peters[4], (1)Center for Science, Technology, and Public Policy, University of Minnesota, Minneapolis, MN, (2)Earth and Environmental Sciences, Los Alamos National Laboratory, Los Alamos, NM, (3)Center for Science, Technology, and Public Policy, University of Minnesota, (4)Department of Civil and Environmental Engineering, Princeton University

Carbon dioxide capture, utilization, and storage (CCUS) is a process that involves capturing carbon dioxide (CO_2) from large stationary point sources and transporting that CO_2 through pipelines to locations where it is injected into deep geologic reservoirs. For CCUS to make meaningful contributions to producing economically valuable products (utilization) or climate change mitigation (storage), large amounts of CO_2 must be transported through dedicated pipeline networks and injected into multiple reservoirs. While CO_2 transportation is considered to be a mature technology with little physical or financial risk, the CO_2 pipeline transportation network is coupled to the geologic formation into which CO_2 is being injected and thus dependent upon reservoir risk. CO_2 or displaced brine may leak from these reservoirs through natural or manmade pathways (e.g., faults, existing wells) and such leakage will incur physical and financial consequences. These consequences may reprioritize reservoir choices over time, and thus the planning, performance, and efficiency of the CO_2 pipeline transportation network thus depends on the performance of the geologic reservoir.

Reservoir leakage risk depends on the physical characteristics (e.g., permeability) of the reservoir and leakage pathways, the three-dimensional proximity of injection to these pathways and other subsurface activities, and the costs that are triggered by the leakage. As fluids escape the injection formation, they may (1) migrate up the geologic sequence and into overlying hydrostratigraphic units, (2) encounter other subsurface activities (e.g., natural gas storage), (3) contaminate underground sources of drinking water (USDW), or (4) escape to the surface. Each of these four potential outcomes incurs costs, and monetizing reservoir leakage risk allows CO_2 pipeline transportation planning to consider the dependence on reservoir performance.

We use the RISCS (Risk Interference of Subsurface CO_2 Storage) model for probablilistic leakage risk assessment of CO_2 injection into subsurface geologic reservoirs, and the SimCCS (Scalable Infrastructure Model for Carbon Capture and Storage) geospatial optimization model to show how leakage risk affects CO_2 pipeline infrastructure deployment. RISCS (e.g., Bielicki et al, 2012) combines probabilistic estimations of geophysical fluid flow and leakage from multiple iterations of the ELSA (Estimating Leakage Semi-Analytically) model (Nordbotten and Celia), three-dimensional data on hydrstratigraphic units and the locations of subsurface activities, and estimates of the financial consequences of leakage using the Leakage Impact Valuation (LIV) method (Pollak et al, 2012). RISCS monetizes leakage risk from geologic CO_2 injection for CCUS. These monetized leakage risks are used by SimCCS (e.g., Middleton and Bielicki, 2009), a cost-minimizing geospatial optimization model for deploying CCUS infrastructure which simultaneously and optimally chooses which CO_2 sources/reservoirs should be deployed, how much CO_2 should be captured from/injected into each source/reservoir, and the pipeline routes, diameters, and CO_2 flows between sources and reservoirs. We conduct this analysis for CCUS options in the Michigan Basin and show how reservoir risk alters the choice of reservoirs and pipeline routes.

References

Bielicki, J., Pollak, M., Wilson, E., Peters, C. and Fitts, J (2012). "Your View or Mine: Spatially Quantifying CO_2 Storage Risk from Multiple Stakeholder Perspectives." *DOE-NETL Project Review Meeting, January 16, 2012.* Pittsburgh PA. Manuscript in preparation for *Environmental Science & Technology.*

Middleton, R. and Bielicki, J. (2009). "A scalable infrastructure model for carbon capture and storage: *SimCCS*," *Energy Policy* 37, 1052-1060.

Pollak, M., Bielicki, J., Dammel, J., Fitts, J., Peters, C., and Wilson, E. (2012). "Estimating Financial Consequences from Geologic Sequestration." Submitted to *Environmental Science & Technology*.

A PHASED APPROACH TO DEVELOPING A PIPELINE NETWORK FOR CO_2 TRANSPORT DURING CCUS

Melanie D. Jensen, Peng Pei, Anthony C. Snyder, and Loreal V. Heebink, University of North Dakota Energy & Environmental Research Center, Grand Forks, ND

Introduction

The concept of carbon capture, utilization, and storage (CCUS) continues to receive considerable attention as an approach to reduce the United States' carbon dioxide (CO_2) emissions. Substantial capital investment will be required to capture, compress, and transport the CO_2 to storage targets if the concept is deployed on a large scale. Because of the lack of national policy or regulatory drivers, the utilization of CO_2 for enhanced resource recovery (enhanced oil recovery [EOR] or enhanced coalbed methane [ECBM] recovery) is likely to provide the impetus for the early deployment of CCUS, as this approach will provide a means to cover the enormous capital cost. National carbon management policies (i.e., carbon regulation by Congress and/or emission standards enacted by the U.S. Environmental Protection Agency) likely would expand this deployment. Unfortunately, many of the large CO_2 sources are not located near appropriate geologic storage areas (i.e., either saline formations or enhanced resource opportunities) and it is likely that some type of pipeline network would be needed to transport the CO_2 from the sources to the storage sinks.

Methodology

Development of a Phased Approach to Pipeline Network Design

Three approaches could be taken to planning a CO_2 pipeline network that supports widespread CCUS (1) One approach is a nationwide network that would transport CO_2 from large industrial sources in geographically diverse areas to large-scale geologic storage sites. A second model consists of a gradual build-out of regional networks in which large CO_2 point sources are connected with existing pipeline infrastructure that serves EOR operations with local storage. In a third scenario, shorter pipelines would directly link large CO_2 sources (such as power plants) with nearby storage locations. Because there are only a few thousand utility and industrial CO_2 emission sources and even fewer large geologic storage targets, it is more likely that the third approach will be the one that is implemented. It should be noted that the CO_2 from smaller industrial sources is unlikely to be captured and transported in a pipeline network because the compression of small amounts of CO_2 for pipeline transport would make such a system cost-prohibitive (1).

Which entities might fund all or part of a CO_2 pipeline network is a subject of considerable interest. A blend of private- and public-sector participation may be required to develop CCUS as a viable industry. The degree to which each of the two sectors provides funding would depend on the specific circumstance. For example, if the economics are positive, private funding may be sufficient to construct and operate a pipeline from a particular source or cluster of sources to an enhanced resource opportunity. Other pipelines may require government funding to defray part of the costs, while still other pipelines that may be mandated by the government to meet an emission reduction target will never be economically viable and may need government funding for the life of the project.

A pipeline network probably would not be built quickly since the drivers for rapid implementation of CCUS are not in place. Instead, a network likely would be built in stages or phases, with the first phase consisting of pipeline segments connecting sources with EOR opportunities, followed by the addition of other sources and sinks as dictated either by the marketplace (in the case of EOR) or national or regional carbon management policy.

An effort was undertaken by the Plains CO_2 Reduction (PCOR) Partnership to estimate how a CO_2 pipeline network might be built out in the PCOR Partnership region, over what time frame it might be built, and how much it might cost. The PCOR Partnership is one of the U.S. Department of Energy's seven regional carbon sequestration partnerships. The PCOR Partnership region is expansive, including the states of Iowa, Minnesota, Missouri, Nebraska, North Dakota, South Dakota, Wisconsin, Montana, and Wyoming and the provinces of Manitoba, Saskatchewan, Alberta, and British Columbia. The geographic and socioeconomic diversity of the region is reflected in the variable nature of the CO_2 sources, which include facilities associated with electricity generation; energy exploration and production activities; agriculture; fuel, chemical, and ethanol production; and various manufacturing and industrial activities.

The first step in determining the timing of the pipeline phases requires a presumption of how aggressively CCUS will be pursued in a region based on various endpoints. The International Energy Agency's (IEA's) BLUE Map scenario has put forth the concept of a 50% emission reduction (compared to levels from the year 2000) by 2050 (2). This reduction falls between the two approaches outlined by Dooley and others (3) when they examined the effects on the U.S. electricity generation assets of stabilizing atmospheric concentrations of CO_2 at 450 ppmv and at 550 ppmv. Although the reductions needed to stabilize the atmospheric CO_2 concentration at 450 ppmv appear to be so stringent as to be nearly unattainable, the timing seems to appropriately delineate the breaks between phases of pipeline development and was therefore adopted for this study. As a result, Phase I was defined as lasting from about 2015 to 2035, with Phase II running from 2035 to 2050, and Phase III beginning in 2050.

PCOR Partnership Methodology

The PCOR Partnership's three-step, phased pipeline planning methodology was developed to compare routes by relatively quickly estimating the amount of CO_2 that could be stored as well as the length and cost of the trunk pipelines required to store that CO_2. The approach is not intended to provide a method for developing a detailed pipeline network design.

The first step in the methodology consists of identifying clusters of CO_2 sources by noting which sources are located proximally to each other on a map. The CO_2 emission rate for each of the sources is taken from one of the many online emission databases, keeping in mind the appropriate capture level for the source type. Expected emission trends are then determined and applied to the known emission values to estimate future CO_2 emission rates.

After the CO_2 source clusters have been identified, the CO_2 storage capacity of each sink or sink cluster must be researched. Some data sets containing this information are available online.

Finally, pipeline routes can be determined using a pipeline routing software, or they can be roughly determined by measuring the distances between the centroids of clusters of CO_2 sources and the centroids of sink clusters. Pipeline costs can be estimated using preliminary engineering estimates or the Carnegie Mellon University IECM (Integrated Environmental Control Model), a free product that is available online.

Timing of the phases is then incorporated into the plan. While the easiest sources from which to capture CO_2 are ethanol plants and gas-processing facilities, the earliest storage (i.e., Phase I) likely will be in areas in which the CO_2 can be profitably used, such as during enhanced resource development activities (i.e., EOR or ECBM production). Many gas-processing facilities are situated on or near oil fields, making them ideally located for this type of activity, assuming that the product from several facilities can be gathered to form a large enough stream to supply an EOR project. Ethanol plants, on the other hand, are usually more widely distributed and may not be located proximally to storage sinks. Therefore, the majority of the ethanol plants may not come into play until late in Phase II or during Phase III network development because the value of the CO_2 volumes, even when consolidated, may not exceed the cost to dehydrate, compress, and build a lengthy pipeline to transport the CO_2 to a storage target. In addition to larger gas-processing plants and well-situated large ethanol plants, other Phase I sources that would be included in a Phase I network would be any power plants having corporate reasons for being an early adopter (e.g., government grants, etc.). It is expected that Phase I pipelines will be a combination of judiciously sited pipelines linking several sources (i.e., a source cluster) to a sink (or a cluster of sinks in a localized area) and one-to-one pipelines transporting CO_2 between one specific CO_2 source and a specific storage target. Whenever possible, existing pipelines were incorporated into a preliminary pipeline network.

Phase II of a network would incorporate more power plants, some of the larger industrial facilities (particularly cement kilns), and the rest of the ethanol and gas-processing facilities, if it makes economic sense to do so. Target geologic sinks would include the rest of the EOR opportunities as well as nearby saline formations. Some of the pipelines in this phase will be the branch lines as well as trunk lines.

Phase III of a network will come into play if sufficiently stringent climate policy and regulations have been put into place so as to force more widespread adoption of CCUS. This phase will include the remainder of the larger coal-fired power plants that must capture CO_2 as well as larger industrial facilities. Target geologic sinks added to the network at this point would consist primarily of saline formations. During this phase, the trunk lines could be connected to other trunk lines in the network and feeder lines could be added from large facilities to connect them with the branch lines. It is also possible that it might not make economic sense to connect all of the pipeline segments together to form a single network during Phase III. In this case, there might be multiple pipeline segments connecting specific source and sink clusters as well as smaller pipeline networks serving specific areas.

Results – Case Study Application of the Methodology to the PCOR Partnership Region

The development methodology was applied to the PCOR Partnership region to estimate a preliminary pipeline network that could be implemented in phases over the next 40 to 50 years. The complete results of this study are presented in Jensen et al., 2012 (4). The volume of CO_2 that would be available from each cluster of sources in the region was determined for three time periods (the present until 2035, from 2035 to 2050, and after 2050), and the most likely storage targets for each source cluster were identified. Preliminary pipeline routes connecting the sources and sinks were determined. Finally, when viewed as a regional whole, the routes were optimized for each network phase.

It was found that a pipeline network of trunk lines roughly 10,780 km in total length could transport sufficient quantities of CO_2 such that the IEA BLUE Map scenario could be met for the PCOR Partnership region by 2050. (The IEA BLUE Map scenario reduction would equal roughly 403.4 Mtonnes CO_2/yr for the PCOR Partnership region.) Meeting this target is dependent upon two critical assumptions. The first, put forward by the Energy Research Group at Dalhousie University (5), is that the output from Canada's electricity generation fleet will increase dramatically until at least 2050. The second assumption is that the Canadian government's goal of CO_2 emission capture (which equates to a 98% CO_2 emission capture rate in 2050 using values included by the Energy Research Group [5]) actually would be attained. These assumptions result in the storage of 335 Mtonnes/yr of CO_2 in the Canadian portion of the PCOR Partnership. When coupled with the expected U.S. CO_2 storage of 220.8 Mtonnes/yr, the overall reduction for the PCOR Partnership region is about 555.6 Mtonnes/yr by 2050. Table 1 summarizes the estimates of length, cost, and amount of CO_2 transported during the three pipeline network development phases. The pipeline routes are shown in Figure 1.

Table 1. Summary of the Preliminary Phased Pipeline Network for the PCOR Partnership Region

	U.S.			Canada				
Phase	I	II	III	I	II	III	Total of Phases I and II	Total, Including Phase III
km of New Pipeline	1078	4184	0	2520	2969	1384	10,751	12,135
CO_2 Transported by New Pipeline, Mtonnes/yr	52.6	168.2	0	145.6	184.5	82.4	551	633.3
Capital Cost of New Pipeline, \$M (2009 US\$)	676.5	2965.6	0	1251	1887	1136	6780.1	7916.1
O&M Cost of New Pipeline, \$M (2009 US\$)	3.4	13.6	0	8.1	9	4	34.0	38.0
Levelized Annual Cost of New Pipeline, \$M (2009 US\$)	79.6	348.0	0	145.6	222	132	795.2	927.2

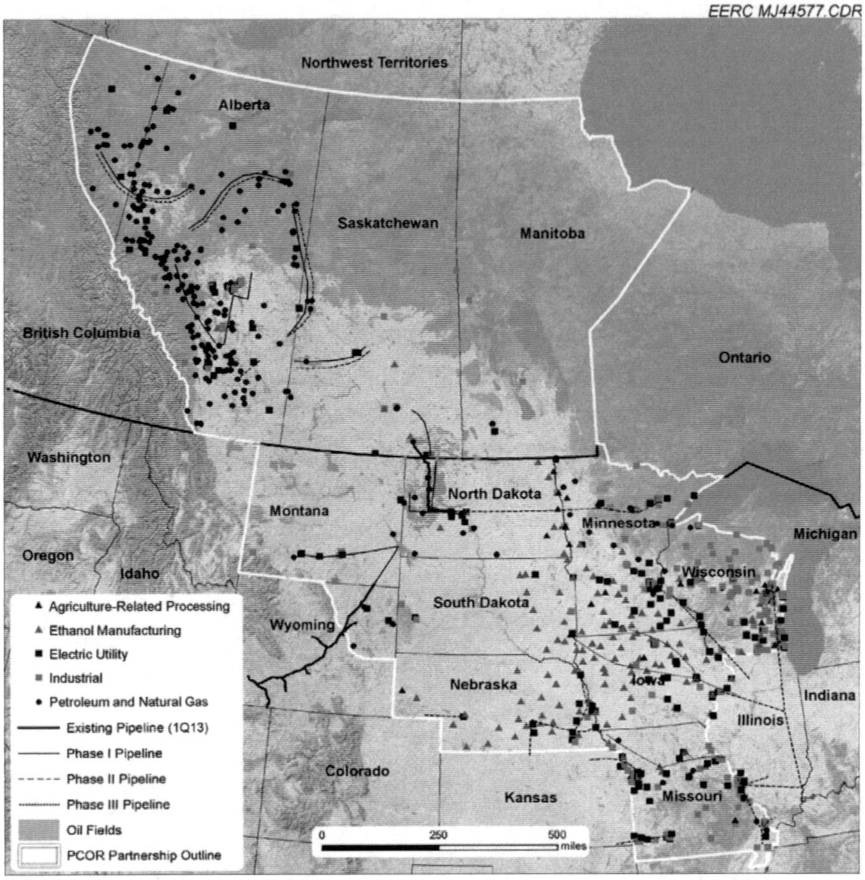

Figure 1. The Phase III Pipeline Network as Determined for the PCOR Partnership

Conclusions

Dooley and others (6) estimated that about 45,000 km of pipeline would be needed in the United States to meet the scenario in which the atmospheric CO_2 is stabilized at 450 ppmv by 2050. The pipeline estimates obtained using the PCOR Partnership methodology indicate that the length required for the U.S. portion of the region totals 5260 km. While this may at first glance seem low, when the number and distribution of regional storage targets is considered, it is obvious that the average pipeline segment would be shorter in the PCOR Partnership region than in many other areas.

According to IEA, long-term strategies are needed to cluster CO_2 sources and develop CO_2 pipeline networks such that source-to-sink transmission of CO_2 is optimized. The preliminary phased pipeline routing methodology developed by the PCOR Partnership is one approach that can help to address this challenge.

Acknowledgments

This material is based upon work supported by the Department of Energy National Energy Technology Laboratory under Award Number DE-FC26-05NT42592. The authors wish to thank not only the U.S. Department of Energy for its support but the many members of the

PCOR Partnership for their support as well. The efforts of the EERC's office services, editing, and graphics teams are greatly appreciated. Finally, the authors gratefully acknowledge the support and assistance of Katherine Anagnost, Janelle Ensrud, and Angela Morgan in the preparation of this document.

Disclaimer

This paper was prepared as an account of work sponsored by an agency of the United States Government. Neither the United States Government nor any agency thereof, nor any of their employees, makes any warranty, express or implied, or assumes any legal liability or responsibility for the accuracy, completeness, or usefulness of any information, apparatus, product, or process disclosed, or represents that its use would not infringe privately owned rights. Reference herein to any specific commercial product, process, or service by trade name, trademark, manufacturer, or otherwise does not necessarily constitute or imply its endorsement, recommendation, or favoring by the United States Government or any agency thereof. The views and opinions of authors expressed herein do not necessarily state or reflect those of the United States Government or any agency thereof.

References

1. Bliss, Kevin et al. (2010), "A Policy, Legal, and Regulatory Evaluation of the Feasibility of a National Pipeline Infrastructure for the Transport and Storage of Carbon Dioxide," Oklahoma City, OK, Interstate Oil and Gas Compact Commission.
2. International Energy Agency (2010), "Energy Technology Perspectives 2010—Scenarios and Strategies to 2050," Paris, International Energy Agency.
3. Dooley, James J., C. Davidson, M. Wise, R. Dahowski (2004), "Accelerated Adoption of Carbon Dioxide Capture and Storage Within the United States Electricity Utility Industry—The Impact of Stabilizing at 450 ppmv and 550 ppmv," in Proceedings of the Seventh International Conference on Greenhouse Gas Control Technologies (GHGT-7), Amsterdam, Elsevier.
4. Jensen, Melanie et al. (2012), "A Phased Approached to Designing a Pipeline Network for CO_2 Transport During Carbon Capture, Utilization, and Storage," report prepared for the U.S. Department of Energy National Energy Technology Laboratory, Grand Forks, ND, Energy & Environmental Research Center.
5. Hughes, Larry and Nikhil Chaudhry (2010), "The Challenge of Meeting Canada's Greenhouse Gas Reduction Targets," presented at the World Energy Congress, September 13–16, Montreal.
6. Dooley, James J. et al. (2009), "Comparing Existing Pipeline Networks with the Potential Scale of Future U.S. CO_2 Pipeline Networks," Energy Procedia 1: 1595–1602.

(422a) Micro-Channel Fuel Processors As Hydrogen Source for Fuel Cells in Distributed Energy Supply Systems (keynote)

Gunther Kolb[1], S. Keller[2], M. O'Connell[3], Jochen Schuerer[4], David Tiemann[4] and Martin Wichert[4], (1)Energy Technology and Catalysis Department, Institut für Mikrotechnik Mainz GmbH , D-55129 Mainz , Germany, (2)Institut für Mikrotechnik Mainz GmbH (IMM),, Mainz, Germany, (3)Institut für Mikrotechnik Mainz GmbH (IMM), Mainz, Germany, (4)Chemical Process Technology, Institut für Mikrotechnik Mainz GmbH, Mainz, Germany

Future sustainable and distributed energy generation will rely in many cases on fuel cell technology. Especially for stationary applications of the smaller scale, mobile and portable power generation systems a compact hydrogen supply is required[1]. Microchannel reactors provide significant potential for the reduction of system size, system complexity and catalyst costs[2]. Scaled-up micro-channel reactors containing thousands of channels create less scale-up issues compared to conventional (fixed bed) systems. They allow a careful heat management during start-up and normal operation, which increases the service life of the catalyst.

At IMM, the development of complete fuel processors for alcohol and hydrocarbon fuels is ongoing, applying microstructured plate heat-exchanger technology. Complete fuel processors are under development applying steam reforming, partial oxidation and partial de-hydrogenation reactions on a large variety of fuels such as methanol[3], ethanol [4,5], natural gas, LPG [6], gasoline, kerosene and diesel [7]. The systems size varies between 100 W and about 20 kW thermal power of the hydrogen produced. All development steps required for building a fuel processor have been addressed i.e. reactor and system simulation and design, catalyst development, catalyst stability against impurities in the fuel and set-up/testing of complete fuel processors. The choice of fuel converted to hydrogen mainly depends on the application and existing infrastructure. While trucks have diesel fuel readily available, LPG is the right choice for caravans. Having demonstrated the feasibility of compact fuel processing, the aspects of future mass production for such devices for consumer applications have been addressed. Complete fuel processor systems (e.g. a 13 kW_{th} diesel fuel processor, a 250W fuel processor/ fuel cell system running on LPG as fuel which will be commercialized in 2012 and a 20kW_{th} methanol fuel processor operated with a high temperature PEM fuel cell[8]) and novel components such as a reactor for the partial de-hydrogenation of kerosene will be presented and discussed.

References

[1] Kolb, G.; *Fuel Processing for Fuel Cells,* 1 Ed.; Wiley-VCH, Weinheim (2008).

[2] Kolb, G.; *"Review: Micro-structured Reactors for Distributed and Renewable Production of Fuels and Electrical Energy"*, Chem. Eng. Proc. **submitted for publication,** (2012).

[3] Kolb, G., Schelhaas, K.-P., Wichert, M., Burfeind, J., Hesske, C., Bandlamudi, G.; *"Development of a Microstructured Methanol Fuel Processor coupled to a High Temperature PEM Fuel Cell"*, Chem. Eng. Technol. **32,** 11 (2009) pp. 1739-1747.

[4] Men, Y., Kolb, G., Zapf, R., Hessel, V., Löwe, H.; *"Ethanol steam reforming in a microchannel reactor"*, Process Saf. Environ. Prot. **85,** B5 (2007) pp. 413-418.

[5] Kolb, G., Men, Y., Schelhaas, K. P., Tiemann, D., Zapf, R., Wilhelm, J.; *"Development Work on a Microstructured 50 kW Ethanol Fuel Processor for a Small Scale Stationary Hydrogen Supply System "*, Ind. Eng. Chem. Res. **50,** (2011) pp. 2554-2561.

[6] Wichert, M., Men, Y., O'Connell, M., Tiemann, D., Zapf, R., Kolb, G., Butschek, S., Frank, R., Schiegl, A.; *"Self-*

sustained operation and durability test of a 300 W-class micro-structured LPG fuel processor", Int. J. Hydrogen Energy **36,** (2011) pp. 3496-3504.

[7] O'Connell, M., Kolb, G., Schelhaas, K. P., Schürer, J., Tiemann, D., Ziogas, A., Hessel, V.; *"Development and evaluation of a microreactor for the reforming of diesel fuel in the kW range"*, Int. J. Hydrogen Energy **34,** (2009) pp. 6290-6303.

[8] Kolb, G., Keller, S., Tiemann, D., Schelhaas, K. P., Schuerer, J., Wiborg, O.; *"Design and Operation of a Compact Micro-channel 5 $kW_{el,net}$ Methanol Steam Reformer with Novel Pt/In_2O_3 Catalyst for Fuel Cell Applications"*, Chem. Eng. J. **submitted for publication,** (2012).

(422b) Tri-Generation of Hydrogen, Heat and Power From a High Temperature Fuel Cell

Jack Brouwer and Roxana Bekemohammadi, National Fuel Cell Research Center, University of California, Irvine, Irvine, CA

Tri-Generation of Hydrogen, Heat, and Power from a
High Temperature Fuel Cell

Jack Brouwer*, Roxana Bekemohammadi, Pere Margalef

National Fuel Cell Research Center
University of California, Irvine

*corresponding author: jb@nfcrc.uci.edu; 949-338-5953 <>Abstract

The current project involves collaborative development and application of a high temperature fuel cell (HTFC) tri-generation system for the first time in the world. In addition, the current tri-generation system is fueled with anaerobic digester gas that is renewably produced at a wastewater treatment plant. The National Fuel Cell Research Center is working with Air Products and Chemicals, Inc. and FuelCell Energy to demonstrate HTFC tri-generation at the Orange County Sanitation District facilities in Fountain Valley, California. The current project is successfully demonstrating local highly efficient production of power, heat and hydrogen from a renewable fuel. The system under test at OCSD can produce up to 125 kg of hydrogen per day, which is sufficient to fuel a fleet of approximately 200 – 300 vehicles. In addition to having zero emissions, fuel cell vehicles are substantially more efficient than internal combustion engine vehicles and can displace, as a result, more petroleum than the equivalent amount of hydrogen (on an energy basis). As a result, the current project will displace between 300 and 360 gallons of gasoline per day. <>Introduction

The National Fuel Cell Research Center (NFCRC) of the University of California, Irvine (UCI) has led research, development and demonstration projects focused upon High-Temperature Fuel Cell systems development. In parallel, the NFCRC is engaged in studies of the hydrogen economy from the generation to the utilization of hydrogen. In recent years, the NFCRC has led the development of a novel and attractive strategy for hydrogen fueling that includes: (1) the conceptualization of the "Energy Station," and (2) HTFC systems for tri-generating hydrogen, heat and power.

One of the most exciting recent developments in integrated energy conversion systems is the concept of poly-generation or tri-generation of power, heat and hydrogen from a high temperature fuel cell. The concept deserves attention and investment because of the significant efficiency improvement, emissions reductions, and resource conservation potential it portends. In recent years attention to this concept has become especially pronounced with the demonstration of the concept as operated on anaerobic digester gas at the OCSD. This abstract briefly summarizes the evolution of the concept and the basic features of the concept to aid in understanding its history and potential. <>Benefits of the High Temperature Fuel Cell Hydrogen Co-Production Concept

The basic concept is to capture and purify hydrogen produced by a high temperature fuel cell for use as a fuel in fuel cell automobiles. The processes of fuel processing and high temperature fuel cell electrochemical production of heat and power are synergistically integrated as shown in REF _Ref323118790 \h Figure 1
08D0C9EA79F9BACE118C8200AA004BA90B02000000080000000E0000005F005200650066003300320033003100310038003700390030000000
, for example, for a molten carbonate fuel cell. Thus, a high temperature fuel cell can be used to co-produce electricity for a building and hydrogen for a local refueling station, avoiding the energy and environmental impacts of hydrogen transport and distribution. Key technical and strategic benefits of this concept are:

(1) Local hydrogen production avoids the negative energy and environmental impacts of traditional hydrogen transport, distribution and dispensing means,

(2) HTFC can theoretically produce up to three times as much hydrogen as is required for electricity production using heat and steam that it naturally produces and would otherwise have to reject,

(3) If all of the reformate/hydrogen stream is sent through the fuel cell its efficiency increases,

(4) Co-generated hydrogen is a much higher value product than thermal energy,

(5) Generation of varying relative amounts of hydrogen and electricity allows additional operator control to increase value added, and reduce operating costs,

(6) Additional hydrogen production provides needed cooling to the fuel cell,

(7) A cooler fuel cell can operate with less excess air, which reduces the main system parasitic load (air blower power) to additionally improve efficiency,

(8) Stationary fuel cell systems are already designed to include integrated natural gas reformation technology, and

(9) Stationary high temperature fuel cell systems with reformers have proven near-zero emissions and high fuel-to-electricity efficiency.

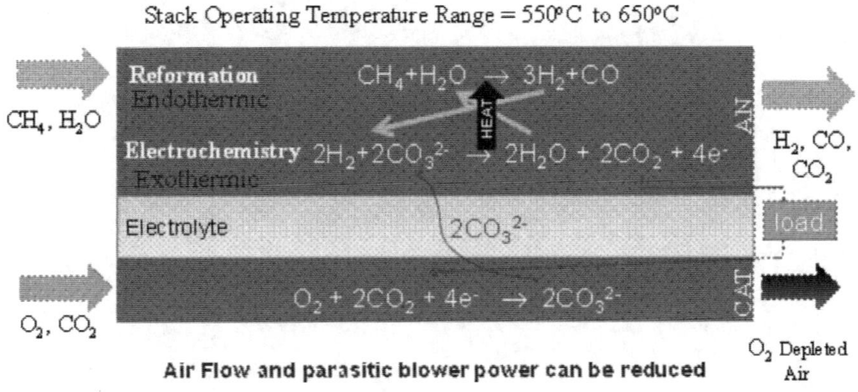

Figure SEQ Figure * ARABIC 1. Tri-generation concept as applied to an internally reforming molten carbonate fuel cell <>History of the Tri-Generation Concept

Some of the earliest presentations of HTFC tri-generation were published in 2001 [e.g., Brouwer et al., 2001]. From 2001 to 2003, the tri-generation concept was advanced through: (a) cycle conceptualization, (b) thermodynamic cycle analyses, and (c) dynamic systems analyses. In 2003, NFCRC worked to develop a collaboration between Air Products and Chemicals Inc. (APCI) and FuelCell Energy (FCE), which led to successful analyses and development of tri-generation with funding support from the U.S. DOE.

NFCRC presented Tri-Generation research results at several technical conferences in 2005 [Leal and Brouwer, 2005a; 2005b; 2005c]. In addition, NFCRC leadership worked with California Governor Arnold Schwarzenegger to develop the California Hydrogen Highway Blueprint Plan, into which they introduced the "Energy Station" concept which includes HTFC tri-generation. [Samuelsen, 2005] In the analyses that accompanied the Blueprint Plan, tri-generation was found to be the most efficient means to produce, deliver, and dispense hydrogen to vehicles.

In 2006, NFCRC researchers published "A Thermodynamic Analysis of Electricity and Hydrogen Co-Production using a Solid Oxide Fuel Cell" [Leal and Brouwer, 2006]. From 2007 to 2009, APCI worked with FCE to advance the concept with U.S. DOE funding support. The NFCRC independently advanced the technology by developing dynamic simulation capabilities. [Shaffer et al., 2008; Shaffer and Brouwer, 2009] NFCRC continued active tri-generation research from 2009 to present, which included completing a dissertation [Margalef, 2010], writing a book chapter [Brouwer and Margalef, 2012], and publishing several papers that analyze tri-generation [Margalef et al., 2011; Shaffer and Brouwer, 2012; Margalef et al., 2012].

From 2011-present, APCI, FCE and NFCRC have partnered to experimentally and theoretically advance tri-generation technology with funding from the DOE and California Air Resources Board. The tri-generation system that has been installed and that is undergoing testing is shown in REF _Ref323119018 \h Figure 2
08D0C9EA79F9BACE118C8200AA004BA90B02000000080000000E0000005F0052006500660033003200330031003100390030003100380000000
. The research being accomplished is proving the tri-generation concept, measuring gas clean-up, hydrogen separation technology performance, hydrogen purity, and overall efficiencies for producing the co-products from anaerobic digester gas, compressing and dispensing hydrogen to on-road fuel cell vehicles in California. In addition, the current effort is providing valuable data for verifying thermodynamic and dynamic models of Tri-Generation technology.

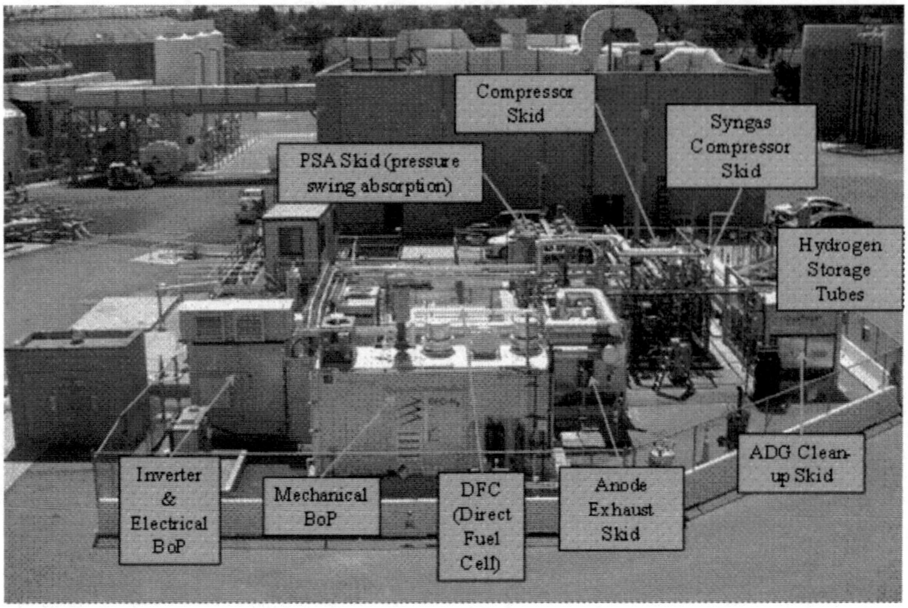

Figure SEQ Figure * ARABIC 2. Photograph of the Tri-Generation system at OCSD indicating major system components <>References ADDIN EN.REFLIST

[1] Brouwer, J., Samuelsen, G.S., Lee, S.W., and O'Connor, T., "Power Park Application of Fuel Cells and Advanced Energy Technologies," Proceedings of the 92nd International District Energy Association Conference, Las Vegas, NV, May 14, 2001.

[2] Samuelsen, G.S., "Energy Station Concept," in California Hydrogen Blueprint Plan, Volume 2, 2005, available on-line at: http://www.hydrogenhighway.ca.gov/plan/reports/volume2_050505.pdf.

[3] Leal, Elisângela Martins, and Jacob Brouwer, "A Thermodynamic Analysis of Electricity and Hydrogen Co-Production Using a Solid Oxide Fuel Cell," Proceedings of the 3rd International Conference on Fuel Cell Science, Engineering and Technology, ASME Paper Number FC2005-74136, May, 2005a.

[4] Leal, E.M., and Brouwer, J. "Thermodynamic Analysis of Production of Hydrogen Using High Temperature Fuel Cells," 2005 ASME International Mechanical Engineering Congress and Expo, Paper Number IMECE2005-81912, November 5-11, 2005b.

[5] Leal, E.M., and Brouwer, J., "Production of Hydrogen Using a High-Temperature Fuel Cell: Energy and Exergy Analysis," Proceedings of 18th International Congress of Mechanical Engineering, Paper Number COBEM05-1514, November 6-11, 2005c.

[6] Leal, E.M., and Brouwer, J., A Thermodynamic Analysis of Electricity and Hydrogen Co-Production using a Solid Oxide Fuel Cell, ASME Journal of Fuel Cell Science and Technology, Volume 3, Issue 2, pp. 137-143, May, 2006.

[7] Shaffer, Brendan P., Hunsuck, Michael, and Jacob Brouwer, "Quasi-3-D Dynamic Model of an Internally Reforming Planar Solid Oxide Fuel Cell for Hydrogen Co-Production," Proceedings of the 6th International Conference on Fuel Cell Science, Engineering and Technology, ASME Paper Number FC08-65193, May, 2008.

[8] Shaffer, Brendan and Jacob Brouwer, "Dynamic model for understanding spatial temperature and species distributions in internal-reforming solid oxide fuel cells," ASME Paper FuelCell2009-85095, June, 2009.

[9] Margalef, Pere, "On the poly-generation of electricity, heat and hydrogen with high temperature fuel cells," PhD dissertation, University of California, Irvine 2010.

[10] Margalef, Pere, Brown, Tim, Brouwer, Jacob, and Samuelsen, Scott, Short communication: Efficiency of poly-generating high temperature fuel cells, Journal of Power Sources, Volume 196, Issue 4, Pages 2055-2060, 15 February 2011.

[11] Margalef, Pere, Brown, Tim, Brouwer, Jacob, Samuelsen, Scott, Conceptual design and configuration performance analyses of poly-generating high temperature fuel cells, International Journal of Hydrogen Energy, Volume 36, Issue 16, Pages 10044-10056, August, 2011.

[12] Shaffer, Brendan and Jacob Brouwer, "Dynamic Model for Understanding Spatial Temperature and Species Distributions In Internal-Reforming Solid Oxide Fuel Cells," Journal of Fuel Cell Science and Technology, accepted for publication, March, 2012.

[13] Margalef, Pere, Tim Brown, Jacob Brouwer, and Scott Samuelsen, Efficiency Comparison of Tri-generating HTFC to Conventional Hydrogen Production Technologies, International Journal of Hydrogen Energy, accepted for publication, March, 2012.

[14] Brouwer, Jacob, and Pere Margalef, " Hydrogen Production by High Temperature Fuel Cells," in Encyclopedia of Sustainability Science and Technology, Robert A. Meyers, ed., Springer, 2012.

(422c) Effect of Storage Time On the Catalytic Activity of Gold for Selective CO Oxidation

Fabio B. Passos, Departamento de Engenharia Química e de Petróleo, Universidade Federal Fluminense, Niterói - RJ, Brazil and Thiago S. Mozer, Departamento de Química, Universidade Federal Fluminense, Volta Redonda - RJ, Brazil

Gold supported alumina catalysts were studied for selective CO oxidation, in order to provide appropriate H_2 for PEM fuel cell. The catalysts were prepared by deposition-precipitation method using $HAuCl_4$ as precursor of gold. The catalysts showed high activity at lower temperatures. The fresh catalysts were tested and after 1 year the activity tests were repeated. A considerable loss in the catalytic activity was observed, the maximum CO conversion decreased and the temperature to reach this conversion was higher. The loss of gold catalytic activity was attributed to the agglomeration of gold particles, which occurred spontaneously at room temperature, and may be favored to the presence of residual chloride from the preparation of the catalysts. The catalysts promoted with copper showed higher activity and a smaller loss of activity after 1 year of storage time. The interaction between gold and copper seemed to decrease to gold mobility, avoiding the agglomeration of gold particles.

NATIONAL ENERGY TECHNOLOGY LABORATORY

Hydrogen and Methane Rich Syngas from Catalytic Coal Gasifier with *In Situ* Capture of H_2S/CO_2 and *Ex Situ* Regeneration of Catalysts

Nicholas Siefert[1,2], Dushyant Shekhawat[1], David Berry[1], Shawn Litster[2]

October 31, 2012

2012 AIChE Annual Meeting, Pittsburgh, PA

[1] DOE/NETL [2] Carnegie Mellon University

U.S. DEPARTMENT OF ENERGY

Coal Gasification w/ *In Situ* H_2S & CO_2 Capture

Approximate Composition:
20%vol Methane
80%vol Hydrogen

On Energy Basis:
50% H_2, 50% CH_4

Lime is Regenerated from Limestone by Oxy-Combusting Unburnt Carbon.
Multiple routes for regenerating alkali hydroxides: either lime or electrodialysis

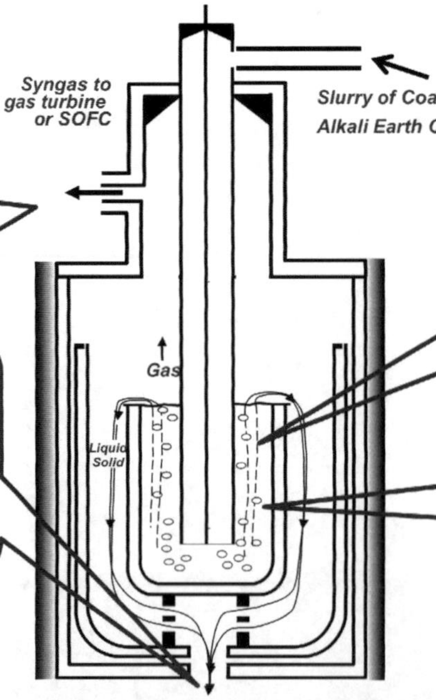

Alkali Hydroxides are Catalyst s. Alkali Hydroxides and Lime Are Capture Agent for Acid Gases, such as CO_2, H_2S and HCl

No oxygen required: exothermic capture & methanation reactions offsets steam-coal gasification

2 NATIONAL ENERGY TECHNOLOGY LABORATORY

Overall Goal: Generate Cheap Electricity at an Overall System Efficiency of 60% HHV

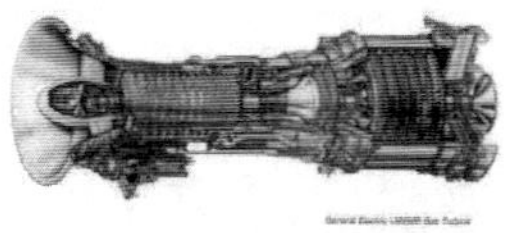

Overall Goal

SECA SOFC Systems

60% Efficiency

- **High Efficiency Power Generation**
  - 60% or greater System Efficiency when Integrating Catalytic Gasification with Solid Oxide Fuel Cells (Li *et al.*, Thijssen)
  - Greater than 50% System Efficiency when high hydrogen & methane content syngas is integrated with gas turbine combined cycle
- **Near-Zero Emissions**
  - Virtually no emissions of HCl, NO_x, SO_x, or particulates
  - Near-pure CO_2 stream produced after electrodialysis unit is ready for sequestration or enhance oil recovery
- **Cost Effective Power Generation**
  - Expected electricity costs on par with building a new non-CO_2 capture PCC power plant, i.e. less than 10% increase in COE

Process Flow Diagram: Gasifier & Hydroxide Regen.

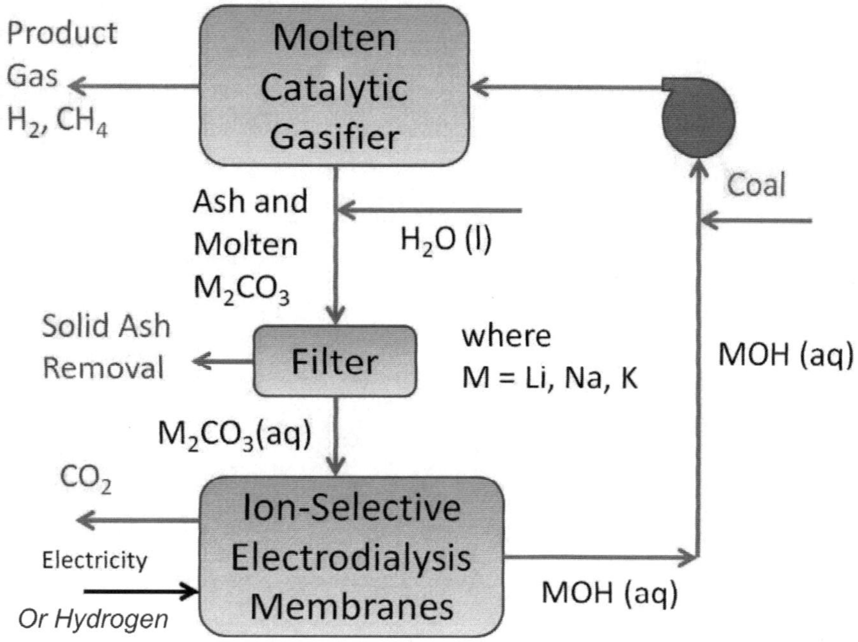

Process Flow Diagram: Gasifier & Lime Regen.

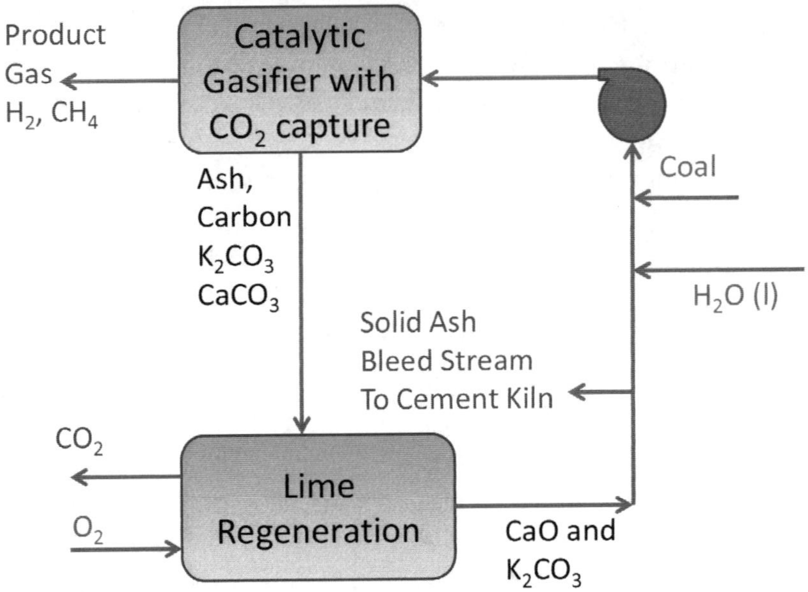

Chemical Reactions of *In Situ* CO_2 Capture

$C(s) + 2\ H_2O(g) \leftrightarrow CO_2(g) + 2\ H_2(g)$ ΔG_{1000K} = -10 kJ/mol

$C(s) + 2\ H_2(g) \leftrightarrow CH_4(g)$ ΔG_{1000K} = +19 kJ/mol

$2\ KOH + CO_2(g) \leftrightarrow K_2CO_3 + H_2O(g)$ ΔG_{1000K} = -91 kJ/mol

Overall: $C(s) + H_2O(g) + 2\ KOH(l) \leftrightarrow K_2CO_3(l) + 2\ H_2\ (g)$

And: $2\ C(s) + H_2O(g) + 2\ KOH(l) \leftrightarrow K_2CO_3(l) + CH_4(g)$

- Other *In Situ* CO_2 capture reactions include

$C(s) + 2\ H_2O(g) + CaO(s) \leftrightarrow CaCO_3(s) + 2\ H_2\ (g)$

ΔG_{1000K} = -33 kJ/mol ΔH_{1000K} = -68 kJ/mol

$2C(s) + 2\ H_2O(g) + CaO\ (s) \leftrightarrow CaCO_3(s) + CH_4\ (g)$

ΔG_{1000K} = -14 kJ/mol ΔH_{1000K} = -157 kJ/mol

Overview of Experiments

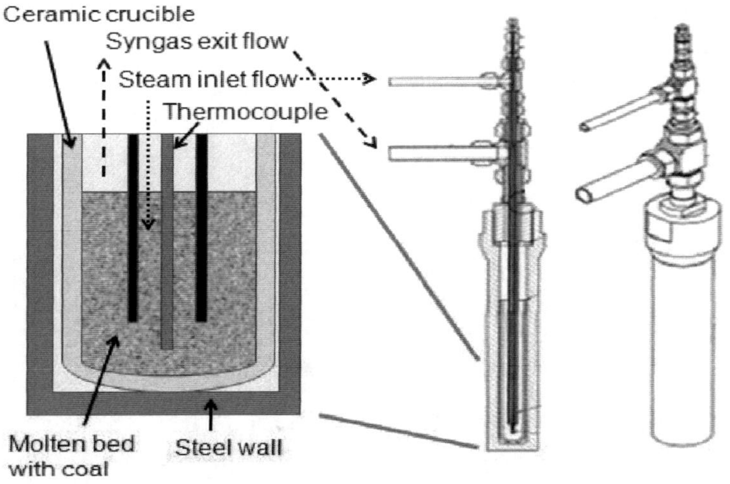

Batch: Coal & Catalyst **Continuous flow: Ar & H_2O**

Syngas analyzed by a GC and/or MS

Steam-Coal Gasification Initial Results

- **Conducted baseline studies in a batch reactor**
  - With eutectic alkali hydroxide catalyst (LiOH:NaOH:KOH)
  - With no catalyst and/or non-catalyst: calcium silicate ($CaSiO_3$)
  - With known catalyst and capture agent, lime (CaO)
  - With KOH, CaO and H_2O added to coal, and then dried
  - Catalytic capture agents were added on an equal CO_2 capture basis
    - MOH=1 : C=1, CaO=0.5 : C=1, $CaSiO_3$=0.5 : C=1

- **Conclusions:**
  - Gasification with alkali hydroxide yields the highest rate of steam-coal gasification while capturing significant amounts of CO_2 and H_2S
  - Gasification with small amount of KOH along with CaO for CO_2 capture provides fast kinetics while providing a simple means of releasing captured CO_2

Effect of Catalyst Type

Pittsburgh#8 Bit., 700 °C, 2.1 MPa,

	H_2	CH_4	C_2HC	CO_2	CO	H_2S	Rate
Catalyst	%vol	%vol	%vol	%vol	%vol	[ppm]	[1/hr]
Alkali Hydroxide	74	17	2.9	2.9	3.3	13	0.7
Lime	51	26	4.8	3.6	15	17	0.4
Calcium Silicate	47	21	3.4	22	7	5530	0.3
No Catalyst	36	27	5.1	17	15	5763	0.2

Wyoming Sub-Bit., 700 °C, 2.1 MPa

	H_2	CH_4	C_2HC	CO_2	CO	H_2S	Rate
Catalyst	%vol	%vol	%vol	%vol	%vol	[ppm]	[1/hr]
Alkali Hydroxide	73	16	2.3	0.8	7.9	2	1.9
Lime	70	18	3.3	0.7	7.9	8	0.7
KOH + Lime	73	13	4.0	0.2	8.0	40	2.6
No Catalyst	38	20	3.0	28	12	722	0.2

The fastest kinetics occurred when 0.2 g-KOH/g-coal and 1.75 g-CaO/g-coal were added to water and Wyodak coal, and then dried before operation.

Current studies are analyzing the degradation rate of the catalyst and capture agents over multiple cycles of gasification and regeneration.

9 NATIONAL ENERGY TECHNOLOGY LABORATORY

(422e) Molten Salt Gasification (MSG) of Refinery Wastes to Produce Hydrogen

Lyman Frost, Western Hydrogen Ltd, Calgary, AB, Canada

Western Hydrogen Limited is developing a process called Molten Salt Catalyzed Gasification (MSG) that allows the production of high-pressure synthesis gas at ~138 bar (~2000 psig) or higher without the need for compression. This makes the process very efficient since the compression of synthesis gas can require a significant amount of energy. The very high thermal efficiency (i.e. theoretical 86%, expected 61%) of the process results in efficient use of resources and a low per unit cost for the synthesis gas produced. This project will use a mixture of algae and water.

The aggressiveness of the MSG molten salt reaction will decompose any carbon-bearing material into a gaseous stream. The reactions can be operated at pressure by simple pressurization of the input water stream and feedstock. The MSG process occurs at pressure in a single reactor but occurs in three steps: 1) reaction of sodium carbonate with water and carbon generating sodium, carbon dioxide, and hydrogen; 2) instantaneous reaction of sodium with water generating sodium hydroxide and hydrogen; and 3) reaction of sodium hydroxide with carbon and water generating sodium carbonate and hydrogen. Experimental data indicates the feed stock ratios can be adjusted so the net of the reactions can be either slightly exothermic or slightly endothermic. The output can be adjusted by changing the operating conditions to generate hydrogen, synthesis gas, or methane. The net chemical reactions with the system operated for synthesis gas are shown below.

$Na_2CO_3 + 3\,C + H_2O \text{ -- > } 2\,Na + 4\,CO + H_2$ *Equation 1*

$2\,Na + 2\,H_2O \text{ --> } H_2 + 2\,NaOH$ *Equation 2*

$2\,NaOH + C + H_2O \text{ --> } Na_2CO_3 + 2\,H_2$ *Equation 3*

The paper will discuss the status of the technology and show data from the laboratory experiments and discuss the status of the pilot plant currently in fabrication.

(422f) Highly Efficient Integrated Chemical Looping Gasification Fuel Cell System

Liang Zeng[1], Mandar Kathe[1], Siwei Luo[1], Fanxing Li[2] and Liang-Shih Fan[1], (1)William G. Lowrie Department of Chemical and Biomolecular Engineering, The Ohio State University, Columbus, OH, (2)Chemical and Biomolecular Engineering, North Carolina State University, Raleigh, NC

Chemical looping process is designed for indirect reactions between fuel and oxidants via exchanging a stream of oxygen carrier particles among separate reactors. It is a promising process that can effectively increase the exergy efficiency for carbonaceous fuel conversion. The chemical looping concept has been widely used for combustion applications in light of its inherent CO_2capture capability. However, its development towards gasification applications, such as hydrogen production, is limited.

Chemical looping gasification (CLG) process can effectively convert carbonaceous fuel such as coal and biomass into hydrogen, and flexibly unite with fuel cell for highly efficient electricity generation with nearly 100% CO_2 capture. This study first generalizes several chemical looping process configurations for hydrogen production. A novel integration between the CLG and the solid oxide fuel cell (SOFC) technology is proposed. The CLG-SOFC system consists of a closed loop between the chemical looping oxidizer and the SOFC anode through the circulation of the gaseous mixture of steam and H_2, which can significantly enhance the electricity generation. Process simulation and exergy analysis confirm the advantages of the proposed process. The results indicate that, through process intensification, the chemical looping gasification strategies have the potential to notably increase process efficiency in the fuel cell system.

Experimental Investigation On the Effects of Trace Contaminants in Coal Derived Syngas On Performance of SOFC Ni-YSZ Anode

Dr. Koji Kuramoto, National Institute of Adv. Ind. Sci. Tech.(AIST)
Onogawa 16-1, Tsukuba, Ibaraki 305-8569, Japan

Introduction

The integrated gasification combined cycle (IGCC) is being developed to meet high-efficiency power generation from coal and the demonstration projects of the pilot scale IGCC power production are ongoing in Spain, United States, The Netherlands and Japan.[1,2] As an extension of the current R&D regarding the IGCC process, we could consider the incorporation of fuel-flexible high-temperature fuel cell into the coal gasification-based power generation process, which is often called integrated gasification fuel cell combined cycle (IGFC) and would have the great possibility to increase the plant efficiency[e.g. 3]. The reason is that the high temperature fuel cells including a solid oxide fuel cell (SOFC) can, in principle at least, directly convert the chemical potential contained in the coal-derived syngas to electricity at high efficiency. So far, a number of studies have been conducted to determine the feasibility of operation of SOFCs on hydrogen and natural gas because such fuel gases are clean and deliverable energy carrier in our daily life and have little chemical contaminants except for the sulfur-based odorants added before using. However, currently a relatively small number of active research works can be seen on the direct use of coal-derived syngas for SOFC. Coal is a complex heterogeneous mixture of organic and inorganic constituents with allothigenic or authigenic origin and the syngas obtained from high temperature coal gasification which predominantly composed of H_2 and CO contains small amounts of various chemical impurities such as condensable hydrocarbons (tar), soot, acidic and halide compounds and heavy metal vapors. In practice, according to the reports [e.g. 1,2], the produced syngas is subjected to the subsequent wet gas cleaning and most of the impurities in the syngas such as dust, soot, heavy metal vapors, and acid and halide compounds are eventually lowered below the acceptable values for avoiding the corrosion of blades of gas turbine in an IGCC system. However, in IGFC system which includes SOFC for power generation, as already pointed out [e.g. 4-7], the degradation of SOFC materials could occur even with the ppm-level or ppb-level contamination by heavy metal vapors in the syngas leading to the performance loss for anodic reaction. Performance loss of SOFCs due to the contaminations by acidic and halide compounds such as H_2S, and HCl were also studied [e.g. 8-10]. These were clearly due to the chemical interactions between Ni-YSZ cermets and trace chemical impurities in the gas under the polarization conditions. To ensure the reliability as well as the durability of coal-syngas use as fuel for SOFC, we need to clarify the mechanism of chemical interactions of these trace impurities with the materials of SOFC anode and electrolyte. Moreover, nickel is considered to be susceptible to poisoning by multiple impurities of volatile acids and halides and heavy metal vapors that are naturally present in the coal syngas. The purpose of the present experimental study is to see the influence of few selected single/multiple impurities in the fuel gas on the performance as well as the chemical/physical constitution of SOFC anode.

Experimentals

Button-shaped single cell: fabrication and preparation

In the power generation test, a button-shaped single SOFC cell with anode of Ni-YSZ (Yttrium stabilized zirconia) cermets was used. An electrolyte disk with a thickness of 1mm was fabricated using the 8YSZ (8 mol% Y_2O_3-92 mol% ZrO_2) powder. A 60%vol Ni – YSZ cermets was used as anode material. Mixed powder of NiO and YSZ dispersed in 2-propanol was painted on the electrolyte disk and calcined at 1673 K for 5 h. Pt mesh was attached to the electrolyte disk directly as a current collector for anode and cathode sides. The active electrode area was 0.785cm^2. Pt wire with diameter of 0.3mm was used as reference electrode (RE) and attached on the side edge of electrolyte. A ring shaped PYLEX grass was used as a gas seal between the cell and the fuel supply tube.

Cell testing fueled by simulated coal syngas with/without impurity doping

The effects of the selected poisonous gases or condensable vapors on the performance of Ni-YSZ anode were studied using a button-shaped half cell with feed of the simulated coal syngas with/without doping of selected poisonous compounds. All tests presented here were conducted under the potentiostatic condition to observe the degradation behavior due to the chemical interaction between the SOFC anode and doped impurities in the fuel gas. **Figure 1** shows a schematic diagram of the experimental apparatus.

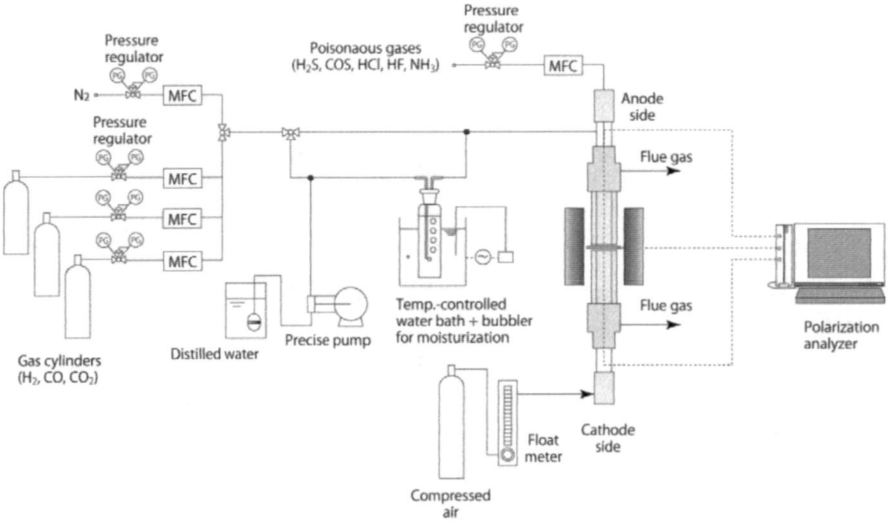

Cell testing with simulated coal syngas

For the tests with the simulated coal syngas, the fuel gas was prepared by mixing the cylindered CO, H_2, CO_2, and N_2 gases with composition of $CO/H_2/CO_2/N_2 = 50/20/4/26$ [vol-%] by a mass flow controlling system and injected to anode chamber at 150ml/min. The composition of the fuel gas was determined from the reported data obtained in the oxygen-blown coal gasification plant in Japan [1]. Dry air was fed to the cathode chamber at 200ml/min. Prior to the fuel cell test, the reduction treatment was conducted for the Ni-YSZ anode. N_2 and dry air were fed to SOFC anode and cathode respectively during heating, and then the NiO-YSZ anode was reduced at 1263K under dry H_2 flow for several hours. After the reduction, cell test was started using 8.7%-humidified H_2 at 1173K at an applied voltage of -0.75 volts versus the reference electrode (RE) which was positioned in the ambient air. When a steady current was attained, the fuel stream was switched from the wet H_2 to the simulated coal syngas. During the tests, the gas concentrations of CO, H_2, CO_2 and H_2 at inlet and outlet of anode chamber were intermittently measured with a TCD-equipped gas chromatograph. When the gaseous impurities were introduced to the anode chamber, a part of the N_2 in the fuel gas was replaced with impurity-containing N_2. The variation of current density with time under the constant applied voltage of -0.75V (vs. RE(air)) was monitored with a polarization analyzer. The AC impedance for the anode with amplitude of 10mV in the frequency range from 0.01Hz to 500kHz under the voltage of -0.75V vs. RE(air) was measured every five hours. In general, depending on the frequency range of voltage perturbation, we obtain some semicircles on the Nyquist diagram of impedance by the complex impedance plot (Cole-cole plot: see Fig.3). The intercept of the high-frequency semicircle with real axis corresponds to the ohmic resistance (Z'_{Ohmic}): the sum of resistances of electrolyte, electrodes and current collectors. The diameter of the semicircles would correspond to the polarization resistances ($Z'_{Polarization}$) caused by electrochemical reaction as well as transport phenomena of reactants, products, and ionized oxygen etc., depending on the frequency range. All tests presented here were conducted using a Solartron SI1287 potentiostat / galvanostat and 1255B frequency response analyzer. Microscopic change in the morphology of anode due to the exposure to the simulated coal syngas with or without impurities doping was observed with SEM-EDX system (VE7800, KEYENCE Co., Japan).

Impurity-doping test: the gaseous compounds

To investigate the effect of the gaseous impurities which possibly present in the practical coal syngas, we selected five gaseous compounds to be doped to the simulated fuel gas. As possible gaseous impurities, we selected H_2S, COS, HCl, HF, and NH_3 and doped them to the fuel gas to see if the chemical degradation of anode will occur or not. In the doping test of these gaseous impurities, N_2 in the fuel gas was partially replaced to the cylindered impurity-containing gas (N_2-balanced) to inject the poisonous gases with a prescribed concentration (about 2-4ppm) without change of gas composition of the main fuel gas.

Impurity-doping test: the condensable compounds (metal vapors and alkali metals)

In the present study, we attempted to dope some condensable compounds such as Na, K and volatile compounds contained in the actual coal to the fuel gas. For doping of the condensable species to the fuel gas, we used reagents ($NaCO_3$, K_2CO_3) or coal ash as the source of volatile impurities from coal. As a source of alkali metal vapors, the mixtures of equal amounts of Na_2CO_3 and K_2CO_3 powders were used which was pelletized into cylindrical shape (2mm-diameter and 2mm height) and dropped into the anode chamber during the fuel cell test. The Na_2CO_3/K_2CO_3 pellet was held just above the anode surface by the Pt mesh and retained in the controlled high temperature region. The TGA pre-exam of Na_2CO_3/K_2CO_3 pellet decomposition at 1173K in the simulated syngas exhibited the nearly constant weight decrease about 4 $\times 10^{-3}$mg/min, which corresponds roughly to the presence of 5-6% of K and Na compounds (e.g. KOH, NaOH, Na(g)) in the simulated syngas. As a source of volatile species contained in the actual coal, we used pellets of low temperature ash (LTA). The LTA sample was prepared by combusting carbonaceous portion of a selected coal sample at temperature below ca. 473K in an oxygen-plasma atmosphere so that the change in the chemical associations of mineral matter of coal ash can be avoided. The preliminary analysis of the LTA composition with ICP-MS spectroscopy detected more than 40 trace elements including Na, K, Zn, Se, Pb, Cd, Sb, Cr (from tens to several thousand ppm level in dry coal ash) but it should be noted that this analytical result could contain missing of components and/or substantial margin of error because we could not have the complete digestion/dissolution of the mineral matter with some acid solutions (e.g. HNO_3 and HF acid) which is necessary to ensure the mass balance of the chemical constitutions of the mineral matter in the ICP-MS measurement. The obtained mineral matters were pelletized without any chemical binder and subjected to the fuel gas atmosphere at the prescribed temperature in the fuel cell test. Note that the volatilization rates of alkali metal or other metallic vapors from the pellets was completely autonomic so the present test should be recognized as a sort of uncontrolled and accelerated impurity-exposure test.

Effects of gaseous impurities on SOFC performance

Figure 2 shows the variation in the current density with time generated by the half cell with the simulated fuel gas without impurities under the constant voltage of -0.75V (vs. RE(air)) and temperature of 1173K. The simulated gas was humidified at ca.20% with temperature-controlled impinger and was fed to the anode chamber at 150cc/min while the dry air to the cathode chamber. Here, the obtained current density was expressed as the normalized way [4] based on the following equation: $I^* = 1 - (I_o - I_{meas}) / I_o$ where I^*: normalized current density, I_o: the current density initially obtained or at the time when impurities are introduced, I_{meaus}: current density at any time. We obtained stable current

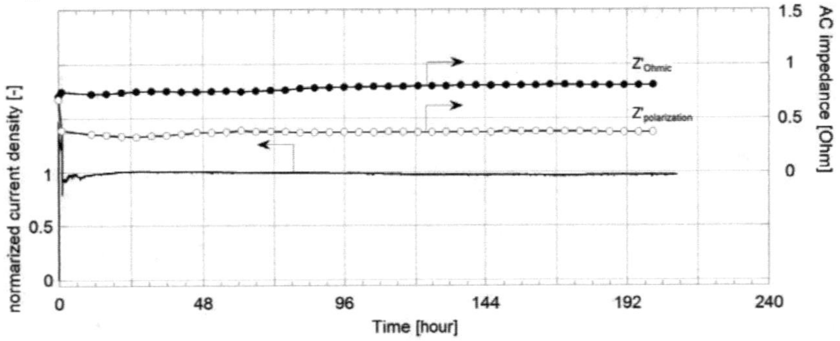

Figure 2 Variation of normalized current density and ohmic and electrode resistances with time obtained with a button-shaped SOFC single cell with fuel gas of $CO/H_2/CO_2/N_2$ = 50/20/4/26vol% with no impurity addition at T_{cell} = 1173K and constant imposed voltage of -0.75V (vs.RE(air))

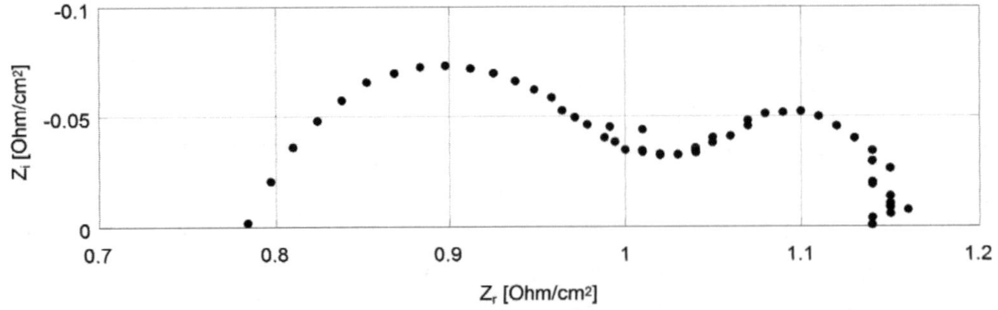

Figure 3 Example of the electrochemical impedance spectra typically obtained in the present study with a button-shaped SOFC single cell with fuel gas of $CO/H_2/CO_2/N_2$ = 50/20/4/26vol% without contamination at T_{cell} = 1173K and constant imposed voltage of -0.75V (vs.RE(air))

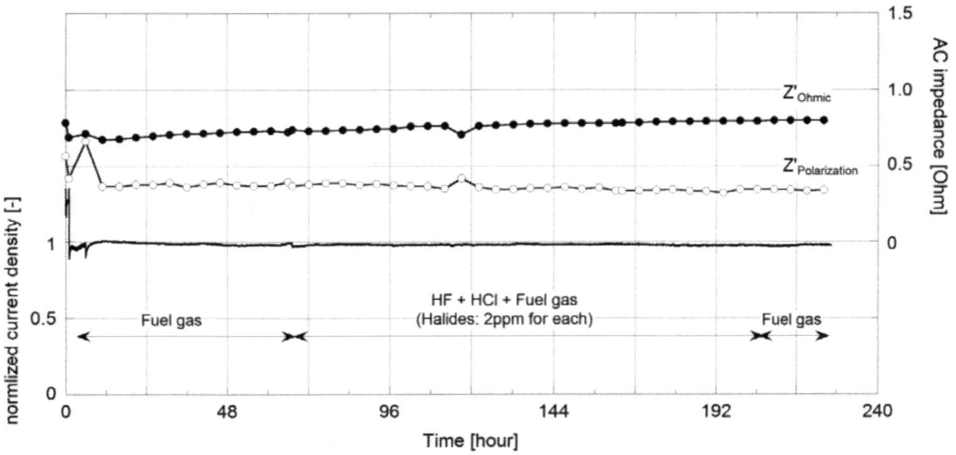

Figure 4 Variations with time of normalized current density and ohmic and electrode resistances obtained with a button-shaped SOFC single cell with fuel gas of $CO/H_2/CO_2/N_2$ = 50/20/4/26vol% with contamination by 2ppm HCl and HF at T_{cell} = 1173K and constant imposed voltage of -0.75V (vs.RE(air))

density (I_o was about 180mA/cm^2 in this case) and the rate of current loss was approximately 20% for 1000hours. Typical cole-cole plots obtained from AC impedance analysis during our test is shown in **Figure 3** and no remarkable change with time in ohmic (Z'_{Ohmic}) and polarization ($Z'_{Polarization}$) resistances were also seen in the case of the fuel gas feed without the impurities.

Figure 4 shows the variation with time of normalized current density, Z'_{Ohmic} and $Z'_{Polarization}$ during the fuel cell test with simultaneous doping of selected halide compounds of HCl and HF with 2ppm for each. So far, concerning the poisoning by the halide compounds (e.g. HCl and CH_3Cl), several experimental results have been presented in the literature. Trembly et al. [10] have examined the effects of HCl on the electrolyte-supported Ni-YSZ anode at the constant voltage of 0.7V and temperature of 1073 and 1173K with simulated coal syngas containing 0, 20, and 160 ppm HCl. They reported that injection of 20 or 160 ppm HCl to the system caused the performance losses in the SOFCs but HCl does not cause irrecoverable degradation to SOFC performance: the cell performance was found to be recoverable upon the removal of HCl from the fuel. Haga et al.[8] reported the slow degradation in cell voltage for H_2-based fuels with 5ppm Cl_2, and the injection of the fuels containing 100 and 1000ppm Cl_2 caused apparent degradation of cell voltage with an almost constant degradation rate. Their FESEM observation confirmed that significant micro-scale morphological change of anode surfaces with exposed 5ppm-Cl_2 for 150hours which might be due to the partial volatilization of chlorinated Ni such as $NiCl_2$. In addition, in a conventional metal recovery process, so called chlorination leaching treatment [e.g. 11] has been adopted where the valuable target metals including Ni in the ashes are chlorinated by exposure to HCl gas and then volatilized to separate the metals from the parent ash. [12] Thus, it was expected that HCl injection to the system would cause deactivation as well as chemical instability of Ni in Ni-YSZ anode due to the partial chlorination of Ni. In our result, as can be seen in the figure, there was almost no difference in the variation in current density and impedances between the tests with or without injection of halide compounds to the system. So, we could conclude that at low concentration of HCl and HF we could not have apparent chemical degradation of Ni-YSZ or accelerated sintering of Ni under the present operating conditions for 120hours. We injected selected poisonous gases such as H_2S, COS, HCl, HF and NH_3 to the system to confirm whether we could have the synergistic performance loss due to the presence of these poisonous gases.

Figure 5 shows the variation with time of normalized current density, Z'_{Ohmic} and $Z'_{polarization}$ during the fuel cell test with doping of five gaseous impurities at different conditions. When H_2S or H_2S/COS mixture was doped to the fuel gas (Figure 5(a)), we observed a sudden current drop at the beginning of the sulfur doping and then the current becomes steady with constant value. Similar observation was also reported by others [8,9]. We also observe that the extent of voltage drop at 4ppm-H_2S doping is larger than that at 2ppm-H_2S doping while no significant difference was observed between the 4ppm-H_2S doping and H_2S/COS mixture doping. From these results, the initial current drop is considered to be only affected by the sulfur concentration rather than the kind of S-compounds for our fuel cell system. In all cases, the current, ohmic and polarization resistances were recovered to the initial level after switching from contaminated fuel to uncontaminated one. As shown in Figure 5(b), when the fuel gas was switched to the gas contaminated with five poisonous gases the current density generated was suddenly dropped and then gradually decreased with time. About 50 hours passed, an unstable oscillation in current density was observed. Switching from contaminated fuel gas to the fuel gas, the current density started increasing and eventually recovered to about the current level nearly equal to that before doping the impurities. The AC impedances at high frequency range (Z'_{ohmic}) and low frequency ($Z'_{Polarization}$) were also varied when the impurities were injected: $Z'_{polarization}$ clearly increased when the fuel gas was switched to the contaminated one. This result suggests that some impurity gases interfered directly with the catalytic and electrochemical reactions in the anode layer as well as transportation behavior (e.g. blockage in adsorption/diffusion of H_2 to three phase boundary) process and main plausible mechanism would be the selective adsorption of sulfur-compounds to the active site or three phase boundaries of anode, as proposed by elsewhere [13]. As reported previously [e.g. 8, 13], when the fuel gas contaminated with the ppm-level sulfur-compounds such as H_2S and COS are introduced to anode chamber we could observe the sudden but slight output drop in the beginning of the impurity introduction and then the lowered performance was maintained without further loss of the performance. At higher concentration of sulfur, on the other hand, effects were different; the unrecoverable damages occurred in the anode due to the interaction of sulfur with Ni and the effects become more significant in the operation under the use of the carbonaceous fuel compared with hydrogen. [14] This kind of enhanced degradation effects by sulfur compounds on Ni or three phase boundary in SOFC anode were also theoretically deduced. [15] In our case, the irrecoverable degradation of Ni-YSZ anode by the injection of five selected poisonous impurities was not observed, maybe because the degradation due to low concentration impurities progressed very slowly over time. As shown in figure 6, SEM observation of the anode surfaces with/without injection of poisonous impurities confirm no drastic morphological change but show that surface of Ni particles in anode layer after exposure to poisonous gases (Figs. 4 and 5) become course-textured surface which might be due to the chemical interaction between Ni and impurities. To elucidate the mechanism as well as the practical influence of the chemical interaction with respect to the durability and stability of the power generation of SOFC fueled by coal syngas the long-term behavior of the low concentration impurities on the micro structural change of Ni as well as deactivation of Ni for electrochemical reactions are needed to be studied further.

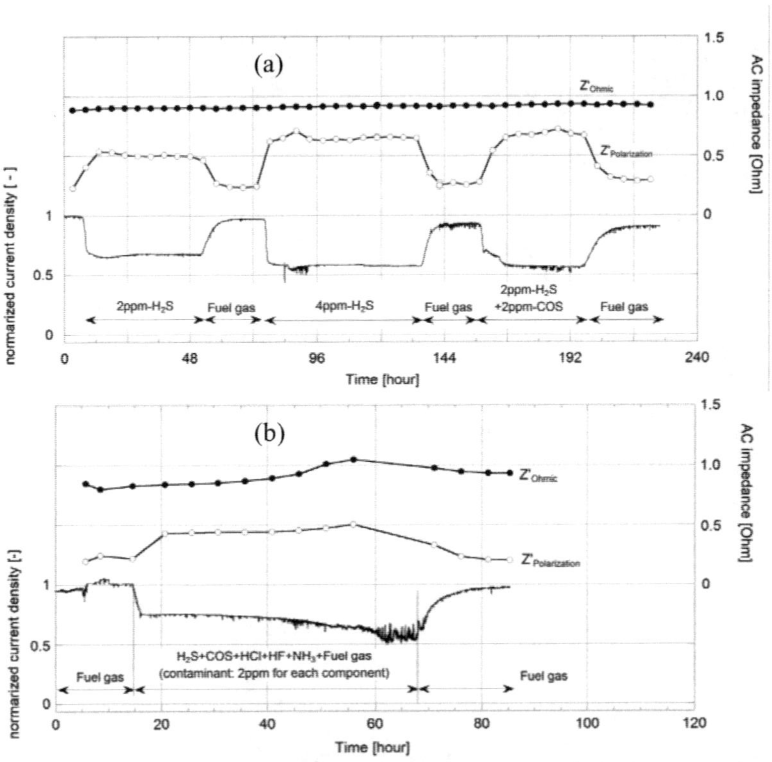

Figure 5 Variations of normalized current density and ohmic and polarization resistances with time obtained with a button-shaped SOFC single cell with fuel gas of $CO/H_2/CO_2/N_2$ = 50/20/4/26vol% with the presence of poisonous impurities at T_{cell} = 1173K and constant imposed voltage of -0.75V (vs.RE(Air)):(a) doping of H_2S and COS with different conditions, (b) doping of H_2S+COS+HCl+HF+NH_3 mixture.

Injection of condensable compounds: metal vapors contained in LTA and Na and K

Figure 6 shows the SEM image and result of an EDX mapping of Ni on the surface of Ni-YSZ anode which was subjected to the exposure of the nascent volatile chemical compounds released from pelletized LTA. The SEM observation confirmed that formation of the spherical and large Ni particles with diameter of about 15-20mm over the anode surface. The initial particle size of NiO is about 1 mm suggesting that the growth of the particle is significant. The formation of the large Ni particles might be caused by eutectic melting of Ni through the chemical interaction between Ni and volatile compounds released from LTA. This apparent growth of Ni has been observed in the fuel cell test with direct feed of coal-derived fuel gas in our previous work [16]. The post-test local chemical composition analysis using EDX confirmed that the larger particles contained only nickel and any other chemical compounds were not detected in the particles. This result suggests that the compounds causing this eutectic Ni melting probably will be interactive with Ni under the condition of polarization, high temperature or equilibrium partial pressure of vapor of the compound(s) but not detectable in the post-test analysis.

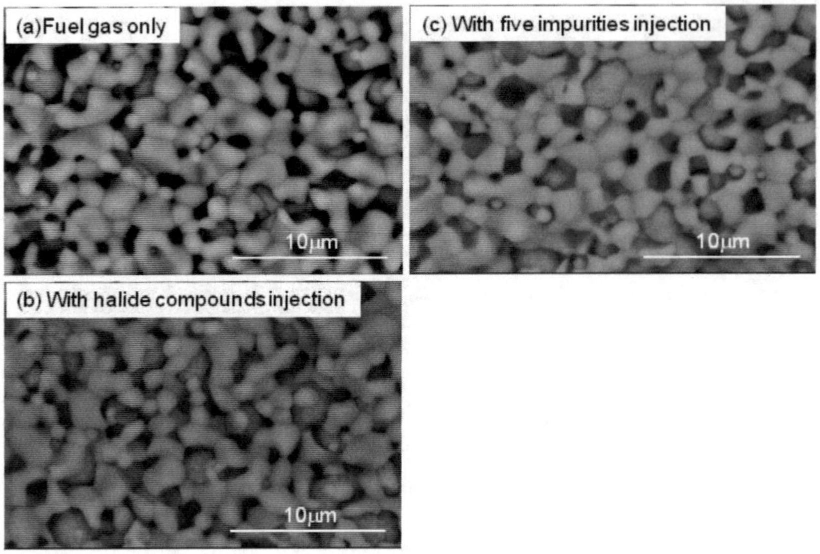

Figure 6 SEM images of anode surfaces with different contaminant-exposure condition: (a) the simulated fuel gas without doping of contaminant, (b) with halide components (HCl and HF) exposure, (c) with exposure of mixture of H_2S, COS, HCl, HF and NH_3

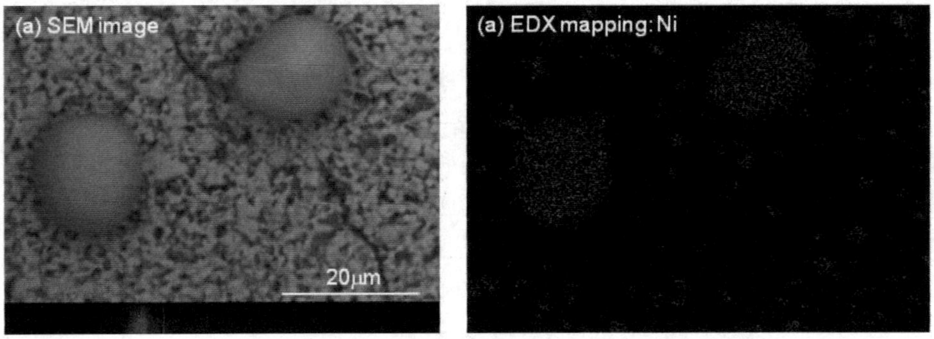

Figure 7 SEM and EDX mapping images of surface of Ni-YSZ anode after the exposure with nascent volatile chemical compounds released from pelletized LTA derived from coal sample: (a) SEM snap shot and (b) Ni distribution over the area in the anode surface.

Sodium and potassium vapors were also introduced into the anode chamber to investigate if the alkaline metal species which are generally present in biomass or brown coal (lignite) would have some negative influence on the performance of SOFC. Figure 8 shows SEM image of anode surface after the exposure with Na and K vapors released from Na_2CO_3/K_2CO_3 pellet and distribution of Si and Na over the spent cell obtained from EDX mapping analysis. SEM observation clearly shows the formation of needle-like deposits in the anode surface and the EDX analysis over the deposits confirmed that the deposits contained Si and Na. Note that the no Si vapor was intentionally doped to the fuel gas during the experiments and one possible source of Si would be a PYREX grass ring which was used for gas sealing between the button cell and gas supply tube. Though not shown here, we observed a stepwise increase in the polarization resistance ($Z'_{Polarization}$) after the injection of the pellets to the anode chamber during the test which might be due to the formation of obstacles for the supply of reactants to the anode layer. Concerning the relevant reaction of Si and Na, Haga et al. [8] reported that Si is possibly present in the form of $Si(OH)_4$ in the coal syngas and will tend to deposit on the anode surface followed by the decomposition into SiO_2 and H_2O. Cayan et al. [4] pointed out in their review paper that SiO_2 may form a grass like layer with Na_2O at the electrode and electrolyte interface, though, SiO_2 itself may not have severe impact on the performance. The phenomena relevant to deposition/condensation caused by the condensable trace species contained in the fuel gas would be expected in the utilization of solid fuel such as coal and biomass and further studies are needed to establish a quantitative assessment about the influences of the gaseous and condensable trace impurities under the accurately-controlled impurity composition of the simulated coal derived fuel gas.

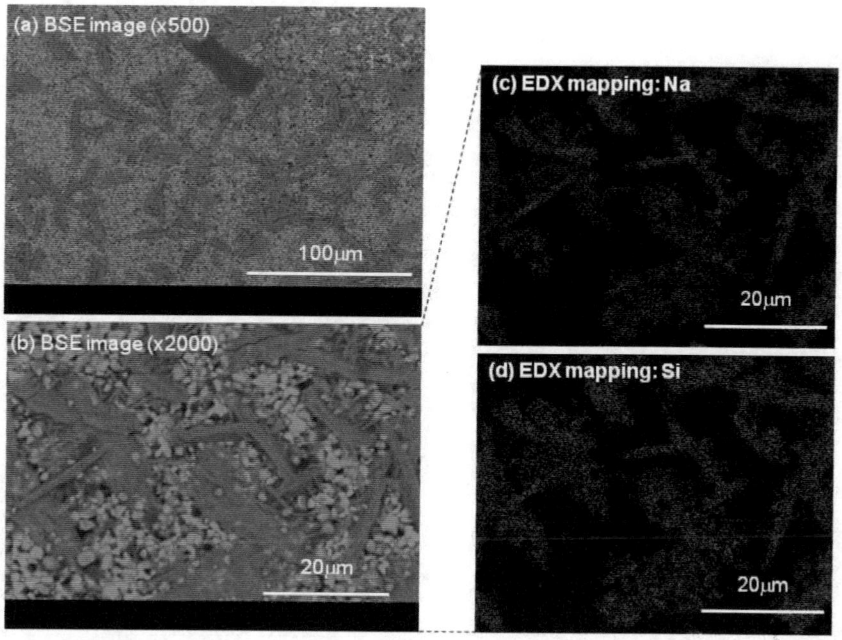

Figure 8 BSE images with different magnifications of anode surface after the exposure with Na and K vapors released from Na_2CO_3/K_2CO_3 pellet (a & b) and distribution of Na and Si over the anode surface obtained from EDX mapping analysis (c & d).

References

1. Ohtsuka et al., *Powder Tech.*, **190**, 340 (2009)
2. Liu et al., *Energy Policy* **36**, 1(2008)
3. Kivisaari et al., *Chemical Engineering Journal* **100**(1-3), 167 (2004)
4. Cayan et al., *Journal of Power Sources* **185**, 595(2008)
5. Trembly et al., *Journal of Power sources* **171**, 818 (2007)
6. Bao et al., *Journal of Power Sources* **193**, 617 (2009)
7. Bao et al., *Journal of Power Sources* **195**, 1316 (2010)
8. Haga et al., *Solid State Ionics* **179**, 1427(2008)
9. Zhang et al., *Int. J. Hydrogen Energy* **35**(22), 12359 (2010)
10. Trembly et al., *J. Power Sources* **169**, 347 (2007)
11. Shimada et al., *Ind. Eng. Chem. Res.* **37**, 894 (1998)
12. Fraissler et al., *Chem. Eng. and Processing: Process intensification* **48**(1), 152 (2009)

13. Sasaki et al., *J. The Electrochem. Soc.* **153**, A2023 (2006)
14. Mukerjee et al., *ECS trans.* **7**, 59-65(2007)
15. Kishimoto et al. *J. The Electrochem. Soc.*, **157**, B802 (2010)
16. Kuramoto et al. *ECS Trans.*,**25**, 2149 (2009)

A MICROCHANNEL WATER-GAS SHIFT REACTOR SIMULATION AND OPTIMIZATION FOR A MOBILE APPLICATION

M.H. Akbari , S.A. Asaee, Shiraz University, Shiraz, Iran

Introduction

Considering the high growth rate in energy consumption and the fact that hydrocarbon fuel emissions are harmful to the environment, the search for renewable energy resources has significantly expanded during the last few decades. In recent years atmospheric greenhouse gas concentrations (particularly CO_2) have increased considerably because of hydrocarbon fuel combustion. Consequently, certain crucial steps have been taken to manage greenhouse gas emissions; these include: enhancement of energy efficiency, utilizing low carbon (or carbon-free) energy sources, and storage of carbon such as CO_2 in geologic formations [1].

Fuel cells are efficient devices for energy conversion with extremely low pollutant emissions in comparison with conventional power generation methods. Due to their simplicity, low temperature operation, and relatively high efficiency, proton exchange membrane (PEM) fuel cells are more appealing for utilization in portable devices than other types of fuel cells. The preferred fuel for a PEM fuel cell is hydrogen; however, hydrogen does not exist in large amounts in nature. Presently, hydrogen is mainly produced from other sources such as hydrocarbon fuels through steam reforming (SR), partial oxidation (POX) or autothermal reforming (ATR). During these processes carbon monoxide (CO) is also generated, while its amount should be below 10 PPM for a hydrogen-rich stream to be usable in a PEM fuel cell. Therefore, a series of reactions are necessary to reduce the amount of CO concentration to sufficiently low levels.

The water-gas shift (WGS) is an important industrial reaction with numerous applications. During this process carbon monoxide and water vapor react to produce carbon dioxide and hydrogen, as follows:

$$CO + H_2O \leftrightarrow CO_2 + H_2, \qquad \Delta H^{\circ}_{298} = -41.1 \quad kJ\ mol^{-1} \tag{1}$$

Thermodynamic equilibrium dictates higher formation of the products at lower temperatures; however, the reaction rate is considerably reduced at such conditions. As a result, the water-gas shift reactor is usually the bulkiest component of a fuel processor [2]. Since for mobile applications size and weight are of crucial importance, a compact and efficient fuel processor (e.g., a microreactor) is more essential in such cases.

Several authors have studied the behavior of water gas-shift reactors using one- or two-dimensional models. Some investigators have developed kinetic models for the water-gas shift reaction on various catalyst types.

Giunta *et al.* [3] used a one-dimensional heterogeneous model for the simulation of a fixed-bed water-gas shift reactor. In that study a commercial $Cu/Zn/Ba/Al_2O_3$ was used as the catalyst. Catalyst deactivation due to thermal factors (sintering) was considered. Their simulation results showed that an isothermal condition yields higher conversions than an adiabatic condition and that using smaller pellet size results in a smaller reactor size.

In another investigation Choi and Stenger [4] studied the water-gas shift kinetic and suggested an equation for the kinetic rate. They utilized a microreactor with commercial $Cu/ZnO/Al_2O_3$ catalyst in the temperature range of 120-250°C. The investigation showed that management and control of the reactor temperature and water addition results in constant CO concentration at the reactor outlet.

Ding and Chan [5] developed a two-dimensional, unsteady model to study the performance of a water-gas shift reactor. The simulation results demonstrated that most of the carbon monoxide conversion occurs in the first 50% length of the reactor. In addition, at higher temperatures less CO conversion was achieved. A gas space velocity range of 107 to 213 h^{-1} was proposed as the optimum operating condition for the simulated reactor.

Sun *et al.* [6] exploited Langmuir–Hinshelwood (LH) kinetic rate model at ambient pressure for noble metal catalysts. They used this model to simulate a water-gas shift reactor for fuel processing from natural gas. The simulation results revealed that LH kinetic rate model yields more acceptable results than power law kinetic rate equations. They also established that reactors with noble metal catalysts are of smaller size by an order of magnitude compared with those that employ copper-zinc as catalyst.

In this paper, a single-channel water-gas shift (WGS) microreactor is simulated using a three-dimensional model with the power law kinetics for a Pt/TiO_2 catalyst [7]. The rate equation is valid in the temperature range of 200-270°C as discussed in detail later. The channel walls are assumed to be isothermal and the reactions occur only on the wall surfaces (channel walls are coated with 0.5 wt% Pt/TiO_2 catalyst texture). Effects of the major contributing parameters on the performance of the microreactor are investigated. These parameters include the feed gas temperature and moisture content, channel geometry (cross section aspect ratio) and length, and gas space velocity. The feed gas composition is taken as the outflow of a typical ATR reactor, and the operating conditions are set in a range such that the device would be paired with a typical low-power PEM fuel cell for mobile applications.

A series of parametric studies have been carried out to obtain the optimum operating conditions and geometry of the microchannel for maximum CO conversion. A matrix of enough number of such optimized microchannels will be required to treat enough synthesis gas to power a low-power PEM fuel cell (e.g., 100 W) in a mobile application.

Model summary

Problem statement

In the present study a rectangular channel microreactor with a hydraulic diameter of 100 μm and a length of around 20 mm is considered. These dimensions were selected based on available fabrication limitations and practical requirements. The flow regime is assumed to be laminar, steady and incompressible. Moreover, the gas mixture is taken as an ideal gas with variable properties based on the local gas composition. The flow Knudsen number at the baseline conditions is 0.0007; this is below 0.001, hence the continuum assumption is valid. Fig. 1 represents a schematic model of the channel geometry.

A typical autothermal reactor outlet gas composition is used which consists of 7 species: CH_4, O_2, CO, CO_2, H_2, H_2O and N_2 [8]. The gas composition at the baseline is given in Table 1.

The main purpose of the present study is to design an optimized microreactor for the intended application. To achieve a set of optimal channel geometry and operating conditions a parametric study is performed. The criterion to be satisfied in these simulations is to reduce CO concentration to below 2% in the outlet gas mixture. Such CO level is in a proper range for the treated gas to be fed to a PROX reactor for further reduction of CO concentration.

Governing equations

The governing equations for a reacting flow in a three-dimensional rectangular duct are given as the conservation of mass, momentum, energy and chemical species in a Cartesian coordinate system. Since the continuum assumption is valid, the Navier-Stokes equations are applicable.

The continuity equation for a 3D steady-state flow is given by

$$\partial(\rho V_i)/\partial x_i = 0 \qquad (2)$$

The Navier-Stokes equations, stated below, represent the conservation of momentum for the given conditions.

$$\rho V_j(\partial V_i/\partial x_j) = -\partial p/\partial x_i + \mu\, \partial^2 V_i/(\partial x_j \partial x_j) \qquad (3)$$

The mass balance for the mth species is presented in Eq. (4), and the molar production rate, ω, is given by Eq. (5) in which r is the WGS reaction rate and ξ is the stoichiometric coefficient of the mth species in the reaction.

$$\frac{\partial(\rho V_j Y_m)}{\partial x_j} = \frac{\partial}{\partial x_j}\left(D_{eff,m}\,\rho\,\frac{\partial Y_m}{\partial x_j}\right) + \omega_m MW_m \qquad (4)$$

$$\omega_m = \xi_m\, r \qquad (5)$$

Eq. (4) is applied for 6 species and nitrogen mass fraction is balanced. Eq. (6) states the conservation of energy.

$$\frac{\partial(\rho c_p V_j T)}{\partial x_j} = \frac{\partial}{\partial x_j}\left(K\frac{\partial T}{\partial x_j}\right) + \rho_{cat}\,\Delta H\, r \qquad (6)$$

Since the chemical reaction occurs only on the internal walls of the microchannel, the source terms in Eqs. (4) and (6) are zero for internal computational cells.

Fluid density is computed from the ideal gas equation for a multi-component mixture using the working pressure and temperature of the reactor. The effective mass diffusion coefficient is computed from the binary diffusion equation [9]. The constant pressure specific heat, dynamic viscosity and thermal conductivity of the gas mixture are computed using the local mixture properties [10].

Chemical kinetics

Various authors have reported kinetic rate equations of the water-gas shift reaction for diverse catalysts and operating conditions. Phatak *et al.* [11] presented a kinetic rate on Pt supported on ceria and alumina which have the operating temperature range of 180-345°C and is applicable at ambient pressure. Also Radhakrishnan *et al.* [12] reported a kinetic rate equation for Pt/Re catalysts supported on ceria-zirconia oxides for hydrogen purification which is generated from natural gas. In another investigation Jacobs *et al.* [13] proposed kinetic rate equation for Pt/CeO_2 fuel processing catalysts for fuel cell applications.

In the present article the WGS kinetic rate on 0.5 wt% Pt/TiO_2 catalyst is used that was developed by Kalamaras *et al.* [7] and given by Eq. (7). The kinetic rate is an empirical power law which is fitted to experimental reaction rates and is valid in the temperature range of 200-270°C.

$$r = k \times P_{CO}^{a} \times P_{H_2O}^{b} \times P_{CO_2}^{c} \times P_{H_2}^{d} \times (1-\beta)$$
$$k = k_0 \times \exp\left(\frac{-E_a}{R_u \times T_{work}}\right) \qquad (7)$$

Here r is the reaction kinetic rate (mol s^{-1} g_{cat}^{-1}), k is the rate constant, and P_i is the partial pressure of species i (atm). Also a, b, c, and d in this equation are the reaction orders with respect to CO, H_2O, CO_2, and H_2, respectively, β is the approach to equilibrium which is presented in Eq. (8), and k_0 is the pre-exponential factor. Moreover, E_a is the apparent activation energy of reaction (kJ $kmol^{-1}$) and T_{work} is the reaction temperature (K).

$$\beta = \frac{P_{CO_2} \times P_{H_2}}{K_{equil} \times P_{CO} \times P_{H_2O}} \qquad (8)$$

$$K_{equil} = \exp\left(\frac{5693.5}{T} + 1.077\ln T + 5.44\times10^{-4}T - 1.125\times10^{-7}T^2 - \frac{49170}{T^2} - 13.148\right) \qquad (9)$$

Reaction orders are tabulated in Table 2. The catalyst specific area is taken at 41 $m^2\,g^{-1}$ [7].

Solution method

In order to solve the non-linear coupled equations of mass, momentum, energy and chemical species conservation, a CFD code is developed based on the finite volume method, in which the SIMPLE algorithm is applied for pressure–velocity linkage. Further information regarding the solution method can be found in [8].

The following boundary conditions are implemented for the solution of the equations. At the microreactor inlet, velocity, pressure, temperature and composition of the gas mixture are given. The reactor walls are permeable due to surface reactions; thus, source terms are used for species at the walls. Furthermore, the walls are assumed to be isothermal. Considering the Knudsen number, the flow regime is continuum and so the no-slip condition is applied at the microreactor walls. The outflow mixture is discharged to a PROX reactor at ambient pressure and the gradients of all other dependent variables are set to zero at the microreactor outlet.

Results and discussion

The present model is developed to investigate the behavior of a water-gas shift surface microreactor for a mobile fuel cell application. For the conditions and dimensions considered in the present study, no compatible data, either numerical or experimental, is available for direct validation of the results. A parametric study is performed to investigate the effect of certain parameters on CO conversion and to obtain optimum geometry and operating conditions of the microreactor.

Baseline simulation

In this section a single microchannel is considered with the geometrical characteristics and operating conditions as tabulated in Table 1.

Contours of the mole fractions of carbon monoxide and hydrogen along the channel mid-plane are depicted in Fig. 2. It should be noted that the length scale in the x-direction is different than that in the z-direction in these figures due to large difference between the channel length and channel width (L / H = 160). It is also worth mentioning that the gradients in species mole fractions along the z-direction are not visible in these figures due to the difference in length scales used in the two directions. It is clear that CO molar consumption is equal to H_2 molar formation in this reaction. Considering the fact that the water-gas shift reaction is molar balanced, the molar mass of the mixture will remain constant through the microchannel. Hence, the N_2, CH_4 and O_2 mole fractions are constant along the channel.

As seen in these figures, H_2 generation and CO removal occur mainly in the primary half of the microchannel length. Reduction of water and carbon monoxide concentrations along the microchannel will lead to a reduction in the water-gas shift reaction rate. Consequently, CO conversion decreases considerably towards the end of the microchannel in comparison with the first half.

Plotted in Fig. 3 are the variation of the mole fractions of CO and other species along the channel. It is observed that CO mole fraction gradient does not approach zero at the channel exit; therefore, this microchannel length will not suffice for reducing the CO concentration to its minimum thermodynamic equilibrium value.

Parametric study

Numerous simulations are performed to inspect the influence of significant parameters on CO conversion. These parameters include channel geometry (cross-sectional aspect ratio) and length, gas space velocity, and feed gas temperature and steam content. To investigate the effect of each parameter, that parameter is varied while the others are held constant.

The aspect ratio of the micro-channel cross section is first varied from 1 (square channel) to 2, while the area of the channel is held constant at 10^{-8} m^2. Plotted in Fig. 4(a) is CO conversion for several aspect ratios with other parameters at their baseline values (given in Table 1). It demonstrates that an increase in the channel aspect ratio will increase CO conversion. This is a sound result, because increasing the aspect ratio of the channel cross section (while keeping the cross-sectional area constant) will increase its surface area. Since the surface density of catalyst is constant, therefore the wider channel provides more reaction sites which leads to an increase in the conversion. However, it should be noted that an increase in the channel surface area also means more catalyst usage. Therefore, it is rational to inspect CO conversion per catalyst area for different aspect ratios. The calculated CO conversion is divided by the catalyst surface area for each case. These results are shown in Fig. 4(b). It is observed that the maximum CO conversion per unit catalyst surface area is obtained for a square cross section (aspect ratio of unity). Therefore, the square cross section is used in the following simulations.

The present water-gas shift reaction rate equation is applicable in the temperature range of 200-270°C, as stated previously. Thus, to study the effect of feed gas temperature on CO conversion, the temperature is varied in this range. In Fig. 5 CO conversion is plotted versus reaction temperature for 20 mm reactor length and five different gas space velocities. It is observed that for all gas space velocities, CO conversion increases significantly with increasing the temperature. The water-gas shift rate equation also shows that the reaction rate will increase with the temperature. However, chemical equilibrium dictates higher CO levels at higher temperatures, leading to lower reactor efficiency.

As mentioned before, the variation of CO conversion is a function of gas space velocity, which is now discussed in greater detail. Fig. 5 reveals that as the space velocity is decreased, the conversion increases. By reducing the space velocity (by decreasing the inlet flow rate) from 5000 h^{-1} to 1000 h^{-1}, conversion of CO improves considerably. This is a reasonable outcome considering the fact that by decreasing the gas space velocity the residence time will increase. It is also observed that at lower temperatures the effect of reducing gas space velocity on CO conversion increase, is greater; on the other hand, as the temperature is increased, CO conversion approaches the same value for different gas space velocities.

The effect of channel length on CO conversion is next investigated. For this purpose, simulations are performed for three different channel lengths of 16, 18 and 20 mm. Fig. 6 illustrates CO conversion as a function of feed gas temperature for three different channel lengths and three gas space velocities. It is observed that by increasing the channel length, CO conversion in increased. However, the effect of increasing the channel length on CO conversion shows little dependence on the temperature.

It is obvious that increasing the channel length will increase the catalyst surface area. Consequently, CO conversion is expected to increase (up to a limit). This is a similar effect to that of reducing gas space velocity by reducing the reactant flow rate. However, it must be noted that increasing the channel length of the WGS reactor in a micro-fuel processing device may not be the best option for mobile applications, since volume is of great concern. Therefore, to properly design a water-gas shift microreactor in such applications, both of these effects should be considered concurrently to obtain the optimum conditions for CO conversion.

It is also noted that for any given reactant flow rate, a particular channel length exists beyond which no considerable improvement of CO conversion is observed. This effect is represented by Fig. 7 in which CO conversion at 270°C and GHSV of 1000 hr^{-1} is plotted versus channel length in the range of 16 to 26 mm. It is observed that as the channel length is increased from 22 to 26 mm, CO conversion changes only slightly. More specifically, increasing the channel length by 4 mm will result in an increase of less than 0.5% in CO conversion, while the overall reactor volume and catalyst usage are increased by 18%. It would therefore be reasonable to make a compromise between conversion and reactor volume, and limit the channel length to 22 mm for the present design.

The effect of feed gas steam content is next studied. The inlet steam mole fraction is varied from 14% to 26%, while other conditions are held at their baseline values (including other species' molar flow rates). The calculated CO conversion for these cases is plotted in Fig. 8(a). It is observed that by increasing the steam content of the mixture, CO conversion is increased. This is in agreement with that expected from thermodynamic equilibrium. However, as the water vapor concentration is increased, the hydrogen concentration is decreased. Fig. 8(a) demonstrates that increasing the water content beyond 24% will have little effect on CO conversion, while as a result of the added water hydrogen concentration will decrease (due to dilution). This is while hydrogen concentration must be kept as high as possible in PEM fuel cell applications. However, it should be noted that the fuel cell operates at much closer to ambient temperatures than 270°C, where much of the water will have condensed out. Thus, considering hydrogen concentration on a dry basis is more reasonable here. Fig. 8(b) reveals that H_2 concentration increases slightly by increasing the inlet steam content. Nonetheless, this change in H_2 concentration (on a dry basis) is very small; thus, one can assume that increasing the inlet steam content will have virtually no effect on the outlet hydrogen concentration on a dry basis. One should note that an increase in the inlet steam content would require more thermal energy upstream of the WGS reactor to vaporize the required water.

Optimum conditions

In order to obtain the optimum conditions, a matrix of 32 case studies for 5 main factors with 3 central points is constructed [14]. As shown in Table 3, the effects of major parameters on CO conversion are considered in this optimization procedure. The steam content has been varied here in 21-26% range because of the desirable effect of higher steam contents on CO conversion.

An analysis of variance for the 32 case studies is given in Table 4. Based on percentage contribution relative to the total sum of squares the relative importance of these five main effects and the one interaction (AC) are discovered. This analysis shows that A (91.03%) has a dominate effect on CO conversion, followed by C (4.02%), the AC interaction (2.30%), D (1.65%), B (0.72%), and E (0.26%), where the letters indicate the coded factors (given in Table 4) and the values given in parentheses denote the percentage of total variability.

CO conversion can be expressed by the following regression equation (in terms of coded factors):

$$\text{CO conv.} = 0.452 + 0.22\,A + 0.02\,B - 0.046\,C + 0.03\,D + 0.012\,E + 0.035\,AC \tag{10}$$

Note that coded values are between -1 and +1. It is clear that CO conversion is increased when increasing the effects with positive regression coefficients (reaction temperature, channel length, steam content, and aspect ratio of channel), and is reduced when increasing the effects with negative value (gas space velocity).

Once the effects of major contributing factors were examined using the 32 case studies, a statistical procedure based on two-independent factors is implemented in order to achieve optimization of CO conversion. The two major factors must be chosen from the five main factors based on their important variability. Thus, the reaction temperature and gas space velocity, with the highest percent contributions, are selected.

The values of the other factors are set based on the results from our previous parametric study described in section 3.2. More specifically, the values of B, D, and E are set to 1, 0, and 0, respectively, in Eq. (10). Then the values of A and C are set to -1, 0, and 1 in turn, to obtain a three-dimensional response surface curve as plotted in Fig. 9. The statistical model is used to understand the behavior of CO conversion and to determine optimum value of the factors required for optimal CO conversion. The optimal conditions for CO conversion in the WGS reactor, estimated by the surface analysis, is 24% H_2O in the feed stream at a reaction temperature of 270°C, and gas space velocity of 1000 h^{-1} in a square channel with length of 22 mm. The predicted CO conversion under the above conditions is about 70%.

In order to verify the predicted conversion, the optimal reaction conditions are applied to the factors. A comparison between the predicted conversion and the calculated one from the actual numerical simulation reveals that the estimated values of the statistical analysis are well fitted to the simulation results.

The geometry and operating conditions of the optimized case are presented in Table 5. The outlet CO mole fraction in this case is 1.4% (dry basis) which is suitable for the treated gas to be fed to a typical PROX reactor [15].

Nonetheless, the optimum conditions should be developed with regards to the desired application and other fuel processor main components. For complete hydrogen generation and purification a reactor train is required. The reactor train consists of an ATR reactor, a WGS reactor and a PROX reactor.

Conclusions

In this investigation a three-dimensional CFD code is developed to model a single-channel water-gas shift surface microreactor for a mobile PEM fuel cell application. The reaction kinetics implemented is for 0.5 wt% Pt/TiO_2 catalyst in the temperature range of 200-270°C. The effects of key parameters are investigated, which include feed gas temperature and water content, channel geometry (cross-section aspect ratio) and length, and gas space velocity. The results of the simulation are as follows:

• A square cross section is the optimum geometry for the channel (based on Pt catalyst usage).

• Raising the temperature in the range of 210-270°C results in more carbon monoxide removal and hydrogen generation.

• Reducing the gas space velocity in the range of 1000-5000 h^{-1} will increase carbon monoxide conversion.

• Increasing the channel length in the range of 16-22 mm yields more CO conversion and hydrogen purification; however, increasing the channel length beyond 22 mm has no significant effect on the water-gas shift microreactor performance.

• Although increasing the steam content in the influent composition in the range of 21-26% may results in more CO conversion, hydrogen concentration in the treated mixture will decrease due to dilution.

In order to obtain the optimum conditions for the water-gas shift microreactor, a statistical model is used. The optimum operating and design conditions are found as follows: feed gas temperature at 270°C with 24% steam content and gas space velocity of 1000 h^{-1} in a square cross section microchannel with the hydraulic diameter of 100 μm and 22 mm length. An optimized water-gas shift microreactor is a vital part of the fuel processing device for low-power PEM fuel cells in mobile applications.

Nomenclature

a	CO reaction order
A	coded value for effect of temperature
AC	coded value for composite effects of temperature and gas space velocity
b	H_2O reaction order
B	coded value for effect of channel length
c	CO_2 reaction order
C	coded value for effect of gas space velocity
c_p	constant pressure specific heat (kJ kg^{-1} K^{-1})
d	H_2 reaction order
D	coded value for effect of steam content
D_{eff}	effective mass diffusion coefficient
E	coded value for effect of channel aspect ratio
E_a	apparent activation energy (kJ $kmol^{-1}$)
ΔH	reaction enthalpy (kJ $kmol^{-1}$)
K	thermal conductivity (W m^{-1} K^{-1})
k	reaction rate constant
k_0	pre-exponential factor
K_{equi}	equilibrium constant
MW	molar mass (kg $kmol^{-1}$)
p	pressure (Pa)
P	partial pressure
r	total reaction rate (kmol m^{-3} s^{-1})
R_u	universal gas constant (J $kmol^{-1}$ K^{-1})
T	temperature (K)
V	velocity component (m s^{-1})
x	coordinate (m)
Y	species mass fraction

Greek symbols

ω	species surface reaction rate (kmol m^{-2} s^{-1})
μ	kinematic viscosity (N s m^{-2})
ξ	stoichiometry coefficient
ρ	Density (kg m^{-3})
β	approach to equilibrium

Sub or superscripts

cat	catalyst
eff	effective
equi	equilibrium
i	index
j	index
m	m^{th} species or reaction
mix	gas mixture
work	working operation

References

[1] C. Song, Fuel processing for low-temperature and high-temperature fuel cells: Challenges and opportunities for sustainable development in the 21st century, Catalysis Today 77 (1-2) (2002) 17-49.

[2] J.M. Zalc, D.G. Löffler, Fuel processing for PEM fuel cells: transport and kinetic issues of system design, J. Power Sources 111 (1) (2002) 58-64.

[3] P. Giunta, N. Amadeo, M. Laborde, Simulation of a low temperature water gas shift reactor using the heterogeneous model/application to a PEM fuel cell, J. Power Sources 156 (2) (2006) 489-496.

[4] Y. Choi, H.G. Stenger, Water gas shift reaction kinetics and reactor modeling for fuel cell grade hydrogen, J. Power Sources 124 (2) (2003) 432-439.

[5] O.L. Ding, S.H. Chan, Water–gas shift reaction – A 2-D modeling approach, Int. J. Hydrogen Energy 33 (16) (2008) 4325-4336.

[6] J. Sun, J. DesJardins, J. Buglass, K. Liu, Noble metal water gas shift catalysis: Kinetics study and reactor design, Int. J. Hydrogen Energy 30 (11) (2005) 1259-1264.

[7] C.M. Kalamaras, P. Panagiotopoulou, D.I. Kondarides, A.M. Efstathiou, Kinetic and mechanistic studies of the water-gas shift reaction on Pt/TiO_2 catalyst, J. Catal. 264 (2) (2009) 117-129.

[8] M.H. Akbari, A. Sharafian Ardakani, M. Andisheh Tadbir, A microreactor modeling, analysis and optimization for methane autothermal reforming in fuel cell applications, Chemical Engineering J. 166 (3) (2011) 1116-1125.

[9] R.C. Reid, J.M. Prausnitz, B.E. Poling, The Properties of Gases and Liquids, McGraw-Hill, New York, USA, (1987).

[10] S.R. Turns, An Introduction to Combustion: Concepts and Applications, McGraw Hill, New York, USA, (2000).

[11] A.A. Phatak, N. Koryabkina, S. Rai, J.L. Ratts, W. Ruettinger, R.J. Farrauto, G.E. Blau, W.N. Delgass, F.H. Ribeiro, Kinetics of the water–gas shift reaction on Pt catalysts supported on alumina and ceria, Catalysis Today 123 (1-4) (2007) 224-234.

[12] R. Radhakrishnan, R.R. Willigan, Z. Dardas, T.H. Vandersput, Water gas shift activity and kinetics of Pt/Re catalysts supported on ceria-zirconia oxides, Appl. Catal. B: Environmental 66 (1-2) (2006) 23-28.

[13] G. Jacobs, L. Williams, U. Graham, G.A. Thomas, D.E. Sparks, B.H. Davis, Low temperature water–gas shift: in situ DRIFTS-reaction study of ceria surface area on the evolution of formats on Pt/CeO_2 fuel processing catalysts for fuel cell applications, Appl. Catal. A: General 252 (1) (2003) 107-118.

[14] D.G. Montgomery, Design and Analysis of Experiments, Wiley, New York, USA, (2001).

[15] M. Vahabi, M.H. Akbari, Three-dimensional simulation and optimization of an isothermal PROX microreactor for fuel cell applications, Int. J. Hydrogen Energy 34 (3) (2009) 1531-1541.

Table 1. Operating conditions and inlet mole fractions for the baseline case.

Parameter	Value
Space velocity (h^{-1})	1,000
Channel length (mm)	16.0
Channel hydraulic diameter (μm)	100.0
Inlet temperature (°C)	270
Inlet pressure (Pa)	101,325
CH_4 mole fraction (%)	10.0
O_2 mole fraction (%)	0.0
CO mole fraction (%)	4.0
H_2O mole fraction (%)	21.5
H_2 mole fraction (%)	30.5
CO_2 mole fraction (%)	17.0
N_2 mole fraction (%)	balance

Table 2. Kinetic parameters.

Catalyst	a	b	c	d	k_0	E_{app} (kJ mol^{-1})
0.5 wt% Pt/TiO_2	0.5	1.0	0.0	-0.7	0.31	45.2

Table 3. Variables in coded and actual units for the 32 case studies.

Factors	Variables	Low (-1)	Medium (0)	High (+1)	Unit
A	Temperature	210	240	270	°C
B	Channel length	16	19	22	mm
C	Gas space velocity	1000	3000	5000	h^{-1}
D	Steam content	21	23.5	26	%
E	Aspect ratio	1	1.5	2	-

Table 4. Analysis of variance of significant terms for the 32 case studies.

Parameter	Effect estimate	Sum of squares	Degrees of freedom	Mean square	Percent contribution
A	0.2202	1.551	1	1.551	91.03
B	0.0196	0.0122	1	0.0122	0.72
C	-0.0463	0.0685	1	0.0685	4.02
D	0.0297	0.0281	1	0.0281	1.65
E	0.0119	0.0044	1	0.0044	0.26
AC	0.0350	0.0392	1	0.0392	2.30
Residual		0.0016		0.000064	
Total		1.7050			

Table 5. Operating conditions and outlet mole fractions for the optimum case.

Parameter	Value
Channel length (mm)	22.0
Channel cross-sectional aspect ratio	1.0
Channel hydraulic diameter (μm)	100.0
Inlet temperature (°C)	270
Inlet steam mole fraction (%)	24
Space velocity (h^{-1})	1,000
Inlet pressure (Pa)	101,325
Outlet CO mole fraction (% dry basis)	1.4

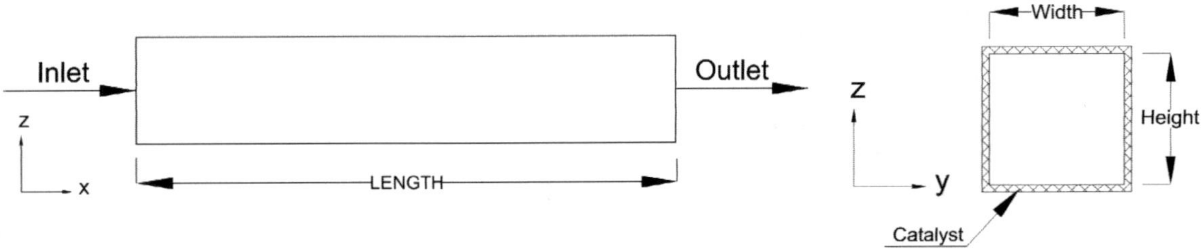

Fig. 1. Schematics of the microchannel reactor.

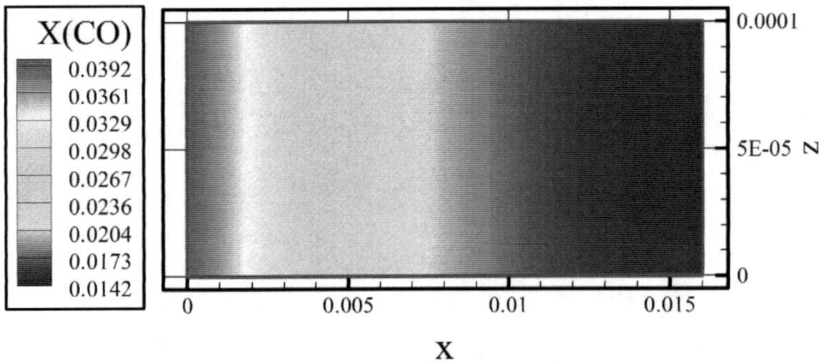

(a)

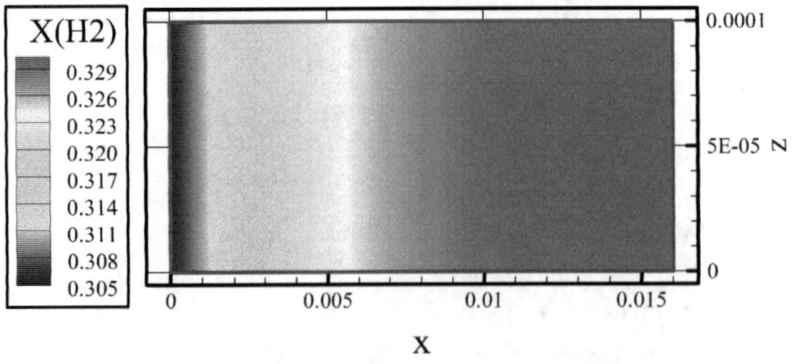

(b)

Fig.2. Contours of species mole fractions along the microchannel mid-plane for the baseline conditions: (a) carbon monoxide (b) hydrogen.

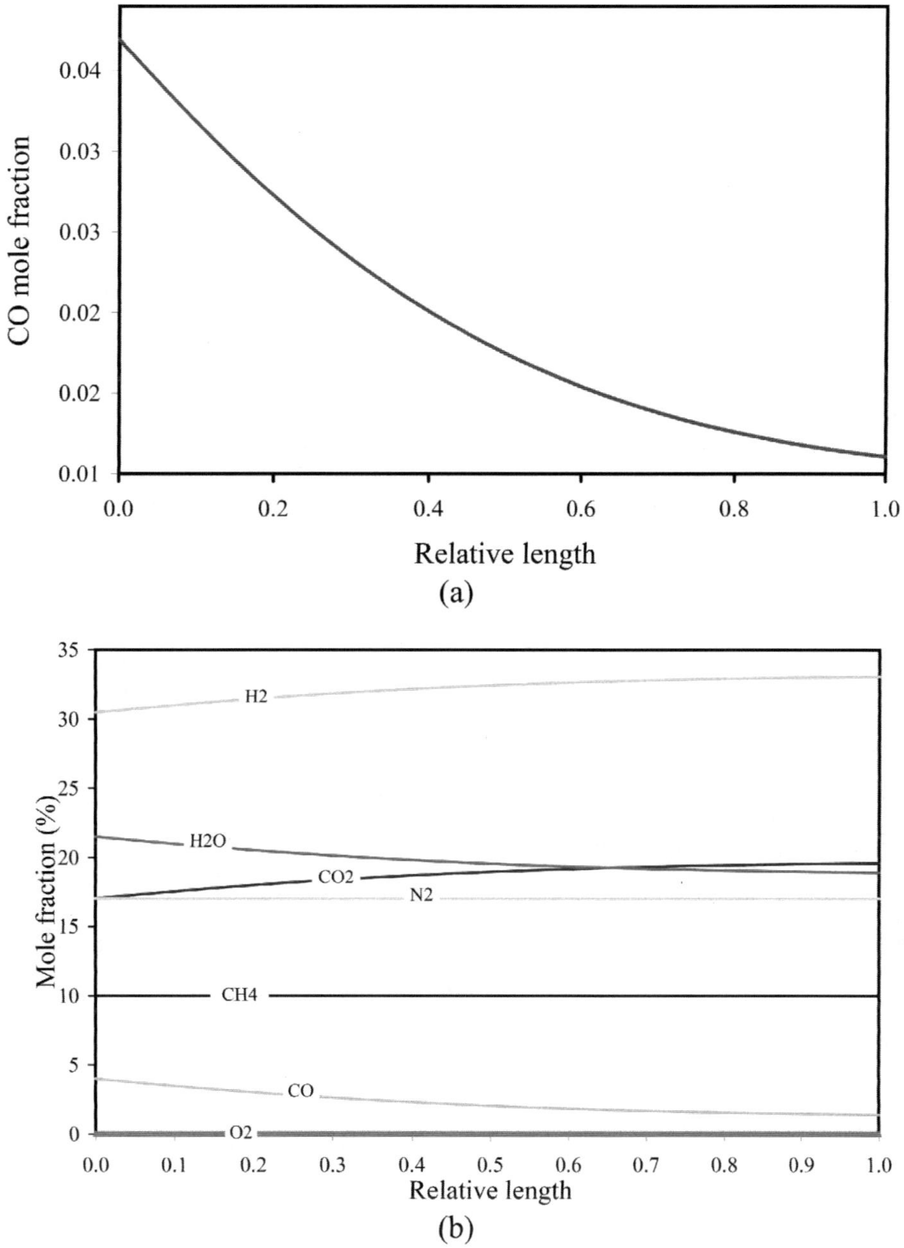

Fig. 3. (a) CO mole fraction, and (b) species mole fraction variations along the microchannel for the baseline conditions.

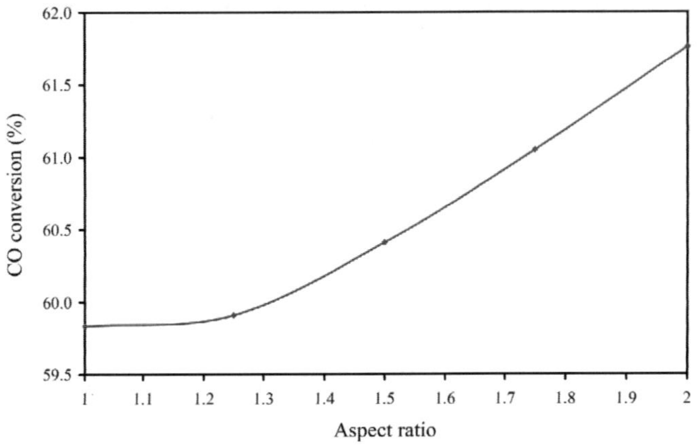

(a)

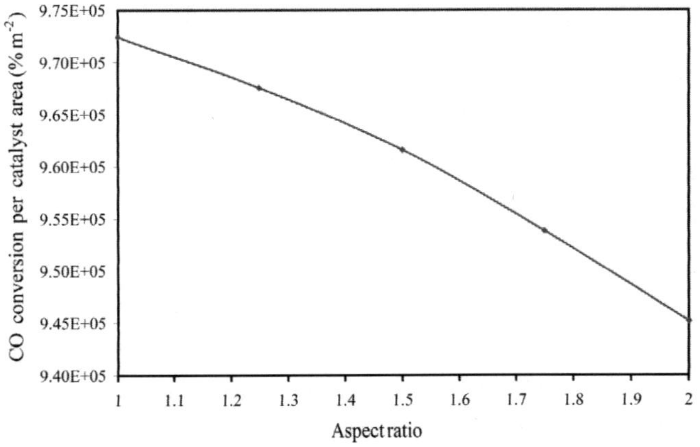

(b)

Fig. 4. Effect of channel aspect ratio on (a) CO conversion, and (b) CO conversion per catalyst area.

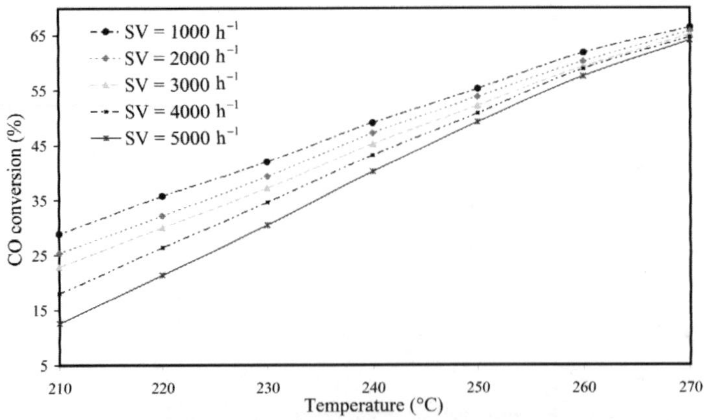

Fig. 5. Effect of inlet temperature and gas space velocity on CO conversion for 20 mm channel length.

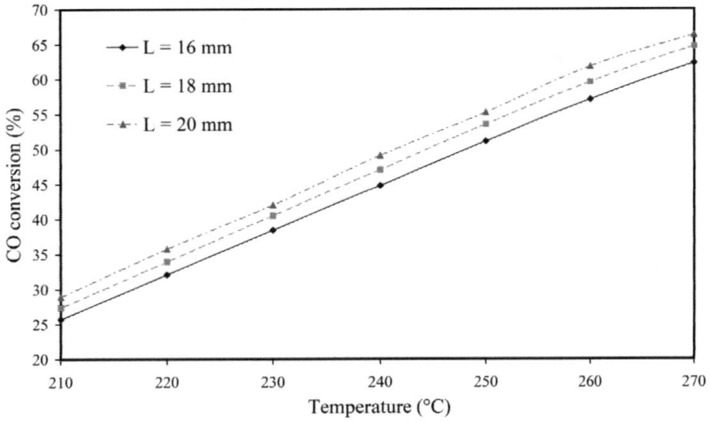

(a)

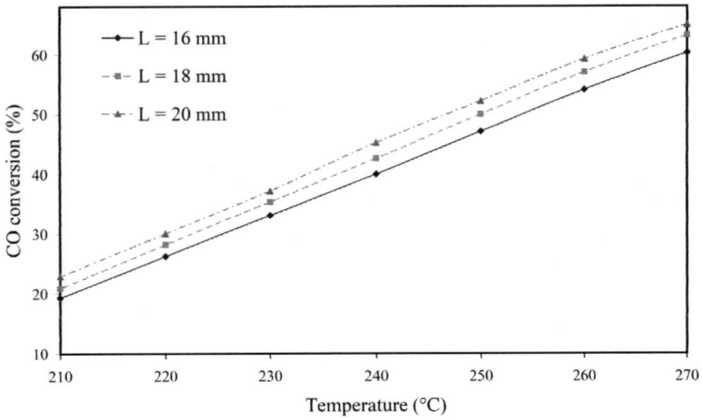

(b)

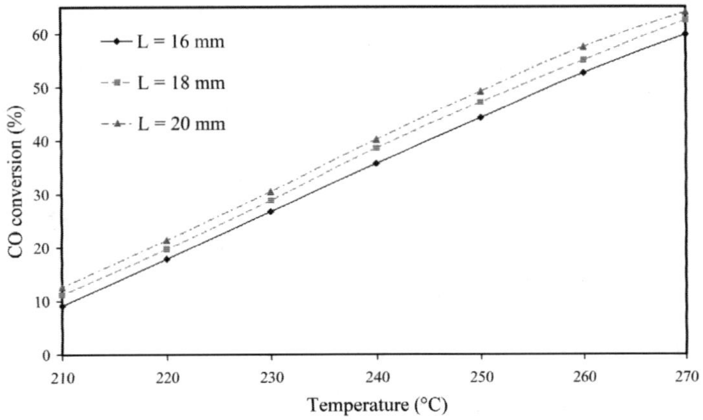

(c)

Fig. 6. Effect of inlet temperature and channel length on CO conversion for different gas space velocities: (a) 1000 h^{-1}, (b) 3000 h^{-1}, (c) 5000 h^{-1}.

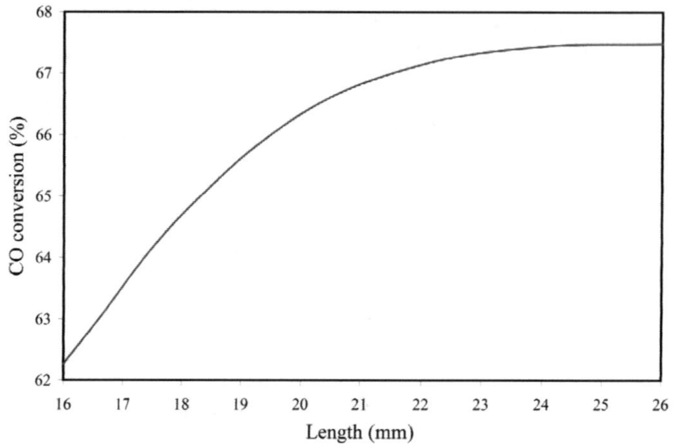

Fig. 7. Effect of channel length on CO conversion.

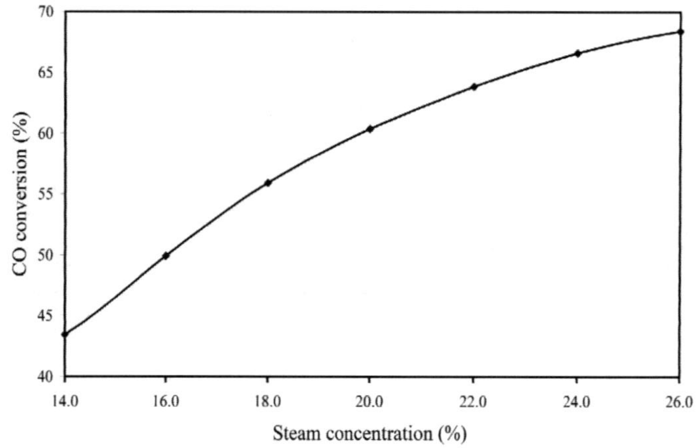

(a)

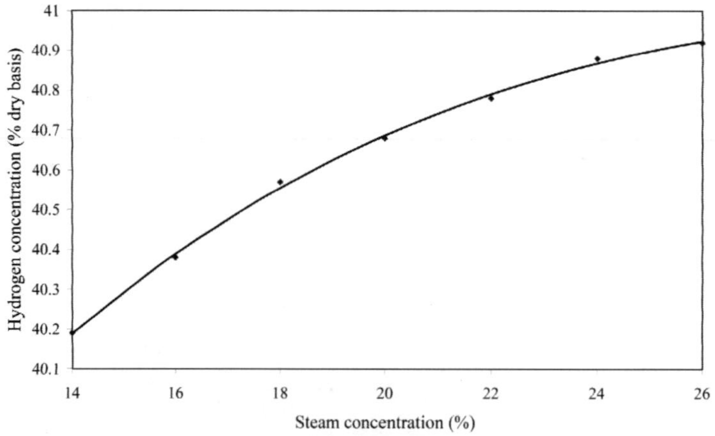

(b)

Fig. 8. Effect of steam concentration on (a) CO conversion, and (b) H_2 concentration.

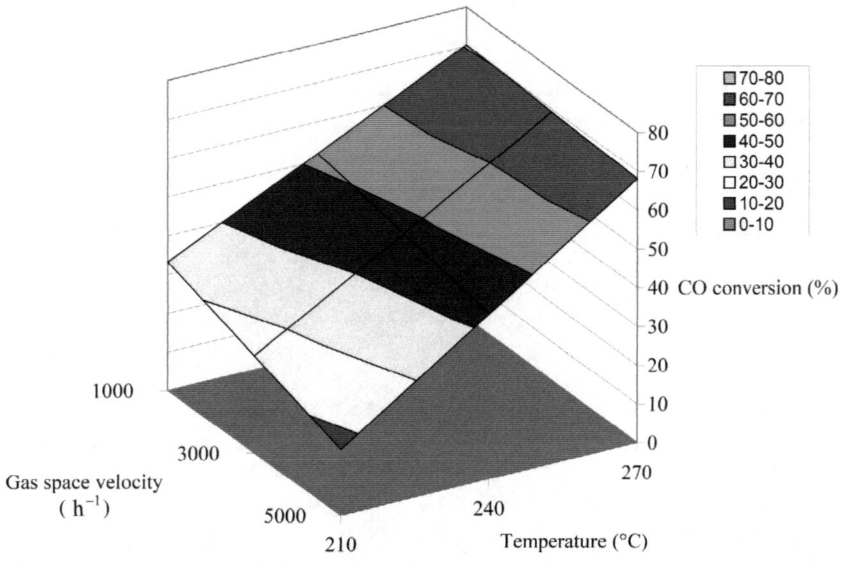

Fig. 9. Contour plot of the response for optimal CO conversion.

Microplasma Reforming for Fuel Cell Feed

Peter J. Lindner[1,2] *and R. S. Besser*[2]

(1)Chemical Engineering, Manhattan College, Riverdale, NY
(2)Chemical Engineering and Materials Science, Stevens Institute of Technology, Hoboken, NJ

Abstract: Microchemical reactor systems are known to intensify chemical reactions by significantly improving heat and mass transfer when compared to conventional sized reactors. Plasmas have been used for chemical reactions for many years, but novel microplasmas (confined with a sub-mm critical dimension) offer strong hydrocarbon conversions and provide a favorable reaction environment due to high densities of reactive ions, radicals, and electrons. Operation at atmospheric pressure and near-room temperature make microplasmas convenient for implementation. We present a novel approach to further enhance chemical reaction intensification through the use of a microplasma reforming reactor.

The results of reforming experiments with a microfabricated microplasma reactor for producing hydrogen-rich product stream are reported. These experiments demonstrate the potential of this approach for energy applications such as combustion or fuel cells where concerns for efficiency and portability are important. To produce the microplasma reactors silicon microfabrication was used. Photolithography and advanced deep reactive ion etching was used to produce microplasma reactors with microchannels widths varying from 25 to 250 μm. A micro hollow cathode discharge (MHCD) configuration was designed with a void space in the cathode region with depths varying from 25 to 250 μm. In these reactors, the void space or hollow was provided by the microchannel, which served as the center of plasma generation and the reaction zone within the reactor.

We examined the conversion of various simple hydrocarbons and alcohols into hydrogen. It was found, consistent with literature reports, that an inert carrier such as nitrogen promotes decomposition of the organic species, encouraging the formation of hydrogen gas. The experimental approach taken involved DC or pulsed DC high voltage power supplies interfaced to the reactor chip for plasma generation. The gaseous feed was set using mass flow controllers. Gas chromatography and mass spectrometry were implemented for analyzing the composition of the stream. A computer interfaced with LabView software allowed simultaneous control and acquisition of data from the experiment.
Hydrocarbon conversions up to 50% were observed, as were various products depending on reactant and the conditions of reaction. Under certain conditions, we observed selectivity of the reaction toward CO over CO_2; this behavior provides a compelling research direction for processing with less global environmental impact than conventional catalytic reforming processes which routinely yield CO_2.

Experimental data will show that our reaction model can explain the key mechanisms that occurred; analysis of the data shows that reactions rates in the microplasma environment surpass those of conventional plasma reactors. The realization of an energetically self-sustaining reforming process through extrapolating system modifications is possible with this model. We conclude that microplasma activation overcomes inherent limitations imposed by catalytically-activated reactions such as catalysis deactivation through poisoning or carbon formation and the need for extended startup times to reach steady state. Microplasma chemical processing is a viable approach for hydrocarbon reforming and merits consideration for a variety of chemical reaction applications.

Keywords: Microplasma, Reforming, Hydrocarbons, Portable Fuel Cell, Carbon Dioxide

1. Introduction

Limitations of batteries have confined the ability of portable technology. Since many devices require small, compact, and portable power sources, seeking an alternative to traditional lithium ion batteries will enhance and extend our technological capabilities. Portable fuel cell technologies have higher energy densities and longer operating lifetimes than current battery technologies [1]. Despite this advantage, fuel cells have not taken over as the primary source of portable power due to the lack of readily available hydrogen. Since the mass production and transportation of hydrogen is impractical and dangerous; hydrogen can be made available by reforming hydrocarbon fuels near the fuel cell location. The development of a microscale reactor to perform reforming on site with a portable fuel cell can aid in developing a portable power device with more capabilities than current technology.

Reforming hydrocarbons is commonly done through catalysis in large scale facilities. For portable devices using a catalyst becomes troublesome due to issues such as sulfur poisoning of catalyst sites, carbon deposits blocking catalyst sites, and long start up times. Blocked and poisoned catalyst sites are solved by high temperature regeneration processes (>600°C) which are impractical for a portable application. Plasma chemical systems have shown promise in reforming various hydrocarbons, but these devices often require high input electrical power and continue to have issues with system efficiencies [2]. These devices are also cumbersome, as they require additional vacuum equipment.

A more streamlined approach that may improve efficiencies is promoting the reforming reaction through the use of microplasma devices. Microplasmas improve on current plasma reactors as they can operate at atmospheric pressure, require much lower power, and have higher electron densities. This work aims to study an all in one microplasma reformer by developing mass and energy balance data for various hydrocarbons under multiple conditions. Due to the increase number of parameters and variability in the reactions a series of carbon dioxide experiments were first done to determine some of the device characteristics.

2. Methods and Materials

The microplasma reactors developed for these reactions are microelectormechanical systems (MEMS). They were produced using common nano/microfabrication techniques such as photolithography and deep reactive ion etching. The type of microplasma devices developed were micro hollow cathode discharge (MHCD) devices. MHCDs were chosen for fabrication because of the high electron density, an indicator of reactivity, measured from MHCDs even when compared to other microplasma devices [3-5]. A schematic and image of the device is shown in figure 1.

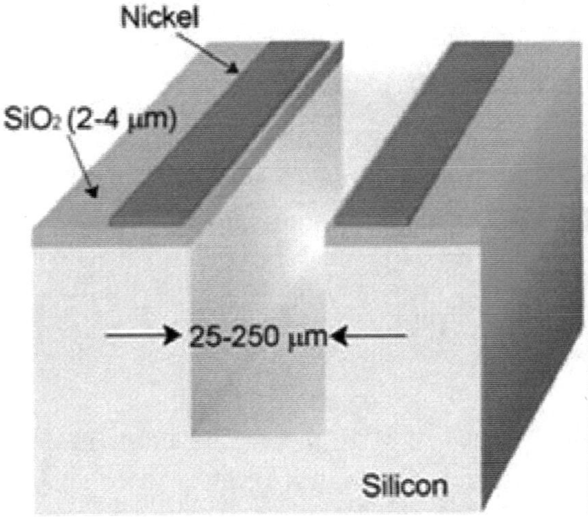

Figure 1. Cross section of microplasma device (left) and image of the reactor (right) [6, 7].

Individual devices were placed into a custom made acrylic/lexant holder which had fluidic connections. Inlet reactants and carrier gases were fed from mass flow control(s). The outlet connected to a 100 microliter sample loop, which sent a sample to a Varian, Inc CP-4900 gas chromatography (GC) unit approximately every two minutes. After the GC, the outlet stream continuously fed a Stanford Research Systems quadruopole mass spectroscopy (QMS) model QMS200. The microplasma device was powered using a Stanford Research System high voltage power supply (model PS310). The cathode of the device was grounded and DC or pulsed DC power was supplied to the anode. The current on the power supply was limited in a range of 1 to 20 mA for various experiments. In order to develop the microplasma, the voltage from the anode to the cathode was increased till a breakdown voltage was reached. The breakdown voltage (~500 V) is a function of the gas, pressure, and electrode distance. After reaching the breakdown voltage and the establishing the microplasma, the voltage generally drops to a much lower value (~150 V), and the current increased to the set limit.

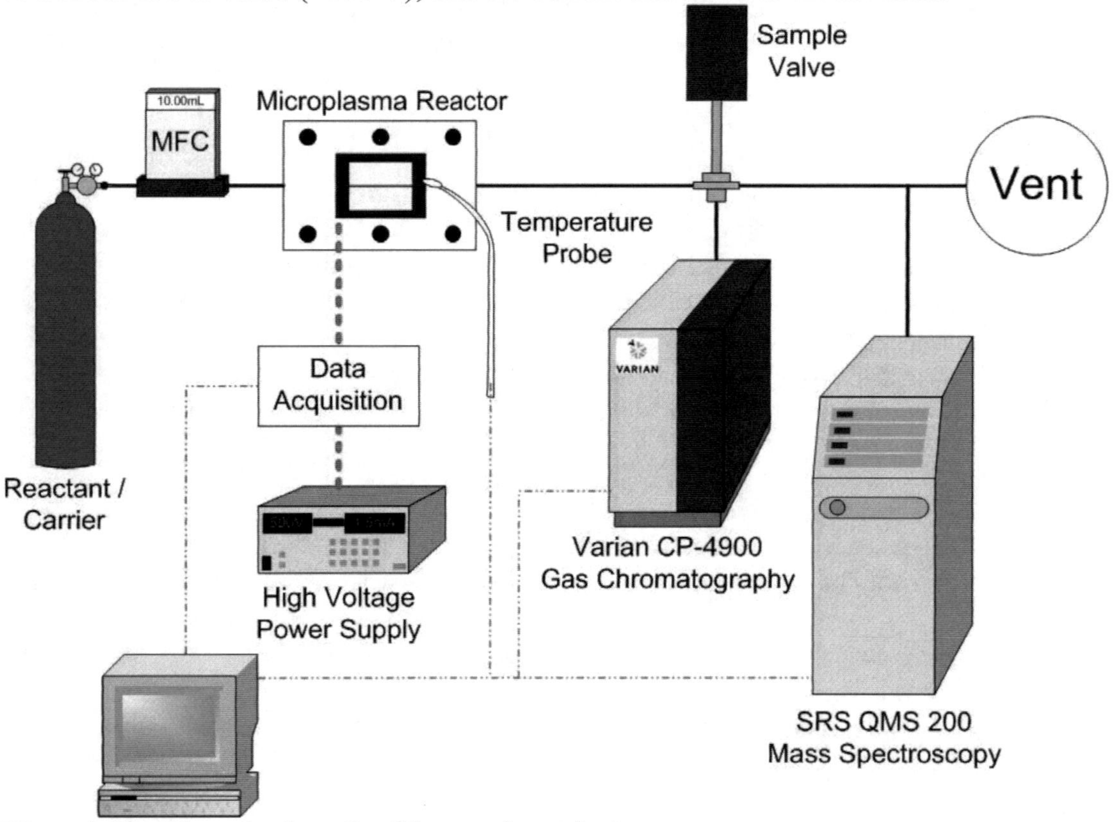

Figure 2. Flow through schematic of the experimental setup.

This setup has been tested using hydrocarbon fuels such as methane, butane and methanol and studies have been published determining reforming capabilities and mathematical models of the reforming reactions [6, 8, 9]. Current experiments using carbon dioxide will be discussed further because of the related findings help increase understanding and avenues of improvement for the devices themselves.

3. Results and Discussion

For our hydrocarbon experiments, a carrier gas was needed to establish a reactive environment. For these experiments nitrogen was used due to its availability in the air. Pure air was not used, to minimize side reactions with oxygen. Nitrogen has shown to facilitate chemical reactions because excitation of nitrogen into higher vibrational and metastable states aids activation of reforming processes [10].

For methane and methanol reactions we have reported the overall plasma reaction shown in equations 1 and 2.

$$1.0CH_4 + 0.05O_2 \rightarrow 1.7H_2 + 0.7C_{(S)} + 0.1C_2H_2 + 0.1H_2O + 0.05C_2H_6 \tag{1}$$

$$1.0CH_3OH \rightarrow 1.4H_2 + 0.6CO + 0.4H_2O + 0.2C_{(S)} \tag{2}$$

Both reactions exhibited minor traces of various hydrocarbons, which is why they do not balance exactly. All hydrocarbon reforming experiments with the tested devices have had issues with device lifetime, stability and consistency. To determine issues with device consistency a study was conducted running various microplasma chemistry experiments decomposing carbon dioxide into carbon monoxide and oxygen through the following reaction:

$$1CO_2 \rightarrow 1CO + 0.5O_2 \tag{3}$$

A series of experiments were tested using only carbon dioxide to establish an understanding of the device. By running with carbon dioxide the reaction was simplified. The system parameters were also reduced by removing the need for a carrier gas. The lifetime of the devices varied from 10 seconds to 180 minutes for these series of experiments. Inlet flow rate, current limit (power), and device dimensions (width, depth, oxide thickness, enhanced perimeter) were tested in this study. Mass and energy balances were determined for each experiment and the findings were compared. Experiments were also tested to understand the repeatability of an experiment. Individual devices were run multiple times under the same experimental conditions. Identical devices were also run under identical conditions to understand variability among equivalent devices

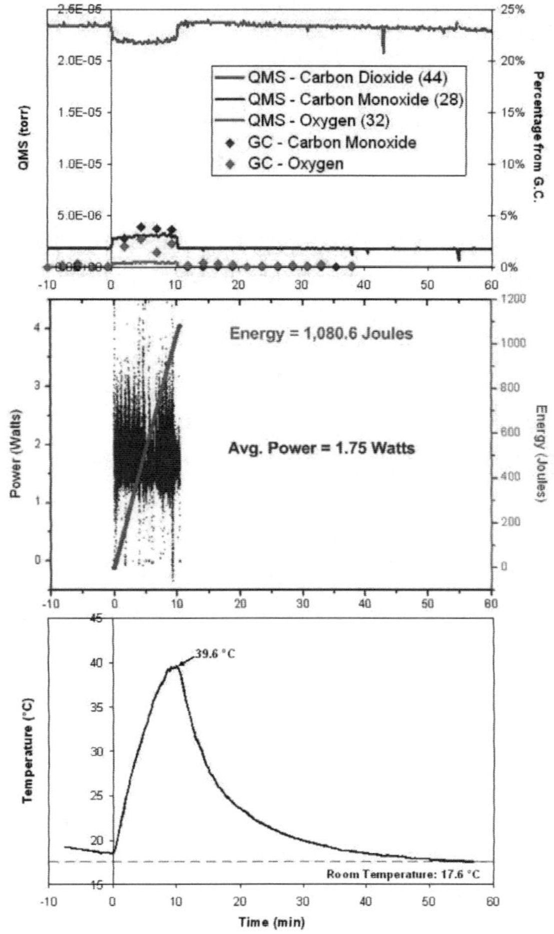

Figure 3. QMS, GC, power and temperature profiles for an example experiment of a microplasma reactor decomposing carbon dioxide for 10 minutes.

The example experiment in figure 3 shows the profile of the temperature, inlet electrical power, QMS and GC data. For this reaction an average of 3.6% of carbon monoxide was produced at a flow rate of 0.20 sccm, as well as 1.8% of oxygen with a flow rate of 0.09 sccm. The device was operated on an average with 1.2 watts of power. Although only a small fraction of the carbon dioxide was decomposed, the microplasma was found to be highly reactive.

For all of the experiments tested a linear trend was found relating the input electrical power into the microplasma device to the amount of carbon monoxide produced (shown in figure 4).

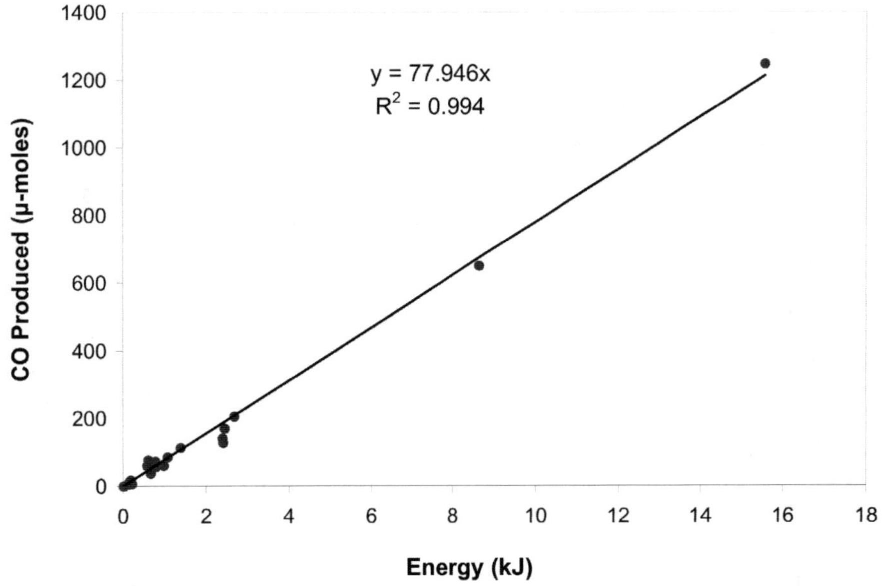

Figure 4. Input electrical energy verses the moles of carbon monoxide produced for various reactions under various conditions.

This linear trend shows that the microplasma device can control the reaction. This also shows that only a small fraction of the energy is being used for the chemical reaction as the inverse of the slope is much greater than the heat of reaction for equation 3 (283 kJ/mol). Approximately 2 to 3% of the input electrical energy is being used to perform the chemical reaction. Most of the energy is being wasted as joule heating of the device. This indicates that by reducing the effects of electrical heating the efficiency of the device could be improved.

Along with the device, the system itself can be improved and engineered. For example one major issue is that the microplasma device holder has an excess bulk volume of 0.3 mL where the reactor volume is at most 0.5 μL (600 times smaller). Reducing this bulk volume can improve the efficiency of the device by allowing more of the reactant to come in contact with the highly reactive microplasma environment. Assuming the calculated reaction rate for a set experiment, a mathematical model was developed allowing us to vary key parameters. Figure 5 shows the effect of conversion and efficiency when varying the bulk volume and the inlet flow rate.

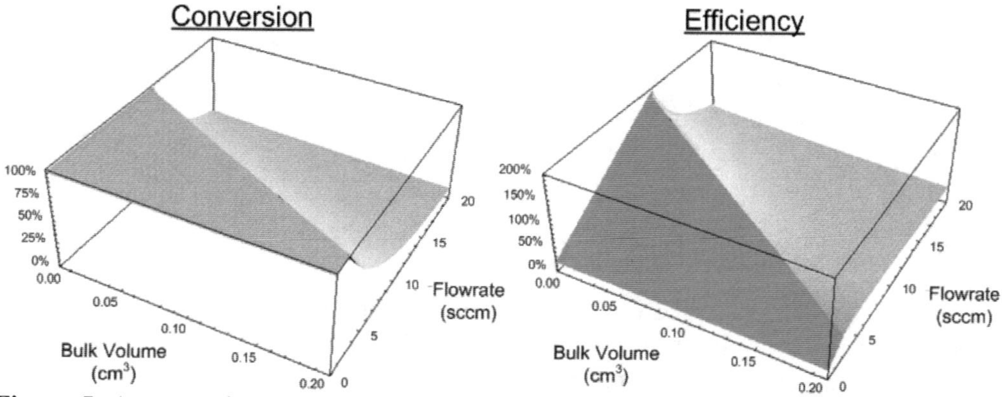

Figure 5. An example mathematical model of varying the flowrate and excessive bulk volume for a methanol reforming reaction[8].

The efficiency depicted in figure 5 is calculated using equation 4 directly below.

$$\eta = \frac{N_{H_2,Produced} * LHV(H_2) + N_{CO,Produced} * LHV(CO)}{E_{in}} \qquad (4)$$

Figure 5 shows that by reducing the flow rate the conversion increases, as the residence time would increase. It also shows that by reducing the bulk volume the efficiency would increase, as more of the reactant would come into the reactive zone. This also demonstrates that there is an ideal flow rate based on the bulk volume, due to the reaction rate. This plot demonstrates the untapped potential of microplasmas as a reforming device.

4. Conclusions and Future Work

Experiments with methane, methanol, and butane showed the feasibility of a microplasma device as a hydrocarbon reformer. These experiments each suffered from inconsistencies in power data, microplasma stability, and issues reaching steady state. Experiments with carbon dioxide have allowed us to calculate experiment variations as well as demonstrate the effects of varying the device dimensions, the devices current limit and the inlet flow rate (affecting residence time). Using mathematical models presented the further potential of these devices to be setup to run more efficiently.

To improve on microplasma reforming devices, further fabrication of new generations of devices should be developed, specifically with a sealed microchannel to remove the effects caused by the bulk volume in the device holder. These devices could also be fabricated to minimize joule heating effects by increasing the oxide thickness layer, and various fabrication improvements [Eden ref, cnf] which are key to minimizing the inlet power. Furthermore, the reactions with carbon dioxide have shown that this device could also be used for a variety of other chemical reactions, in this case, reforming carbon dioxide, a combustible fuel, from a greenhouse pollutant.

Acknowledgements

The authors wish to acknowledge the assistance and support of Sang Youp Hwang, George Wohlrab and Bruce Fraser. The support of the National Science Foundation under grant no NSF DGE-0742462 and the Army Research Office under grant no. W911NF-07-1-0118 are gratefully acknowledged.

References

[1] Kundu A, Jang JH, Gil JH, Jung CR, Lee HR, Kim SH, et al. Micro-fuel cells - Current development and applications. Journal of Power Sources. 2007;170(1):67-78.

[2] Petitpas G, Rollier JD, Darmon A, Gonzalez-Aguilar J, Metkemeijer R, Fulcheri L. A comparative study of non-thermal plasma assisted reforming technologies. International Journal of Hydrogen Energy. 2007;32(14):2848-67.

[3] Schoenbach KH, Verhappen R, Tessnow T, Peterkin FE. Microhollow cathode discharges. Applied Physics Letters. 1996;68(1):13-5.

[4] Kim GJ, Y.J. H, Iza F, Lee JK. Evolution of the electron energy distribution function in pulsed micro hollow cathode discharge. Computer Physicis Communications. 2007;177(131).

[5] Kim GJ, Iza F, Lee JK. Electron and ion kinetics in a micro hollow cathode discharge. Journal of Physics D: Applied Physics. 2006;39:4386-92.

[6] Besser RS, Lindner PJ. Microplasma Reforming of Hydrocarbons for Fuel Cell Power. Journal of Power Sources. 2010;196(21):9008-12.

[7] Besser RS, Najem A. Compact Hydrogen Generation through JP-8 Fuel Processing Using Microplasmas. Cornell NanoScale Facility Research Accomplishments. 2007;2006-2007.

[8] Lindner PJ, Besser RS. A Microplasma Reactor for Chemical Process Intensification. Chemical Engineering & Technology. 2012;35(7):1249 - 56.

[9] Lindner PJ, Besser RS. Hydrogen porudction by methanol reforming in a non-thermal atmospheric pressure microplasma reactor. International Journal of Hydrogen Energy. 2012;37(18):13338 - 49.

[10] Indarto A, Choi JW, Lee H, Song HK. Effect of additive gases on methane conversion using gliding arc discharge. Energy. 2006;31(14):2986-95.

(491b) Dry Reforming Assisted Nonthermal Plasma Catalytic Reactor

Mahammadunnisa Shaik[1], Manoj Kumar Reddy P[2], B. Ramaraju[2] and **Surbahmanyam Ch**[2], (1)IIT Hyderabad, Hyderabad, India, (2)Chemistry, IIT Hyderabad, Hyderabad, India

Dry reforming assisted nonthermal plasma catalytic reactor

Shaik Mahammadunnisa, P. Manoj Kumar Reddy, B. Ramaraju, Ch.Subrahmanyam

Department of Chemistry, IIT Hyderabad-502205, India

csubbu@iith.ac.in

The negative effect of global warming of carbon dioxide (CO_2) is well established and there is an immediate need to develop large scale CO_2 utilization techniques. Catalytic CO_2 reforming of natural gas also has drawbacks such as high reaction temperature and clogging caused by coke formation. Hence, alternative technologies for CO_2 utilization are much warranted [1]. In this context, non-thermal plasma generated under ambient conditions seems to be the best choice, where CO_2 can act as an oxidant due to the fact that its deoxygenation proceeds rapidly in the absence of O_2. CO_2 reforming of aliphatic hydrocarbons, especially methane also contributes to the co-processing of two potential global warming compounds [2].

The experimental set up used for CO_2 reforming consists of plasma generation in a quartz tube of length 30 cm and diameter was 21 mm and effective discharge area was 15 cm. Stainless steel rod was used as an inner electrode and the outer electrode was Cu wire wound around the quartz tube. The discharge gap was 3.5 mm. In order to understand the influence of catalyst, 500 mg NiO/Al_2O_3 was placed at the end of plasma zone. The AC voltage in the range 10 to 22 kV at 100Hz was applied. The discharge power (W) of the DBD reactor was calculated by V-Q Lissajous method [3]. Products were identified by using an online GC-2014 and confirmed with a GC-MS. The influence of various parameters like input power, total feed flow rate, dilution and catalyst on the conversion and product selectivity were studied.

Typical results indicated that NiO catalyst partially reduced to Ni showed highest conversion and methane conversion was always higher than that of CO_2. As seen from Fig. 2, increasing voltage showed higher conversion and selectivity towards syngas (Fig. 2). At 22 kV and 11 W, selectivity to H_2 and CO was 30% and 55%, respectively with plasma, whereas, plasma combining NiO catalyst increased the selectivity to 55% and 65%. It was observed that partial reduction of NiO further improved the conversion and selectivity.

THE GRAPHIC ON THIS PAGE WAS NOT AVAILABLE ON THE E-MEDIA VERSION OF THIS PAPER.

Fig. 1. Effect of applied voltage on conversion

Fig. 2. Effect of Applied voltage on H_2 and CO selectivity

1. Xinli Zhu, PeipeiHuo, Yue-ping Zhang, Dang-guo Cheng, Chang-jun Liu, Applied Catalysis B: Environmental 81 (2008) 132–140.

2. Chang-junLiu , Gen-huiXu , Timing Wang, Fuel Processing Technology 58 1999 119–134.

Dry (CO_2) reforming of methane using Rh substituted pyrochlores ($La_2Zr_2O_7$)

Devendra Pakhare [1], Victor Abdelsayed [2], Daniel Haynes [2], Dushyant Shekhawat [2], James Spivey [1]

NETL NATIONAL ENERGY TECHNOLOGY LABORATORY

[1] – Department of Chemical Engineering, Louisiana State University, Baton Rouge, LA 70803

[2] – National Energy Technology Laboratory, Collins Ferry Rd., Morgantown, WV 26507, USA

U.S. DEPARTMENT OF ENERGY
Center for Atomic-Level Catalyst Design
an Office of Basic Energy Science
Energy Frontier Research Center

Outline

1. **Dry reforming background**
2. **Catalysts**
3. **Catalyst characterization**
   – XRD
   – CH_4 TPR
4. **Conclusion**

1. Dry (CO_2) Reforming of Methane (DRM) background

Dry Reforming of Methane (DRM)

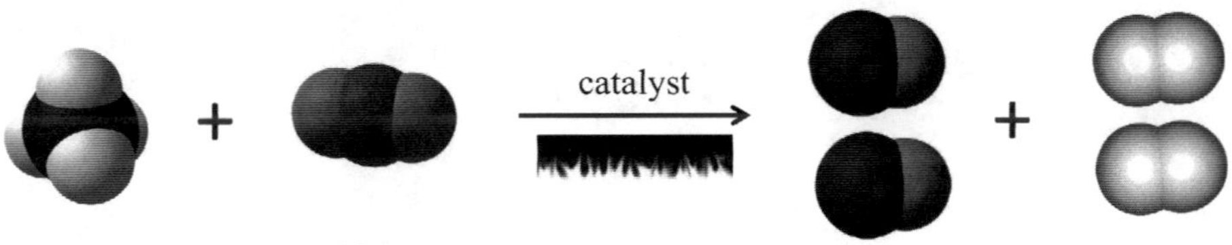

CH_4 CO_2 2 CO 2 H_2

- Involves two extremely stable reactant molecules.
- Highly endothermic reaction ($\Delta H_{298K} = +59.1$ kcal mol^{-1})
- Thus operates at high temperatures.

U.S. DEPARTMENT OF ENERGY | Center for Atomic-Level Catalyst Design | *an Office of Basic Energy Science* *Energy Frontier Research Center*

Introduction

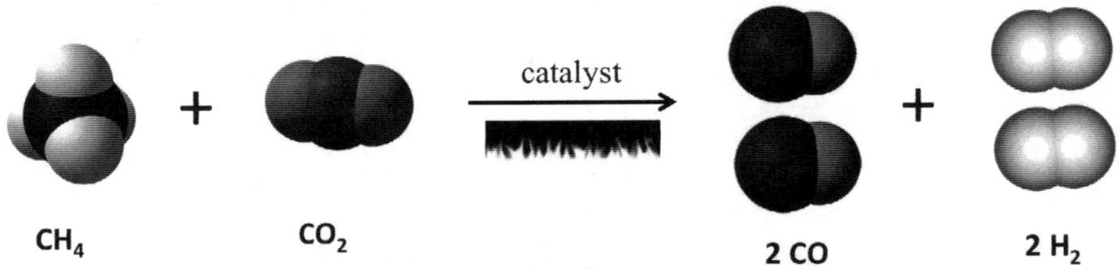

- Other side reactions; carbon formation.

1)

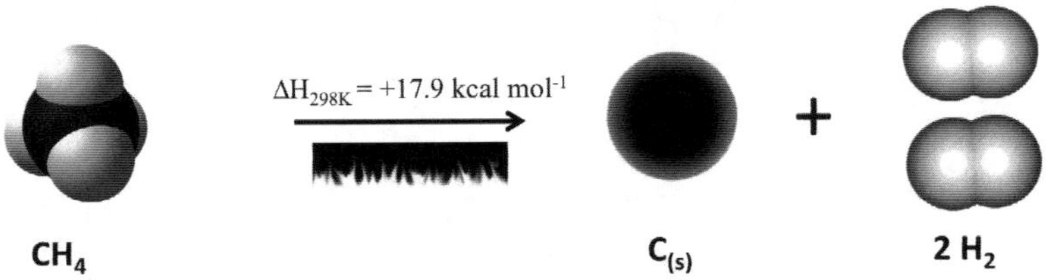

- Methane decomposition

October 3, 2012 5

U.S. DEPARTMENT OF ENERGY | Center for Atomic-Level Catalyst Design | *an Office of Basic Energy Science* *Energy Frontier Research Center*

Introduction

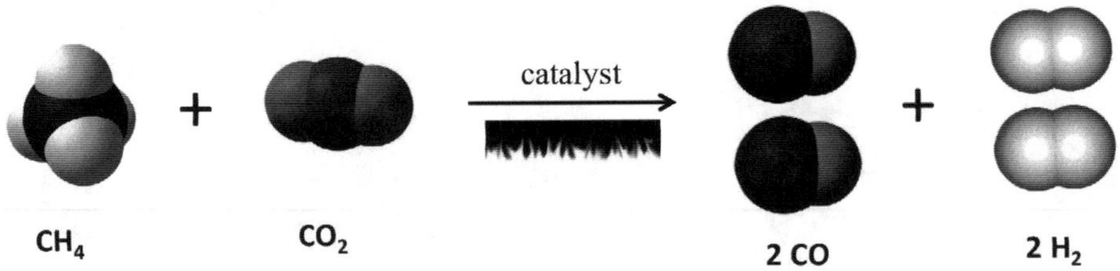

- Other side reactions; carbon formation.

2)

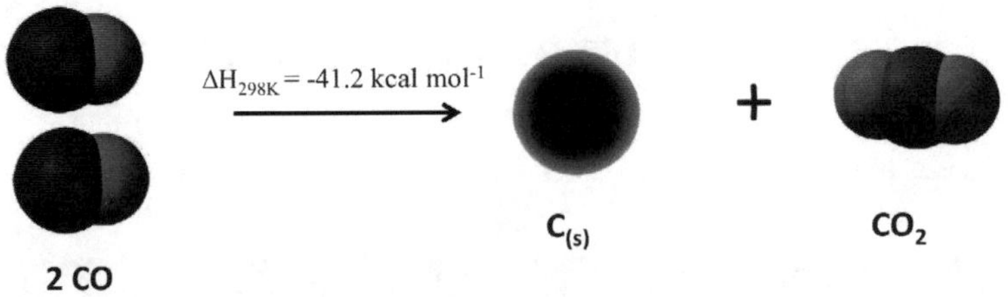

- CO disproportionation

October 3, 2012 6

U.S. DEPARTMENT OF ENERGY | Center for Atomic-Level Catalyst Design | an Office of Basic Energy Science Energy Frontier Research Center

2. Catalysts

October 3, 2012 7

U.S. DEPARTMENT OF ENERGY | Center for Atomic-Level Catalyst Design | an Office of Basic Energy Science Energy Frontier Research Center

Catalyst used in the study

- Pyrochlores : metal oxide crystals.
- General formula:

  $A_2B_2O_7$
- We use $La_2Zr_2O_7$ with Rh on (Zr) B-site

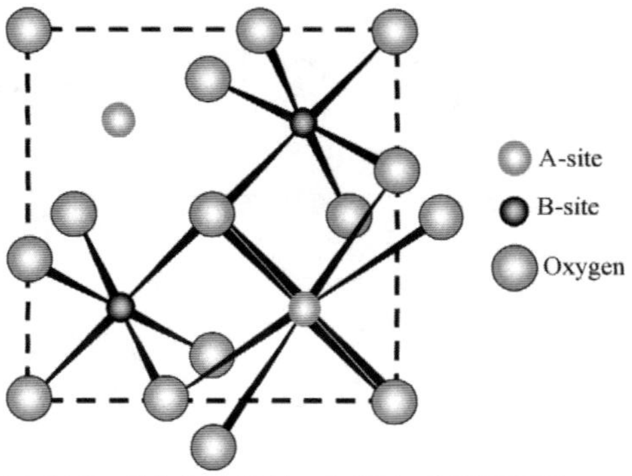

D.J. Haynes, D.A. Berry, D. Shekhawat, J.J. Spivey, Catal. Today, 136 (2008) 206-213.

October 3, 2012 8

Why Pyrochlores??

- Thermally stable.
- Oxygen conductivity inhibits carbon formation.
- No literature on pyrochlores for DRM except for that by Ashcroft et al (1993).
  - A-site composed of Nd/ Sm/ Eu/ Gd
  - B-site composed of Ir or Ru.
  - eg., $Eu_2Ir_2O_7$
  - Pyrochlores used in their work decomposed.
- We partially substitute B-site with Zr and Rh/Ru.

U.S. DEPARTMENT OF ENERGY
Center for Atomic-Level Catalyst Design
an Office of Basic Energy Science
Energy Frontier Research Center

Catalyst used in this study

- 0% by wt Rh represented as **[LZ]**
- 2% by wt Rh represented as **[LRhZ (2%)]**
- 5% by wt Rh represented as **[LRhZ (5%)]**

3. Catalyst characterization

(a) X-Ray Diffraction (XRD)

U.S. DEPARTMENT OF ENERGY
Center for Atomic-Level Catalyst Design
an Office of Basic Energy Science
Energy Frontier Research Center

Perovskite peak for LRhZ (5%)

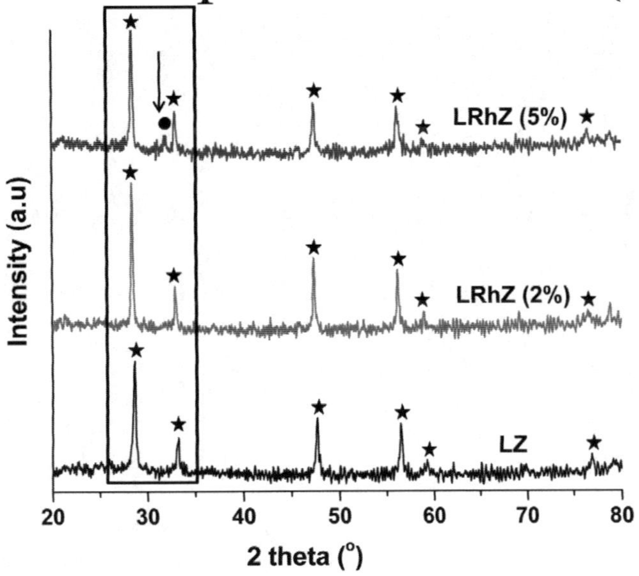

D.J. Haynes, D.A. Berry, D. Shekhawat, J.J. Spivey, Catal. Today, 136 (2008) 206-213.

Possible perovskite peak for LRhZ (2%)

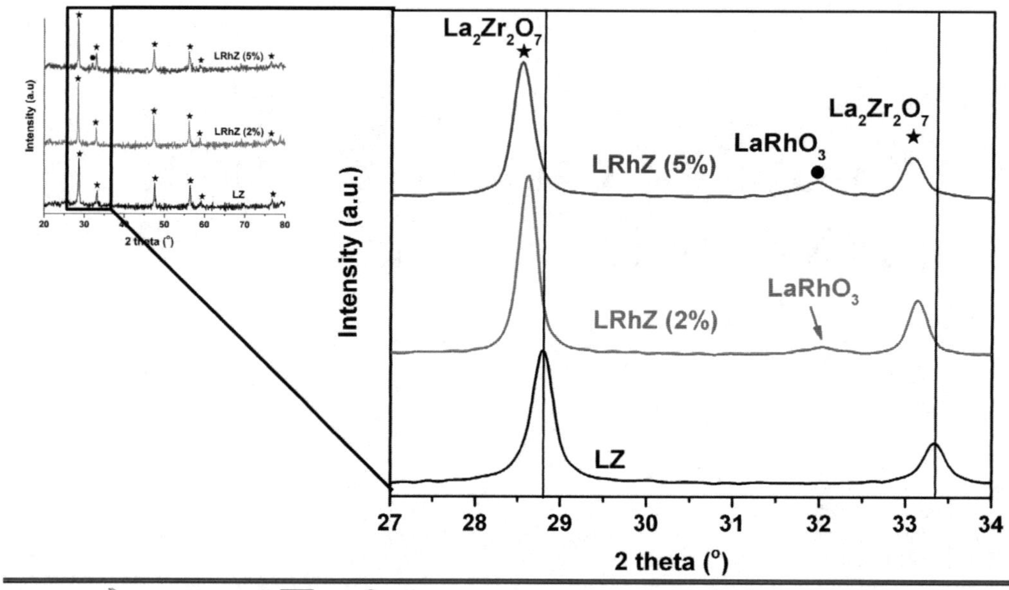

3. Catalyst characterization

(b) CH_4 TPR

Catalyst pre-treatment for CH_4 TPR

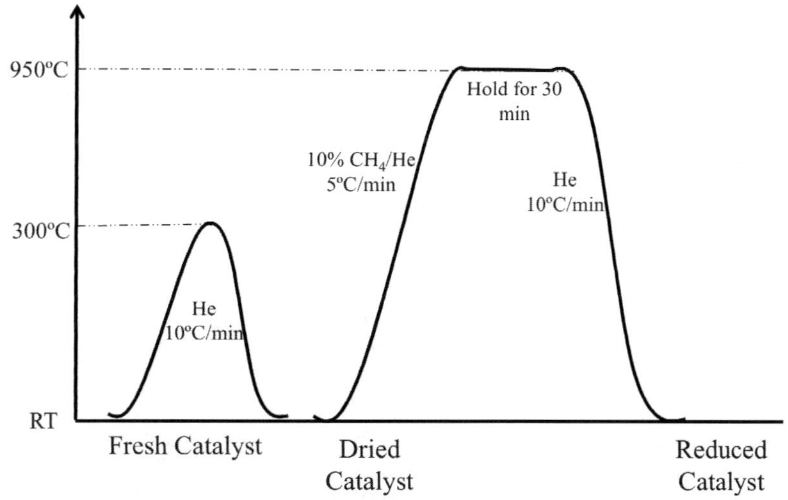

U.S. DEPARTMENT OF ENERGY
Center for Atomic-Level Catalyst Design
an Office of Basic Energy Science
Energy Frontier Research Center

LZ reduces at 605ºC by conducting O_2

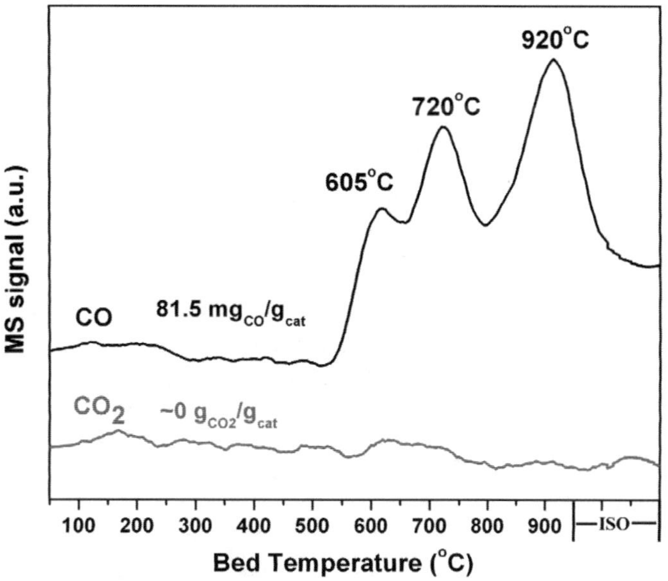

LRhZ (2%) reduces at (570°C) lower temperature than LZ

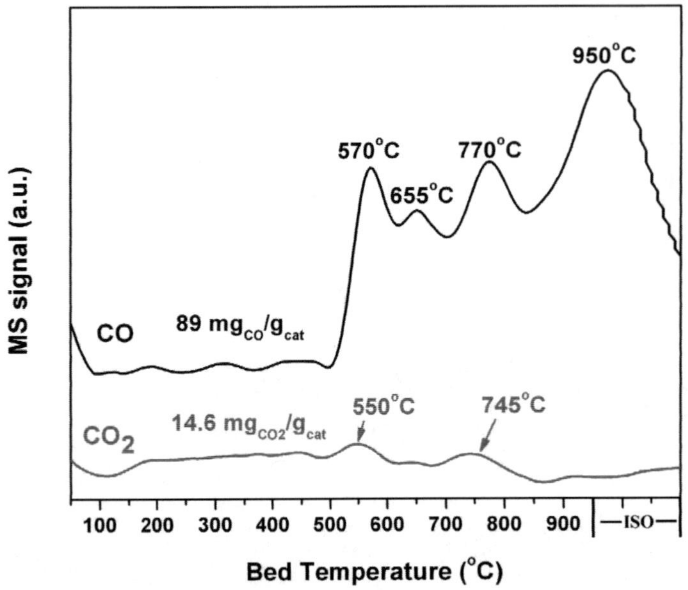

LRhZ (5%) has the lowest reduction temperature

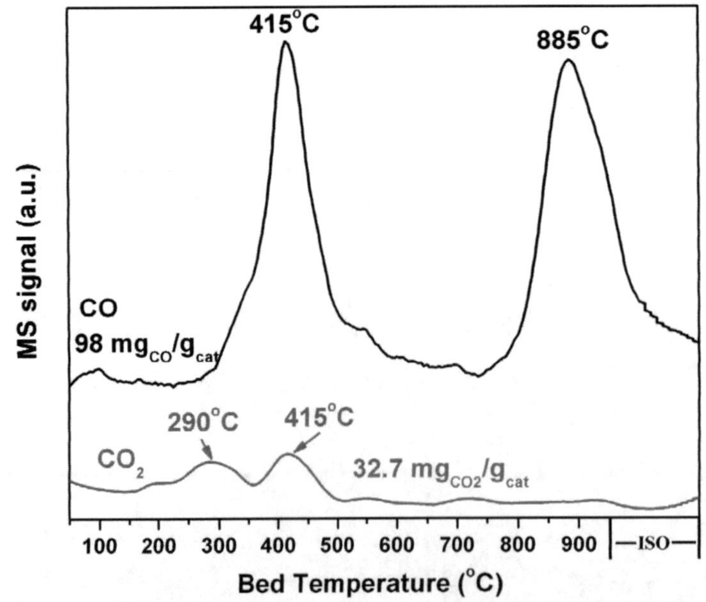

Summary from CH_4 TPR

Catalyst	Amount of CO formed (mg CO/gcat)	Amount of CO2 formed (mg CO2/gcat)	Amount of reactive lattice O2 (mg O2/gcat)	Increase in O2 conductivity wrt LZ [§X/¶LZ]
LZ	81.5	0	46	1
L2RhZ	89	14.6	61.4	1.33
L5RhZ	98	32.7	79.7	1.73

- Increasing Rh substitution increases oxygen conductivity of pyrochlores.
- Rh substitution increases the reducibility.

Possible scenario of $\mathbf{CH_4}$ interaction with **lattice oxygen**

CH$_4$ TPR of LZ

2 species of O for LZ

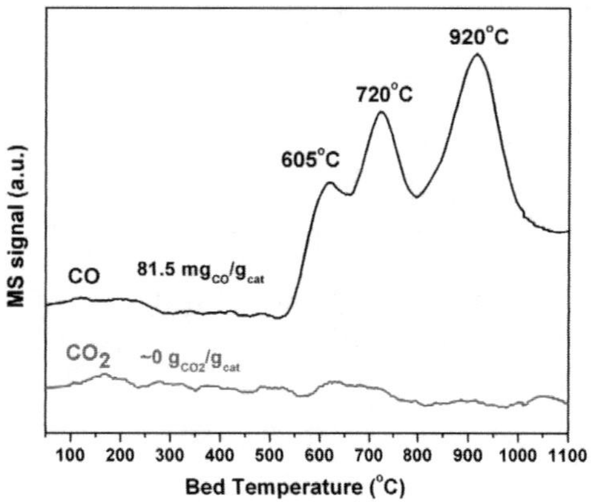

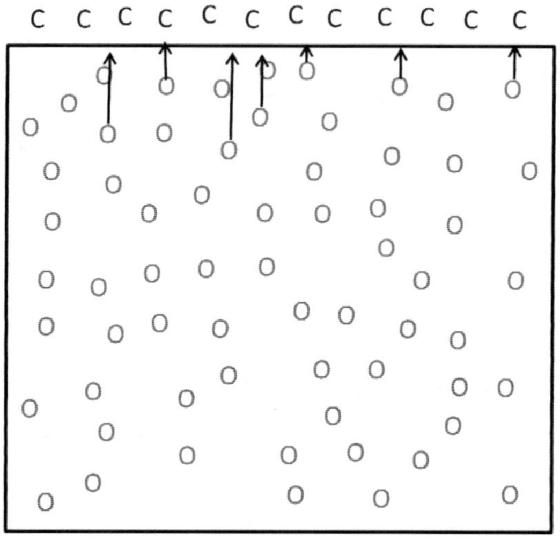

U.S. DEPARTMENT OF ENERGY
Center for Atomic-Level Catalyst Design
an Office of Basic Energy Science
Energy Frontier Research Center

CH$_4$ TPR of LZ

2 species of O for LZ

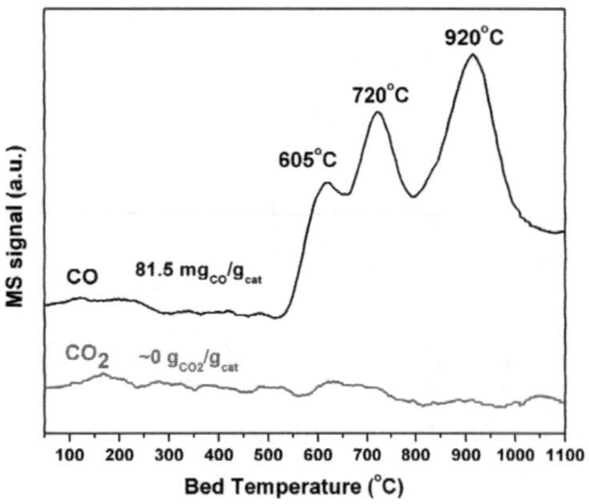

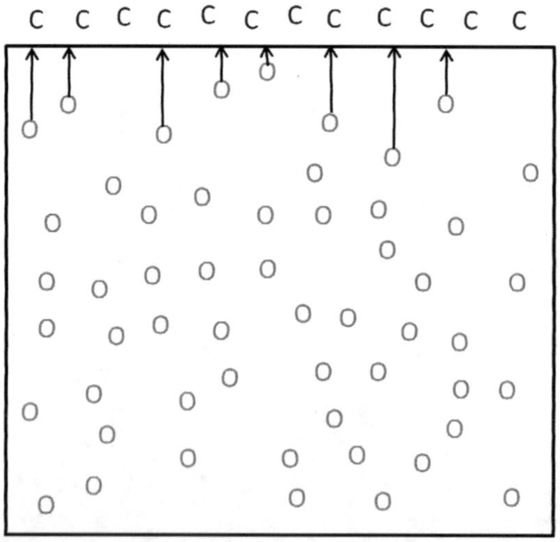

CH_4 TPR of LRhZ (2%)

Low temperature could be Rh-O species

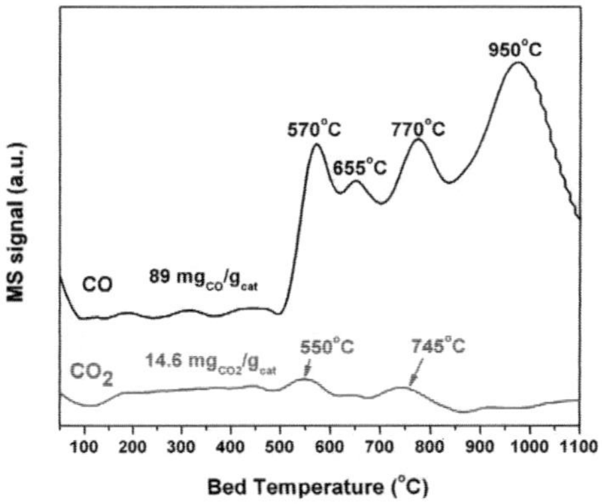

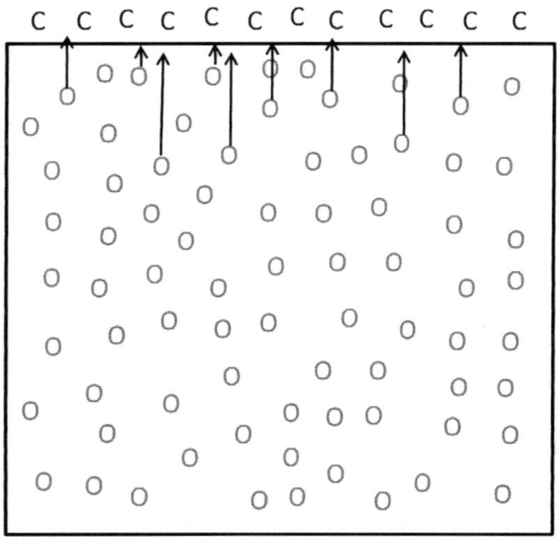

CH_4 TPR of LRhZ (2%)

Low temperature could be Rh-O species

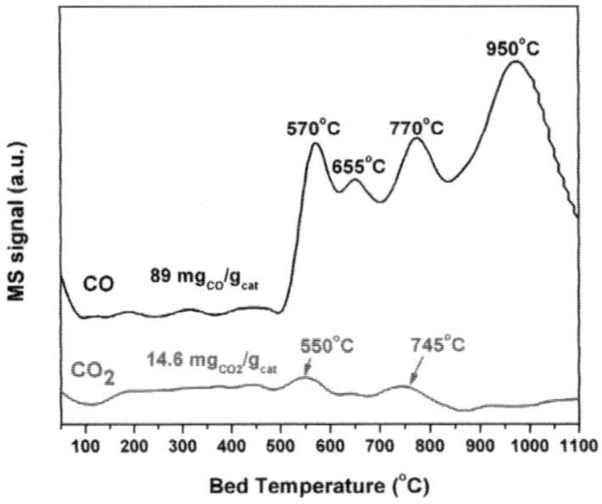

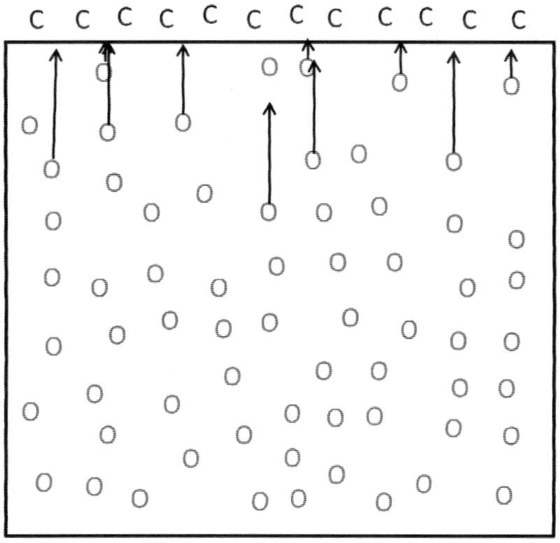

U.S. DEPARTMENT OF ENERGY | Center for Atomic-Level Catalyst Design | an Office of Basic Energy Science Energy Frontier Research Center

Low temperature could be Rh-O species

CH_4 TPR of LRhZ (5%)

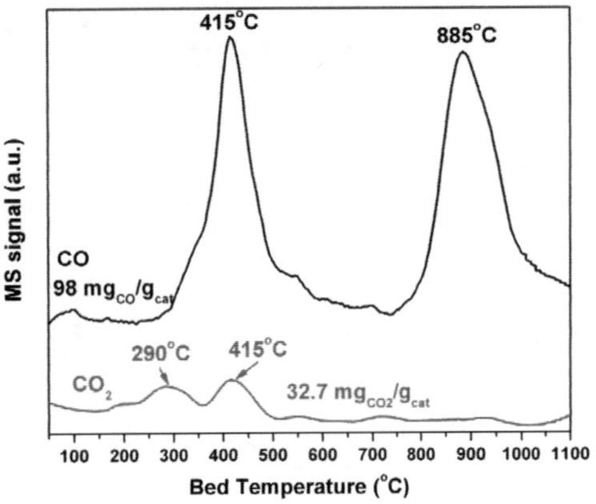

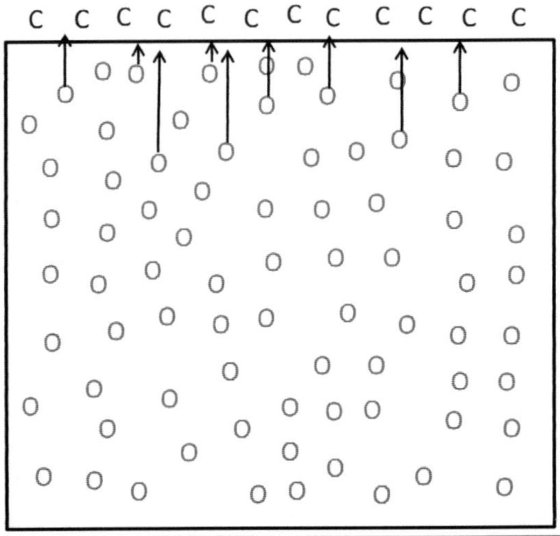

U.S. DEPARTMENT OF ENERGY | Center for Atomic-Level Catalyst Design | an Office of Basic Energy Science Energy Frontier Research Center

Low temperature could be Rh-O species

CH_4 TPR of LRhZ (5%)

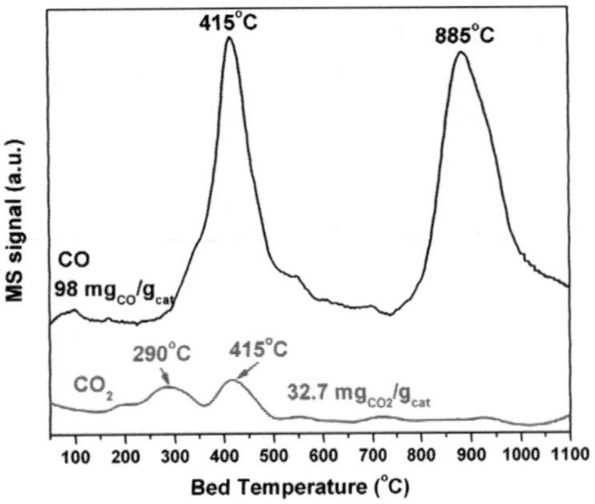

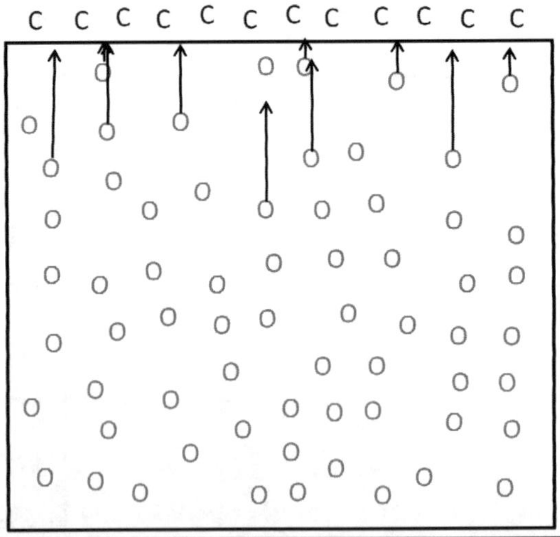

4. Conclusion

Conclusions

- Limited substitution on the B-site else forms perovskites.
- Rh substitution catalyzes the oxygen conductivity of the pyrochlores.
- Increased oxygen conductivity of LRhZ (5%) could also be due to the perovskite phase.
- Oxygen associated with Rh is more reactive than that associated with lanthanum zirconate.

Acknowledgement

- This material is based upon work supported as part of the Center for Atomic Level Catalyst Design, an Energy Frontier Research Center funded by the U.S. Department of Energy, Office of Science, Office of Basic Energy Sciences under Award Number DE-SC0001058.

U.S. DEPARTMENT OF ENERGY
Center for Atomic-Level Catalyst Design
an Office of Basic Energy Science
Energy Frontier Research Center

Thank you

3. Catalyst characterization

(e) Temperature Programmed Oxidation of catalysts spent for CH_4 TPR

Carbon accumulation on LZ during CH_4 TPR

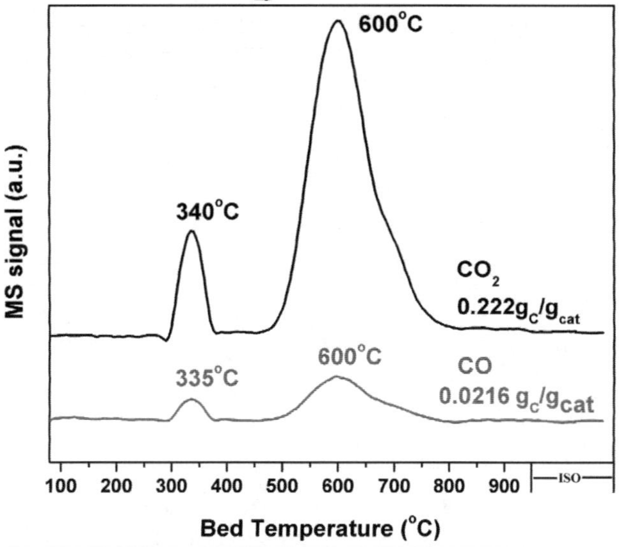

U.S. DEPARTMENT OF ENERGY | Center for Atomic-Level Catalyst Design | an Office of Basic Energy Science Energy Frontier Research Center

Carbon accumulation on LRhZ (2%) during CH_4 TPR

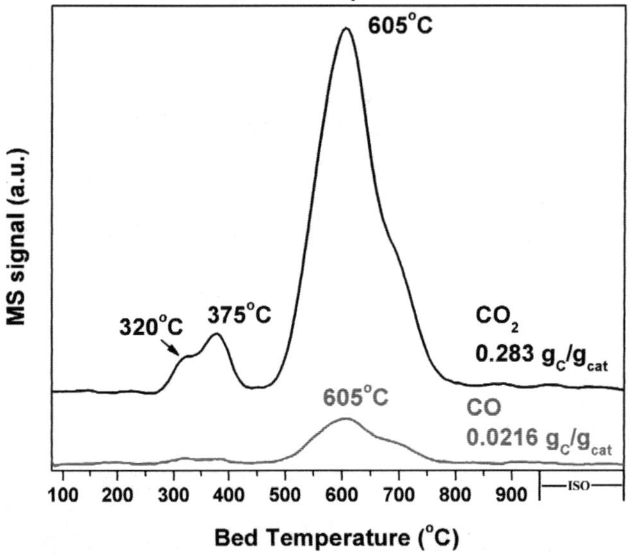

U.S. DEPARTMENT OF ENERGY | Center for Atomic-Level Catalyst Design | an Office of Basic Energy Science Energy Frontier Research Center

Carbon accumulation on LRhZ (5%) during CH_4 TPR

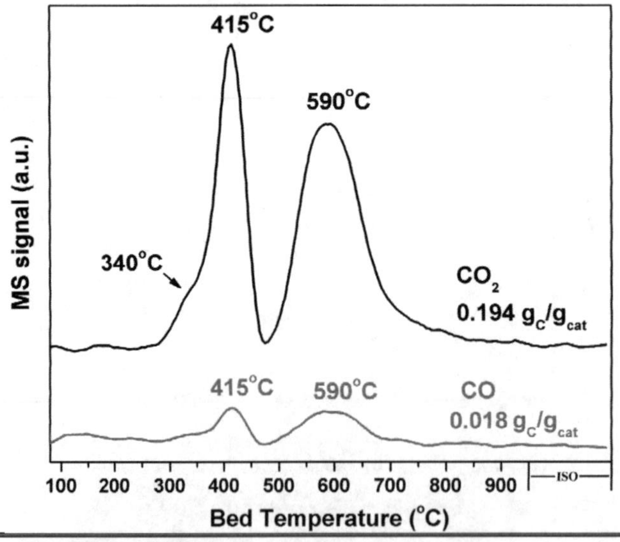

Summary of carbon formed during CH_4 TPR

Catalyst	Amount of carbon oxidized at lower temperature [<500 ºC] (g_C/g_{cat})	Amount of carbon oxidized at high temperature [>500 ºC] (g_C/g_{cat})	Total amount of carbon formed during TPR by CH_4 (g_C/g_{cat})
LZ	0.027	0.217	0.244
LRhZ (2%)	0.028	0.277	0.305
LRhZ (5%)	0.085	0.128	0.213

- High oxygen conductivity of LRhZ (5%) resulted in lower carbon accumulation.
- Higher carbon formation on LRhZ (2%) than LZ could be due to greater activation of CH_4 for LRhZ (2%) than LZ.

Simulation of methane steam reforming enhanced by *in situ* CO_2 sorption utilizing K_2CO_3 promoted hydrotalcites for H_2 production

Naruewan Chanburanasiri, Chulalongkorn University, Bangkok, Thailand.

Ana M. Ribeiro, University of Porto, Porto, Portugal
Alirio E. Rodrigues, University of Porto, Porto, Portugal*
Navadol Laosiripojana, King Mongkut's University of Technology Thonburi, Bangkok, Thailand
Suttichai Assabumrungrat, Chulalongkorn University, Bangkok, Thailand.

**Corresponding author (arodrig@fe.up.pt)*

Abstract

The hydrogen production performance of sorption enhanced methane steam reforming (SEMSR) was investigated in this study. Three different K_2CO_3 promoted hydrotalcites (HTC) including HTC A: industrial K_2CO_3 promoted HTC reported in the work of Ding and Alpay (1), HTC B: commercial HTC from SASOL impregnated with K_2CO_3 in the work of Oliveira et al. (2), and HTC C: commercial K_2CO_3 promoted HTC from SASOL, were considered. A set of experiments was carried out to measure of CO_2 adsorption on HTC C and a 1-D heterogeneous dynamic fixed bed reactor mathematical model was developed to simulate the performance of SEMSR. It was observed that the CO_2 adsorption characteristics were different among the HTCs, resulting in different sorption enhanced characteristic curves. The reaction period that can be operated to produce the high purity hydrogen (>98%) depends on the sorbent type and operating condition. The system utilizing HTC B offers the best performance with the pre-breakthrough period of 116.67 min at the operating condition: $F_{CH_4} = 0.1656$ mmol/min, $T = 773$ K, $P = 0.2$ MPa, $S/C = 11.5$ and catalyst/total solid = 0.05 while the system with HTC A offers 33.33 min that is better than 8.33 min of HTC C. For HTC B, the increase of steam to methane (S/C) ratio leads to the increase of the pre-breakthrough period. The increase of pressure results in the increase of the time before breakthrough when the S/C value is higher than 11.5 at 773 K and 6.143 at 863 K. The temperature of 863 K is suitable for the operation at a low S/C value while the temperature of 773 K is appropriate for higher S/C values.

Introduction

The motivation to switch the fossil fuels to an alternative fuel is an emerging area of research and development during the past decade. Hydrogen is a clean energy carrier for sustainable consumption that can be used in various goals such as raw materials in ammonia and methanol manufactures (1) as well as energy source for electrical power generation and transportation fuel (2). Sorption enhanced methane steam reforming process (SEMSRP) is an interesting multifunctional reactor that is used to produce high-purity hydrogen in a single step (3). This process is based on the addition of a sorbent into the reaction system for selective separation of a product, and thereby shifting the equilibrium of reversible reaction according to Le Chatelier's principle (4). The advantage of this hybrid reactor is the reduction (or even elimination) of the separation process while obtaining high yield and purities of the desired product and conversion (5).

Hydrotalcite is a promising CO_2 adsorbent used in high temperature adsorption and sorption-enhanced reaction process with some main advantages such as high stability and resistance to steam in cyclic operation (6-8). The CO_2 adsorbent feature is shown through CO_2 adsorption capacity represented by CO_2 equilibrium isotherm, and CO_2 adsorption kinetics characterized by CO_2 adsorption kinetic constant. There are a number of papers studying CO_2 adsorption by hydrotalcite sorbents. Most CO_2 adsorption experiments determined CO_2 equilibrium isotherm at different conditions such as with/without presence of steam (wet or dry condition), CO_2 partial pressure, temperature and type of sorbents. From various works, there is a conclusion that the hydrotalcite with K_2CO_3 offers higher CO_2 adsorption capacity than that without K_2CO_3 (9-13).

For this work, the commercial K_2CO_3 promoted hydrotalcite from SASOL denoted by HTC C was tested in a fixed-bed reactor to observe the CO_2 adsorption capability. The performance of HTC C was subsequently compared with the other two HTC sorbents: industrial K_2CO_3 promoted HTC reported in the work of Ding and Alpay (1) (HTC A) and commercial HTC from SASOL impregnated with K_2CO_3 in the work of Oliveira et al. (2) (HTC B). The effect of the operating parameters, including temperature, pressure and steam to methane ratio, on the pre-breakthrough period, the period that offers high purity hydrogen (> 98%), were studied by a 1-D heterogeneous dynamic fixed bed reactor model. The model composes multi-component and overall mass balances, Ergun relation for pressure drop and energy balance for the bed-volume element. The linear driving force (LDF) model is used for the mass transfer rate of CO_2 in the adsorbent.

Experimental

The adsorbent was commercial K_2CO_3 promoted hydrotalcite from SASOL ($MG30w/K_2CO_3$) in cylindrical shape of 4.8 mm diameter and 4.63 mm length. A fixed bed reactor was used for CO_2 adsorption experiment. The feed stream contained helium, carbon dioxide and water while helium and water were used for adsorbent regeneration. The outlet stream of the column was analyzed by gas chromatography (GC 1000, Dani Chromatographs) equipped with an on-line multiport 16-loop valve system for sample injection (Valco Instruments Company Inc.), a capillary column (Carboxen 1010 Plot, Supelco) and a thermal conductivity detector.

Numerical simulation

The model was solved in gPROMS modeling software (Process System Enterprise). The centered finite difference method of second order was used with at least 100 intervals for the axial coordinate of the reactor and 10 intervals for the radial coordinate of the catalyst extrudates.

Results and discussion

CO_2 sorption equilibrium isotherms of commercial K_2CO_3-hydrotalcite (HTC C)

The purpose of performing CO_2 breakthrough experiments is to determine CO_2 adsorption equilibrium isotherm of HTC C, that is, the equilibrium amounts of CO_2 chemisorbed as a function of CO_2 partial pressure. The carbon dioxide sorption experiment of HTC C in wet condition (27% H_2O) was carried out at 633, 689 and 740 K. The CO_2 equilibrium isotherm at each temperature was determined at 5 levels of CO_2 partial pressure (0.005, 0.01, 0.02, 0.035 and 0.045 MPa). Figure 1 shows a breakthrough curve for the experiment using CO_2 partial pressure of 0.01 MPa and temperature of 633 K. The area above the breakthrough curve is the source of equilibrium CO_2 chemisorbed amount. The CO_2 equilibrium isotherms of HTC C at three temperatures (633, 689 and 740 K) under wet feed condition are shown in Figure 2. The bi-Langmuir model which is a simple model that allows the consideration of 2 different and independent types of sites (physical sorption and chemical reaction) is the first model used to fit these experimental results. The best fitted model parameters (q_{max1}, q_{max2}, k_{0eq1,CO_2}, k_{0eq2,CO_2}, $-\Delta H_{sorb}$ *and* $-E_{sorb}$) are reported in Table 1 and the fitting of the data is represented by the solid lines in Figure 2.

From the best fitting parameters obtained, it can be assessed that mostly of the CO_2 adsorption capacity on HTC C is from the physical adsorption. For example, for the adsorption at P_{CO_2} of 0.08 MPa and 753 K, the amount of physical adsorption is 0.695 mol/kg whereas the amount of chemical reaction is only 0.00001 mol/kg. Consequently, the Langmuir model is adequate to describe the CO_2 equilibrium isotherm of HTC C.

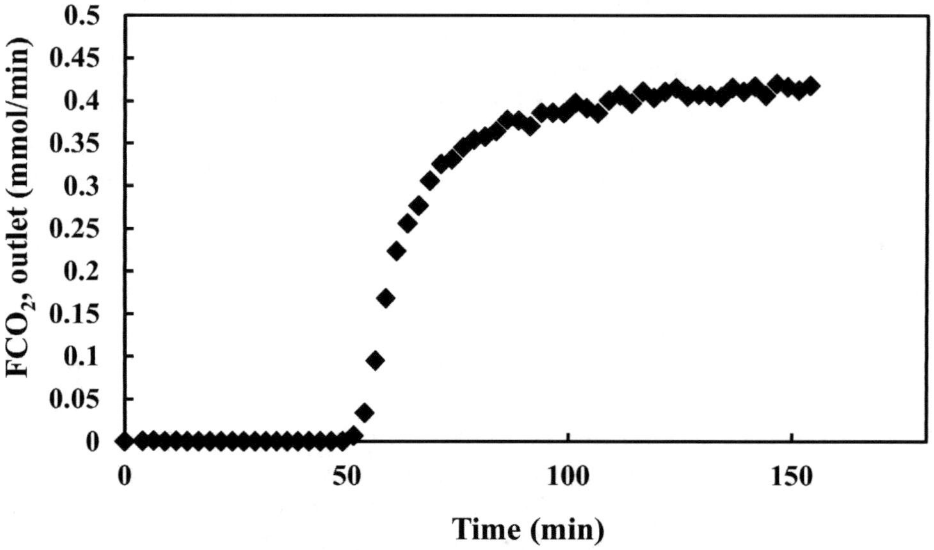

Figure 1. Molar flowrate at the exit of the column for breakthrough experiments of CO_2 (10% CO_2) at 633 K and 0.1 MPa with presence of steam.

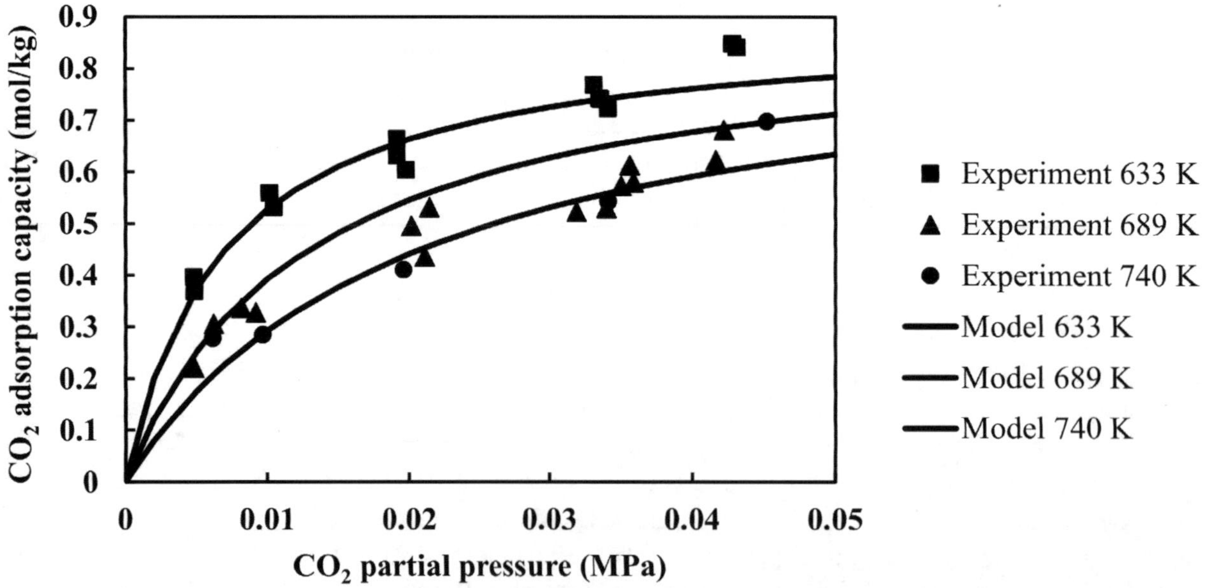

Figure 2. CO_2 sorption equilibrium isotherms of K_2CO_3 promoted hydrotalcites from SASOL (HTC C) at 633, 689 and 740 K.

Table 1. Fitting parameters of the bi-Langmuir models for K_2CO_3 promoted hydrotalcites from Ding and Alpay (1) (HTC A), Oliveira et al. (2) (HTC B) and this work (HTC C).

Parameter	Ding and Alpay (HTC A)	Oliveira (HTC B)	This work (HTC C)
q_{max1} (mol/kg)	0.65	0.423	0.8938
q_{max2} (mol/kg)	0	0.351	0.146
k_{0eq1,CO_2} (bar^{-1})	1.1309	9.07×10^{-3}	8.13×10^{-3}
k_{0eq2,CO_2} (bar^{-1})	0	1.01×10^{12}	1.0×10^{7}
$-\Delta H_{sorb}$ (kJ/mol)	17	40	39.360
$-E_{sorb}$ (kJ/mol)	0	-130.831	-157.314

Comparison of CO_2 sorption capacity of K_2CO_3 promoted hydrotalcites from different sources

The performance of HTC C obtained from the experiments is compared with those of HTC A and HTC B through CO_2 equilibrium isotherm. The equilibrium isotherms of the three materials at 773 K are shown in Figure 3. It is clear that K_2CO_3 promoted hydrotalcites from different sources offer different CO_2 adsorption characteristics as represented by the differences in model types and model parameters. The equilibrium isotherms of HTC A and HTC C can be described by the Langmuir model with different model parameters while the bi-Langmuir model is used to describe the adsorption of HTC B. The fitting parameters of Ding and Alpay's work (1) and Oliveira et al.'s work (2) are also shown in Table 1. The values of CO_2 adsorption capacity of the different materials are compared in the range of 0-1 bar of CO_2 partial pressure (P_{CO_2}). From this figure, the adsorption capacity of HTC B is higher than HTC C as well as HTC A at all ranges of CO_2 partial pressure. Moreover, it seems that HTC C can adsorb CO_2 better than HTC A at higher CO_2 partial pressure while HTC A can adsorb more CO_2 than HTC C at lower CO_2 partial pressure.

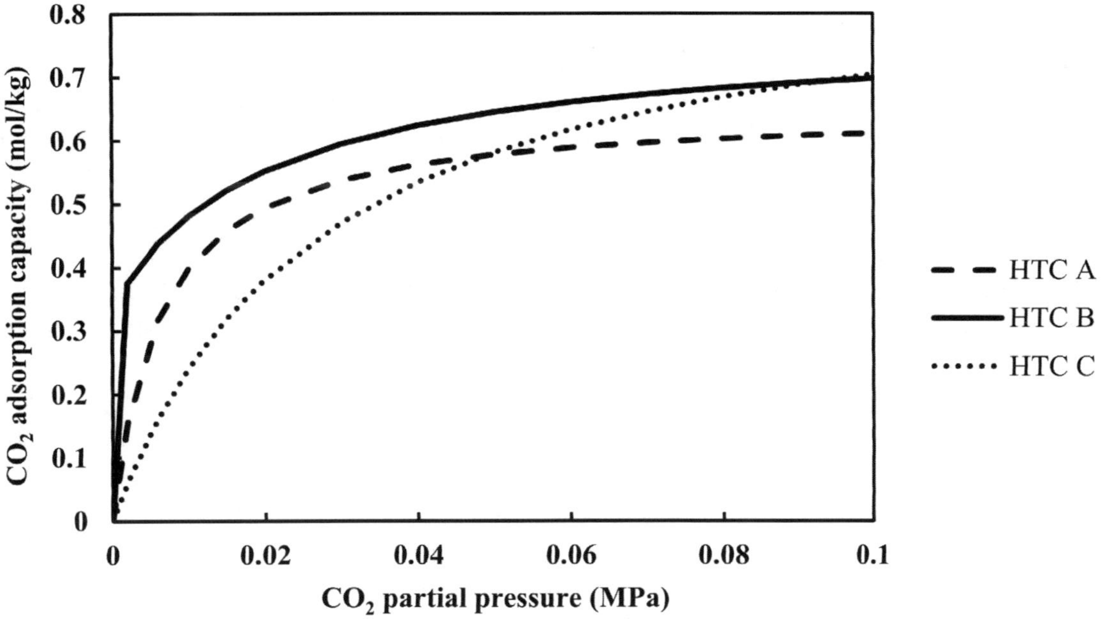

Figure 3. CO_2 sorption isotherms of HTC A, HTC B and HTC C at 773 K.

Sorption enhanced methane steam reforming performance using different K_2CO_3 promoted hydrotalcites

The sorption enhanced methane steam reforming process (SEMSRP) of the system containing HTC C is first simulated to observe the performance. Figure 4 shows a typical product composition profiles (dry basis) when operating SEMSRP at the operating condition: F_{CH_4}=0.1656 mmol/min, T=773 K, P=0.2 MPa S/C=11.5 and catalyst/total solid = 0.05. This profile can be divided into 3 regions. In the first region named pre-breakthrough period, hydrogen concentration is high because the sorbent removes the CO_2 produced, displacing the reaction equilibrium towards the formation of more H_2 product according to Le Chatelier's principle. When the sorbent becomes saturated, the hydrogen concentration decreases from the maximum value to the equilibrium value (breakthrough period). After that the product composition is constant (post-breakthrough period) and equal to the value obtained in a conventional process with no sorbent.

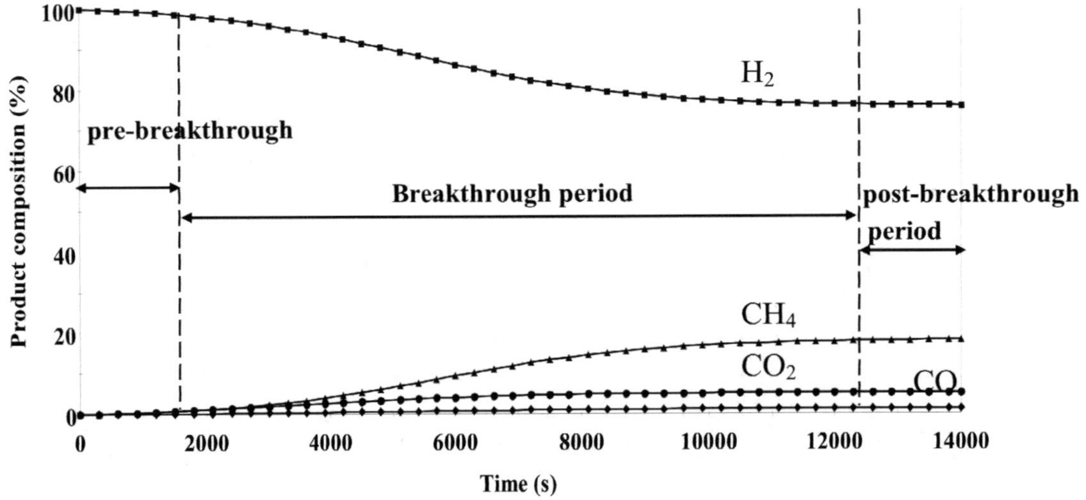

Figure 4. The product composition (dry basis) utilizing HTC C as a function of time at the operating condition: F_{CH_4}=0.1656 mmol/min, T=773 K, P=0.2 MPa, S/C=11.5 and catalyst/total solid = 0.05.

In Figure 5, the hydrogen purity (dry basis) of the systems with different K_2CO_3 promoted hydrotalcites (HTC A, HTC B and HTC C) are compared at the same operating condition: F_{CH_4} = 0.1656 mmol/min, T=773 K, P=0.2 MPa, S/C=11.5 and catalyst/total solid = 0.05. Two adsorption kinetic models of Ding and Alpay (1) and Oliveira et al (2) are considered. The kinetic model of Ding and Alpay (1) is a function of equilibrium isotherm and of temperature while the other one, from Oliveira et al. (2), is a function of temperature only. The similarity of the response curves calculated with the two kinetic models implies that the equilibrium isotherm affects insignificantly the CO_2 adsorption kinetic parameter. Consequently, the term that affects the CO_2 adsorption rate and therefore affects sorption enhanced methane steam reforming curve when using the different K_2CO_3-hydrotalcites is the term of CO_2 sorption equilibrium isotherm. The response curve of HTC B has long pre-breakthrough period followed by fast breakthrough. In the case of both HTC A and HTC C, short pre-breakthrough period followed by slow breakthrough are observed. The system utilizing HTC B offers the best performance with the pre-breakthrough period of 116.67 min at the operating condition: F_{CH_4} = 0.1656 mmol/min, T = 773 K, P = 0.2 MPa, S/C = 11.5 and catalyst/total solid = 0.05 while the system with HTC A offers 33.33 min that is better than 8.33 min of HTC C.

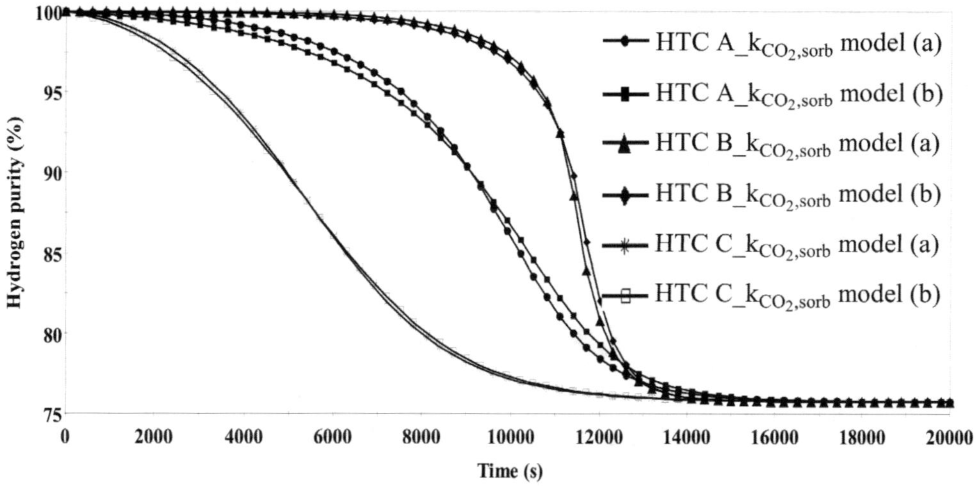

Figure 5. H_2 purity (dry basis) from SEMSR with different HTCs as a function of time simulated by using two CO_2 adsorption kinetic models, (a) Oliveira model (1) and (b) Ding and Alpay model (2) (F_{CH_4}=0.1656 mmol/min, T=773 K, P=0.2 MPa, S/C=11.5 and catalyst/total solid = 0.05).

The effect of the operating parameters including pressure, temperature and S/C ratio on the pre-breakthrough period – the period that offers the high purity hydrogen (>98%) were studied using HTC B. Figures 6 and 7 show the pre-breakthrough period in the system using HTC B at the operating conditions: F_{tot}=2.07 mmol/min and catalyst/total solid = 0.05. The difference of these two figures is the operating pressure, the pressure of 0.1 MPa used for Figure 6 and the pressure of 0.2 MPa used for Figure 7. The operation with higher temperature (863 K) gives longer pre-breakthrough period than at lower temperature (773 K) in low S/C range; however, the pre-breakthrough period of 773 K is longer than that of 863 K when the steam in the process is increased. The increase of steam to methane (S/C) ratio leads to the increase of the pre-breakthrough period. The increase of the pressure from 0.1 MPa (Figure 6) to 0.2 MPa (Figure 7) results in the increase of the time before breakthrough in all ranges of S/C value at 863 K, unlike the results at 773 K. At the lower temperature, the pre-breakthrough period increases with the increasing of pressure only when the S/C value is higher than 11.5.

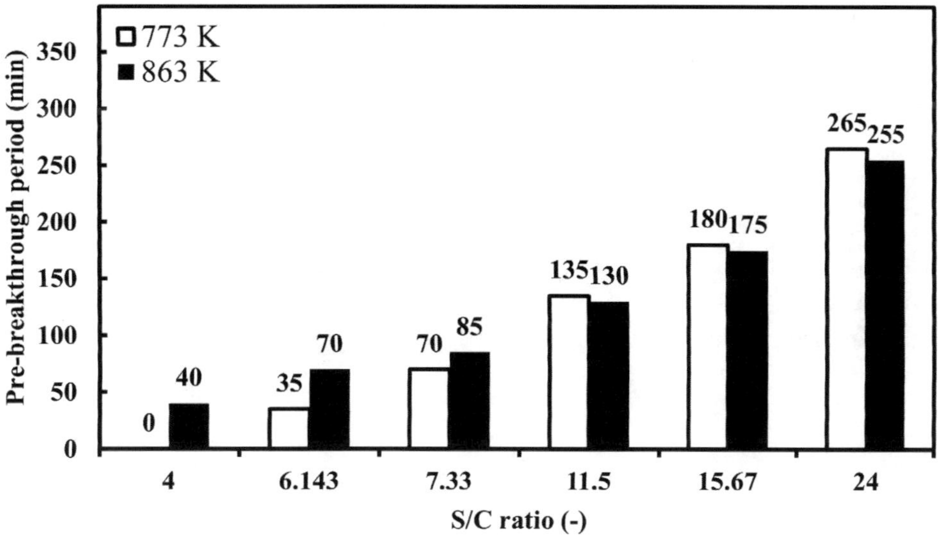

Figure 6. Pre-breakthrough period (min) as a function of S/C ratio utilizing K-hydrotalcites from HTC B at the operating condition (F_{tot}=2.07 mmol/min, P=0.1 MPa and catalyst/total solid = 0.05).

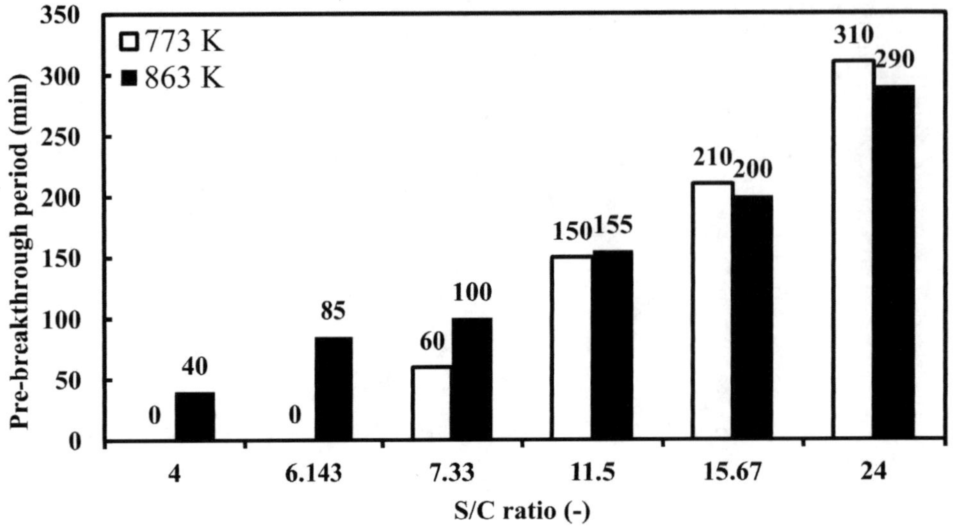

Figure 7. Pre-breakthrough period (min) as a function of *S/C* ratio utilizing K-hydrotalcites from HTC B (F_{tot}=2.07 mmol/min, *P*=0.2 MPa and catalyst/total solid = 0.05).

Conclusion

This work focused on the sorption enhanced methane steam reforming processes (SEMSRP) utilizing different types of K-hydrotalcites for high purity hydrogen production. The commercial K_2CO_3 promoted hydrotalcites from SASOL (HTC C) that was tested in this work is compared with the industrial K_2CO_3 promoted hydrotalcite material in the work of Ding and Alpay (HTC A) and commercial hydrotalcite from SASOL impregnated with K_2CO_3 in the work of Oliveira et al. (HTC B). The CO_2 adsorption characteristics were different among the HTCs, resulting in different sorption enhanced characteristic curves. The period of time that can be operated to produce high purity hydrogen (>98%) depended on the sorbent type and operating conditions. The system utilizing HTC B gives the best performance while the system with HTC A is better than HTC C. For HTC B, the increase of steam to methane (*S/C*) ratio leads to the increase of the pre-breakthrough period. The increase of pressure results in the increase of the time before breakthrough when the S/C value is higher than 11.5 at 773 K and 6.143 at 863 K. The temperature of 863 K is suitable for the operation at a low *S/C* value while the temperature of 773 K is appropriate for higher *S/C* values.

Acknowledgement

The authors gratefully acknowledge the support from the Thailand Research Fund, the Royal Golden Jubilee PhD Program and the project PEst-C/EQB/LA0020/2011, financed by FEDER through COMPETE - Programa Operacional Factores de Competitividade and by FCT - Fundação para a Ciência e a Tecnologia.

References

(1) Ding, Y., Alpay.E. (2000), "Equilibria and kinetics of CO_2 adsorption on hydrotalcite adsorbent", *Chemical Engineering Science*, 55, 3461-3474.

(2) Oliveira, E. L. G (2009), "Sorption enhanced steam methane reforming process for continuous production of hydrogen in pressure swing adsorptive reactors", PhD *dissertation of University of Porto*, 1-285.

(3) Rawadieh. S., Gomes. V. G. (2009), "Steam reforming for hydrogen generation with in situ adsorptive separation", *International Journal of Hydrogen Energy*, 34, 343-355.

(4) Li. Z. S., Cai. N. S. (2007), "Modeling of multiple cycles for sorption-enhanced steam methane reforming and sorbent regeneration in fixed bed reactor", *Energy & Fuels*, 21, 2909-2918.

(5) Barelli. L., Bidini. G., Gallorini. F., Servili. S. (2008), "Hydrogen production through sorption-enhanced steam methane reforming and membrane technology: A review", *Energy*, 33, 554-570.

(6) Lee. K. B., Beaver. M. G., Caram. H. S., Sircar, S. (2008). "Reversible chemisorbents for carbon dioxide and their potential applications", *Industrial & Engineering Chemistry Research*, 47, 8048-8062.

(7) Molga. E., Cherbanski. R. (2012), "Hydrogen production integrated with simultaneous CO_2 sequestration on fly ashes from power plants", *Chemical Engineering & Technology*, 35, 539-546.

(8) Lee. K. B., Verdooren. A., Caram, H. S., Sircar, S. (2007), "Chemisorption of carbon dioxide on potassium-carbonate-promoted hydrotalcite", *Journal of Colloid and Interface Science*, 308, 30-39.

(9) Hutson. N. D., Attwood. B. C. (2008), "High temperature adsorption of CO_2 on various hydrotalcite-like compounds", *Adsorption-Journal of the International Adsorption Society*, 2008, 14, 781-789.

(10) Choi. S., Drese. J. H., Jones. C. W. (2009), "Adsorbent materials for carbon dioxide capture from large anthropogenic point sources", *Chemsuschem*, 2, 796-854.

(11) Oliveira, E. L. G., Grande. C. A., Rodrigues. A. E. (2008), "CO_2 sorption on hydrotalcite and alkali-modified (K and Cs) hydrotalcites at high temperatures", *Separation and Purification Technology*, 62, 137-147.

(12) Lee. J. M., Min. Y. J., Lee. K. B., Jeon. S. G., Na. J. G., Ryu. H. J. (2010), "Enhancement of CO_2 sorption uptake on hydrotalcite by impregnation with K_2CO_3", *Langmuir*, 26, 18788-18797.

(13) Halabi. M. H., De Croon. M. H. J. M., Van Der Schaaf. J., Cobden. P. D., Schouten, J. C. (2012), "High capacity potassium-promoted hydrotalcite for CO_2 capture in H_2 production", *International Journal of Hydrogen Energy*, 37, 4516-4525.

(491e) Bi-Reforming of Methane and Natural Gas Under High Pressure

Alain Goeppert[1], Miklos Czaun[1], Robert May[1], Surya Prakash[2] and George Olah[1], (1)Loker Hydrocarbon Research Institute, University of Southern California, Los Angeles, CA, (2)Loker Hydrocarbon Research Institute - Chemistry Department, University of Southern California, Los Angeles, CA

Bi-Reforming of Methane and Natural Gas Under High Pressure

Alain Goeppert, Miklos Czaun, Robert B. May, G. K. Surya Prakash, George A. Olah, Loker Hydrocarbon Research Institute, Department of Chemistry, University of Southern California, Los Angeles, CA 90089, USA

Introduction

By choosing the right proportions between water, CO_2 and CH_4, the combination of steam and dry reforming of methane can generate syn-gas with a H_2/CO ratio of 2 ideal for the synthesis of methanol.[1,2] This combination of steam and dry reforming has been named bi-reforming.

$$\text{Steam reforming} \quad CH_4 + H_2O \rightleftharpoons CO + 3H_2 \qquad \Delta H_{298K} = 49.1\ \text{kcal.mol}^{-1}$$

$$\text{Dry reforming} \quad CH_4 + CO_2 \rightleftharpoons 2CO + 2H_2 \qquad \Delta H_{298K} = 59.1\ \text{kcal.mol}^{-1}$$

$$\text{Bi-reforming} \quad 3CH_4 + 2H_2O + CO_2 \rightleftharpoons 4CO + 8H_2 \longrightarrow 4CH_3OH$$

Bi-reforming could be advantageous in the use of natural gas sources containing substantial amounts of CO_2. This CO_2 would, otherwise, have to be separated to allow further processing of the natural gas. Some natural gas sources contain CO_2 concentration from 5% up to 70%. In most cases, once separated, the CO_2 is then released into the atmosphere. Only in few locations is CO_2 sequestered. The natural gas at the Sleipner platform in Norway containing 9% CO_2 is for example currently separated and sequestered beneath the North Sea in a deep saline aquifer.

Experimental

To mimic conditions closer to commercial operations, the bi-reforming reaction was conducted at pressures up to 35 bars in a tubular reactor system specially built for this purpose and able to withstand both high pressure and high temperature in a gas mixture with a high carbon activity. The catalysts used were based on Ni deposited on various supports including alumina, alkali earth oxides and combinations thereof. Reaction temperatures ranging from 700°C to 870°C were investigated. A gas feed composition of $CH_4/CO_2/H_2O$ with a molar ratio of 3/1.2/2.4 was typically used. The reaction was followed by an online GC equipped with a TCD.

Results

In general the catalysts were tested for at least 50 hours to determine their stability as a function of time. The conversion of methane as well as carbon dioxide was stable over the length of the experiments (Figure 1). The obtained H_2/CO ratio of the reaction gases was close to the desired value of 2. Increasing the amount of water and CO_2 compared to methane increased the methane conversion but did not significantly change the H_2/CO ratio.

As expected from a thermodynamic point of view the methane conversion decreased with increasing pressure. To some extent this effect could be compensated by a higher reaction temperature and/or higher water and CO_2 content in the feed gas.

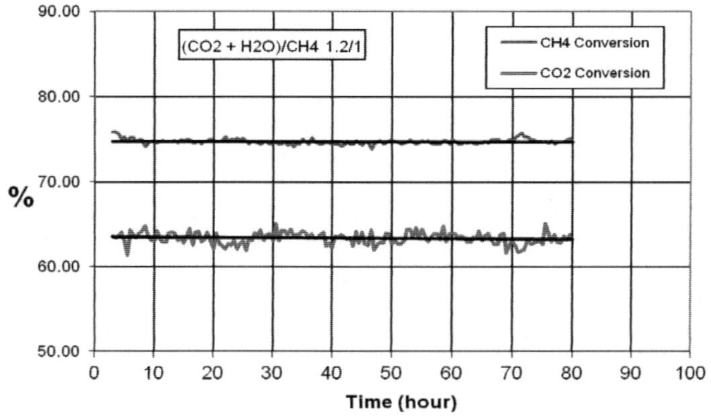

Figure 1. Bi-reforming at 7 bars. Conversion of methane and carbon dioxide.

Acknowledgements

Support of our work by the Loker Hydrocarbon Research Institute and the United States Department of Energy is gratefully acknowledged.

References

[1] G. A. Olah, G. K. S. Prakash, A. Goeppert, *J. Am. Chem. Soc.* **2011**, *133*, 12881.

[2] G. A. Olah, A. Goeppert, G. K. S. Prakash, Beyond Oil and Gas: The Methanol Economy, 2nd ed., Wiley VCH, Weinheim, Germany, **2009**.

(491g) Production of Hydrogen and Nanocarbon From Catalytic Decomposition of Methane Over a Ni-Fe/Al2O3 Catalyst

Yi Jin, School of Chemical Engineering, Tianjin Key Laboratory of Catalysis Science and Technology, Tianjin, China, Gaowei Wang, Tianjin Key Laboratory of Catalysis Science and Technology, Tianjin, China and **Yongdan Li**, Tianjin Key Laboratory of Applied Catalysis Science and Technology, Tianjin University, Tianjin, China

Methane catalytic decomposition reaction (MCD) has received much attention in recent years as a potential route for the direct production of COx (CO/CO_2)-free hydrogen because of its environmentally benign, energy efficient and process simple characteristics, which produces an especially suitable feed for low temperature hydrogen fuel cells, e.g. proton exchange membrane fuel cells (PEMFC) and alkaline fuel cells. Besides of generating COx-free hydrogen, MCD produces valuable carbon nanotubes (CNTs) and nanofibers (CNTs) with great application potentials. The catalysts used for MCD are usually group 8-10 transition metals, such as Ni, Fe and Co. The co-precipitated nickel-based catalysts are very active and stable at low temperatures (e.g. 773-873 K). MCD is endothermic and high reaction temperature is advantageous for obtaining high concentration hydrogen. However, the durability of the Ni/Al_2O_3 catalyst declines swiftly with the increase of the reaction temperature. To improve its durability and delay its deactivation at expected reaction temperatures, the third component, such as metals, or oxides were introduced.

In this work, the influence of the doping of iron to co-precipitated Ni-Al catalysts on the stability and activity of the reaction was investigated. The mixed oxide was prepared from the precursor which possesses a well-crystallized hydrotalcite-like anionic clay structure. In the mixed metal oxides obtained from calcination of the co-precipitated precursors, NiO phase dominates and the iron oxide is uniformly dispersed in NiO phase. The constant temperature reaction result indicates that the addition of iron enhances the stability of the Ni/Al_2O_3 catalyst greatly. Since the diffusion coefficient of carbon atoms through iron metal is three orders of magnitude higher than that through nickel metal. Modification of the catalyst with doping iron increases the carbon diffusion rate thus enhances the stability of the catalyst at high temperatures. The carbon yield of the prepared $Ni-Fe/Al_2O_3$ catalyst (mole ratio of Ni-Fe-Al=2:1:1) at 923 K was 851.3 g C/ g cat which is the highest yield in the literature reported and the reaction lasted more than 200 h. This work provides a new catalyst which is extremely stable for MCD reaction and is promising for the H_2 production for PEMFC.

(491h) Natural Gas Direct Carbon Fuel Cell for Hydrogen and Electricity Production (NGDCFC)

Meyer Steinberg, HCE, LLC, Melville, NY

Combining the decomposition of methane (natural gas) to hydrogen and carbon in molten carbonate with the direct carbon fuel cell (DCFC) is ideally suited for co-production of hydrogen and electricity for supplying hydrogen and electrical driven vehicles at fuel filling stations, A louvered section in the anode compartment of the cell allows hydrogen to be produced separately from the CO2 produced in the anode compartment The cathode compartment provides the carbonate ion which migrates through a membrane to the main anode compartment where it reacts with the carbon formed by the decomposition of the methane to produce CO2.. The electron circuit is then completed for the cell designated as NGDCFC. At the demonstrated 80% DCFC thermal efficiency for electricity production from the carbon, there is enough waste thermal energy to allow the endothermic decomposition of the methane fuel feed. Further heat exchange of the effluent gasses with the entering methane fuel and air allows the cell to operate auto-therma lly. Thus, the overall energy efficiency of the NGDCFC of natural gas to hydrogen and electricity approaches 100%. The NGDCFC is at least 20% more efficient than the highly developed steam reforming of methane process for hydrogen production, which does not produce any electricity. Conversely, it is 40% more efficient than the natural gas combined cycle power process for electricity. Furthermore, the CO2 is obtained from the cell at 100% concentration thus not requiring any further capture and the CO2 can either be directly sold or sequestered. Preliminary cost estimates indicate that the hydrogen can be produced in the NGDCFC at less than $1.00/gge. This is based on the current price of $7.00/MSCF for natura l gas feed and selling the electricity for only 60 mills/kWh) which is the lowest cost from coal in the country. Depending on whether hydrogen fuelled vehicles or electrically driven vehicles will dominate the market in the future, the NGDCFC provides the flexibility of supplying either of these fuels from one processing unit. Additionally, the electricity can always be sold back to the grid at the lowest marginal cost of power generated from fossil fuels.

(515a) CFD Simulation of Pressurized, Staged Oxyfuel Combustion

Benjamin M. Kumfer, Fei Xia, Bhupesh Dhungel and Richard L. Axelbaum, Energy, Environmental & Chemical Engineering, Washington University in St. Louis, St. Louis, MO

Presently, the costs associated with the capture of carbon dioxide from coal-fired power plants using first-generation oxyfuel combustion technology are prohibitively high. The U.S. Department of Energy has set a goal of developing technologies that can lead to 90% capture of carbon dioxide, with an increased cost of electricity of no more than 35%, as compared to a similar plant without carbon capture. One approach that has shown promise to both reduce capital costs and improve power plant efficiency, as compared to first-generation oxyfuel technology, is pressurized oxy-combustion.

In this work, a pressurized fuel-staged combustion approach is evaluated utilizing CFD simulation. By staging the combustion, the temperature and heat transfer can be controlled. This allows for the elimination of other temperature control processes, such as flue gas recycle or water/steam injection. The potential benefits of fuel staging which may lead to improved cycle efficiency are: reduced process gas volume, increased radiative heat transfer, reduced oxygen demands, increased CO_2 purity entering the carbon compression and purification unit (CPU), and reduced auxiliary power demands.

The principle of staged combustion is that combustion is carried out initially at very high stoichiometric ratio, enabling control of temperature even with very little or no flue gas recycle. To enable combustion under non-stoichiometric conditions, fuel is brought into the combustion chamber in stages. The concept is illustrated in Figure 1 below. Three stages are shown for illustration purposes only, and more stages may be optimum. In the first stage, the coal is burned in pure oxygen. However, the amount of coal is small and the amount of excess O2 is very large. With proper mixing, the temperature of the combustion products of the first stage can be controlled since there is a large amount of excess O2, which effectively acts as a diluent. Also, the amount of coal consumed in the first stage is relatively small so that while the stoichiometric flame temperature near the burner may be high, the rapid mixing with excess oxygen and large radiative heat transfer causes the temperature to drop quickly. The products of the first stage, a mixture of O2, CO2 and H2O, enter stage 2, where more fuel is injected and more O2 is consumed. Since the O2 is diluted with products and the products have been cooled, the peak flame temperature in this stage can be lower than the first stage. This process continues in multiple stages until nearly all of the O2 is consumed.

Results from a CFD simulation of a fuel-staged oxycombustion process are presented. The simulation is performed utilizing ANSYS FLUENT 13. The objective of this work is to determine the radiative and convective heat transfer characteristics that are required to design a unique boiler to accommodate staged combustion. In addition, consideration is given to the optimum distance between stages, optimum placement of excess oxygen in the first stage (i.e. burner injection vs. in furnace injection), and the number of stages and the thermal input at each stage.

Figure 1. – Illustration of fuel-staged oxycombustion process

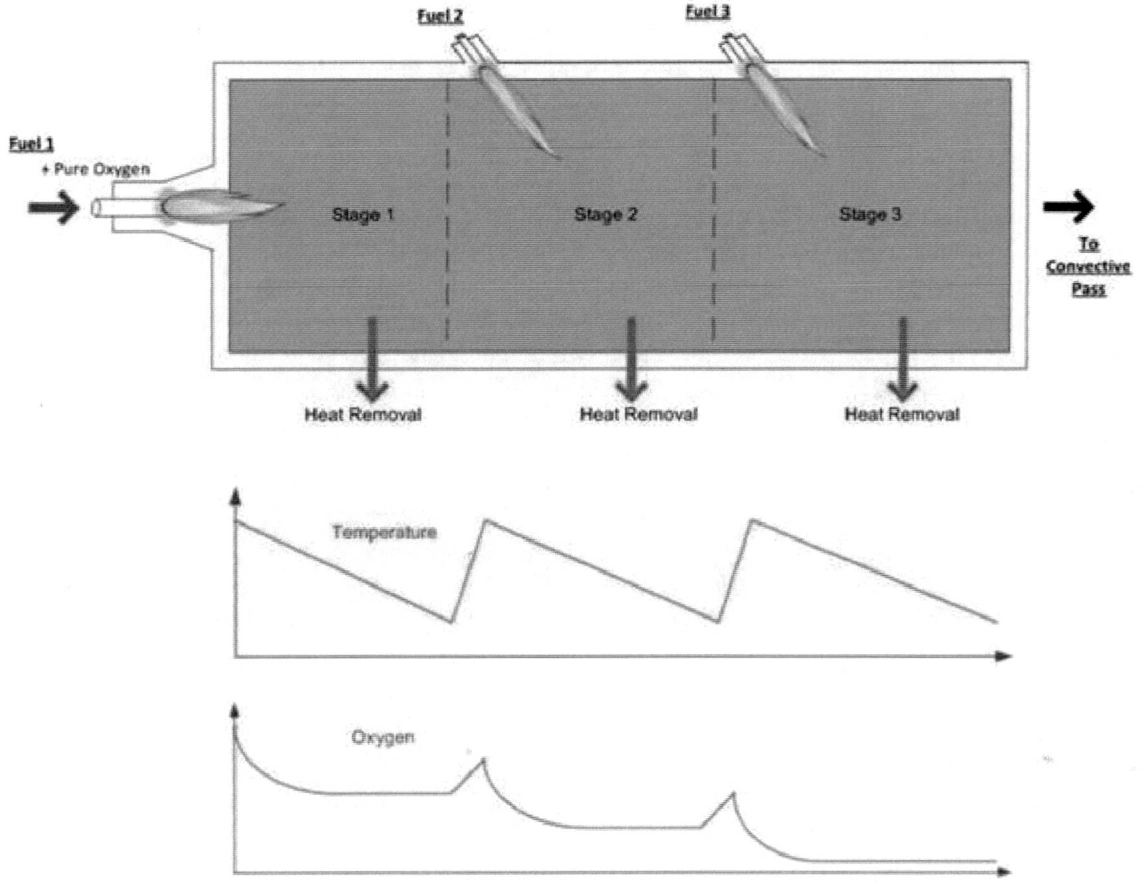
Fuel 2
Fuel 3
Fuel 1
+ Pure Oxygen
Stage 1
Stage 2
Stage 3
To
Convective
Pass
Heat Removal
Heat Removal
Heat Removal
Temperature
Oxygen

(515b) Double Perovskite Materials for Applications in Chemical Looping and Oxygen Storage

Jonathan W. Lekse[1], Sittichai Natesakhawat[2], John P. Baltrus[1] and Christopher Matranga[1], (1)U.S. Department of Energy, National Energy Technology Laboratory, Pittsburgh, PA, (2)US Department of Energy, National Energy Technology Laboratory, Pittsburgh, PA

Chemical looping combustion is a process currently being studied for its potential to reduce the energy penalty associated with carbon capture. To reduce this energy penalty, chemical looping makes use of metal oxides as the source of oxygen for combustion of a fuel. The resulting waste stream consists only of CO_2 and water. The water can then be condensed resulting in a pure CO_2 stream ready for sequestration. One interesting class of materials for chemical looping applications are double perovskite oxides due to their ability to rapidly store and release oxygen. Previously, the double perovskite $BaYMn_2O_{5+x}$ was shown to rapidly and reversibly store and release oxygen with unprecedented kinetics. The ability of these materials to store and release oxygen is derived from the fact that the cations in these materials can vary in oxidation state which allows the oxygen stoichiometry to vary. Altering the cation composition in these materials will allow for the optimization of both the amount of oxygen stored and the storage/release kinetics. Four double perovskite materials were synthesized and studied for this work, $BaLaFe_2O_{5+x}$, $BaLaCo_2O_{5+x}$, $BaYCo_2O_{5+x}$ and $BaYFe_2O_{5+x}$. Characterization of the materials using X-ray diffraction and X-ray photoelectron spectroscopy will be presented. We will also present adsorption/desorption studies and basic kinetic data as well as initial results of testing with methane as a model fuel. The Fe-containing compounds have proven to be the most interesting as they are stable for multiple adsorption/desorption cycles with both nitrogen/air and hydrogen/air at multiple temperatures.

Using $Ca_xLa_{1-x}Mn_{1-y}M'_yO_{3-\delta}$ as oxygen carriers for chemical-looping combustion

Tobias Mattisson[1], Ali Hedayati[1], Abdul-Majeed Azad[3], Magnus Ryden[1], Henrik Leion[2]

[1]Department of Energy and Environment and [2]Department of Chemical and Biological Engineering, Chalmers University of Technology, S-412 96, Göteborg, Sweden

[3]Department of Chemical Engineering The University of Toledo, OH 43606-3390, Toledo, USA

Abstract

Oxygen carriers of the type $Ca_xLa_{1-x}Mn_{1-y}M'_yO_{3-\delta}$ (CLMM') were investigated as oxygen carrier materials. A matrix of potential substituted $Ca_xLa_{1-x}Mn_{1-y}$ $M'_yO_{3-\delta}$ (M' = Fe, Ti, Cu, Mg) perovskites were synthesized using mechanical homogenization of primary solids in a rotary evaporator followed by drying and calcination. The resulting powders were subjected to particle fabrication via extrusion followed by crushing and sieving. The oxygen release properties and the reactivity with methane and syngas was investigated using a batch fluidized batch reactor made of quartz. A sample of 15 g of particles and a fuel flow of 450 Nml/min was used for all reactivity experiments. All of the materials showed high rates of oxygen release in gas phase at 900°C, exhibiting so called oxygen uncoupling behavior. In addition, the reactivity with methane was high, approaching 100% gas yield for all of the materials at 950°C, except the Cu-doped material which defluidized during reduction. All materials also exhibited high rate of oxidation. It can be concluded that the CLMM' materials reacted with fuel through both conventional chemical-looping (direct gas-solid reactions) and via chemical-looping with oxygen uncoupling (with the oxygen released from the particles). The stability of the materials was excellent with very little or no attrition. No agglomeration was noticed for most of these materials. High reactivity and very good oxygen uncoupling properties in combination with the low cost of the primary materials, make this system very interesting for chemical-looping combustion of different types of fuel.

Introduction

Chemical-looping combustion (CLC) is an innovative method to oxidize fuels with inherent CO_2 sequestration that uses well-established boiler technology similar to circulating fluidized bed boilers (CFB). [1] Because of those characteristics it could potentially play a significant role in achieving large reductions in global CO_2 emissions A CLC system is composed of two interconnected reactors with a circulating oxygen carrier which transfers oxygen between combustion air and the fuel, see Fig. 1. When fuel and air are introduced into these reactors, the following overall reactions occur:

$$(2n+m)Me_xO_y + C_nH_{2m} \rightarrow (2n+m)Me_xO_{y-1} + nCO_2 + mH_2O \qquad (1)$$

$$2Me_xO_{y-1} + O_2 \rightarrow 2Me_xO_y \qquad (2)$$

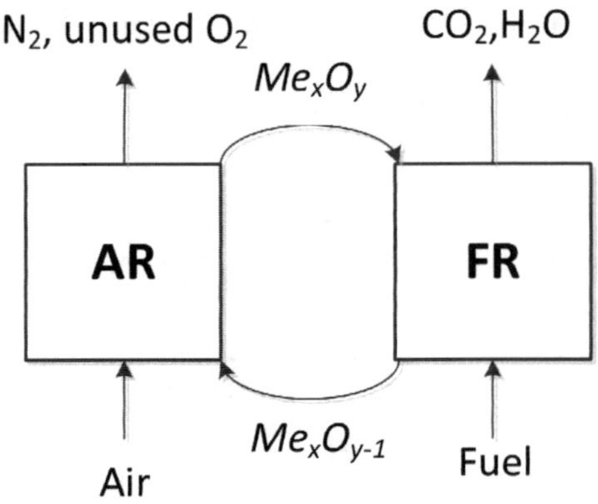

Figure 1. Chemical-looping combustion

Here Me_xO_y and Me_xO_{y-1} are the fully oxidized and reduced forms of the oxygen carrier and C_nH_{2m} is the fuel, and could in gaseous, liquid or solid form. However, in the fuel reactor itself it is likely that the fuel reacts with the particles through a gas-solid reaction. This means that for solid fuels there will need to be a gasification step in order to transform char to reactive gases CO and H_2. [2] Commonly proposed oxygen carrier materials include transition metal oxides such as NiO, Fe_2O_3, CuO or Mn_3O_4, with or without inert support materials such as for example Al_2O_3 or ZrO_2. [3]

With some oxygen carrier materials gas-phase O_2 can be released directly in the fuel reactor according to reaction (3): [4]

$$Me_xO_y \rightarrow Me_xO_{y-1} + 0.5O_2 \qquad (3)$$

The oxygen partial pressure is dependent upon the temperature and oxide system used. If there is a fuel present it will react directly with released O_2 through normal combustion, which will facilitate further O_2 release until all available fuel is consumed. The reduced oxygen carrier can then be recirculated to the air reactor where it is reoxidized according to reaction (2). This reaction scheme described above is referred to as chemical-looping with oxygen uncoupling (CLOU). Here, some common metal oxide systems based on Cu, Mn and Co could be potential candidates, [4], but recently a number of combined metal oxides have also been proven to have CLOU properties see review by Ryden et al. [5]

One such family of possible CLOU materials are the ABO_3-type perovskites, which have interesting properties by virtue of their oxygen non-stoichiometry, which in certain cases can be altered by doping at the A- or B-site. This class of materials is interesting for chemical-looping combustion (CLC) and chemical-looping with oxygen uncoupling (CLOU) applications because of their ability to release a significant amount of active oxygen through,

$$ABO_{3-\delta_{ar}} \leftrightarrow ABO_{3-\delta_{fr}} + \tfrac{1}{2}(\delta_{fr}-\delta_{ar})\, O_2 \qquad (4)$$

The oxygen released in the gas phase, given by (δ_{fr}-δ_{ar}), can be used during CLOU process to burn gas, liquid or solid fuel. The term 'δ' gauges the degree of oxygen non-stoichiometry and 'ar' and 'fr' refer to the air and the fuel reactor, respectively. The actual amount of active oxygen can be increased or reduced by changing parameters such as the ambient temperature, pressure or the O_2 fugacity. [5] For example, in the oxidizing environment of the air reactor, δ_{ar} is likely given by the partial pressure of oxygen at the outlet of the reactor, which is typically around 5% . [1] In the overall reducing fuel reactor the perovskite is at equilibrium at a lower oxygen content given by a δ_{fr} which is larger than δ_{ar} meaning that there will be a release of gaseous oxygen.

One particular promising group of perovskite-type oxygen carrier material is the calcium manganate perovskite $CaMnO_{3-\delta}$, which has been shown to be excellent for gaseous fuel combustion. [6, 7] For instance, Rydén et al. examined $CaMn_{0.875}Ti_{0.125}O_{3-\delta}$ in a small circulating fluidized bed reactor and reported good fluidization behavior, stable operation, O_2 release in inert atmosphere and that complete conversion of CH_4 was achievable. [7] Recently, Källen et al., investigated a similar type of material, $CaMn_{0.9}Mg_{0.1}O_{3-\delta}$, in a 10 kW natural gas fired CLC unit at Chalmers. The results were very promising, with a high rate of oxygen release and full conversion of the natural gas at certain conditions, superior to that of a NiO based oxygen carrier in the same unit. [8]

This paper reports the results from a study on the performance of a series of A and B-site doped $CaMnO_{3-\delta}$ oxygen carriers. Several materials were manufactured, where Ca was partially substituted with small amount of La at A-site, and Mn at the B-site was partially substituted with Fe and other relevant metals such as Ti, Mg and Cu. The oxygen carrier particles were investigated with respect to criteria important for CLOU, i.e. oxygen release behavior and reactivity with fuel, such as methane and syngas in this case.

Experimental

Material preparation

The oxygen carrier particles were prepared using a general methodology which has been previously outlined in detail. [9] Here, an aqueous slurry of the fine and pure raw chemical powders was homogenized and mixed in a rotary evaporator at low pressure. Great care was taken in order to ensure that the material was well dispersed. After drying, the resulting powder was turned into a pliable dough using various dispersants and binders and additional mixing. The resulting dough was then processed to form thin strands using a hand-held single screw extruder. These strands were dried and then calcined in a muffle oven at temperatures in the range of 1100 to 1300°C using a dwelling time of 6h. The resulting material was sieved in order to obtain particles in the size range of 125-250 μm, which could be tested in the batch fluidized bed reactor. Table 1 shows the matrix of materials prepared. For the reactivity tests it was decided to focus on the materials sintered at 1300°C.

Table 1. Materials prepared by extrusion

System	Thermal history and ID	Thermal history and ID	Thermal history and ID
$CaMnO_{3-\delta}$	1100°C/6h: CM1100	1200°C/6h: CM1200	1300°C/6h: CM1300
$Ca_{0.9}La_{0.1}MnO_{3-\delta}$	1100°C/6h: CLM1100	1200°C/6h: CLM1200	1300°C/6h: CLM1300
$Ca_{0.9}La_{0.1}Mn_{0.9}Mg_{0.1}O_{3-\delta}$	1100°C/6h: CLMMg1100	1200°C/6h: CLMMg1200	1300°C/6h: CLMMg1300
$Ca_{0.9}La_{0.1}Mn_{0.9}Ti_{0.1}O_{3-\delta}$	1100°C/6h: CLMT1100	1200°C/6h: CLMT1200	1300°C/6h:CLMT1300
$Ca_{0.9}La_{0.1}Mn_{0.9}Fe_{0.1}O_{3-\delta}$	1100°C/6h: CLMF1100	1200°C/6h: CLMF1200	1300°C/6h:CLMF1300
$Ca_{0.9}La_{0.1}Mn_{0.9}Cu_{0.1}O_{3-\delta}$	1100°C/6h:CLMCu1100	1200°C/6h:CLMCu1200	1300°C/6h:CLMCu1300

Experimental procedure and data evaluation

The CLOU and reactivity experiments were performed in a fluidized bed quartz reactor which had a length of 820 mm and a porous 22 mm in diameter quartz plate placed 370 mm from the bottom. A sample of 15 g of oxygen carriers with size of 125-250 μm was placed on the porous plate and the reactor was heated to the temperature of interest in a flow of 5% O_2 in N_2. When the intended target temperature was achieved, the particles were fluidized with pure N_2 and the outlet oxygen

Table 2. Details of fluidized bed tests

No of cycles	Gas stream	F_{ox} (mL_n/min)	F_{in} (mL_n/min)	t_{in} (s)	F_{red} (mL_n/min)	t_{red} (s)	T_{ox} (°C)	T_{red} (°C)
3	nitrogen	900	600	360	-	-	900	900
1	nitrogen	900	600	360	-	-	900	900 → 1000
3	methane	900	600	60	450	20	950	950
3	syngas	900	600	60	450	80	950	950
3	nitrogen	900	600	360	-	-	900	900
3	nitrogen	900	600	360	-	-	900	900 → 1000

concentration was measured during the inert period in order to establish how well the oxygen carrier particles released gas phase oxygen. The particles were exposed to consecutive cycles of oxidizing and inert periods at 900°C. They were also subjected to oxidation cycles at 900°C but with the temperature raised to 1000°C during the oxygen uncoupling period; see Table 2 for the details of the experiments.

The reactivity of the oxygen carrier was tested with methane and syngas (50% CO + 50% H_2). The reactive gases and gaseous products were expelled from the reactor by introduction of N_2 in between the active periods, thereby avoiding the mixing of reactive gases.

The gas from the reactor was led to an electric cooler for removing water and then to a Rosemount NGA 2000 Multi-Component gas analyzer, for monitoring the gas flow and the concentration of CO, CO_2, CH_4 and O_2. The temperature was measured at a distance of 5 mm under and 10 mm above the porous quartz plate using Pentronic CrAl/NiAl thermocouples with inconel-600 enclosed in quartz shells. The temperature presented in the paper is the set-point temperature, i.e. the temperature at the beginning of reduction. From high frequency measurements of the pressure drop it was possible to determine if the bed was fluidized or not.

The degree of conversion, X, describes the extent to which the oxygen carriers are oxidized and is defined as follows:

$$X = \frac{m - m_{red}}{m_{ox} - m_{red}} \tag{5}$$

where m is the actual mass of the sample, m_{ox} is the mass of the fully oxidized sample, and m_{red} is the mass of the sample in its fully reduced form. The degree of conversion of oxygen carriers as a function of time during reduction with methane and syngas is calculated from the outlet gas concentrations using eq. 6 and 7, respectively.

$$X_i = X_{i-1} - \int_{t_0}^{t_1} \frac{1}{M_0 P_{tot}} \dot{n}_{out} (4p_{CO_2,out} + 3p_{CO,out} - p_{H_2,out}) dt \tag{6}$$

$$X_i = X_{i-1} - \int_{t_0}^{t_1} \frac{1}{M_0 P_{tot}} \dot{n}_{out} (2p_{CO_2,out} + p_{CO,out} - p_{H_2,out}) dt \tag{7}$$

Here, X_i is the conversion as a function of time for a period i, X_{i-1} is the degree of conversion after the foregoing period; t_0 and t_1 are the time at the beginning and end of the period, respectively; M_0 is the molar mass of active oxygen in the unreacted oxygen carrier; n_{out} is the molar flows of dry gas entering the analyzer; P_{tot} is the total pressure; $p_{CO2,out}$, $p_{H2,out}$, and $p_{CO,out}$ are the outlet partial pressures of CO_2, H_2, and CO after removal of water vapor.

In order to better compare oxygen carrier materials which contain different amounts of oxygen, a mass-based conversion, ω, is defined as follows:

$$\omega = \frac{m}{m_{ox}} = 1 + R_0(X - 1) \tag{8}$$

For analysis of gas conversion, the fraction of CO_2 in the outlet gas flow was calculated on dry basis as follows:

$$\gamma = \frac{p_{CO_2}}{p_{CH4} + p_{CO_2} + p_{CO}} \tag{9}$$

Results and Discussion

Oxygen uncoupling properties

All the prepared particles showed reasonable mechanical strength, as determined from the crushing strength (CS) tests on them prior to the reactivity investigations. All particles (in the size range of 180-250 μm) had a CS above 1 N, see Table 3. Although the CS may not correlate linearly with attrition and fragmentation behavior in a real CLC system, another perovskite ($CaMn_{0.9}Mg_{0.1}O_{3-\delta}$) tested by Källen et al. in a 10 kW natural gas fired reactor had a similar CS and maintained stable fluidization with almost no attrition for more than one hundred hours in hot conditions. [8]

All the tested materials showed CLOU behavior, releasing a substantial amount of oxygen at 900°C. This can be seen in Fig 2, where the oxygen concentration during an inert period is shown as a function of time. The oxygen release trend is similar for all the particles, which showed a steady decline in the amount of oxygen release as a function of time. This behavior is typical for this kind of materials, where the partial pressure of oxygen is related to the oxygen stoichiometry in the perovskite, which continually changes due to oxygen release through reaction (4). The pure, undoped $CaMnO_{3-\delta}$, has the highest oxygen partial pressure throughout most of the inert period, while the titanium doped formulation (CLMT1300) has the lowest partial pressure of oxygen among all the investigated materials. The Cu-doped material has a different type of release behavior, in comparison to the rest, and the oxygen concentration is almost constant for a good part of the reaction, and this material

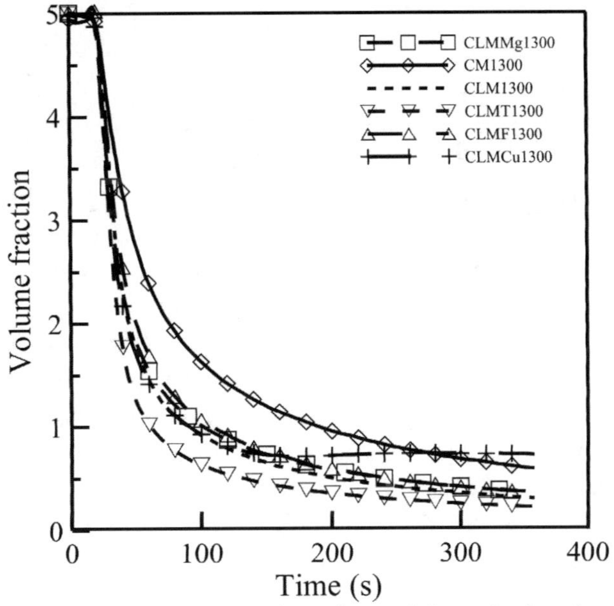

Figure 2. Oxygen volume fraction as a function of time during inert period at 900°C.

actually had the highest concentration at the end of the inert period, see Table 3. There was very little change in the oxygen release rate as a function of cycle number, indicating that no deactivation of the material was occurring in successive cycles. The oxygen concentration after 360 s in inert atmosphere at 900°C and 1000°C, both before and after the fuel cycles, is given in Table 3. As can be readily seen, there is no or very little loss in oxygen release activity even at higher temperature, confirming the excellent stability of these oxygen carrier particles. The CLMCu1300, however, defluidized at the higher temperature, and thus no relevant data could be obtained.

Table 3. Oxygen concentration after 360 s in inert before and after fuel cycles

Sample	Crushing strength (N)	O_2 at 900°C before fuel	O_2 at 900°C after fuel	O_2 at 1000°C before fuel	O_2 at 1000°C after fuel
		[%]	[%]	[%]	[%]
CM1300	1.4	0.59	0.59	0.96	0.89
CLM1300	1.2	0.30	0.25	0.77	0.72
CLMMg1300	1.3	0.36	0.33	0.69	0.56
CLMT1300	1.5	0.22	0.21	0.61	0.57
CLMCu1300	2.8	0.78	0.65	1.06	Defluidized
CLMF1300	1.7	0.35	0.34	0.87	0.82

Reactivity with methane and syngas

The reactivity experiments with methane were performed at 950°C using 15 g of particles in the bed. A typical concentration profile for the reducing period with CLMF1300 material is shown in Figure 3a. Initially, the particle is oxidized in 5% O_2. When the oxidizing stream is replaced with nitrogen the oxygen concentration decreases steadily in the same way as was shown in Fig. 2 for the CLOU tests. When the fuel is added, there is a peak of carbon dioxide, with no CO or CH_4 in the outlet. At the same time the oxygen concentration decreases to zero. It is not believed that there is actually oxygen present during any part of the reduction period, and the overlap of carbon dioxide and oxygen is likely due to the phenomenon of back-mixing. After a few seconds of complete combustion, there is some methane released from the outlet of the reactor. As the reduction period ends, there is no oxygen released from the oxygen carrier, which indicates that the active oxygen for CLOU has been removed during the reduction process. Thus it is likely that the reaction of methane with oxygen carrier material is a combination of direct or indirect reaction of the methane with the solid particles in addition to a reaction with the gaseous oxygen released from the oxygen carrier via CLOU. Figure 2b shows the oxygen concentration profile during the subsequent oxidation period, when 5% O_2 is introduced at the same temperature. During the initial period, all of the oxygen in the inlet gas reacts with the depleted oxygen carrier. Subsequently (after ~250 s) oxygen breaks through and approaches the optimum inlet concentration (5%). The fact that no oxygen is released initially suggests that either the perovskite has been reduced to such a high degree of reduction that the equilibrium partial pressure of oxygen is very low, or new non-perovskite phases are formed, which do not release oxygen in the gas phase.

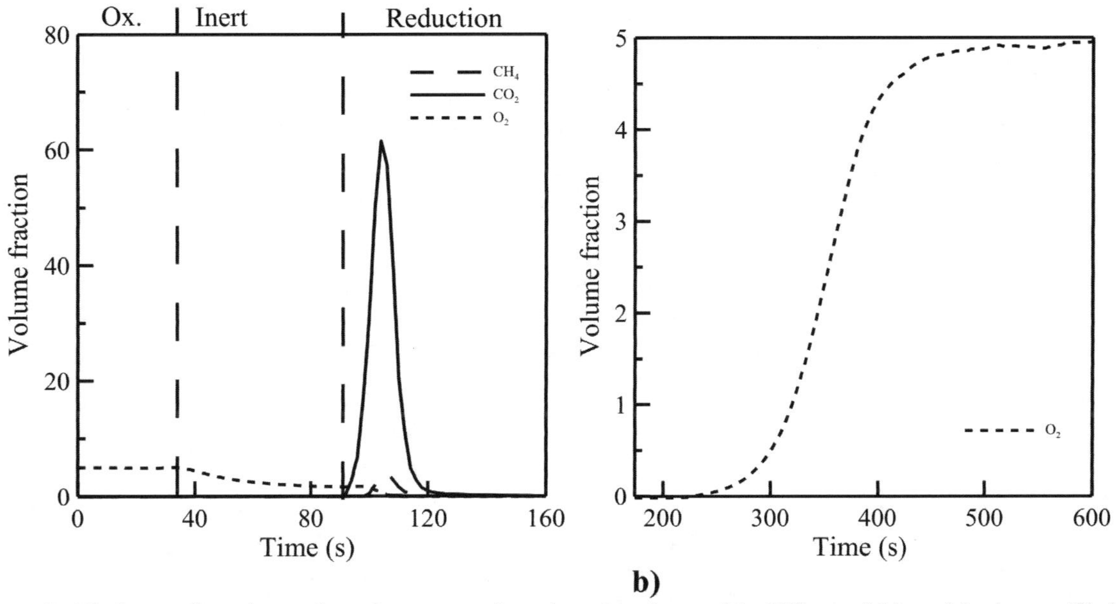

Figure 3. Volume fraction of outlet gases for a) reduction with CH_4 and b) oxidation utilizing the CLMF1300 at 950°C.

Figure 4 shows the gas yield as a function of the mass-based degree of conversion, ω, for the second methane cycle for all the oxygen carrier particles investigated here, the exception being CLMCu1300, where defluidization took place during reduction. Clearly, all materials showed high gas yield; it was complete initially and remained above 80% during the entire reducing period.

With the exception of the Cu-doped material, there was no defluidization seen for any of the particles during testing. Also, very little or no attrition was seen during the experimental sequence. Of course the total number of cycles and the gas velocities are relatively low here, and further long-term tests will be needed in order to confirm the mechanical stability of these materials.

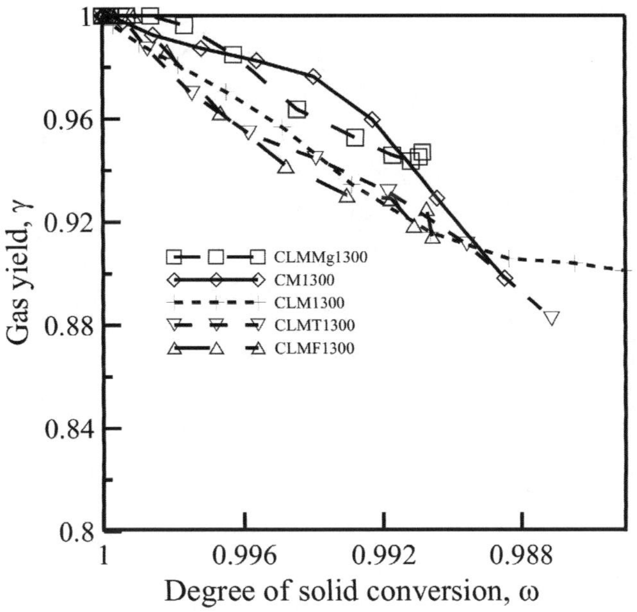

Figure 4. Gas yield, γ, as a function of ω for all particles when reduced with CH_4 at 950°C.

The reactivity was also investigated with syngas composed of 50% CO and 50% H_2. As H_2 was not measured, only the gas yield with respect to CO could be established. For all samples there was complete gas yield at 950°C, independent of the conversion level of the solids.

Conclusions

A series of new perovskite-type oxygen carriers have been investigated with respect to oxygen uncoupling properties and reactivity with methane and syngas. With the exception of Cu-doped material, almost complete methane gas yield could be obtained at 950°C. With syngas, complete gas yield was seen for all materials at the same temperature. No or very small degree of attrition was noted during the experiments; there were no agglomeration tendencies either, the exception again being the Cu-doped material. The particles have CLOU properties, which is an advantage with respect to fuel conversion. As the oxygen carriers were prepared with relatively cheap raw chemicals, i.e. $Ca(OH)_2$ and Mn_2O_3 in addition to small amounts of dopants, these type of oxygen carriers could be very interesting for chemical-looping applications with all types of fuel.

Acknowledgments

This study was partly funded by Stiftelsen Olle Engkvist Byggmästare.

References

1. Lyngfelt, A., B. Leckner, and T. Mattisson, *A fluidized-bed combustion process with inherent CO2 separation; application of chemical-looping combustion.* Chem. Eng. Sci, 2001. **56**: p. 3101-3113.
2. Leion, H., et al., *Solid fuels in chemical-looping combustion using oxide scale and unprocessed iron ore as oxygen carriers* Fuel, 2009. **88**: p. 1945-1954.
3. Adanez, J., et al., *Progress in chemical-looping combustion and reforming technologies* Progress in Energy and Combustion Science, 2011. **38**: p. 215-282.
4. Mattisson, T., A. Lyngfelt, and H. Leion, *Chemical-looping oxygen uncoupling for combustion of solid fuels* International Journal of Greenhouse Gas Control, 2009. **3**: p. 11-19.
5. Ryden, M., et al., *Combined oxides as oxygen carrier material for chemical-looping with oxygen uncoupling* in *2nd International Conference on Chemical-Looping*. 2012: Darmstadt.
6. Leion, H., et al., *Use of CaMn0.875Ti0.125O3 as oxygen carrier in chemical-looping with oxygen uncoupling.* Energy and Fuels, 2009. **23**: p. 5276-5283.
7. Rydén, M., A. Lyngfelt, and T. Mattisson, *CaMn0.875Ti0.125O3 as oxygen carrier for chemical-looping combustion with oxygen uncoupling (CLOU)-Experiments in a continuously operating fluidized-bed reactor system.* International Journal of Greenhouse Gas Control 2010. **5**: p. 356-366.
8. Källen, M., et al., *CaMn0.9Mg0.1O3-δ as Oxygen Carrier in a Gas-Fired 10 kWth Chemical-Looping Combustion Unit.* Submitted for publication, 2012.
9. Hedayati, A., et al., *Evaluation of Novel Ceria-Supported Metal Oxides As Oxygen Carriers for Chemical-Looping Combustion.* Industrial & Engineering Chemistry Research, 2012. **In press**.

(515d) Mechanism of Methane Chemical Looping Combustion with Fe2O3 Promoted with CeO2

Duane Miller, URS, Morgantown, WV and Ranjani V. Siriwardane, National Energy Technology Laboratory, U.S. Department of Energy, Morgantown, WV

Chemical-looping combustion (CLC) is a promising technology for fossil fuel combustion that prevents CO_2 dilution reducing the energy penalty of CO_2 separation from flue gases. A good oxygen carrier for CLC will readily react with the fuel gas and will be reoxidized upon being in contact with oxygen. This study investigates the development of a CeO_2 promoted Fe_2O_3-hematite oxygen carrier suitable for the methane chemical-looping combustion process. Composition of CeO_2 is between 5-25% and is lower than what is generally used for supports in Fe_2O_3 carrier preparations. Temperature programmed reaction studies showed that the addition of even 5 wt,% of CeO_2 enhances the reaction capacity of the Fe_2O_3 oxygen carrier by influencing the decomposition and combustion of methane. Fixed bed reactor data showed that the 5 wt. % cerium oxides with 95% iron oxide produce two times as much carbon dioxide per unit time in comparison to the sum of carbon dioxide when the oxides are tested separately. This effect is likely due to the reaction of CeO_2 with methane forming intermediates which are reactive for extracting oxygen from Fe_2O_3 at a considerable higher rate than the rate of direct reaction of Fe_2O_3 with methane. These studies reveal that 5%-CeO_2/Fe_2O_3 gave stable conversions over 15 cycles. Lab scale reactor studies (pulse mode) suggest the mechanism of CO_2 formation begins by the methane reaction with CeO_2 lattice oxygen to form reaction intermediates which continues to react with oxygen from Fe_2O_3 leading to its complete oxidation to form CO_2. This mechanism is supported by the characterization studies, which also suggest that the formation of carbonaceous intermediates may affect the reaction rate and selectivity.

OXYGEN CARRIERS FOR CHEMICAL-LOOPING COMBUSTION OF SOLID FUELS

Tobias Mattisson[1], Anders Lyngfelt[1], Henrik Leion[2], Magnus Ryden[1]

[1]Department of Energy and Environment and [2]Department of Chemical and Biological Engineering, Chalmers University of Technology, S-412 96, Göteborg, Sweden

Abstract

The main economic advantage of CLC is likely realized for combustion of solid fuels such as coal, petroleum coke and biomass. In a process where the fuel and air reactor are designed as interconnected fluidized beds and where the solid fuel is well-mixed with the oxygen carrier particles there are two main mechanisms for char and volatile conversion: i) Normal CLC where the volatiles and gasification intermediates react directly with the oxygen carrier particles through a gas-solid reaction and ii) through CLOU or chemical-looping with oxygen uncoupling, where the oxygen carrier particles release gaseous oxygen which reacts with the solid fuel components through normal combustion. In addition, several oxygen carriers have been identified to have some CLOU properties, and for these so-called semi-CLOU materials, it could be speculated that the solids and volatiles react through both CLC and CLOU processes in parallel. The role of the oxygen carrier is thus different and depends upon the process which is used above. For route i) the oxygen carrier needs to convert the intermediate syngas components, mainly H_2 and CO, as well as volatiles at a sufficient rate to achieve high gaseous yield to carbon dioxide. Here the oxygen carrier may have a limited effect on the char conversion. For route ii), or CLOU, the oxygen carrier needs to supply oxygen to the fuel at a sufficient rate to achieve a high rate of char and volatile conversion. For semi-CLOU materials, the role of the oxygen carrier material could depend upon a variety of factors, including the rate of oxygen release, and where it is released. The paper discusses the three processes described above and proposes a simple generic analytical model to evaluate the three types of processes above. From the model, quantitative reactivity criteria for oxygen carrier particles can be obtained. It is suggested that it is likely a substantial advantage to utilize oxygen carriers with CLOU properties to process solid fuels, both with respect to char and volatile components. The advantage of semi-CLOU materials is less clear.

Introduction

Chemical-looping combustion (CLC) and Chemical-looping with oxygen uncoupling (CLOU)

Chemical-looping can be used to burn solid fuels through two main mechanisms. In normal CLC with solid fuels the volatile part of the fuel react directly with the oxygen carrier (OC) through reaction (1), to CO_2 and H_2O. The remaining char is gasified according to reaction (2), and the intermediate gasification products, mainly CO and H_2, react with the oxygen carrier directly via (3),

$$\text{Volatiles } (CH_4, CO, H_2) + MeO \rightarrow Me + CO_2 + H_2O \quad (1)$$

$$\text{Char (mainly C)} + H_2O/CO_2 \rightarrow CO/H_2 \quad (2)$$

$$CO/H_2 + MeO \rightarrow Me + CO_2/H_2O \quad (3)$$

a)

The rate of reaction (2) determines how well the solid fuel is converted in the fuel reactor, and determines the residence time needed in order to achieve a high degree of carbon capture. Reaction (1) and (3) gives a measure of the functionality of the oxygen carrier, and a high rate would be beneficial.

In CLOU the solid fuel is oxidized according to the overall reactions (4)-(6) below. The metal oxides release oxygen in the gas phase according to reaction (4). Both the volatile and char components react directly with the released oxygen according to reaction (5) and (6) respectively.

$$2MeO \leftrightarrow 2Me + O_2 \quad (4)$$

$$Volatiles\ (CH_4,\ CO,\ H_2) + O_2 \rightarrow CO_2 + H_2O \quad (5)$$

$$Char\ (mainly\ C) + O_2 \rightarrow CO_2 \quad (6)$$

In order to avoid unreacted oxygen being released from the fuel reactor, the overall rate of oxygen production through reaction (4) should equal the rate of oxygen consumption through reaction (5) and (6). If the rate of oxygen release exceeds the combustion rate, unreleased oxygen will be present in the outlet stream, which may be a disadvantage with respect to flue gas processing. On the other hand, if the combustion rate exceeds the overall release rate, there will be unburnt components, either gas or char, from the outlet of the fuel reactor. Hence, the functionality of the oxygen carrier is determined by the rate of reaction (4), and the release rate governs how much oxygen carrier will be needed in the fuel reactor at full combustion conditions.

There are certain materials which have only a small amount of oxygen which can be released via CLOU and where it can be expected that both CLC and CLOU reactions proceed in parallel. Some combined metal oxide systems and natural manganese ores could be examples of such materials. [1]

This paper will present a simple analytical generic model for the gas conversion in the fuel reactor when solid fuels is utilized. Here, the gas conversion can be obtained as a function of only a few easily determined parameters. From the model, rate criteria for oxygen carriers in order to achieve a certain level of gas conversion at a given oxygen carrier inventory can be determined.

Generic analytical model of the fuel reactor for CLC, CLOU and semi-CLOU

In order to determine the reactivity needed for oxygen carriers in a fuel reactor, a generic fuel reactor model was derived which can be used for the different conversion mechanisms, i.e. i) CLC, ii) CLOU and iii) semi-CLOU materials. In deriving the model the following constraints were used:

- The model should be simple with an analytical solution
- Employs effective rate constants for the different reaction rates of the OC, and these rate constants should be easily obtained from laboratory measurements.
- Result in gas conversion as a function of readily determined parameters.

Generic equation

In the following analysis, it will be assumed that both volatiles and char gasification products are composed of CO and H_2 only. Figure 1 shows a schematic of the two general cases studied. The analysis will be performed using a mass balance over a section *dm* of the bed solid material for two general cases: a) gaseous reactants (H_2 or CO) released from the bottom of the bed, which then flow in a plug-flow pattern through the bed where these components react with the oxygen carrier, see Fig. 1a, and b) gaseous reactant components CO and H_2 are released evenly throughout the bed, Fig 1b. The first case would represent a likely reaction pattern for volatile components if they are released at the bottom of the fuel reactor. Case b) is representative gasification of the char fraction to a syngas, consisting of CO and H_2, which then reacts with the oxygen carrier particles. Case b) results in a gas expansion throughout the bed due to constant production of gases from the solid char, which

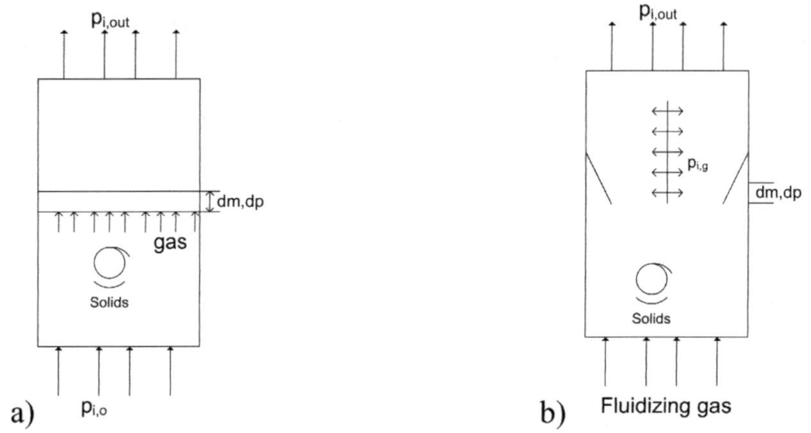

Figure 1. Reaction model for fuel reactor for Case a) where CO or H_2 are introduced from the bottom of the bed, and Case b) where gases are generated evenly throughout the bed.

should be taken into account. This is not the case when the CO and H_2 are released at the bottom of the bed, for instance, from the volatiles, as the reactions with the oxygen carrier particles to fully oxidized products does not constitute a change in flow. Of course oxidation of hydrocarbons would result in gas expansion when reacting, but this has not been taken into account here.

Assuming a well mixed solid phase (char and OC particles) and gases moving according to plug flow, a general mass balance over a layer *dm* of the bed can be performed. The difference in partial pressure dp of the reactant i, i.e. either CO and H_2, can be written as:

$$dp_i = \left[\frac{-k_F m_{tot}}{F_0} p_i - \kappa p_i - \frac{bk_C m_{tot} RT}{F_0} + p_{i,g}(1+\kappa)\right] \cdot \frac{dm}{m_{tot}\left(1+\kappa \frac{m}{m_{tot}}\right)}. \qquad (7)$$

The definition of the parameters in eq. (7) are given in the List of symbols section, and the details of the derivation will be given in the extended version of this paper. The first term on the r.h.s reflects the consumption of syngas reactant (CO or H_2) by direct reaction with the OC, i.e. reaction (3) above. The second term is related to the dilution of the reactive component by release of syngas. The third term is due to the reaction with gas phase oxygen released from the particles, i.e. CLOU and the finally the fourth term is due to the production of gaseous component i, i.e, $p_{i,g}$ corresponds to the source of reactant through char gasification in Case b. As is evident from the general equation, it is assumed that the syngas components react via a first order reaction with respect to component i. Integrating equation (7) provides the general solution for the partial pressure of gaseous component i,

$$p_i = \frac{1}{(\alpha+\kappa)}\left\{p_{i,g}(1+\kappa) - p_{i,c} + \left[(\alpha+\kappa)p_{i,o} - p_{i,g}(1+\kappa) + p_{i,c}\right]\left(\frac{1+\kappa\frac{m}{m_{tot}}}{1+\kappa\frac{m_0}{m_{tot}}}\right)^{-\frac{\alpha}{\kappa}-1}\right\} \qquad (8)$$

In the above equation, $\kappa = F_r/F_o$, where F_r is the gas flow of released gas components due to gasification of the char and F_o is the flow of the fluidizing gas, $p_{i,o}$ is the partial pressure of reactant at the bottom of the bed and α a dimensionless number defined as:

Table 1. Model assumptions and simplified expressions for the gas conversion of reactive component i, in the fuel reactor.

Case	**Assumptions in eq. (8)**	**Expression**
Chemical-looping combustion (CLC)		
a) CO or H_2 introduced from bottom of fuel reactor	$p_{i,c}=0$, $m_o=0$, $m=m_{tot}$, $p_{i,g}=0$, $\kappa=0$, $p_{i,o}=p_{i,max}$	$\gamma = 1 - e^{-\alpha}$
b) CO or H_2 generated evenly in the bed	$p_{i,c}=0$, $m_o=0$, $m=m_{tot}$, $p_{i,0}=0$, $p_{i,g}=p_{i,max}$	$\gamma = 1 - \frac{\left[(1+\kappa)-(1+\kappa)^{-\alpha/\kappa}\right]}{\alpha+\kappa}$
Chemical-looping with oxygen uncoupling (CLOU)		
a) CO or H_2 introduced from bottom of fuel reactor	$m_o=0$, $m=m_{tot}$, $p_{i,g}=0$, $\alpha=0$, $\kappa=0$, $p_{i,o}=p_{i,max}$	$\gamma = \zeta_i$
b) CO or H_2 generated evenly in the bed	$m_o=0$, $m=m_{tot}$, $p_{i,0}=0$, $\alpha=0$, $p_{i,g}=p_{i,max}$	$\gamma = \zeta_i$
Semi-CLOU		
a) CO or H_2 introduced from bottom of fuel reactor	$m_o=0$, $m=m_{tot}$, $p_{i,g}=0$, $\kappa=0$, $p_{i,o}=p_{i,max}$	$\gamma = \left[1+\frac{\zeta}{\alpha}\right](1-e^{-\alpha})$
b) CO or H_2 generated evenly in the bed	$m_o=0$, $m=m_{tot}$, $p_{i,0}=0$, $p_{i,g}=p_{i,max}$	$\gamma = 1 - \frac{1}{(\alpha+\kappa)}\left\{\left[(1+\kappa)(1-\zeta)\right]\left[1-(1+\kappa)^{-\frac{\alpha}{\kappa}-1}\right]\right\}$

$$\alpha = \frac{k_F m}{F_0} \tag{9}$$

The rate constant k_F is an effective rate constant, expressed as m_n^3gas/kg OC,s and gauges the reactivity of the oxygen carrier with respect to the direct gas-solid reaction between oxygen carrier and gas. This rate constant can be obtained through simple experimental evaluations, and can also be related to common rate constants, such as the shrinking-core model. [2] Further, $p_{i,c}$ in eq. (8) is the partial pressure of the generated component i which has been consumed via the CLOU reaction, and calculated from,

$$p_{i,c} = \frac{b k_C m_{tot} RT}{F_0} \tag{10}$$

Here, the constant b is related to the moles of species i reacted per mol of oxygen released from the oxygen carrier and depends upon to what extent CO or H_2 reacts with the oxygen, i.e. rate of H_2 and CO combustion. The effective rate constant k_c is the rate constant of O_2 release via reaction (4). In the analysis above it is assumed that there is no equilibrium restraint on reaction (4). Below follows the solution of the general equation (8) for the three different cases explored in this work.

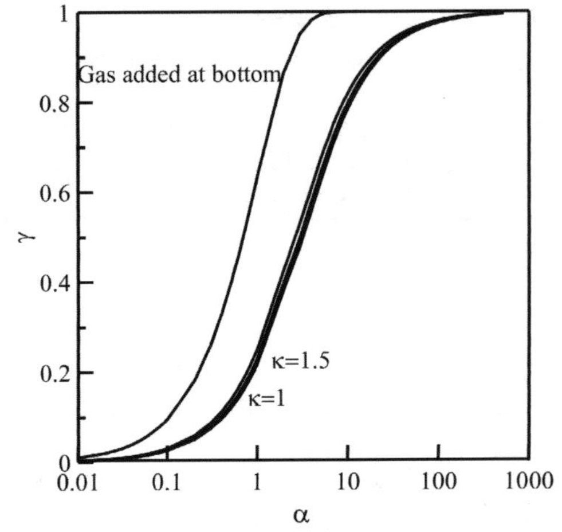

Figure 2. Gas conversion, γ, as a function of α for CLC, both when CO or H_2 is added from the bottom of the fuel reactor and generated throughout the bed.

Chemical-looping combustion (CLC)

Case a) CO and H_2 generated at bottom of bed. It will be assumed that the reactant components are composed of CO or H_2 only. The general equation (8) can be simplified, according to the assumptions given in Table 1. For instance, there is no CLOU effect, and hence $p_{i,c}=0$ in eq. (8) There are no gases generated in the bed, and thus $p_{i,g}=0$. Further, $p_{i,o}=p_{i,max}$ as the inlet partial pressure will be the same as the maximum partial pressure if no reaction with the OC is taking place. Setting $m=m_{tot}$, equation (8) can be simplified and reduced to:

$$p_i(\alpha) = p_{i,max} \cdot e^{-\alpha}. \tag{11}$$

The gas conversion is expressed as,

$$\gamma = 1 - \frac{p_i}{p_{i,max}} \tag{12}$$

and thus the gas conversion as a function of α is simply,

$$\gamma = 1 - e^{-\alpha} \tag{13}$$

Case b) CO and H_2 generated evenly in the bed. When the syngas is produced evenly throughout the bed, there will be a continual increase in flow throughout the bed, which is due to the gases generated by the gasification. In this case the boundary conditions are such that the gas concentration at the bottom are zero, and hence $p_{i,o}=0$. On the other hand, there will be a partial pressure of gaseous species i, due to the gasification reactions, and if no reaction takes place with the OC, $p_{i,g}=p_{i,max}$ in equation (8). With some further assumptions, as given in Table 1, a combination of equation (8) and (12) gives, [2]

$$\gamma = 1 - \frac{\left[(1+\kappa) - (1+\kappa)^{-\alpha/\kappa}\right]}{\alpha + \kappa} \tag{14}$$

Figure 2 shows the gas conversion, γ, as a function of the dimensionless number α for both cases above, i.e. eq. (13) and (14). Clearly, the release of gases throughout the bed of material will mean that a much higher bed mass will be needed in order to achieve high gas conversion. The exponential form is indicative of the first-order dependence assumed with respect to p_i and means that at high degrees of gas conversion a relatively large increase in the effective rate constant or bed mass would be needed in order to achieve a constant incremental increase in gas conversion. Further, the effect of gas expansion is rather limited as seen for the curves with different values of κ.

Chemical-looping with oxygen uncoupling (CLOU)

For the CLOU process the released volatiles and char originating from the fuel will react directly or indirectly with the released oxygen from the oxygen carriers. The amount of OC needed in the fuel reactor is simply related to the oxygen release rate and the oxygen demand of the fuel. Below follows the analysis of this process in a similar way as above.

Case a) CO and H_2 released from bottom of bed. The syngases will be combusted by gas-phase oxygen, generated by CLOU, and hence α=0 in eq. (8). There are no gases generated in the bed, and thus $p_{i,g}=0$. Setting $p_{i,o}=p_{i,max}$ as the inlet partial pressure will be the same as the maximum partial pressure if no reaction is taking place. At $m=m_{tot}$, equation (8) can be simplified and reduced to:

$$p_i(\alpha_2) = p_{i,max} - p_{i,c} \tag{15}$$

The gas conversion can then be expressed as

$$\gamma_i = \zeta_i \tag{16}$$

where ζ_i is the CLOU factor for species i, which is a measure of how much of the required oxygen is supplied in the gas phase by the oxygen carrier, and defined as

$$\zeta_i = \frac{p_{i,c}}{p_{i,max}(1+\kappa)} \tag{17}$$

When ζ_i is unity, the oxygen demanded by the fuel is equal to the supply rate from the oxygen carrier. In relation to a feasible scenario of volatiles released from the bottom, there will likely be a mixture of CO, H_2, and hydrocarbons which will demand oxygen. Hence, in order to determine the actual reaction pattern of component i, the individual reaction rates need to be determined.

Case b) CO and H_2 generated evenly in the bed. Although there is no CO or H_2 generated through gasification during CLOU of solids fuel, the generic equation (8) can still be employed to analyze char combustion with CLOU. With the assumptions given in Table 1 in combination with eq. (12) the solution would be the same as that given in equation (16) above.

In summary, it is clear from equation (16) and (17) that the gas conversion for a CLOU process is lineraly dependent upon the oxygen release rate from the oxygen carrier. This is in contrast to the normal CLC case, where there is a clear logarithmic dependence of the gas conversion on the reactivity and bed mass, see Fig. 2. Hence if there is a sufficient release of oxygen from the particles, it may be easier to obtain full conversion of gas using CLOU particles.

Combined CLC and CLOU (semi-CLOU)

Recently a number of combined oxides have been shown to have CLOU properties. {Ryden, 2012 #424} But experimental evidence suggests that some of these materials only release a part of their active oxygen as CLOU, and in some cases, the rate of release is limited. Here it is likely that the materials react via both CLC and CLOU mechanisms in the fuel reactor. Equation (8) will be used to analyze these semi-CLOU materials in the same way as above.

Case a) CO and H_2 released from bottom of bed. As with the cases above, here it will be assumed that there are no gases generated in the bed, and hence $p_{i,g}=0$. Also, $p_{i,o}=p_{i,max}$ as the inlet partial pressure will be the same as the maximum partial pressure if no reaction is taking place. Setting $m=m_{tot}$, equation (8) and (12) can be combined and reduced to:

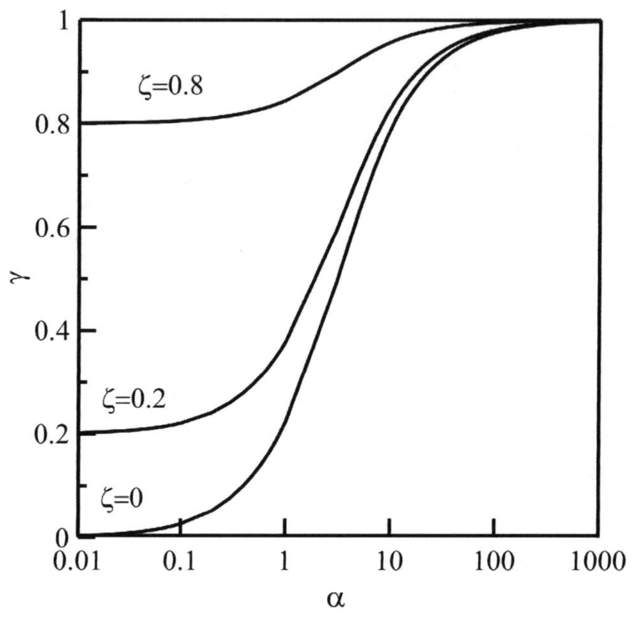

Figure 3. Gas conversion, γ, as a function of α for semi-CLOU with different values of the CLOU-factor ζ.

$$\gamma = \left[1 + \frac{\zeta}{\alpha}\right](1 - e^{-\alpha}) \tag{18}$$

Case b) CO and H_2 generated evenly in the bed. When the gases are generated throughout the bed of particles, the partial pressure at the bottom, $p_{i,o}$, will be zero. On the other hand, there will be a partial pressure of gaseous species i, due to the gasification, and if no reaction takes place with the OC, $p_{i,g}=p_{i,max}$ in equation (8), as was the case with normal CLC. With some further assumptions, as given in Table 1, equation (8) and (12) combine to yield,

$$\gamma = 1 - \frac{1}{(\alpha + \kappa)}\left\{\left[(1+\kappa)(1-\zeta)\right]\left[1 - (1+\kappa)^{-\frac{\alpha}{\kappa}-1}\right]\right\}, \tag{19}$$

Figure 3 shows the gas conversion, γ, as a function of the dimensionless number α for the case of constant generation of CO and H_2 in the bed. Here some different values of the CLOU factor are shown. As can been observed, the initial conversion at low α is dominated by the given CLOU reaction. At higher values of α, the effect of CLOU becomes less prominent as the oxygen released will react with CO and H_2 that would anyway have reacted with the oxygen carrier particles.

Discussion

In a chemical-looping process it is important to try to keep the solids inventory as low as possible, in order to reduce the size of reactors as well as the amount of oxygen carrier material in the system. The analytical expressions obtained in this work gives the gas conversion as a function of α,ζ and κ, see Fig 2 and 3. These expressions then be used to obtain reactivity criteria for oxygen carrier particles which react via one of the three mechanisms discussed. For instance, for a particle reacting only via the CLC mechanism, the α needed to achieve a certain gas conversion is given in Figure 2. The necessary effective rate constant k_F which is needed by a particular oxygen carrier for a given bed mass and fluidization flow can then be obtained through rearrangement of eq. (9),

$$k_F = \frac{\alpha F_0}{m} \quad (9b)$$

A similar procedure can be used for semi-CLOU materials, given a certain CLOU-factor. For pure CLOU oxygen carriers, the solution is linear and trivial, see eq. (16).

Conclusions

This paper presents a general analytical model for the fuel reactor using solid fuels. The model generates the gas conversion as a function of only a few parameters, i.e. α, ζ and κ. The model is applied to three cases for solid fuel conversion, i.e. CLC, CLOU and semi-CLOU processes. Two cases are analyzed for each conversion mechanism a) reactive species CO and H_2 are introduced at the bottom of the fluidized bed and b) the CO and H_2 are generated throughout the bed of oxygen carrier material. From the model it is possible to determine the effective rate constants necessary to achieve specified levels of gas conversion.

List of symbols

b	stoichiometric coefficient (mol species i reacted/mol O_2 released)
F_0	fluidizing flow at the bottom of the fluidized bed (m_n^3/s)
F_r	additional gas released due to the gasification reaction (m_{n3}/s)
k_F	effective reaction rate constant for the CLC reaction (m_n^3/(kg.s))
k_c	effective rate constant for CLOU reaction (mol O_2/(kg.s))
m	mass of oxygen carrier (kg)
m_{tot}	mass of oxygen carrier in the fuel reactor (kg)
p_i	partial pressure of species i (bar)
$p_{i,o}$	inlet partial pressure of species i (bar)
$p_{i,c}$	partial pressure of species i reacted via CLOU (bar)
$p_{i,g}$	partial pressure of species i originating from char gasification (bar)
$p_{i,max}$	maximum partial pressure of species i if no reaction with oxygen carrer occurs (bar)
α	dimensionless number defined in eq. (9)
γ_i	conversion of species i (%)
κ	dimensionless number gauging the extent of volume expansion in the bed
ζ	CLOU-factor

Acknowledgements

This work in this paper was financed by Vattenfall AB as well as Chalmers University of Technology Energy Area of Advance.

References

1. Ryden, M., et al., *Combined oxides as oxygen carrier material for chemical-looping with oxygen uncoupling* in *2nd International Conference on Chemical-Looping*. 2012: Darmstadt.
2. Berguerand, N., et al., *Chemical-Looping Combustion of Solid Fuels in a 10 kWth Unit.* Oil & Gas Science and Technology - Rev. IFP, 2011. **DOI: 10.2516/ogst/2010023**.

(515f) Development and Performance Evaluation of a Cost-Effective Oxygen Carriers

Liangyong Chen, Yi Zhang, Fang Liu, Chunli Han, James K. Neathery and Kunlei Liu, Center for Applied Energy Research, University of Kentucky, Lexington, KY

Chemical looping combustion (CLC) for power generation has the inherent potential of enriching CO_2 from flue gases in-situ with low energy penalty compared to other commercial available options. However, the feasibility of applying CLC to burning coal is heavily depended on the suitability of oxygen carrier (OC). The ideal OCs shall present high reactivity, high resistance to attrition, no agglomeration at high temperature, cost-effective, and low secondary environmental impact. University of Kentucky Center for Applied Energy Research (CAER), which focuses on developing advanced CLC of solid fuel for large-scale power generation, has been devoting to development of cost-effective OCs using Fe-based materials. The purpose of present research is to investigate the potential of red mud [residue (waste) from alumina production] to produce cost-effective OCs for coal-fueled CLC and to obtain its performance for comparison with Fe-based OCs. The OC samples used are made from red mud that contains 33wt% Fe-element by freeze granulation (FG) methods. In preliminary study, the reductivity and oxygen transport capacity of red mud OCs produced from different sintering temperatures (1000 ^{o}C-1200°C) and times were tested in a thermogravimetric analyzer coupled with a mass spectrometer (TG-MS) using gaseous fuel as reduction agent. The BET surfaces, SEM imagines, mechanical strengths and agglomeration temperature of these red muds OCs were obtained. The performance of the red mud OCs was compared with that of OCs from commercially available Fe_2O_3 and α-alumina powders. The results showed that it is feasible to produce cost-effective Fe-based OCs from red mud with similar performance as the OC prepared from 50% $Fe_2O_3$and 50% α-alumina powders.

(515g) Study of Wyoming PRB Coal for Application of Solid Fueled Chemical Looping Combustion with TGA-MS

Yi Zhang, Jeffrey Slone, Aurora M. Rubel, James K. Neathery, Fang Liu, Chunli Han, Liangyong Chen and Kunlei Liu, Center for Applied Energy Research, University of Kentucky, Lexington, KY

Study of Wyoming PRB Coal for Application of Solid Fueled Chemical Looping Combustion with TGA-MS

Yi Zhang, Jeffery T. Slone, Aurora M. Rubel, James K. Neathery, Fang Liu, Chunli Han, Liangyong Chen and Kunlei Liu*

University of Kentucky, Center for Applied Energy Research

Lexington, KY 40511-8410

Abstract: Since Wyoming coal is used to generate electricity in 38 states and represents 55% of the coal used in the US, a feasibility study of the application of chemical looping combustion technology to Wyoming coal would benefit its use in the future carbon constrained world. Wyoming PRB coal was acquired from a local power plant in the State of Kentucky. The coal was first pyrolyzed at 500-600 °C to remove volatiles and tar to obtain char, which was served as the solid fuel in experiments reported in this paper. The purpose to remove volatiles and tar from coal was to minimize the potential agglomeration of oxygen carriers (OCs) inside reducer, and volatiles and tar produced would be introduced into reducer as part of fluidization media in the CLC process. In this work, a variety of oxygen carriers (OC) were premixed with PRB char with char-to-char ratio of 10:90 (weight based), tested in a thermogravimetric analyzer coupled with a mass spectrometer and a water vapor generator (TGA-MS-WVG). The TGA temperature was ramped up to 950 °C at heating rate of 10 °C/min. The weight change and its changing rate were used to quantify the degree of fuel conversion and reactivity of the OCs. Tested OCs include commercially available catalysts such as fused iron, Wustite and pure Fe_2O_3 powder and housed-made synthetic OCs such as freeze-granulated Fe_2O_3/Al_2O_3(50:50), NiO/Al_2O_3(50:50) and CuO/Al_2O_3(50:50). The carbon conversion rate and reactivity were calculated and compared.

(515h) Techno-Economic Evaluation of Chemical Looping Combustion for Industrial-Scale Applications

John T. Ramsey[1], Michael J Gallagher Jr.[1], Stephen Carpenter[1] and George Richards[2], (1)National Energy Technology Laboratory, URS Corporation, Morgantown, WV, (2)National Energy Technology Laboratory, U.S. Department of Energy, Morgantown, WV

Chemical looping combustion (CLC) is a promising approach to capture carbon dioxide from fossil fuel combustion at low cost. The CLC process relies on metal-based oxygen carrier materials to transport oxygen to the combustion zone, thereby eliminating the costs to separate nitrogen from CO_2, as in air-fired systems. Over the last decade, several lab and pilot-scale CLC systems have been constructed and evaluated, but CLC technology has not transitioned into industrial-scale use due to uncertainties in several performance factors still under investigation. The performance factors primarily include the reactivity, cost, and lifetime of the oxygen carrier material, which dictates the overall size and power-generating capacity of large-scale systems. In this study, a reduced order model was developed to compare oxygen carriers, based on their reaction kinetics and thermodynamic properties, in simulated large-scale (10 and 100 MW) CLC systems. The results of the screening model were fed into an ASPEN simulation of a plant-scale system model to indentify conceptual design configurations that have potential for commercialization. The study compares a single industrial configuration of CLC versus a baseline industrial carbon capture and storage case by considering heat and material balances, preliminary process flow diagrams, and estimations of first-order costs. The outcome of this techno-economic evaluation will provide a basis to identify industrial end-use applications and markets that have the greatest potential to benefit from CLC technology.

(529b) CFD Modeling of Commercial-Scale Entrained-Flow Coal Gasifiers

Jinliang Ma, URS Corporation, Morgantown, WV and Stephen E. Zitney, AVESTAR Center, National Energy Technology Laboratory, Morgantown, WV

Optimization of an advanced coal-fired integrated gasification combined cycle system requires an accurate numerical prediction of gasifier performance. Computational fluid dynamics (CFD) has been used to model the turbulent multiphase reacting flow inside commercial-scale entrained-flow coal gasifiers. Due to the complexity of the physical and chemical processes involved, the accuracy of sub-models requires further improvement. Built upon a previously developed CFD model for entrained-flow gasification, the advanced physical and chemical sub-models presented in this paper include a moisture vaporization model with consideration of high mass transfer rate and a coal devolatilization model with more species to represent coal volatiles and the heating rate effect on volatile yield. The global gas phase reaction kinetics is also carefully selected. To predict a reasonable peak temperature of the coal/O_2 flame inside an entrained-flow gasifier, the reserve reaction of H_2 oxidation is included in the gas phase reaction model. The enhanced CFD model is applied to simulate two typical commercial-scale oxygen-blown entrained-flow configurations including a single-stage down-fired gasifier and a two-stage up-fired gasifier. The CFD results are reasonable in terms of predicted carbon conversion, syngas exit temperature, and syngas exit composition. The predicted profiles of velocity, temperature, and species mole fractions inside the entrained-flow gasifier models show trends similar to those observed in a diffusion-type flame. The predicted distributions of mole fractions of major species inside both gasifiers can be explained by the heterogeneous combustion and gasification reactions and the homogeneous gas phase reactions. It was also found that the syngas compositions at the CFD model exits are not in chemical equilibrium, indicating the kinetics for both heterogeneous and gas phase homogeneous reactions are important. Overall, the results achieved here indicate that the gasifier models reported in this paper are reliable and accurate enough to be incorporated into process/CFD co-simulations of IGCC power plants for system-wide design and optimization.

Ultrasound Measurements of Temperature Profile Across Gasifier Refractories

Yunlu Jia and Mikhail Skliar
Department of Chemical Engineering
University of Utah

Coal and biomass gasification is a process characterized by extreme operating conditions of high temperatures and pressures, chemical aggressiveness, and mechanical abrasion. Several technological challenges impact the reliability and economics of coal gasification. The key challenge remains the complete lack of sensors that are capable to reliably perform in harsh environment over an extended period of operation. The conventional approach of developing hardened conventional insertion sensors has proven to have limited effectiveness in the aggressive gasifier environment. Gasifier operators routinely use "indirect measurements", such as methane concentration in the gasifier product gas ("temperature" is reported in ppm of methane!), to estimate the average temperature in the system. Such indirect methods cannot be used to infer temperature if the observed methane concentration is caused by the presence of a region or regions inside the gasifier that are running particularly hot. Note that optical techniques, including combustion specific measurements of temperatures and reaction composition, are hardly suitable for a pressurized, slagging gasifier since the flowing slag blocks optical access ports. They are also not suitable for measuring thickness and other characteristics of non-reflective and non-transparent containments.

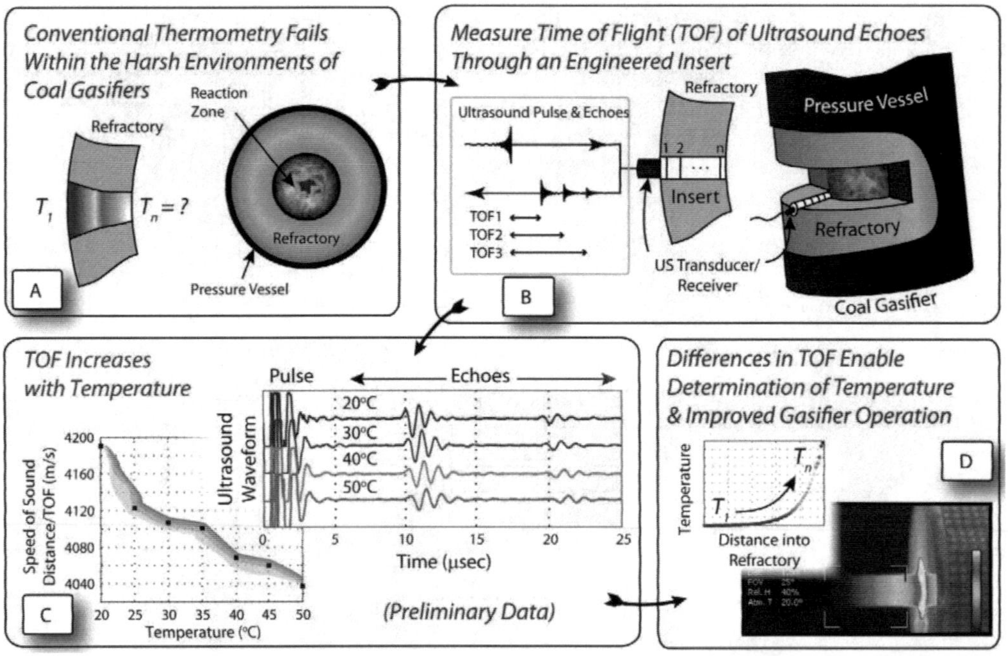

Fig. 1: The pulse propagates through an engineered material which produces multiple partial echoes (panel B). The TOF of each echo is measured and used to calculate the SOP which changes with the temperature (panel C) of the corresponding segment of the refractory. By sequentially estimating the temperature of each segment, the temperature distribution along the entire path of ultrasound propagation is obtained (panel D).

In this presentation, we describe a novel approach that uses noninvasive ultrasound to measure: (1) spatial distribution of temperature across the refractory or other containment; (2) temperature of the hot side of the refractory and (3) refractory thickness. In the simplest implementation of the temperature measurements using ultrasound (US), the time of flight (TOF or return delay) of an US pulse introduced on the cold side of the refractory and reflected from the refractory's hot side is measured. The temperature dependence of the speed of sound can then be used to determine an "average" temperature over the refractory thickness. This simple approach fails when significant thermal gradients are present, as is the case for gasifier refractories. The central idea, summarized in Fig. 1, which enables us to measure the temperature distribution when large gradients are present, is to create an US propagation path inside the refractory with controlled backscattering at predetermined spatial locations. In Fig. 1, the same piezoelectric element serves as the transducer and receiver; modification for the case of a separate receiver is straight forward. The transducer-generated pulse propagates through an engineered material which produces multiple partial echoes (panel B). The time of flight of each echo is measured and used to calculate the speed of sound (SOS) which changes with the temperature (panel C) of the corresponding segment of the refractory. By sequentially estimating the temperature of each segment, the temperature distribution along the entire path of ultrasound propagation is obtained (panel D). We will present experimental validation of this approach and outline options for creating engineered materials that produce partial US reflections from pre-determined positions. The achievable accuracy and spatial resolution of the measured temperature profile will be discussed.

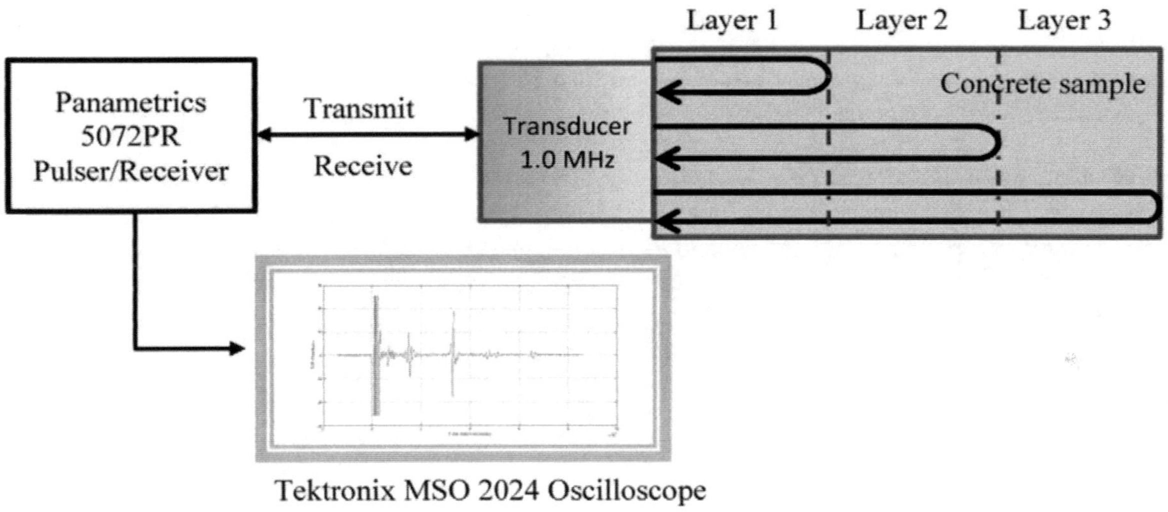

Fig. 2: Experimental setup for ultrasound testing of layered samples in pulse-echo mode.

The key component to implement the proposed approach is the ability to create partial ultrasound reflections from known spatial locations inside the sample. An ideal choice for partial reflectors would be a material with identical thermal expansion, and chemical and mechanical resistances similar to that of the refractory material. Using cementitious materials as refractory model, we investigated different ways to introduce partial internal reflections. It was found that sequential casting of cementitious refractory with time in between to allow for a partial curing creates enough properties variations to introduce partial reflections. The ultrasound tests were carried out using Panametrics pulser/receiver (model 5072PR) and Panametrics immersion transducer (model V302) which has a central frequency of 1 MH. The data were acquired using Tektronix oscilloscope (model MSO 2024) interfaced to a computer. The overall data acquisition setup is shown in Figure 2.

The ultrasound testing revealed that cementitious samples produce clear partial internal reflections from each interface created by layered casting of cementitious refractory. The reflections are clear without waveform overlap for a 3-layered sample shown in Figure 3. This suggests that 1 inch or better spatial resolution of the ultrasound measurements of temperature distribution may be achieved with the proposed approach.

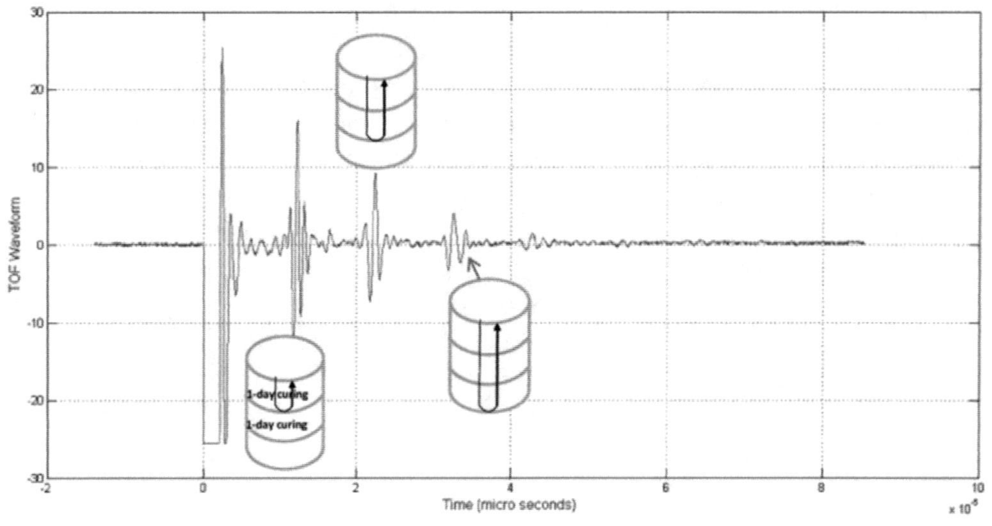

Fig. 3: Ultrasound response for samples with the two internal interfaces obtained by allowing the two initial concrete layers to cure for 1 day each. The last layer was cured for at least one day. Water-cement ratio equal to 0.44 was used. Echoes from both interfaces are well defined, with relatively high signal-to-noise ratio.

The preliminary ultrasound measurements on these cementitious samples were carried out at low temperatures with the water saturated sample kept in a temperature-controlled water bath. All cementitious samples were tested with randomizing temperature order and repeated at least 6 times for each temperature between 20 and 50°C, in 5°C increments. After each change of the water bath temperature, sufficient time was allowed for thermal equilibration to occur before the next TOF measurement.

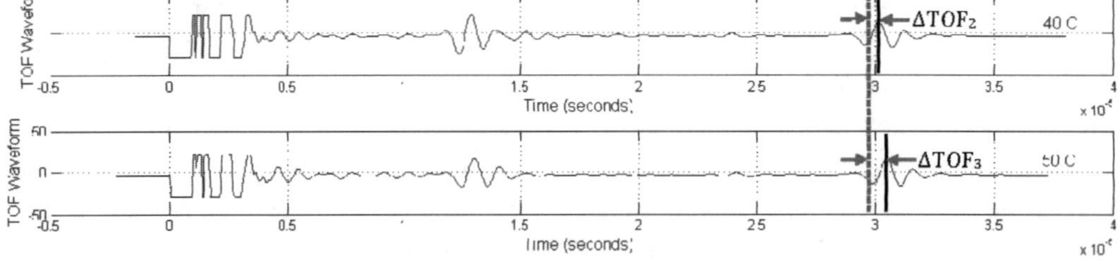

Fig. 4: The change in the TOF between echo waveforms at different temperatures is calculated by crosscorrelation with the reference waveform acquired at 20°C. The results are shown for a refractory sample with a single embedded partial ultrasound reflector located in the middle of the sample and obtained by layered casting of refractory concrete.

Since the speed of sound is calculated as the distance traveled by the ultrasound pulse (in pulse-echo experiments, this distance is equal to twice the length of the sample, L) divided by the time of propagation (TOF),

$$\mathrm{SOS} = \frac{2\mathrm{L}}{\mathrm{TOF}},$$

the method for precise measurements of the TOF is needed. The change ΔTOF in the time it takes for an echo to return to the transducer after reflection from the sample's distal end was measured as a function of temperature by performing cross-correlation analysis between the reference waveform (the first trace in Fig. 4, corresponding to room temperature of 20°C, was used as a reference) and the waveforms obtained at different temperatures. The speed of sound at each temperature is then calculated as

$$SOS = \frac{2L}{Ref\ TOF + \Delta TOF},$$

where the reference TOF at 20°C was obtained using the multiple round-trip method.

Figure 5 summarizes results on the measured speed of sound as a function of temperature, obtained with different experimental and data analysis methods. These preliminary results are consistent and confirmed that there is a strong, measurable relationship between the temperature and the speed of sound in the model refractory in the considered temperature range.

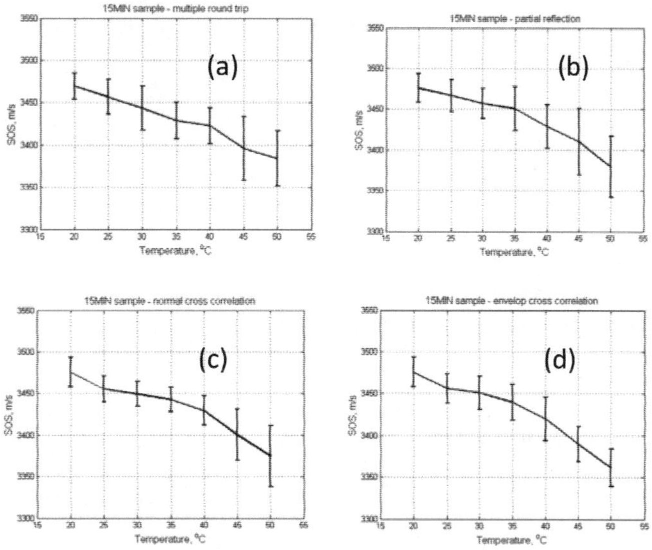

Fig. 5: For the same sample as used in Fig. 4, the calibration curves for the speed of sound as a function of temperature were obtained using different experimental and data analysis methods. Error bars show 95% confidence interval. (a) The calibration curve was obtained using round trip measurements of the TOF at each temperature and the waveform crosscorrelation of echoes from the distal end of the sample. (b) The integrated delay line method was used to obtain TOF at each temperature using waveform crosscorrelation between the distal and internal echoes. (c) ΔTOF at each temperature relative to the reference values of the TOF at 20°C is used to calculate the SOS. ΔTOF is found by crosscorrelation of the waveforms. The reference TOF is found using the round trip method. (d) The same as (c) but the ΔTOF is found by crosscorrelating the waveform envelopes.

(529d) Recent Developments in Carbonaceous Chemistry for Computational Modeling (C3M) with Uncertainty Quantifications (UQ)

Kiran Chaudhari, Chemical Engineering, West Virginia University, Morgantown, WV, Richard Turton, Department of Chemical Engg., West Virginia University, Morgantown, WV, Chris Guenther, National Energy Technology Laboratory, U.S. Department of Energy, Morgantown, WV, Mehrdad Shahnam, Computational Science Division, National Energy technology Laboratory, Morgantown, WV, Aytekin Gel, Alpemi Consulting LLC, Phoenix, AZ, Philip Nicolleti, URS,National Energy Technology Laboratory, Morgantown, WV, Tingwen Li, URS, Morgantown, WV, Dirk VanEssendelft, National Energy Technology Laboratory, Morgantown and Nate Weiland, Mech. & Aerosp. Eng, West Virginia University, Morgantown, WV

The Department of Energy's National Energy Technology Laboratory (NETL) has developed a software platform entitled Carbonaceous Chemistry for Computational Modeling (C3M) that is used to access a variety of kinetic processes and reaction mechanisms typically found in coal gasification, gas clean-up, and carbon capture processes. It improves coal/biomass/ petcoke gasification process simulation by linking computational fluid dynamic (CFD) software codes such as open source software Multiphase Flow with Interphase Exchanges (MFIX) developed at NETL, ANSYS-FLUENT by ANSYS Inc. and BARRACUDA by CPFD Software with chemical kinetics and laboratory data. These kinetic expressions describe the fundamental steps taking place in the gasification of coal/petcoke/biomass, namely, devolatilization, tar-gas chemistry, soot formation, followed by the heterogeneous and homogeneous gasification and combustion reactions. For this purpose, the kinetic data generated through a number of detailed models such as PC Coal Lab (PCCL), Chemical Percolation Model for Coal Devolatilization (CPD), Solomon's Functional-Group, Depolymerization, Vaporization, Cross-linking (FGDVC) model, or experimental data currently being generated at NETL are used.

C3M provides the option of fuel selection such as a wide variety of coal, biomass and petcoke. The C3M graphical user interface (GUI) has been modified to allow users to exercise the various kinetic models to evaluate graphically the effect different fuels and/or gasifier operating conditions have on gasification kinetics along with the yield of product species. After scrutinizing the data, the C3M GUI allows the multiphase CFD user to select kinetic information from a particular kinetic model and have that information properly formatted and seamlessly integrated into the CFD input files ready for immediate use. By incorporating chemical reaction mechanisms into a multiphase computational fluid dynamic (CFD) model, C3M effectively opens a virtual window into the actual operation of a gasifier giving engineers, designers, and plant managers access to information inside a gasification-based energy system, which has previously been unavailable.

Uncertainty Quantification (UQ) for coal gasification processes is a unique feature that can be coupled with C3M. Utilizing a UQ engine and toolbox (e.g., PSUADE, DAKOTA), the user can observe and predict the uncertainties/variations in product yields and reaction rates with the prescribed variability in the operating conditions and fuel properties. This is achieved by conducting systematically designed runs on the kinetic packages available in C3M and analyzing the output with the UQ toolbox. This is a very cheap and cost effective method in terms of time and computer capability. The preliminary results and the manner by which the UQ work can be extended for CFD packages will be discussed.

In all circumstances, C3M provides the user the ability to conduct virtual kinetic experiments using leading kinetic packages and available experimental data to evaluate kinetic predictions as a function of fuel and sorbent type and/or operating conditions.

(529e) A One-Dimensional Transient Model of a Single-Stage, Downward-Firing, Entrained-Flow Gasifier

Job Kasule[1], Richard Turton[1], Debangsu Bhattacharyya[1] and Stephen E. Zitney[2], (1)Department of Chemical Engg., West Virginia University, Morgantown, WV, (2)AVESTAR Center, National Energy Technology Laboratory, Morgantown, WV

The integrated gasification combined cycle (IGCC) technology has emerged as an attractive alternative to conventional coal-fired power plant technology due to its higher efficiency and cleaner environmental performance especially with the option of CO_2 capture and sequestration. The core unit of this technology is the gasifier whose optimal performance must be understood for efficient operation of IGCC power plants. This need has led a number of researchers to develop gasifier models of varying complexities.

Whereas high-fidelity CFD models can accurately predict most key aspects of gasifier performance, they are computationally expensive and typically take hours to days to execute on high-performance computers. Therefore, faster one-dimensional (1D) partial differential equation (PDE)-based models are required for use in dynamic simulation studies, control system analysis, and training applications. A number of 1D gasifier models can be found in the literature, but most are steady-state and have limited application in the practical operation of the gasifier. As a result, 1D PDE-based dynamic models are needed to further study and predict gasifier performance under a wide variety of process conditions and disturbances.

In the present study, a 1D transient model of a single-stage downward flow GE/Texaco-type gasifier has been developed. The model comprises mass, momentum and energy balances for the gas and solid phases. The model considers the initial gasification processes of water evaporation and coal devolatilization. In addition, the key heterogeneous and homogeneous chemical reactions have been modeled. The resulting time-dependent PDE model is solved using the well-known method of lines approach in Aspen Custom Modeler®, whereby the PDEs are discretized in the spatial domain and the resulting differential algebraic equations (DAEs) are then solved to obtain the transient response. The transient response of various gasifier performance parameters to certain disturbances commonly encountered in the real world operation of commercial IGCC plants will be presented. These disturbances include ramp and step changes in input variables such as coal flow rate, oxygen-to-coal ratio and water-to-coal ratio, among others. Comparison of gasifier model predictions to available dynamic data will also be discussed.

CO-PYROLYSIS OF LOW RANK COALS AND BIOMASS AT TRANSPORT GASIFIER CONDITIONS

Nathan T. Weiland, National Energy Technology Laboratory, Pittsburgh, PA
West Virginia University, Morgantown, WV

Nicholas C. Means, National Energy Technology Laboratory, Pittsburgh, PA
URS Corp., South Park, PA

Chris Guenther, National Energy Technology Laboratory, Morgantown, WV

Introduction

Given the EPA's proposed rules on limiting CO_2 emissions from new coal power plants to match those of natural gas, blending of carbon-neutral biomass with coal is an increasingly attractive option for power producers. Further, gasification technologies that can capture and store CO_2 prior to combustion are also being considered, and efforts to reduce the cost of electricity from such plants include the consideration of cheaper, lower rank coals and lignites, a significant and under-utilized resource within the United States. Transport gasifiers hold significant promise for achieving the desired CO_2 emissions reductions in this manner, particularly in regard to low rank fuels. As a result, the United States Department of Energy's National Energy Technology Laboratory (NETL) has undertaken a significant research effort to advance the technology of transport gasification through state-of-the-art multi-phase computational modeling.

In order to most accurately model the effects of co-feeding low rank coal and biomass into such a reactor, an experimental effort has been carried out to obtain kinetic rates and product distributions for co-pyrolysis and co-gasification of coal and biomass feedstocks currently in use at the Transport Integrated Gasifier (TRIG) at the National Carbon Capture Center in Wilsonville, AL. The focus of this study is to evaluate the complex pyrolysis reaction and to provide fundamental co-pyrolysis kinetic data under transport gasifier conditions for inclusion into NETL's Carbonaceous Chemistry for Computational Modeling (C3M) platform, which supplies kinetic information to multiphase CFD codes and other computational models [1].

Experimental Details

Co-pyrolysis testing is performed in the High Pressure Tubular Reactor (HPTR), an isothermal drop reactor shown in Figure 1. A dried sample of pure or blended feedstock is placed in the hopper and purged with argon while the 3 zone heater is heated to the desired temperature. Under a constant flow of argon, the backpressure valve is used to set the desired pressure. Once the reactor temperature and pressure reaches a steady state, the hopper valve is opened and the sample falls to the preheated quartz frit to pyrolyze. The product gases are swept out of the reactor in less than 0.5 seconds and are sampled at 1.4 Hz via an in-situ mass spectrometer, which is calibrated to quantitatively detect argon and the primary syngas constituents CO, CO_2, CH_4, H_2, H_2O, H_2S and C_2H_4. Comparison to the known mass flow rate of argon yields a quantitative gaseous product distribution, and the remaining char is weighed after the reactor cools down, with tar being computed by difference. The experimental test matrix includes temperatures ranging from 600 – 975 °C, and pressures of 1, 4 and 16 atm, similar to transport gasifier conditions.

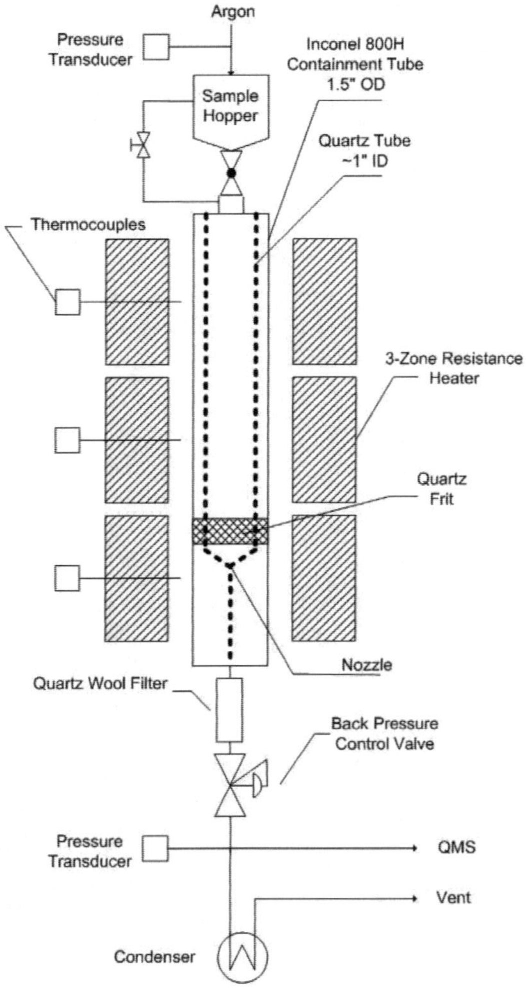

Figure 1: The High Pressure Tubular Reactor

The feedstock test matrix includes sub-bituminous Powder River Basin (PRB) coal, Mississippi lignite (LIG), Southern Yellow Pine (woody biomass, WB), multiple PRB/WB blends, and an 80/20 LIG/WB blend. All samples are ground and sieved to a size range of 106-300 μm, then dried for 24 hours. Proximate and ultimate analyses of the feedstocks are presented in Table 1 on a dry-basis.

Table 1: Proximate and Ultimate Analysis of Feedstocks

Sample	PRB	LIG	WB
Proximate Analysis (wt%, dry basis)			
Volatile	46.18	43.76	85.10
Ash	6.60	25.71	0.64
Fixed Carbon	47.22	30.53	14.26
Ultimate Analysis (wt%, dry basis)			
Carbon	66.21	51.75	48.84
Hydrogen	4.20	3.57	5.78
Nitrogen	1.21	1.27	0.38
Oxygen (diff)	21.30	16.97	44.21
Sulfur	0.48	0.73	0.15

Results and Discussion

Co-pyrolysis of PRB/WB blends at 1 atm was extensively studied as a base case for comparison. Plots of experimental char and gas product distributions are plotted on the left of Figure 2, and are found to be approximately linear with respect to temperature and biomass blend ratio. These results are generalized to all blend ratios, temperatures, and pressures within the ranges studied through four linear curve fit coefficients generated by a regression analysis, as represented by the red mesh on each plot. Some variability in the char data occurs due to incomplete recovery of all of the char in some cases, while variability in the gas data is mostly due to slight pressure variations at the gas sampling location.

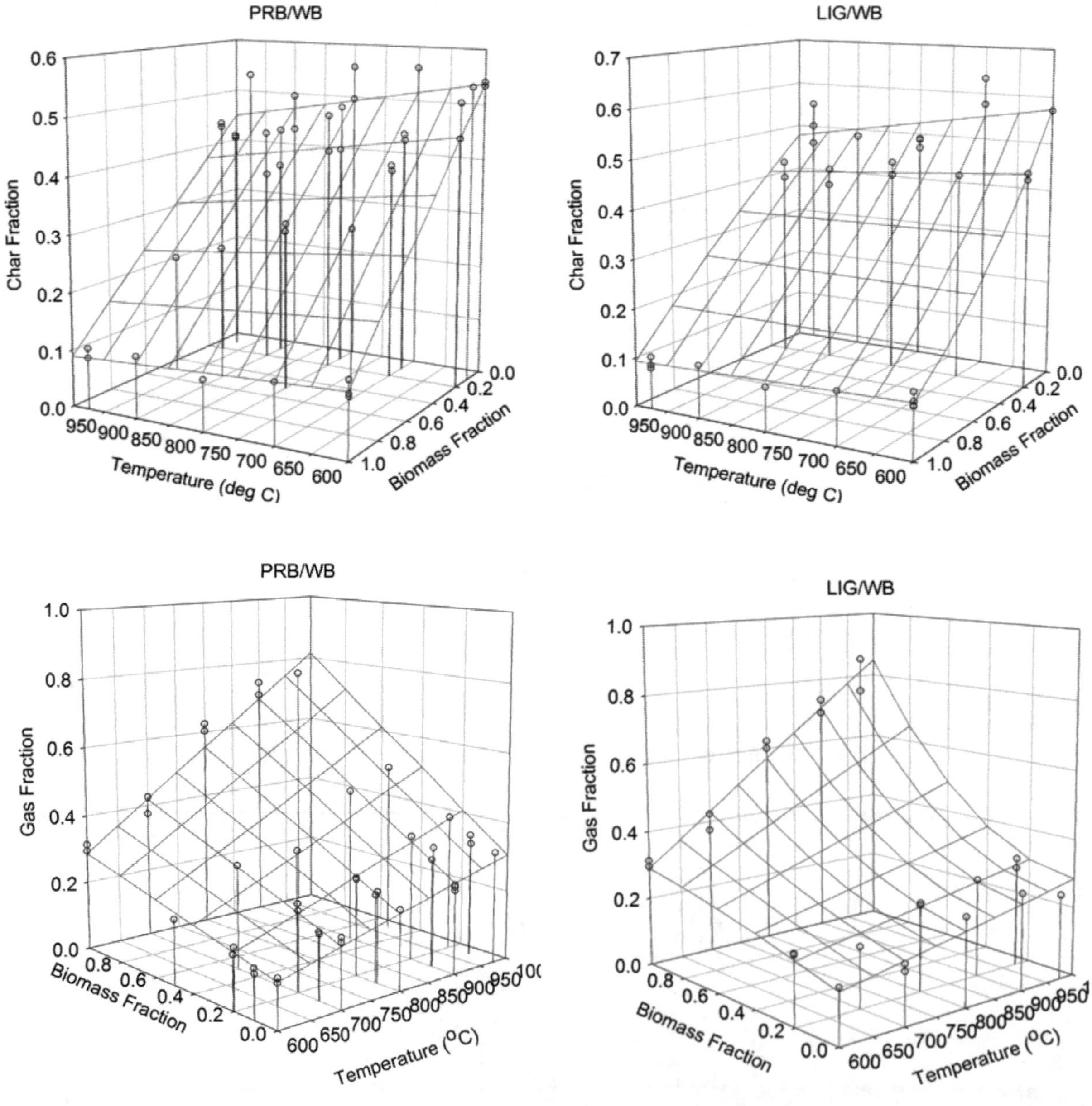

Figure 2: Char (top) and gas (bottom) co-pyrolysis products as fractions of the original feedstock weight for PRB/WB (left) and LIG/WB (right) at 1 atm pressure. Experimental data (blue) and curve fits (red).

Also plotted on the right side of Figure 2 are char and gas product fractions from co-pyrolysis of lignite and woody biomass at 1 atm. LIG/WB char product fractions are largely linear with respect to temperature and biomass feed fraction, however, gas product fractions appear to be somewhat nonlinear with respect to biomass fraction, particularly at temperatures above 700 °C. The char and gas product data are curve fit to a 6-parameter regression that is linear in temperature, but quadratic in biomass weight fraction, as shown by the red mesh. This nonlinearity is consistent with the observations of Haykiri-Acma and Yaman [2], who note that lower rank coals are more susceptible to nonlinear interactions when co-pyrolyzed with biomass. However, their TGA-based experiments can attribute this behavior to overlapping regions of high mass loss at lower temperatures for low rank coals, while higher rank coals pyrolyze at higher temperatures relative to biomass. Isothermal drop-tube experiments performed by Wei et al. [3] show negligible nonlinearities with respect to biomass weight fraction for bituminous and brown coals, though their experiments were only performed up to 700 °C. Fixed bed co-pyrolysis experiments with sawdust and sub-bituminous coal are shown to yield nonlinearities in product yields at lower temperatures (500-700 °C), with linear distributions with respect to biomass fraction and temperature at higher temperatures [4]. The nonlinear nature of the LIG/WB co-pyrolysis results reported here should however be confirmed with tests at intermediate biomass weight fractions (e.g. 50%) to verify these claims.

Kinetic data for the PRB/WB and LIG/WB studies at 1 atm pressure were analyzed to determine the kinetic reaction rate for the first 60-90% of the co-pyrolysis reaction. These were fit to Arrhenius plots and kinetic activation energies, E_a, and pre-exponential factors, k_0, were determined at each biomass blend ratio for a simple first order reaction rate equation. The resulting coefficients are plotted in Figure 3, where they are shown to be non-linear with respect to biomass blend ratio, indicating a significant reaction synergy in co-pyrolysis of low rank coals and biomass that must be accounted for in accurate modeling of transport gasifier systems. Similar results have been presented by Seo et al. [5] for isothermal co-pyrolysis of blends of sub-bituminous coal and refuse derived fuel, composed primarily of biomass.

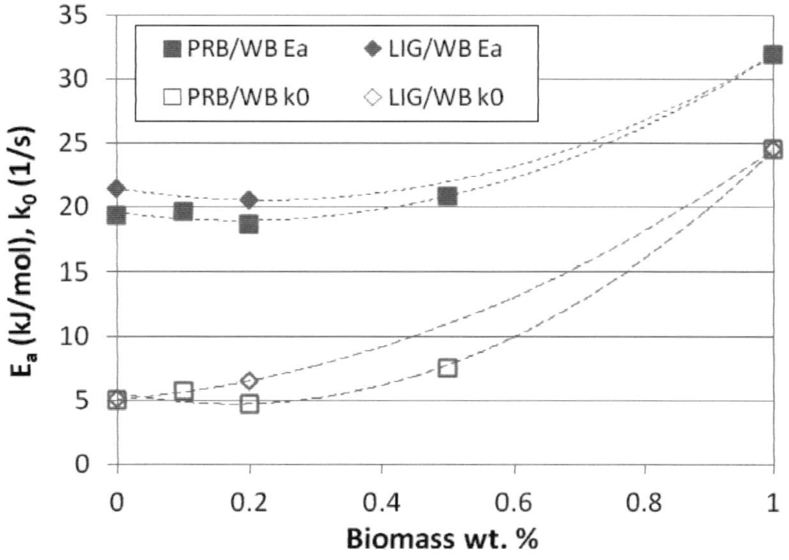

Figure 3: First order rate coefficients for co-pyrolysis of PRB/WB and LIG/WB at 1 atm pressure and temperatures of 600 – 975 °C.

Product distributions for co-pyrolysis of PRB/WB blends at pressures of 1, 4, and 16 atm are presented in Figure 4 for a representative temperature of 700 °C. Char product fractions do not appear to be affected by pressure, though gas yields significantly increase with pressure and biomass weight fraction. Tar yields, though subject to more uncertainty as they are determined by difference, decrease with pressure, particularly for high biomass feed fractions. This is consistent with the analysis of Solomon et al., who observe decreasing tar yields with pressure in coal pyrolysis [6]. This is attributed to an increased residence time of pyrolysis volatiles within the pores of the hot pyrolyzing particle at higher pressures, which leads to heterogeneous cracking of tars to lighter gases.

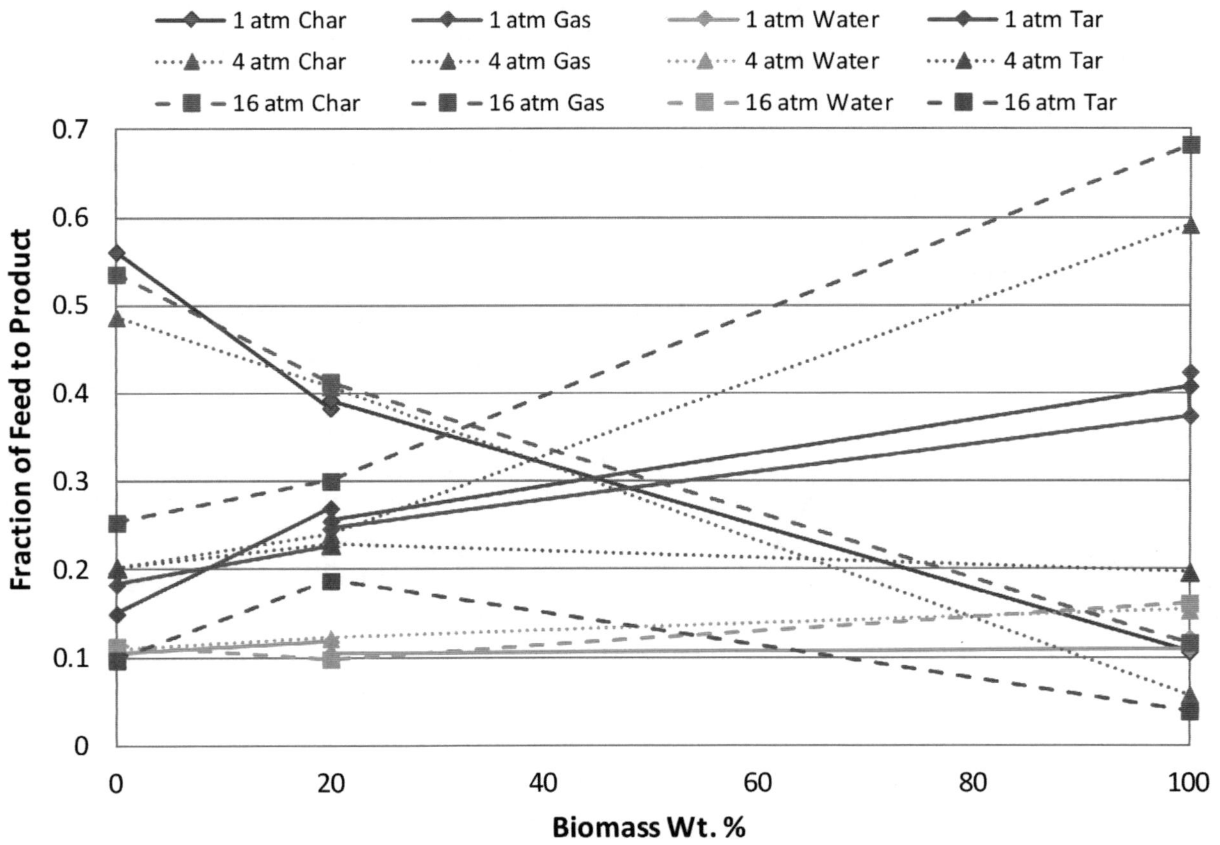

Figure 4: Major product distributions for PRB/WB co-pyrolysis as a function of pressure and biomass weight fraction at 700 °C.

Product gas fractions for the same experiments as in Figure 4 are presented in Figure 5. Of the gases produced during pressurized pyrolysis, CO_2 yields are favored at high pressure across all biomass feed fractions. CO yields increase significantly with biomass weight fraction and temperature (not shown), particularly at higher pyrolysis pressure. Hydrogen is shown to be favored for high coal feed fractions and biomass at high pressure. Methane increases with pressure, consistent with Ref. [6], as well as with biomass fraction. Ethylene trends are similar to methane, but are not shown for clarity. These gas species production results are consistent with the isothermal, atmospheric pressure experiments of Wei et al. [3] and Zhang et al. [7] with respect to biomass feed fraction and temperature, with the exception of hydrogen, which increases with biomass in their experiments.

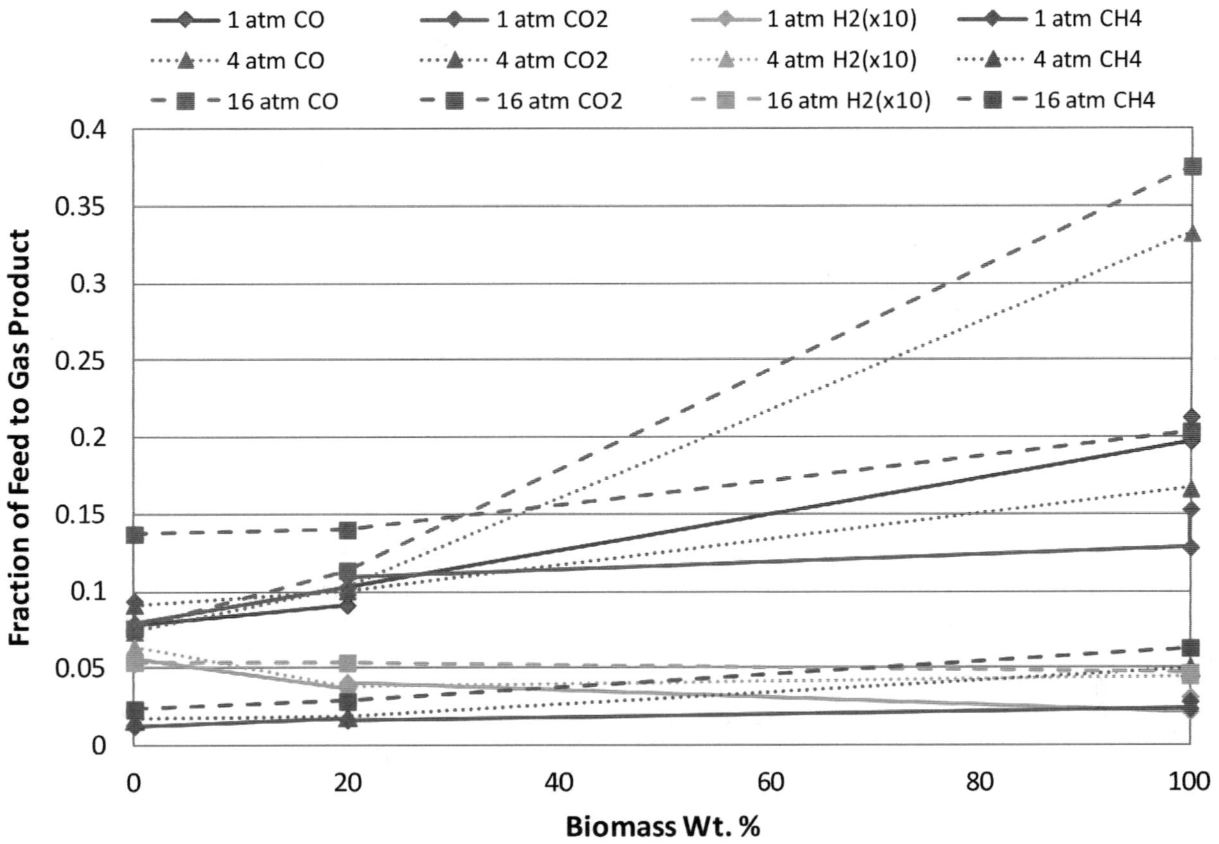

Figure 5: Gas species product distributions for PRB/WB co-pyrolysis as a function of pressure and biomass weight fraction at 700 °C.

Conclusions and Future Work

In atmospheric pressure co-pyrolysis experiments, sub-bituminous coal and woody biomass fuel blends tend to behave in a linear fashion with respect to product distributions, but nonlinearly with respect to kinetic rates. Likewise, lignite and biomass fuel blends also behave nonlinearly in kinetic terms, and also tend to exhibit nonlinear gaseous product distributions at higher temperatures. At higher pressures, tars from sub-bituminous coal and biomass blend co-pyrolysis tend to crack to light gas species, particularly CO_2 for high-coal blends, and CO for high-biomass blends. Non-linearities in product distributions also appear to be more prevalent in high pressure pyrolysis for these feedstocks.

Co-pyrolysis kinetic analyses will be pursued in future work, as will the incorporation of these results into the C3M kinetic database. Afterwards, studies of the kinetics of co-gasification of these feedstocks and their blends at high pressures will be performed, using CO_2, H_2O, and O_2 as the gasification agents.

Acknowledgements

As part of the National Energy Technology Laboratory's Regional University Alliance (NETL-RUA), a collaborative initiative of the NETL, this technical effort was performed under the RES contract DE-FE0004000.

References

1. Guenther, C., D. VanEssendelft, et al. (2012). "Gasification Modeling in the 21st Century." Power Engineering, June, pp. 80-82.
2. Haykiri-Acma, H. and S. Yaman (2010). "Interaction between biomass and different rank coals during co-pyrolysis." Renewable Energy **35**(1): 288-292.
3. Wei, L.-g., L. Zhang, et al. (2011). "Effects of feedstock on co-pyrolysis of biomass and coal in a free-fall reactor." Journal of Fuel Chemistry and Technology **39**(10): 728-734.
4. Park, D. K., S. D. Kim, et al. (2010). "Co-pyrolysis characteristics of sawdust and coal blend in TGA and a fixed bed reactor." Bioresource Technology **101**(15): 6151-6156.
5. Seo, M. W., S. D. Kim, et al. (2010). "Pyrolysis characteristics of coal and RDF blends in non-isothermal and isothermal conditions." Journal of Analytical and Applied Pyrolysis **88**(2): 160-167.
6. Solomon, P. R., M. A. Serio, E. M. Suuberg (1992). "Coal Pyrolysis: Experiments, Kinetic Rates and Mechanisms." Progress in Energy and Combustion Science **18**: 133-220.
7. Zhang, L., S. Xu, et al. (2007). "Co-pyrolysis of biomass and coal in a free fall reactor." Fuel **86**(3): 353-359.

(529g) Dynamic Modeling of Detachment, Transport, and Flow of Slag in an Entrained-Flow Gasifier

Pratik Pednekar[1], Debangsu Bhattacharyya[2], Richard Turton[2] and Raghunathan Rengaswamy[3], (1)Department of Chemical Engineering, West Virginia University, Morgantown, WV, (2)Department of Chemical Engg., West Virginia University, Morgantown, WV, (3)Chemical Engg. Dept., Clarkson University, Potsdam, NY

Dynamic Modeling of Detachment, Transport, and Flow of Slag in an Entrained-Flow Gasifier

Pratik Pednekar, PhD Student, West Virginia University

Department of Chem. Eng., West Virginia University, Morgantown WV 26506

ppedneke@mix.wvu.edu

Debangsu Bhattacharyya, Assoc. Professor, West Virginia University

Department of Chem. Eng., West Virginia University, Morgantown WV 26506

Debangsu.Bhattacharyya@mail.wvu.edu

Tel: 3042939335, Fax: 3042934139

Richard Turton, Professor, West Virginia University

Department of Chemical Engineering, WVU, Morgantown, WV 26506

Richard.Turton@mail.wvu.edu

Tel: 3042939364, Fax 3042934139

Raghunathan Rengaswamy, Professor, Texas Tech University

Department of Chemical Engineering, Texas Tech University, Lubbock, TX 26507

Raghu.Rengasamy@ttu.edu

Tel: 8067421765, Fax 3042850903

Abstract

The harsh environment inside slagging gasifiers causes degradation of the refractory lining. Replacement of the refractory lining is carried out every 1-2 years and is very expensive and leads to significant down-time, which lowers the overall availability of hot-wall slagging gasifiers. Due to the harsh operating condition inside a slagging gasifier, direct, in-situ measurements of either the transport variables or refractory degradation are not possible with current state-of-the-art technology. Therefore, a dynamic model describing the detachment, transport, and flow of slag has been developed for better understanding of the effects of various operating conditions and physicochemical properties on the degradation of the refractory. This can eventually lead to the development of better monitoring and prevention techniques.

Several models exist in the open literature for calculating the thickness of the slag layer flowing along the gasifier wall. However, there is a dearth of models that consider detachment of slag droplets from the char particles and the subsequent transport of the slag particles to the wall. In this work, a detachment model is developed by considering the dominating forces acting on the slag droplets attached to the unconverted char. In addition, a model of the turbulent deposition of the slag to the gasifier wall is developed. Furthermore, a model of the molten slag that flows along the refractory lining has been developed by considering mass, momentum, and energy balance equations. Due to the temperature gradient along the molten slag layer, a solid slag layer exists below the molten slag layer. A model of the solid slag layer is also developed using energy balance equations.

The slag sub-model is integrated with a 1D steady-state model of a single-stage, downward-firing, oxygen-blown, slurry-fed, entrained-flow gasifier developed previously at WVU in the Aspen Custom Modeler® (ACM) environment. The gasifier model includes mass, momentum and energy balance equations for solid and gas phases. The model also includes a number of heterogeneous and homogeneous chemical reactions along with devolatilization and evaporation of the slurry feed. It was assumed that the size of the char particles did not change as the ash remains attached to the unconverted char. Therefore, a shrinking core model was appropriate. However, in the current model, droplets of molten slag are considered to detach from the unconverted char when their diameters exceeds a critical value. Therefore, the shrinking core assumption is replaced with a shrinking particle description that is more appropriate for the current model. A discussion of the effect of various operating parameters on the formation and deposition of the slag will be given in this presentation.

HYDROGEN PRODUCTION FROM RENEWABLE SOURCES: R&D CHALLENGES, TARGETS and PATHWAYS

Sara Dillich[1], Eric Miller[2], Erika Sutherland[3], Sarah Studer[4]
[1]U.S. Department of Energy, Washington, DC, USA
[2]U.S. Department of Energy, Washington, DC, USA
[3]U.S. Department of Energy, Washington, DC, USA
[4] AAAS Science & Technology Policy Fellow, Washington, DC, USA

An important component of the U.S. Department of Energy (DOE) Fuel Cell Technology (FCT) Program is the R&D portfolio for research and development of technologies for low-cost, highly efficient hydrogen production from diverse, renewable sources. The goal is to reduce the cost of produced, delivered, and dispensed hydrogen to \$2.00-\$4.00/kg (untaxed). This cost goal is independent of the technology pathway and takes into consideration a range of assumptions for fuel cell electric vehicles (FCEVs) to be competitive with hybrid electric vehicles (HEVs), such as fluctuations in gas price and differing fuel economies[1].

Several key hydrogen production technologies representing a range of production scales and timeframes to market are being addressed in the FCT portfolio. Examples of production process technologies include advanced electrolysis, reforming of bio-derived liquids, and photo-electrochemical, solar-thermochemical, and photolytic biological water splitting. R&D addresses materials durability, energy efficiency, and capital cost reductions associated with the production pathways. The research portfolio includes a focus on hydrogen from direct water splitting through the advanced technology pathways named above, and also includes R&D of integrated systems for renewable electrolysis utilizing advanced technologies for direct coupling of electrolyzers with wind and solar power. In addition, fermentative and microbial-aided electrolysis technologies are being developed to produce hydrogen from organic matter.

Analyses of hydrogen production cost sensitivities have identified key challenges to further improvements in efficiencies and the realization of the DOE hydrogen cost goal. Capital costs, process and solar-to-hydrogen efficiencies as well as materials durability are key factors to be addressed through R&D. Results of analyses help to inform program planning and are used to set short term and long term portfolio targets, priorities, and directions. Case studies for many of the production pathways, made using H2A v3, a spread sheet tool for modelling process economics can be found on-line[2,3].

New technologies are also needed to reduce the cost of delivery and dispensing of hydrogen. The FCT hydrogen portflio includes R&D to develop technologies for hydrogen transportation, distribution, and end user operations such as compression, storage, and dispensing. Current research addresses tube trailer, tanker truck, and pipeline delivery pathways with a focus on reduction of capital costs and improvement of energy efficiency of hydrogen delivery systems. The research portfolio also includes the development and testing of technologies for low-cost station designs or upgrades; and seeks to improve delivery system reliability and integrity through the testing of advanced materials, including fiber reinforced polymer and polymer composites (to be used in storage vessels, compressor components, pipes, etc.).

Significant improvements in hydrogen production and delivery efficiencies and costs have been made in recent years. R&D progress is summarized in annual presentations and progress reports made by FCT funded project teams and by the FCT progam[4,5].

1. Office of Fuel Cell Technologies Program Record #11007, *Hydrogen Threshold Cost Calculation*, http://hydrogen.energy.gov/pdfs/11007_h2_threshold_costs.pdf
2. Department of Energy Hydrogen and Fuel Cells Program H2A Analysis, http://hydrogen.energy.gov/h2a_analysis.html
3. Department of Energy Hydrogen and Fuel Cells H2A Production Analysis, http://hydrogen.energy.gov/h2a_production.html
4. Department of Energy Hydrogen and Fuel Cells Program Annual Merit Review and Peer Evaluation, http://www.hydrogen.energy.gov/annual_review.html
5. Department of Energy Hydrogen and Fuel Cells Program Annual Progress Reports, http://www.hydrogen.energy.gov/annual_progress.html

Paper #282008

Auto-Thermal Reforming of Biodiesel-Ethanol-Diesel Blends for Solid Oxide Fuel Cell Applications

Jiefeng Lin[+]*, Thomas A. Trabold[+]*, Mark R. Walluk*, and Daniel F. Smith*

*Center for Sustainable Mobility, Rochester Institute of Technology
[+]Golisano Institute for Sustainability, Rochester Institute of Technology
111 Lomb Memorial Drive, Rochester, New York 14623

Introduction

Solid oxide fuel cells operated at a relatively high temperature generally serve as the combined heat and power systems for stationary applications due to their high fuel efficiency [1]. Solid oxide fuel cell-based auxiliary power units (SOFC-APUs) are also used in long-haul diesel heavy duty trucks to supply auxiliary electricity for the drivers during truck idling periods [2]. Rather than keeping the full diesel engine running at rest, trucks utilizing SOFC-APU systems within the range of 2-10 kW to provide secondary electrical power supply and achieve a relatively high fuel efficiency and low environmental impact. Even though many hydrocarbon fuels (like JP-8, jet fuel, gasoline, or diesel) can be fed into the SOFC-APU system by introducing a catalytic reformer to produce a hydrogen-rich reformate [3-5], several challenges (e.g. systematic integration, carbon formation, and catalyst sintering) need to be overcome in order to maintain the stable performances of the systems, particular for transportation applications [4,6]. To avoid carrying an external tank, trucks with SOFC-based APUs would presumably use the same petro-diesel as the main fuel both for diesel engine combustion during traveling and hydrogen production at rest. Our previous study on diesel auto-thermal reformation (ATR) with anode recycled gases suggests that oxygen to carbon ratio (O/C), fuel utilization fraction, and anode recycle fraction all significantly affect soot formation, and there is a strong correlation between soot formation and ethylene concentration in the effluent gas [4]. Rising concerns over volatile crude oil prices and adverse environmental impacts from using fossil fuels are accelerating the transition of fuel consumption from only diesel to blends with biodiesel (e.g., B20). Thus, it is important to conduct a study of practical diesel blends for truck transportation and evaluate the reformation of these diesel blends, which has not yet been widely reported in the open literature.

The major drawbacks in biodiesel-diesel (B-diesel) fuel blends are their poor performance relative to cold flow and fuel kinematic viscosity limits, which make them difficult to atomize into small droplets and effectively deliver to the engine at low temperature [7]. Alcohol based co-solvents have been reported as additives to improve B-diesel performance and also considered as oxygenates to accelerate complete fuel combustion [8]. In this study, ethanol is added to the B-diesel blends as an additive to improve certain fuel properties and these mixtures are referred to as BED blends (biodiesel, ethanol, and diesel). A comprehensive study has been conducted in our research group to establish a model and optimize the blending composition of BED fuel in terms of system profitability with multi-criteria, including practical fuel quality specifications (kinematic viscosity, density, lower heating value, cloud point, cetane number, fuel stability and sulfur contents), economic viability (fuel production cost, market demand, and fuel market price), and fuel policy (bio-fuel tax credit). The proposed model indicates a fuel blend with 25% biodiesel, 10% ethanol, and 65% diesel (B25E10D65, by volume) enables satisfying these multi-objectives and maximizing the profitability of the refinery firms [9]. Thus, B25E10D65 blended fuel is used as a representative BED blend in this study and the objective is to analyze its performances during autothermal reforming, which would contribute to a fundamental understanding of blended hydrocarbons catalytic reformation. Ultra-low sulfur diesel (ULSD, $C_{15}H_{28}$)

and biodiesel ($C_{19}H_{36}O_2$ - based on fuel certifications) were considered as two basic fuels to compare and were individually reformed under the same conditions as BED reformation in this work. Solid carbon deposition during hydrocarbon ATR has been recognized as a primary degradation mode in SOFC-APU systems, but the dynamic carbon formation is difficult to detect and control. To overcome these challenges, this work applies a direct photo-acoustic soot meter to measure the onset of carbon production and quantify its formation in a single-tube reformer. By integrating the carbon measurement with a mass spectrometer to determine the composition of effluent gases from the reformer, the optimum operating environment for BED ATR with carbon-free deposition and peak hydrogen yield can be identified. Thermodynamic analysis based on the method of total Gibbs free minimization is implemented as well to evaluate the equilibrium compositions of effluent products. The experimental investigation complimented with theoretical analysis of BED blend ATR enables effectively optimizing the onboard reforming conditions.

Thermodynamic Analysis

The Gibbs free total energy minimization method was applied in this work and implemented with Matlab® because this approach enables quantifying the presence of condensed species in the reaction products without necessarily knowing the exact chemical reaction involved [4,10,11].

The total Gibbs energy with two phases can be expressed as:

$$G = G_{gas} + G_{condensed} = \sum_{i=1}^{N} n_i (G_i^o + RT \ln \Phi_i y_i P) + \sum_{i=N+1}^{Ns} n_i G_{c(s),i} \qquad (1)$$

where n_i is the number of moles of species i; G_i^o is standard state free energy of species i; y_i mole fraction of species i; T, P, R are temperature, pressure, and gas constant, respectively. When solid carbon is considered in the system, its Gibbs free energy $G_{c(s),i}$ equalizes to zero [10,11]. The gas phase species were assumed to behave ideally because of the high operating temperature and low pressure (slightly above atmospheric) in the reactor, so the fugacity coefficient $\Phi_i = 1$. Also, Eq. (1) is subjected to the constraint of elemental mass balance:

$$\sum_{i=1}^{N} n_i a_{ik} = b_k \qquad k = 1, \ldots, M \qquad (2)$$

where a_{ik} is the number of atoms of element k in molecule i and b_k is the total amount of element k in the effluent gases.

After introducing the Lagrange Multipliers, Eq. (1) and Eq. (2) can be combined and the partial derivatives with respect to n_i are set equal to zero in order to minimize the total Gibb free energy. Thus, ten non-linear equations with a total of ten variables (including six for molar fractions, y_i, of species i and four for Lagrange multipliers λ_k) are generated and can be solved by applying the Newton-Raphson method and employing the fsolve function in Matlab® [10]. Details on the thermodynamic analysis are available in our previous works [4,10,12]. For BED ATR thermodynamic modeling, biodiesel, ethanol, and diesel are assumed to react independently with oxygen-to-carbon ratio (O_2/C) and steam-to-carbon ratio (S/C) based upon the total fuel carbon content. The compositions of the effluent gases are simulated in the temperature range of 400 -1000°C, O_2/C range of 0 - 0.4 with constant S/C=0.605.

Experimental

Fig. 1 provides a schematic diagram of a single-tube reformer system apparatus, which including fuel and water delivery modules, a single-tube reformer with four separate sets of rhodium-stabilized alumina catalyst wash-coated ceramic monolith, AVL photo-acoustic based micro-soot meter, and the mass spectrometer. The reformer is entirely contained within an electrically heated tube furnace. Because high hydrocarbon fuels like diesel and biodiesel tend to self-pyrolyze in the vaporization step and form solid carbon prior entering the reactor [4,6], a customized nozzle was developed and used to improve fuel atomization and promote homogeneous steam/fuel mixing (droplet diameter size range between 10 to 20 μm [10]). Super-heated steam at 380°C generated from a heated stainless steel coil is sprayed out of the nozzle and mixed finely with the fuels and carrier nitrogen gas. For more detailed description of this vaporization process and the customized nozzle, refer to our relevant report [10]. Five thermocouples TC1 to TC5 located along the reactor centerline are used to acquire the temperature profile of the catalyst and a thermocouple is used to monitor the temperature of the front end of the reactor (trumpet-shaped). An AVL micro soot meter (Model No. 483, AVL List GmbH, Graz, Austria) is applied to directly measure the carbon concentration in the reformate stream. This instrument utilizes the photo-acoustic effect to obtain accurate carbon concentration measurement as low as 5 μg/m^3 (0.005 mg/m^3) [13]. By the photo-acoustic effect, periodic heating and cooling of the dispersed carbon particles caused by a modulated laser beam produces pressure fluctuations that can be detected by sensitive microphones. The microphone signal is linearly proportional to the carbon concentration in the sampling volume. This approach has the capability to detect dynamic carbon evolution during fuel ATR at different operating conditions and provides time-resolved results. The mass spectrometer was applied to quantify the effluent gas species, including N_2, H_2, O_2, CO, CO_2, CH_4, C_2H_4, and C_2H_6.

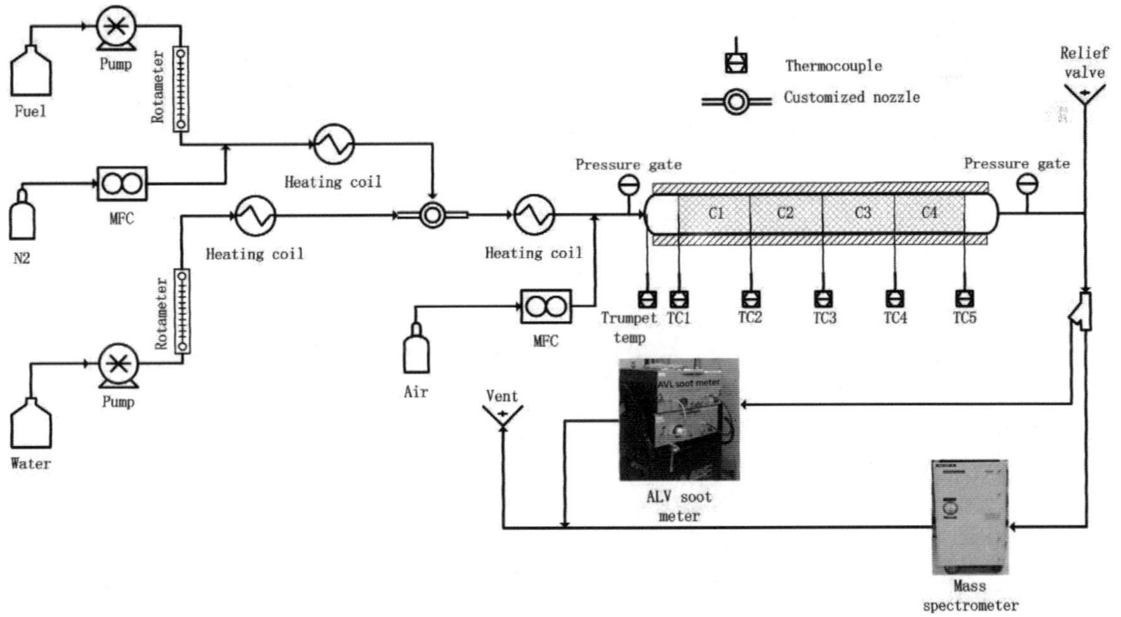

Fig.1 - Experimental diagram of the single-tube reformer test apparatus

Hydrogen yield from reforming each hydrocarbon fuel (diesel, biodiesel, and BED) was derived from thermodynamic analysis described in the above section, and then used to determine the flow rate of the input fuel in order to achieve same amount of hydrogen production rate (3.5 mg/s) under the same operating conditions (total O/C, gas hourly space velocity, initial operating temperature, and S/C). Table

1 provides the experimental test conditions for each fuel. Total oxygen-to-carbon ratio (O/C) is defined as the summation of oxygen from fuel, air, and water to the carbon from the fuel. Gas hourly space velocity was kept constant at 34,120/hr under the condition of standard temperature and pressure (STP) throughout these tests by adjusting nitrogen input flow. This study also included tests to identify the carbon formation boundaries of these hydrocarbon fuels by reducing the reformer operating temperature while monitoring the AVL soot meter to quantify the carbon concentration.

Table 1 - Initial experimental test points for the ATR of diesel, biodiesel, and BED

Fuel	Fuel flow rate (mg/s)	O_2/C	S/C	Total O/C	Theoretical H_2 yield (mg/s) @ 1000°C
Diesel	18.9	0.435	0.605	1.475	3.5
Biodiesel	22.1	0.382	0.605	1.475	3.5
BED	20.4	0.342	0.605	1.475	3.5

Results and Discussion

Equilibrium Analysis

Fig. 2 illustrates the gas compositions of some reformate species and carbon content of BED ATR as a function of operating temperature and O_2/C with constant S/C. The peak hydrogen production rate of BED ATR at 15.6 mol/mol fuel occurs at 725°C average reformer temperature with $O_2/C = 0.3$ (Fig. 2a), which is also within the region of high risk to form solid carbon (Fig. 2b). When the reformer temperature increases above 725°C, the moles of hydrogen slightly decrease due to the water gas shift reaction [4,14]. The temperature range from 500 to 800°C is the main operating region to accelerate CO evolution and solid carbon reduction because the molar concentrations of CH_4 and H_2O are relatively low and the Boudouard reaction (Eq. 3) dominates the coke activity [4,11]. The carbon-free boundary highlighted in Fig. 2b suggests the combination of input air and reformer temperature is crucial to mitigate carbon formation in BED ATR. Methane decreases with increase of air flow and temperature (Fig. 2c), while carbon dioxide monotonically increases with O_2/C increase. Carbon monoxide reaches its peak yield with $O_2/C = 0.34$ at high temperature and then gradually decreases as air flow continues increasing, with a similar trend occurring for hydrogen when air flow increases.

$$CO_2 + C \longrightarrow 2CO \quad (3)$$

Experimental Test: Initial Set Points

Table 1 provides the initial test points for diesel, biodiesel, and BED ATR under the same operating conditions to achieve the same theoretical hydrogen production rates. Table 2 summarizes the measured reformate gas production based on the input fuel flow rates provided in Table 1. The production rates of hydrogen in the experimental tests are all slightly less than the predicted values because thermodynamic modeling in this work considers fuel reforming as an ideal process, namely, the carbon mitigation reactions are assumed to completely occur when the reactor reaches the carbon free boundary and the thermal-transition loss is ignored. However, the differences of hydrogen production

Table 2 - Gas yields from ATR of diesel, biodiesel and BED [S/C=0.605 and total O/C=1.475 at 1000°C]

	Output (mg/s)					
Input fuel	H_2	CO	CO_2	CH_4	C_2H_4	C_2H_6
Diesel	3.41	30.88	9.11	0.02	0	0
Biodiesel	3.33	32.61	9.56	0.43	0.12	0
BED	3.37	32.02	7.17	0.23	0.05	0

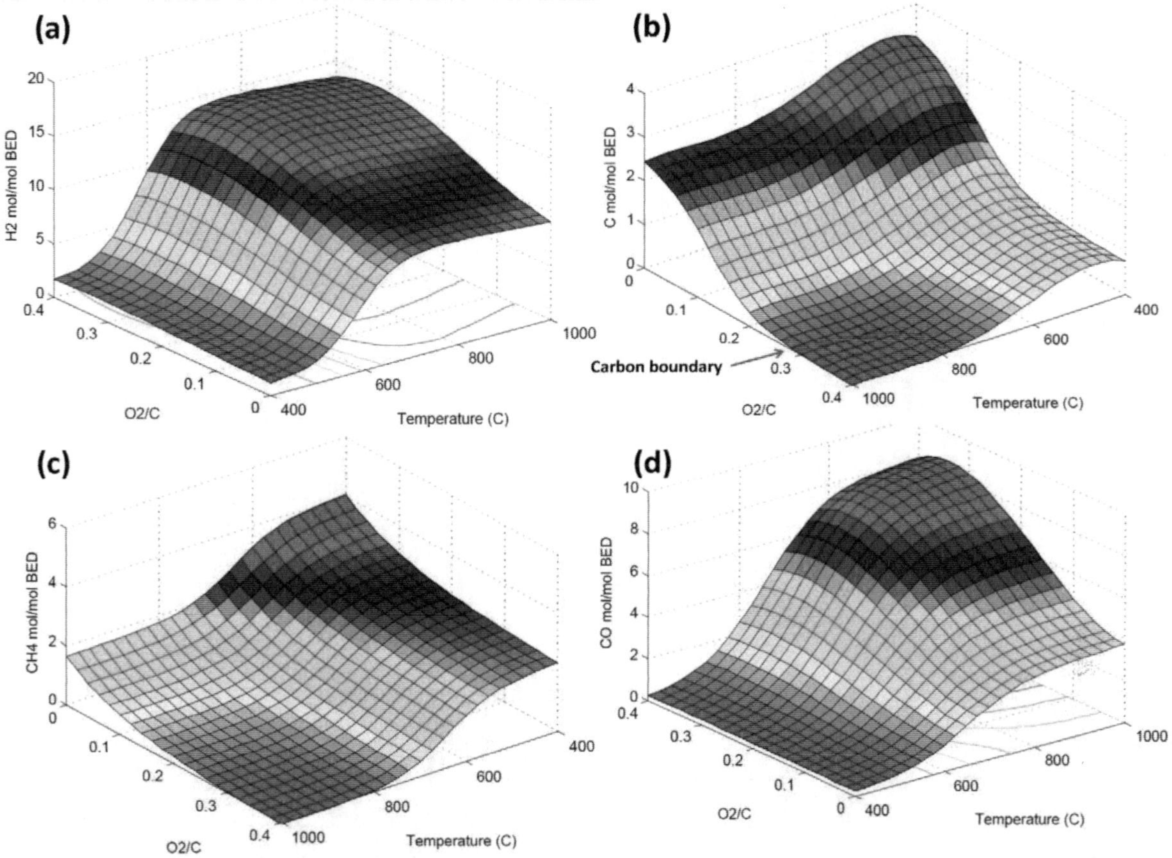

Fig. 2 - Thermodynamic equilibrium gas products of BED ATR (a) H_2, (b) solid carbon, (c) CH_4, and (d) CO with S/C = 0.605 (the scales are altered to provide adequate visualization of the surface plot)

rates between experimental results and those derived from thermodynamic analysis are less than 5.5% by implementing the proposed simulation methodology (diesel, biodiesel, and BED differences of 2.64%, 5.11%, and 3.86%, respectively). Under the conditions of S/C = 0.605 and total O/C = 1.475 at 1000°C, diesel ATR produces negligible methane, ethylene, and ethane, whereas certain fractions of methane and ethylene are present in the effluent from biodiesel ATR. Fig. 3 shows the temperature profiles of the reactor during the ATR of these three types of fuel under the initial conditions. The temperatures throughout the reactor with biodiesel ATR are all higher than those with diesel ATR, which may be due to more fuel input for biodiesel ATR (see Table 1). Temperature profile of BED ATR behaves in as a different manner as either diesel ATR or biodiesel ATR. For instance, because the exothermic reaction of ethanol oxidation occurs in the front end of the reformer and promotes biodiesel/diesel vaporization for the case of BED ATR, its trumpet temperature is higher than the one in other two cases. For BED ATR, temperature near the front of the first catalyst in the reactor (TC1 shown in Fig. 1) stays around 60°C lower than diesel/biodiesel ATR, which helps reduce the risk to create local hot spots and sinter the front end of the catalyst at high temperature.

Experimental Test: Temperature Reduction

To capture the chemical reaction changes inside the reformer as a function of the average reformer temperature (average of TC1-TC5), the reactor temperature was reduced from initial conditions at 1000°C with a 0.1°C/s reduction rate by adjusting the external power supplies to the tube furnace. The mass spectrometer was used to quantify the gas composition, while the AVL soot meter was applied to monitor the soot concentration simultaneously. Fig. 4(a) and Fig. 4(b) show the gas compositions of the effluent mixes from ATR experiments of biodiesel and BED, respectively. Overall, each gas behaves similarly as temperature reduces. For example, H_2 and CO are both gradually reduced and CO_2, CH_4, C_2H_4, and C_2H_6 increase when temperature is reduced from its initial test point at 1000°C. Fig. 4(a) depicts the unstable reaction of biodiesel ATR when operating temperature is <875°C, which may be

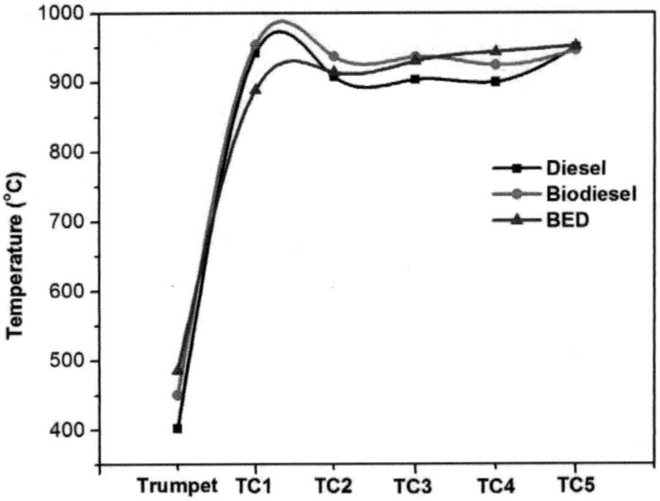

Fig. 3 - Temperature profile of the reactor for diesel, biodiesel, and BED ATR

due to the high fraction of high hydrocarbon fuel decomposed in the biodiesel vaporization process and it increases the possibility to form solid carbon inside the reformer. Significant intermittent fluctuations occur in ethylene composition when the temperature is below 925°C and this experimental phenomenon can be explained with the mechanism of carbon deposition on the rhodium catalyst: the solid carbon formed insider the reformer not only would block the surface electrochemical reaction of the catalyst which leads to the increment of ethylene, but it could also degrade the catalyst support [11]. Thus, it is challenging to reform high hydrocarbon fuels like biodiesel without identifying the optimum operating environment and avoiding irreversible effects on the catalysts. As compared to the gas compositions of biodiesel ATR, BED ATR has a relatively "smooth" gas composition profile (Fig. 4b) which indicates that ethanol serving as the oxygenate may promote biodiesel/diesel fuel decomposition to low hydrocarbon mixtures before entering the reactor.

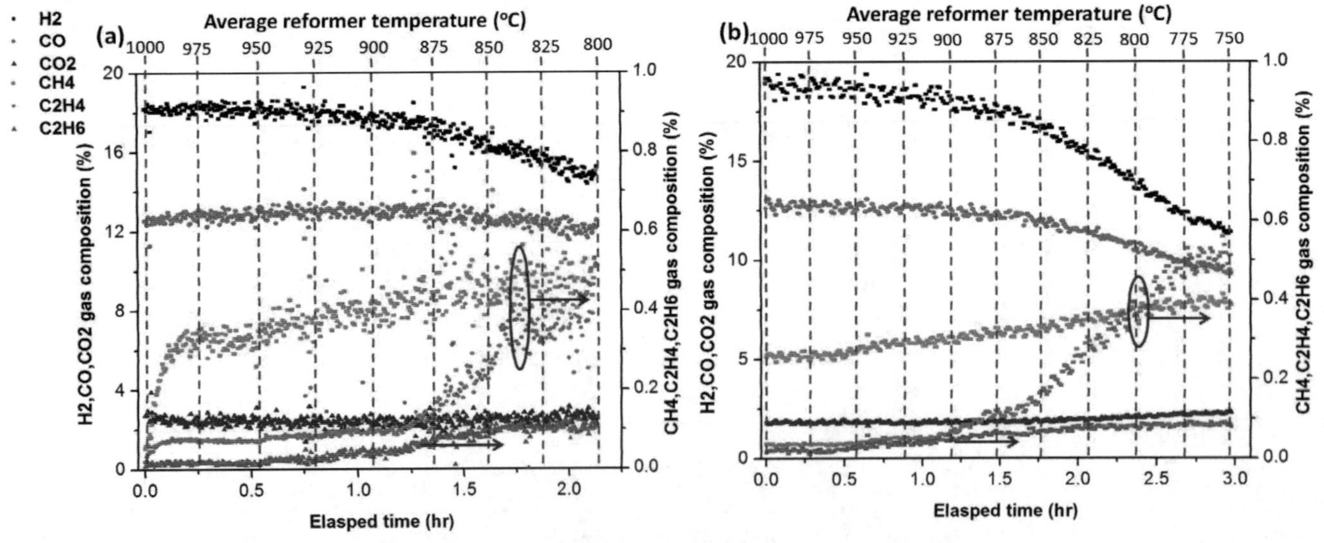

Fig. 4 - Compositions of effluent gas from (a)biodiesel, (b)BED ATR with reformer temperature reduction

Fig. 5 shows the carbon concentrations of diesel, biodiesel, and BED ATR as reformer temperature is reduced. The approximate critical temperatures at the onset of solid carbon formation are 825°C, 920°C, and 865°C for diesel, biodiesel and BED, respectively. Similar to the gas composition profiles, BED ATR has a relatively "smooth" carbon concentration profile as temperature is reduced. Based on our previous experiments, the intermittent "spikes" are believed to be actual indications of high carbon concentration which may result from a process of continual carbon accumulation and release from the internal reformer surfaces.

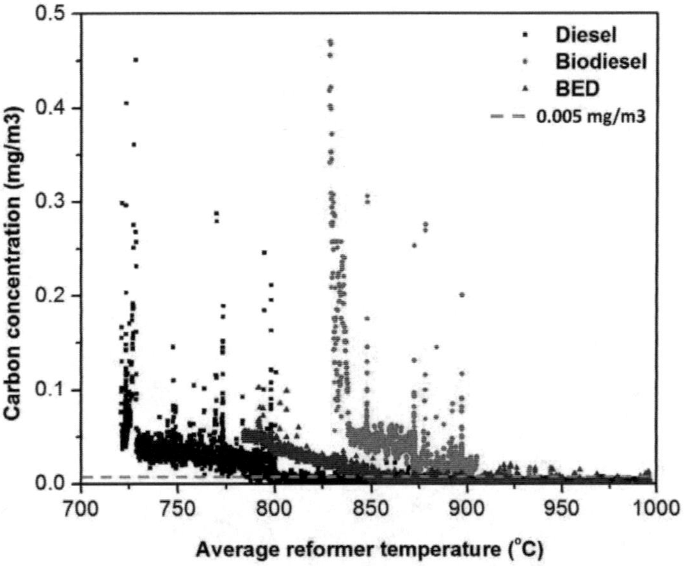

Fig. 5 - Carbon concentrations of diesel, biodiesel, BED ATR with temperature reduction

Conclusions

Biodiesel is considered as a promising alternative fuel for transportation applications. However, many challenges (such as its poor performance relative to cold flow and high fuel viscosity) limit its wide applications in various geographical regions. Thus, certain additives are needed for biodiesel/diesel blended fuels to improve their performance, either for diesel engine combustion during traveling or hydrogen production at rest. This study proposes ethanol as an additive for B-diesel and investigates the autothermal reformation performance of a BED blend (B25E10D65). Thermodynamic analysis based on total Gibbs free energy minimization was applied to extrapolate the optimum operating conditions for BED ATR (S/C, O2/C, and operating temperature) to achieve high hydrogen yield with no carbon formation. A mass spectrometer and AVL soot meter were simultaneously used in the experimental apparatus to quantify the reformate gas composition and carbon concentration of effluent mixtures from the reformer, respectively. To achieve the same hydrogen production rates under the same operating conditions (reformer temperature, S/C, gas hourly space velocity, and total O/C), biodiesel ATR and BED ATR produce measurable amounts of methane and ethylene that serve as potential precursors to carbon formation while diesel ATR produces negligible quantities of these gases [15]. As the reactor temperature is reduced, BED ATR indicates a "smooth" profile during the temperature transition in regards to the measured stability of the reformate concentration. Strong correlation between ethylene and carbon concentration was identified, which is in agreement with our previous studies. Critical temperatures for diesel, biodiesel, and BED to form solid carbon during fuel reforming were determined as 825°C, 920°C, and 865°C, respectively. Thus, BED blended fuel would serve as a promising alternative fuel for SOFC-APU application in transportation sectors. Future works will focus on extrapolating the interaction terms of each hydrocarbon fuel in BED from experimental results and further improve the accuracy of the equilibrium analysis.

Acknowledgements

The authors gratefully acknowledge the support of the Office of Naval Research (ONR), under Contract No. N000141110200, which provided a graduate research assistantship for J. Lin. We also acknowledge the valuable communications with Edward House and Mark Cervi, both with the ONR Laboratory in Philadelphia, PA.

References

[1]. Hajimolana, S.A. and Soroush, M. (2009), "Dynamics and control of a tubular solid-oxide fuel cell," *Ind. Eng. Chem. Res.,* 48 (13), pp. 6112-6125.

[2]. Singhal, S.C. (2002), "Solid oxide fuel cells for stationary, mobile, and military applications," *Solid State Ionics*, 152, pp. 405-410.

[3]. Strohm, J.J., Zheng, J., and Song, C. (2006), "Low-temperature steam reforming of jet fuel in the absence and presence of sulfur over Rh and Rh-Ni catalysts for fuel cell," *J. Catalysis,* 238(2), pp. 309-320.

[4]. Trabold, T.A., Lylak, J.S., Walluk, M.R., Lin, J.F., and Troiani, D.R. (2012), "Measurement and analysis of carbon formation during diesel reforming for solid oxide fuel cells," *Int. J. Hydrogen Energy*, 37, pp. 5190-5201.

[5]. Shekhawat, D., Berry, D.A., Haynes, D.J. and Spivey, J.J. (2009), "Fuel constituent effects on fuel reforming properties for fuel cell applications," *Fuel*, 88, pp. 817-825.

[6]. Borup, R.L., Inbody, M.A., Tafoya, J.I., Guidry, D., Parkinson, W.J. (2004), "Diesel reforming for SOFC auxiliary power units," LA-UR-04-5370, Los Alamos National Laboratory.

[7]. Alptekin, E., Canakci, M. (2008), "Determination of the density and the viscosities of biodiesel-diesel fuel blends; Renewable Energy," *Renewable Energy*, 33, pp. 2623-2630.

[8]. Fernando, S., Hannan, M. (2004), "Development of a novel biofuel blend using ethanol-biodiesel-diesel microemulsions: EB-diesel," *Energy & Fuels*, 18, pp. 1695-1703.

[9]. Lin, J., Gaustad, G., Trabold, T.A. (2012), "Enabling sustainable fuel: optimizing biodiesel-ethanol-diesel blends," submitted to *Applied Energy.*

[10]. Lin, J., Walluk, M.R., Smith, D.F., and Trabold, T.A. (2012), "Mitigation of carbon formation during autothermal reforming of biodiesel for solid oxide fuel cell applications," *Proceedings of the ASME 2012 6th International Conference on Energy Sustainability & 10th Fuel Cell Science*, ESFuelCell2012-91092, July 23-26, 2012, San Diego, CA, USA.

[11]. Parmar, R.D., Kundu, A., Karan, K. (2009), "Thermodynamic analysis of diesel reforming process: Mapping of carbon formation boundary and representative independent reactions", *J. Power Sources*, 194, pp.1007-1020.

[12]. Lin, J., Smith, D.F., Babbitt, C.W., Trabold, T.A. (2011), "Assessment of bio-fuel options for solid oxide fuel cell-based auxiliary power units," *Sustainable Systems and Technology (ISSST), 2011 IEEE International Symposium*, 16-18 May 2011.

[13]. Schindler, W., Haisch, C., Beck, H.A., Niessner, R., Jacob, E., Rothe, D., (2004), "A photoacoustic sensor system for time resolved quantification of diesel soot emissions", *SAE Paper 2004-01-0968.*

[14]. Graschinsky, C., Giunta, P., Amadeo, N., and Laborde, M. (2012), "Thermodynamic analysis of hydrogen production by autothermal reforming of ethanol," *Int. J. Hydrogen Energy*, 37(13), pp. 10118-10124.

[15]. Ahmed, K., and Foger, K. (2010), "Fuel processing for high temperature high efficiency fuel cells," *Ind. Eng. Chem. Res.,* 49, pp. 7239-7256.

(562c) Development of a Butanol Fuel Processor for Person-Portable Fuel Cell Power Systems

Stephen Jolly and Hossein Ghezel-Ayagh, FuelCell Energy, Inc., Danbury, CT

A novel partial oxidation (POX) catalyst and butanol-to-hydrogen fuel processor sub-system concept have been developed to fill the need for person-portable power systems with high energy density for extended duration operation. The system is designed to reform isobutanol through partial oxidation with air. A multi-stage carbon monoxide removal train was developed to convert the CO present in the reformate stream. The resulting hydrogen stream is suitable for integration with a PEM fuel cell for person-portable power system applications.

An integrated system design approach was undertaken to optimize the fuel processing system configuration and operating conditions. Among the available fuel processing options, systems based on both partial oxidation and auto-thermal reforming (ATR) of isobutanol fuel were investigated. The former, i.e. partial oxidation reforming, results in a simpler but less efficient design to implement. The latter, i.e. autothermal reforming, is a more efficient approach for conversion of isobutanol to hydrogen for the fuel cell anode. A systematic design study has determined the optimal operating conditions for the partial oxidation and autothermal-based systems.

Energy density, in terms of W-h/kg and W-h/L, is a critical parameter for person-portable power systems. These systems must be lightweight and compact, especially for military applications to avoid hindering a soldier's movements in the field. The butanol-fueled system under development has an inherent advantage due to the increased energy density of the fuel, as compared to methanol-fueled systems. A complete, integrated sub-system was designed using 3-D CAD modeling software to minimize the size and weight of the reformer. 3-D CAD renderings were created for both the partial oxidation and autothermal-based butanol fuel processor systems. The POX-based butanol reforming system is estimated to achieve an energy density of 810 W-h/L and 850 W-h/kg, while the ATR-based system is estimated to achieve an energy density of 1,000 W-h/L and 2,000 W-h/kg.

The concept feasibility of reforming isobutanol to a hydrogen-rich stream via partial oxidation was experimentally demonstrated. Novel catalyst formulations and catalyst support structures have been developed and tested. A Rh/Ce-based reforming catalyst effectively converted isobutanol to an H_2-rich stream with no coke formation under the desired system operating conditions. The catalyst was also demonstrated to be regenerable, should carbon deposition occur. In summary, the concept feasibility of both a novel butanol reforming catalyst and fuel processor sub-system design to effectively convert isobutanol to hydrogen with the potential to be packaged into a compact, high energy density system has been demonstrated.

(562d) Cartridge-Based Ceramic Microreactor Networks for Process Intensification in Fuels Reforming: Demonstration and Modeling of Complex 2-D Thermal Integration

Bhanu Vardhan Reddy Kuncharam, Artie McFerrin Department of Chemical Engineering, Texas A&M University, College Station, TX, Shalini Damodharan, Texas A&M University, College Station, TX and **Benjamin Wilhite**, Artie McFerrin Dept. of Chemical Engineering, Texas A&M University, College Station, TX

Increasing concern with atmospheric pollution and difficulties associated with nuclear power have caused a surge of interest in using fuels cells as a means of electrical power generation for utility power and also for electrical vehicles[1]. With the difficulty of storing hydrogen, efficient microreformers are developed to convert hydrocarbon fuels into hydrogen on-site for utilization by proton-exchange membrane (PEM) or solid-oxide fuel cells (SOFC's) in portable electronics devices and automotive applications. Methanol has been identified as an attractive fuel to produce hydrogen because of its high hydrogen-to-carbon ratio and lower inter-carbon bonds which therefore can be reformed efficiently at moderate temperatures [2]. Thermally integrated microchannel reactors have been intensively studied as a power technology for realizing all-in-one portable reformers capable of autothermal hydrogen generation via combinations of two or more chemical processes in a compact unit [4].

Our research group has previously demonstrated a new cost-effective approach for realizing heat integrated ceramic microchannel networks capable of coupling several unique chemical processes in complex two-dimensional radial distribution patterns over a broad range of scale without loss in design flexibility [3]. This technique presents the following advantages; (i) ease of catalyst packing, (ii) higher conversion and yield with less amount of catalyst, (iii) increase in Overall hydrogen yield and thermal efficiency, (iv) location of hotspot at the center of the reactor axial length and (v) external insulation to the microchannel reactor resulted in a limited amount of radial heat losses to ambient.

This work presents the design of a ceramic microchannel network, interfaced with brass distributors in order to evaluate the thermal integration issues of a new class of versatile, scalable, integrated ceramic microchannel network. Detail thermal and reaction experiments are performed varying the flowrates of the reaction streams and varying the configuration of catalyst packing in order to find the optimum operating parameters to achieve substantial gain in overall hydrogen yield using this compact, cost-effective, and cartridge based reformer. Experimental results are compared with detailed 3-D COMSOL simulations to demonstrate the ability to flexibly design unique and/or non-intuitive thermal integration schemes.

[1]J.C.Amphlett,K.A.M.Creber,J.M.Davis,R.F.Mann,B.A.Peppley and D.M.Stokes, "Hydrogen Production by Steam Reforming of Methanol for Polymer Electrolyte Fuel Cells",*Int.J.Hydrogen Energy.*, **1994,** *19*, pg 131-137.

[2]Jeong-Se-Suh, Ming-tsang Lee, R.Greif, C.P.GrigoropoulosHan,"A study of steam methanol reforming in a microreactor", *J. Power Sources*. **2007**, 173, pg 458-466.

[3]Angela Moreno, Benjamin A.Wilhite, "Autothermal hydrogen generation from methanol in a ceramic microchannel network", *J. Power Sources.*, **2010,** 195,pg 1964-1970.

[4]A.Moreno,S.Damodharan and B.A.Wilhite, "Influence of Two-Dimensional Distribution schemes upon Reactor Performance in a Ceramic Microchannel Network for Autothermal Methanol Reforming", *Ind. Eng. Chem. Res.*,**2010**,49(21),pg 10956-10964.

(562e) Protonic Membrane Reactors: Converting Hydrocarbon Resources and CO2 to Fuels

Eric D. Wachsman, University of Maryland Energy Research Center, University of Maryland, College Park, MD

Membrane reactor technology holds the promise to circumvent thermodynamic equilibrium limitations by *in-situ* removal of product species, resulting in improved chemical yields. Recent advances in mixed-conducting oxide-membrane technology present the possibility for a dramatic reduction in the cost of converting petroleum, coal and biomass derived feed stocks to hydrogen and other "value added" hydrocarbons. We have developed novel membrane reactor technology, based on high temperature proton conductors, that can convert a wide range of hydrocarbons to pure H_2, and syngas for subsequent Fisher Tropsch synthesis of liquid fuels and chemical feed stocks. By simultaneous H_2 permeation and catalysis, we have demonstrated the ability to increase water gas shift yields >70% over thermodynamic limitations. Similarly we have demonstrated increases in steam reforming yields, and the ability to reform CH_4 with CO_2. The later creates the opportunity for a revolutionary method for carbon sequestration.

(562f) One-Stage Process for Water Gas Shift in a Pd-Based Membrane Reactor

Giuseppe Barbieri[1], Adele Brunetti[1,2], Alessio Caravella[1] and Enrico Drioli[1,2], (1)Institute on Membrane Technology of the National Research Council - (ITM-CNR), Rende, Italy, (2)University of Calabria, Rende, Italy

The water gas shift (WGS) reaction is the upgrading stage in the cycles of hydrogen production by, for example, steam reforming of light hydrocarbons from fossil or renewable sources. This reaction limited by thermodynamics often carried out in industry to convert CO, producing more hydrogen. A WGS industrial process consists of two reactors in series: the first one operates at a high temperature (300-400°C), exploiting the advantages offered by a fast kinetics of FeCr based catalysts; the second one operating in a low temperature range (200-300°C) benefits of the higher thermodynamic conversion.

This work proposes the use of one Pd-based membrane reactor (MR) [1], operating in the same temperature range of the high temperature WGS reactor as suitable alternative to the whole traditional reactor (TR) process. The Pd-based membrane allows the selective removal of H_2 from the reaction volume, thus, favouring the increase of the equilibrium conversion close to the total value. Operated in the high temperature range, the MR can exploit the faster kinetics offered by Fe-Cr based catalyst, together with the higher permeation rate, obtaining CO conversion and H_2 recovery significantly higher than the whole TR system.

The values of gas hourly space velocity (GHSV), temperature, H_2O/CO feed molar ratio, feed composition, etc. used in the simulations are those typical of an industrial application of a WGS upgrading stage. A feed pressure of 15 bar was assumed as reference value, this values being the strength limit of self supported Pd-Ag membranes currently available at market level. However, a feed pressure of 30 bar was also considered as it is the typical pressure used in industrial processes of hydrogen production. The pressure demonstrates to be one of the most interesting variables of MR processing.

MR always exceeded a CO conversion of 80% and 95% for H_2O/CO feed molar ratios of 1 and 3, respectively. These conversion values are always higher than those of the whole traditional process and also exceed significantly the TR equilibrium conversion (TREC). As a first important result, this comparison leads to conclude that only one MR can replace the two reactors of the traditional process. The MR operating at the higher temperature regime allows the kinetically faster Fe-Cr based catalyst to be used, implying, as a consequence, a very large (ca. 10 fold) reduction of the total catalyst required by the traditional process for a feed molar ratio of 1. The low temperature reactor present in the traditional process, where the Cu-Zn catalyst requires a space velocity ca. 10 times lower than the high temperature stage, is anymore necessary in MR proposed solution owing to the higher CO conversion achieved thank to the hydrogen removal through the membrane. Passing from a feed molar ratio of 5 to 3, MR volume corresponds to 33 and 21%, respectively, of the one required by the whole TR process. A further reduction of the feed molar ratio, towards the stoichiometric value of 1, leads to an extra reduction of this parameter. This also has consequence on the reduction of the plant footprint.

In addition, around 90% of the H_2 fed and produced by reaction in the Pd-Ag MR was recovered as permeate at 15 bar and ca. 450°C (outlet temperature). This stream, completely pure in H_2, did not require any separation/purification, contrarily to the one exiting from the traditional process. Furthermore, other hydrogen was also contained in retentate stream. MR retentate concentrated (ca. 60-70%) and compressed in CO_2 can be easily captured.

[1] Barbieri G., **Brunetti A.**, Caravella A., Drioli E., "Pd-based Membrane Reactors for one-stage Process of Water Gas Shift", RSC Adv., 2011, 1 (4), 651-661 http://dx.doi.org/10.1039/C1RA00375E A

Acknolewdgements

The research under this project is co-funded by the European Union Seventh Framework Programme (*FP7/2007 - 2013*) under DEMCAMER project (NMP3-LA-2011-262840)

(562g) Water Gas Shift Reaction in Thin Pd-Layer Supported On PSS Membrane Reactor

Angelo Basile[1], Pietro Pinacci[2], Francesca Drago[2], Simona Liguori[3], Pietro Paolo Morrone[4] and Adolfo Iulianelli[1], (1)Institute on Membrane Technology of the National Research Council - (ITM-CNR), Rende, Italy, (2)RSE SpA, Milan, Italy, (3)CNR-ITM, Milan, Italy, (4)CNR-ITM, Cosenza, Italy

Abstract

In this study, the performances of water gas shift (WGS) reaction performed in a porous stainless steel (PSS) supported thin Pd-layer membrane reactor (MR) have been experimentally investigated in terms of CO conversion, hydrogen recovery and hydrogen purity collected in the permeate stream. The membrane is constituted by a 10 mm thin Pd-layer deposited by electroless plating deposition onto a tubular PSS support and a high temperature Fe-Cr based WGS catalyst is used during the experimental campaign. Prior to reaction tests, the membrane has been characterized to permeation analyses by supplying pure gases of interest such as H_2, CO, CO_2 and their mixtures. As a best result, at T = 400 °C and Dp = 101 kPa, a hydrogen/helium ideal gas selectivity equal to around $1.0 \cdot 10^4$ has been reached. Afterwards, at the reaction temperature of 400 °C, the influence of parameters such as reaction pressure (from 3 to 9 bar), gas space hourly velocity, GHSV, (from 3450 to 14000 h^{-1}) and steam to carbon ratio variation between 1/1 and 4/1 has been investigated. The aim of this paper is, then, to study the potentiality of a composite Pd/PSS MR applied to WGS reaction with the intent of achieving higher CO conversion than conventional reactors and to recover further a highly concentrated hydrogen-rich stream in the MR permeate side.

Techno-economic assessment of membrane reactor technologies for pure hydrogen production for fuel cell vehicle fleets

AIChE 2012 annual meeting - Paper # 262376

Leonardo Roses[1(*)], Giampaolo Manzolini[1], Stefano Campanari[1], Ellart De Wit[2], Michael Walter[2]

[1] Politecnico di Milano, Dipartimento di Energia, Milano, Italy
[2] HyGear B.V., Arnhem, The Netherlands
(*) corresponding author. Tel: +39 022399 3817, e-mail: leonardo.roses@mail.polimi.it

Abstract

In the context of a desirable 'carbon neutral' energy economy, among the most promising solutions for replacing today's GHG emitting vehicles is the use of hydrogen as an energy carrier. In the pathway towards a future infrastructure based on renewable energy sources, a medium term step would rely on the use of fossil fuels for on-site production of hydrogen, feeding small fleets of fuel cell vehicles. Great interest is on natural gas as a primary source due to its high hydrogen to carbon ratio. State of the art technology for the production of hydrogen from natural gas include a series of reacting steps typically involving steam reforming (at 800°C or above), a water gas shift reactor and final purification of hydrogen through pressure swing adsorption (PSA). An alternative that has been subject of growing interest is the use of thin (2-50 micron thick) Pd-alloy materials as hydrogen perm-selective membranes for the embedded extraction of pure hydrogen from the chemical reactor; this system is usually known as 'membrane reactor'.

Within a membrane reactor, the continuous extraction of hydrogen from the reacting mixture shifts the equilibrium conversion of reactants towards higher values, hence production of hydrogen can go beyond the equilibrium limit of traditional reactors. The process intensification obtained with the use of membrane reactors reduces the number of components and eliminates the need for batch operating PSA units, besides allowing a more compact system.

This paper studies the adoption of palladium based membrane reactor technologies for pure hydrogen production from natural gas. In particular, three system layouts are analysed and compared to the traditional option: (i) autothermal reforming membrane reactor, (ii) steam reforming membrane reactor (externally heated) and (iii) water gas shift membrane reactor downstream of a steam reformer. The comparison is made in terms of performances and techno-economic considerations for the design of compact systems for on-site production of hydrogen at filling stations. The systems are designed for 50 m^3/h (1766 cfh) of hydrogen, which corresponds to refilling 25 vehicles a day with 4 kg of hydrogen (approximately 418 km driving range on fuel cell vehicles with 70 MPa storage tank).

Keywords: hydrogen production, membrane reactor, palladium membrane, pure hydrogen, fuel cell vehicles

Introduction

In the challenge of reducing carbon footprint in mobility, the use of hydrogen as an energy carrier has been envisioned since many years as a promising solution for avoiding the emission of carbon dioxide from vehicles. Though, the transformation into a hydrogen based economy raises some challenging obstacles such as the required infrastructure in filling stations and the issue on primary energy source for the production of hydrogen. Diverse regional and national programs are already active in the introduction of hydrogen infrastructure for future transport market with a significant share of hydrogen based technology [1–3].

In the long term the prospective solution consists of using wind and solar or any other renewable energy sources to produce hydrogen by means of electrolysis of water. However in a transition process moving towards this objective, hydrogen is likely to be produced from reforming of light hydrocarbons, as natural gas. The use of natural gas seems as the most feasible option due to the fact that many countries already have a distribution network widely developed, and also because its low carbon to hydrogen ratio leads to lower specific CO_2 emission compared to heavier hydrocarbons.

The benchmark technology for the production of hydrogen from methane would consist in the production of a hydrogen-rich syngas by means of steam reforming at temperature of around 800 °C followed by a water gas shift reactor, and later separation of hydrogen using pressure swing adsorption. An alternative technology that has been focus of increasing attention in the latest decades is the use of perm-selective membranes for the separation of pure hydrogen. This technology can prospectively lead to reduction in costs in allow for a more compact solution leading to equipment savings.

This work compares different configurations adopting membrane reactor technology with the traditional option using pressure swing adsorption. Three system layouts based on membrane reactors are studied and their results contrasted with the benchmark option. The model is implemented in Aspen Plus®.

Membrane reactor

Membrane reactor technology generally refers to the integration of a reacting process with membranes having the characteristics of being selectively permeable to a specific species, allowing the addition or removal of that species from the reacting mixture. In this work, the application of hydrogen selective membranes based on palladium alloys to steam reforming is evaluated: the H_2 is separated from other components as CO_2, CO and steam is and it is sent to the permeate side of the membrane. Figure 1 shows the case of a steam reforming membrane reactor, where the extraction of hydrogen from the reforming reaction shifts the equilibrium conversion of methane towards higher values with respect to the conventional reforming reaction. In this way, higher values of conversion can be achieved at lower reacting temperature. This is in fact one of the key effects that turn the membrane reactor fuel processing attractive when compared to traditional approach. The other advantage consists of performing the hydrogen conversion and separation in just one reactor with potential equipment savings.

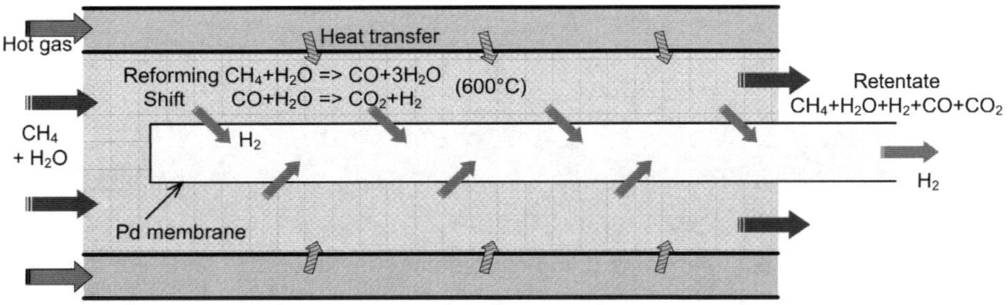

Figure 1: Schematic view of steam reforming membrane reactor (SR-MR).

Separation of pure hydrogen can be achieved through the use of dense palladium-based membranes of reduced thickness (2 to 50 µm) [4–7]. An expression that describes the permeation of hydrogen across the membrane is shown in equation (1).

$$J_{H_2} = P_{H_2}\left(p^{n}_{H_2,ret} - p^{n}_{H_2,prm}\right) \qquad 1$$

were 'P_{H_2}' is the permeance of the membrane measured in $mol/(s\, m^2\, Pa^n)$, the term between brackets is the driving force for permeation depending on the respective partial pressure of hydrogen 'p_{H_2}' on the retentate and permeate side, and 'n' the exponent for the driving force which is experimentally determined to be in the range of 0.5-1.0.

Many factors may affect the permeance of a membrane, as the type of alloy, the manufacturing procedure, type of support, surface quality, heat treatment and thickness of the perm-selective layer are surely the most important. A technological challenge is addressed by the objective of a permeance of $8.52{\cdot}10^{-3}$ $mol/(s\, m^2\, Pa^{0.5})$ set by the DOE as a target for 2015 on its Hydrogen and Fuel Cell Program [8]. In practical applications this value of permeance is still far from the performances achieved by many developers, however some have already proved this permeance to be a feasible objective [5,9]. In this work the model considers a membrane having the permeance set by the DOE 2015 target above mentioned.

System layout

Three innovative configurations and one reference case are considered in this work. They are as follows:

- SR-PSA: Steam reforming with pressure swing adsorption
- SR-MR: Steam reforming membrane reactor
- ATR-MR: Autothermal reforming membrane reactor
- WGS-MR: Water gas shift membrane reactor

Starting from the reference case, the SR-PSA layout is shown in Figure 2. Natural gas is compressed to the operating pressure and preheated prior to the desulfurization unit. The steam reforming reactor is controlled at an outlet temperature which is optimized for performance falling typically within 800 °C and 900 °C. Then, the reformate is cooled down prior to the water gas shift reactor (WGSR). In the reforming reactor, the reaction .R.1 and R.2 (see below) take place, whereas in the WGSR the operation is selective only to R.2 over a Fe-based catalyst.

$$CH_4 + H_2O \Rightarrow CO + 3\,H_2 \qquad \text{R.1}$$

$$CO + H_2O \Rightarrow CO_2 + H_2 \qquad \text{R.2}$$

The syngas is then cooled down to the operating temperature of the pressure swing adsorption unit (PSA) of 50 °C. The impurities from the PSA and possibly additional natural gas are sent to the burner required for sustaining the endothermic reforming reaction and producing the required steam.

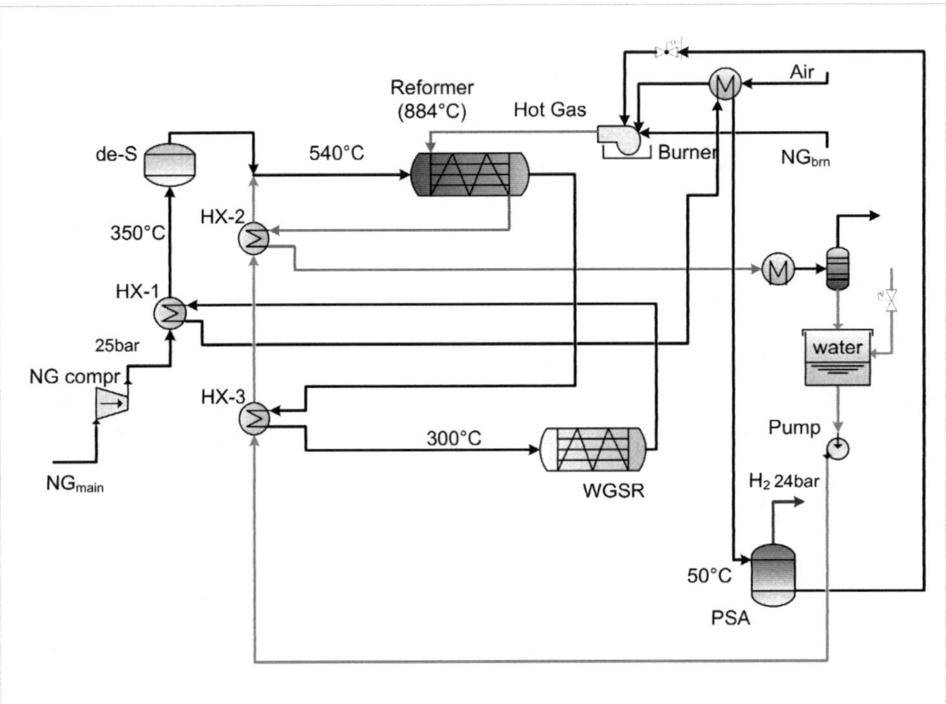

Figure 2: Layout traditional system "SR-PSA"

The configuration involving Steam Reforming Membrane Reactor (SR-MR) is shown in Figure 3. In this architecture, the separation of hydrogen takes place in a single reactor operating at 600 °C. In the SR-MR reactor, the equilibrium of both reactions R.1 and R.2 is shifted towards the products thanks to the simultaneous extraction of hydrogen from the reacting section. The extracted hydrogen is then cooled down prior to the inlet of the compressor. The retentate together with additional natural gas is sent to the burner required for sustaining the endothermic reforming reaction and for steam production.

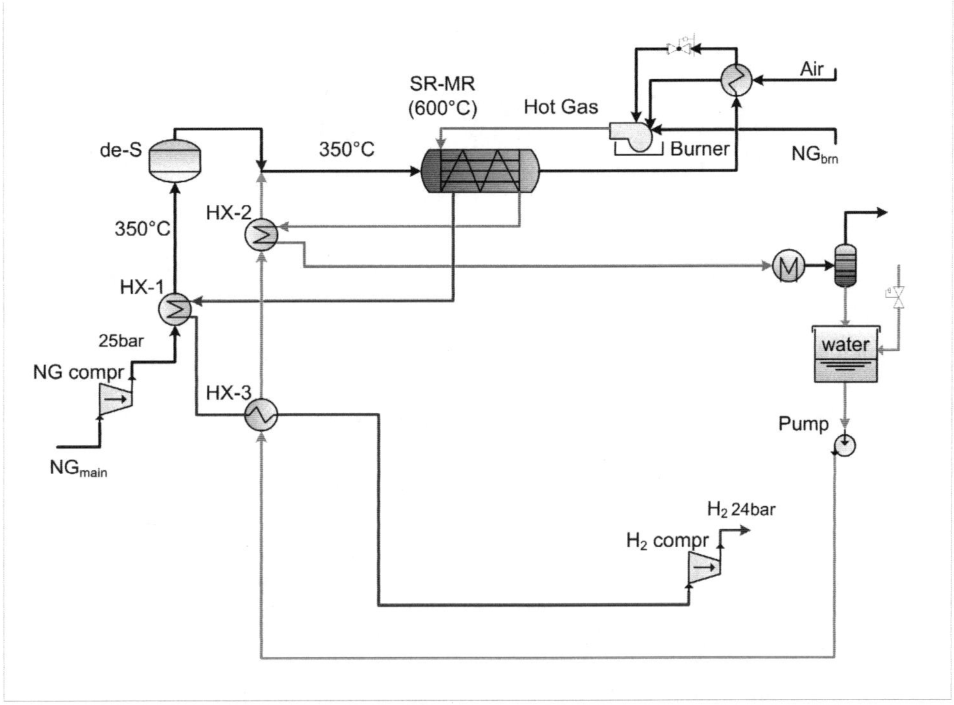

Figure 3: Layout Steam Reforming Membrane Reactor system "SR-MR"

The layout of the system with the autothermal membrane reactor is shown in Figure 4. In this system, the reactions in the ATR are thermally balanced: the exothermal reaction of methane partial oxidation R.3 and partly R.2 supply the heat to support the reforming R.1 which is endothermic. This configuration therefore requires an air compressor. The fraction of stoichiometric air provided to the reactor is determined in order to keep the outlet temperature of the reactor at 600 °C. The retentate which has unconverted species and the remaining hydrogen is combusted with additional air and then cooled down while producing steam.

$$CH_4 + 2\ O_2 => CO_2 + 2\ H_2O \qquad \text{R.3}$$

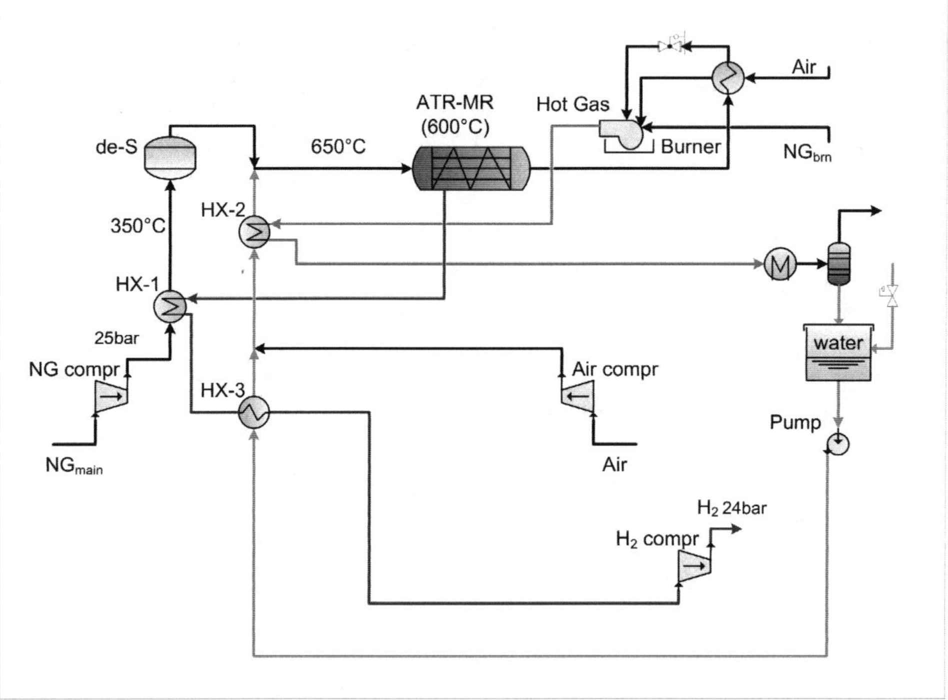

Figure 4: Layout Autothermal Reforming Membrane Reactor system "ATR-MR"

The layout of the system with water gas shift membrane reactor (WGS-MR) is shown in Figure 5. In this configuration a traditional steam reforming reactor is used for methane conversion, with the syngas then cooled down to 300 °C as inlet temperature to a WGS-MR. In the shift reactor only R.2 takes place over a Fe-based catalyst, together with the extraction of hydrogen through the perm-selective membrane. The hot gas produced by combustion of the retentate and additional natural gas is use for sustaining the reforming reaction.

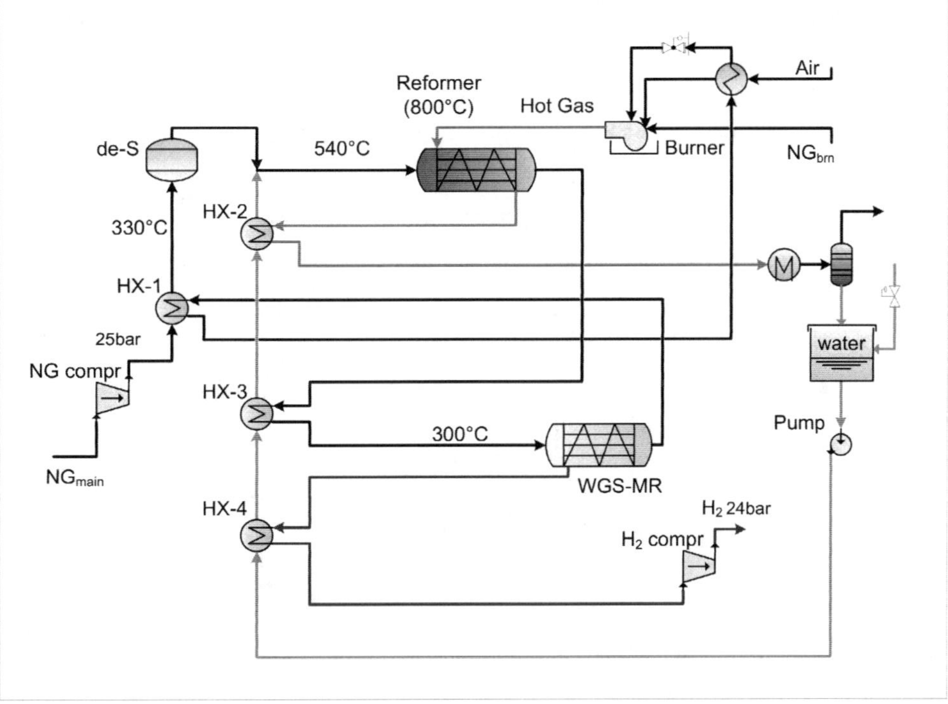

Figure 5: Layout Water Gas Shift Membrane Reactor system "WGS-MR"

Design and modelling considerations

The size of the hydrogen production system is set in order to supply a fleet of 25-30 refills a day of hydrogen fuelled vehicles. This definition of scale of production corresponds to an early market stage where the hydrogen fuelled vehicle is not fully immersed in the market, yet. A fuel cell powered family car performing at approximately 105 km/kg_{H2} [10], assuming a range of autonomy of 483 km requires storage for 4.6 kg_{H2}. Therefore, a plant with a production rate of 4.5 kg_{H2}/h would supply hydrogen for 30 vehicles a day topping up their tanks by 75-80 % of their capacity.

The most traditional technology for separation of pure hydrogen is based on pressure swing adsorption (PSA) unit. The equilibrium-based process of PSA typically operates at pressure ranging 2 to 3 MPa [11]. In this work the operating pressure in the reforming reactor was set to 2.5 MPa_g for all configurations studied, and a 0.1 MPa pressure drop was assumed across the PSA giving pure hydrogen at 2.4 MPa_g. For automotive applications high purity hydrogen is desired (grade 5.0), hence the operation of the PSA would operate at lower recovery factor than for production of technical grade hydrogen (e.g. 2.5) [12]. Of the overall hydrogen flow rate at PSA inlet, a recovery of 75 % was assumed.

The compressors used for natural gas, air and hydrogen are reciprocating multistage type with intercooling stages in order to keep discharge temperature below approximately 150 °C (300 °F). Air and natural gas are compressed to 2.5 MPa_g in the ATR-MR system whereas in the other cases, air is only required at low pressure at the burner.

Regarding the compression of hydrogen, in the models of all four system layouts the outlet pressure of hydrogen was set to 2.4 MPa_g to match the conditions set by SR-PSA. The consumption of a booster which compress the hydrogen from 2.4 MPa_g to 70 MPa_g required by the refilling station was calculated independently being that it is the same (5 stage intercooled) in any of the four cases studied.

Table 1 summarizes the operating parameters of the compressors.

Table 1: Compression parameters

	Units	Natural gas	Hydrogen	Air (ATR-MR case)
Inlet pressure	kPa_g	4	9	0
Outlet pressure	MPa_g	2.5	2.4	2.5
Number of stages		3	5	4
Compression ratio		2.92	1.87	2.25
Isentropic efficiency	%	80	80	80
Mechanical-electrical efficiency	%	98	98	98
Maximum discharge temperature	°C	155	130	155
Maximum temperature after intercooling	°C	50	50	50

With regards to the design of the membrane reactor, all membrane cases were designed with equal permeate pressure set slightly above ambient pressure (9 kPa_g). In linear membrane reactors as the one shown schematically in Figure 1, the fraction of hydrogen on the pressurized side of the membrane decreases towards the exit of the reactor reducing the driving force for permeation. Therefore the flux of hydrogen is the lowest towards the end of the reactor as it shown in Figure 7. The criteria used for membrane reactor design is that of keeping as minimum flux between the 25 and 30 % of average flux.

The reforming reactors are designed for operating at temperatures between 800 and 900 °C in the SR-PSA and WGS-MR systems and at 600 °C in the ATR-MR and SR-MR. In the ATR-MR reactor, a maximum of 50 °C above the operating temperature is admitted at the reactor inlet (see Figure 4). The inlet temperature for the WGS reactors is of 300 °C in the corresponding systems either with or without membranes. In all cases the reactors are considered adiabatic.

For all four investigated cases a steam excess in the reforming reaction of 50 % is considered, leading for example to a steam to carbon ratio of 3.0 at the inlet of the steam reforming reactor. A desulfurization step is required prior to any reactor since sulphur poisons reforming catalyst. For effective sulphur removal, this process typically takes place between 300 and 400 °C over a ZnO bed.

The model is coded using the Aspen Plus® simulation software. The thermodynamic properties were evaluated using a standard Peng-Robinson cubic equation of state [13]. For simplicity the fuel inlet has been considered pure methane. Further design considerations are summarized in Table 2.

Table 2: Design parameters

Parameter	units	value
Steam reforming control temperature in SR-PSA	°C	884
Steam reforming control temperature in SR-MR	°C	800
ATR-MR and SR-MR control temperature	°C	600
WGSR and WGS-MR inlet temperature	°C	300
Desulfurization temperature	°C	330-350
Minimum temperature difference for all heat exchangers	°C	40
PSA hydrogen recovery	%	75
Excess feed water for complete hydrogen production	%	50
Excess air for combustor	%	40
Reactors operating pressure	MPa_g	2.5
Hydrogen output pressure	MPa_g	2.4
Hydrogen production capacity	m^3/h	50
Membrane permeance	$mol/(s\ m^2\ Pa^{0.5})$	$8.52 \cdot 10^{-3}$

Performance comparison

An initial comparison regarding operating performance in terms of efficiency and cost was performed. Table 3 shows the results for energy cost performance. In terms of thermal efficiency, both SR-MR and ATR-MR show higher efficiency than the traditional option. This benefit comes from the lower exergy losses associated with a lower operating temperature of the reformer and simultaneous hydrogen separation which reduces temperature swing losses. The electric power consumption, as expected is higher in any of the membrane reactor systems with respect to SR-PSA, due to the higher demand for hydrogen compression. ATR-MR system is also penalized for the air compression.

After evaluating the thermodynamic performance of the three innovative cases with the reference case, the comparison was then performed in terms of H_2 production costs. In this assessment only the two main operating costs, natural gas and electricity, were considered. The prices of natural gas and electricity were taken as follows: 5.87 $/GJ of gas (6.34 $/Mcf) from US industrial average price in the period January 2001 to June 2011 [14]; and 0.103 $/kWh from US commercial sector 2011 average [15]. Table 3 also shows that the production costs in the traditional system remain lower than in any of the three membrane reactor based configuration. The latter are penalized by the higher electricity consumption and consequently affected by the cost of electricity. However, it is seen that with the SR-MR the production cost is relatively close to the SR-PSA option. The production costs are strongly related to the prices of natural gas and electricity.

Table 3: Performance comparison

	units	SR-PSA	SR-MR	ATR-MR	WGS-MR
NG input	$kW_{(HHV)}$	222.6	203.0	195.9	234.6
Reformer	%	100.0	73.7	100.0	95.2
Burner	%	0.0	26.3	0.0	4.8
Total electric consumption	kW	11.7	18.0	22.8	19.1
NG compression	kW	3.0	2.0	2.6	3.0
Air compression	kW	0	0	4.4	0
H_2 compression 2.4 MPa_g	kW	0	7.3	7.3	7.3
H_2 compression 70 MPa_g	kW	7.9	7.9	7.9	7.9
BOP and controls	kW	0.8	0.8	0.6	0.9
Membrane area	m^2	-	0.308	0.437	0.150
H_2 output	$kW_{(HHV)}$	177.1	177.1	177.1	177.1
Thermal efficiency	% $H_2/NG_{(HHV)}$	79.6	87.3	90.4	75.5
Energy efficiency	%	75.6	80.2	81.0	69.8
Production cost	$/$kg_{H2}$	1.27	1.33	1.41	1.50
NG cost	$/$kg_{H2}$	1.00	0.92	0.88	1.06
Electricity cost	$/$kg_{H2}$	0.27	0.41	0.52	0.44

The natural gas price for instance, shows strong variability linked to location or political reasons, for example the yearly averaged industrial price in 2010 was 4.3 and 11.2 $/GJ in Texas and in Rhode Island respectively. Moreover, natural gas prices are substantially higher than the reference adopted in other countries, for instance in several EU countries as well as Japan. Illustrating this effect, Figure 6 shows the results of the production costs recalculated for 4.3 and 11.2 $/GJ compared to the reference case of 5.87 $/GJ. The figure shows how a relatively high price of natural gas would put the SR-MR solution as convenient over the traditional option in terms of variable production costs.

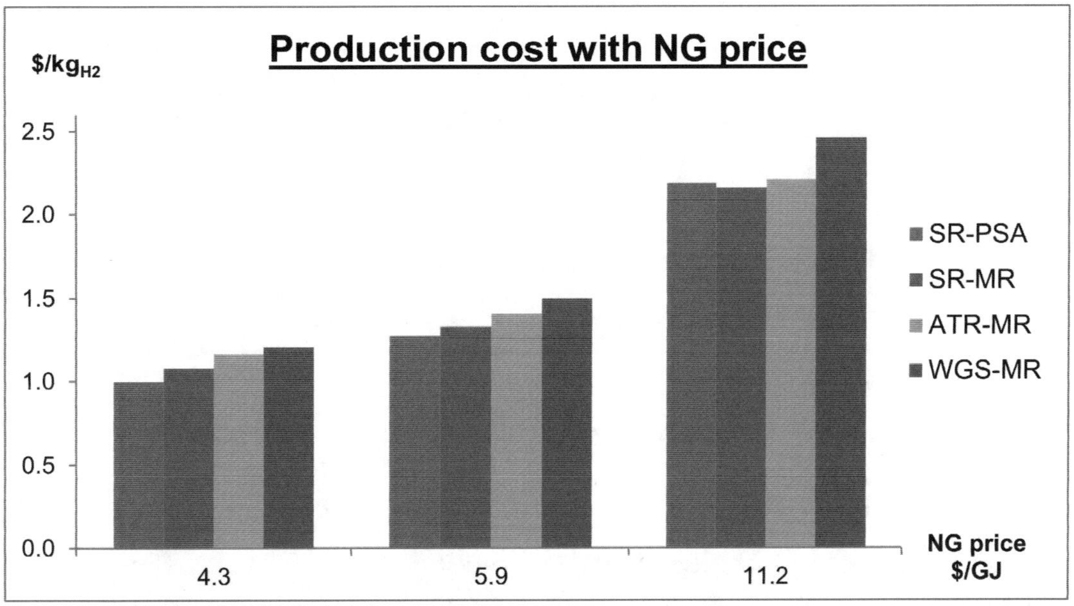

Figure 6: Change in hydrogen production costs with NG price

Cost comparison

Until the technology for hydrogen distribution at filling stations is not a settled market, the investment and operation costs of the system remain uncertain. Based on cost estimations, a detailed study conducted in 2002 for a system similar in configuration and scale to the SR-PSA described in this work, and concluded an investment of $ 253014 from which $ 18075 corresponds to the PSA unit and $ 20780 to the hydrogen compressor [16]. Considering these costs for the traditional system, the deviations from this cost were considered as shown in Table 4 distinguishing the costs associated with key components. In particular the cost of hydrogen compressor varies from the low inlet pressure used in the MR systems and the 2.4 MPa_g inlet pressure of the SR-PSA system. The membrane cost was taken as 5382 $/m^2 (500 $/ft^2) as in the DOE target for 2015 [8], which will be discussed below in a sensitivity analysis.

The table shows how the total cost of hydrogen in $/kg for the SR-MR solution can be competitive with the traditional system, whereas both in ATR-MR and WGS-MR the costs are higher. However, the differences between the total costs are not so significant in relative terms, taking back to the idea of the heavy bond the final costs to those of the energy sources.

Table 4: Comparison of cost of hydrogen

	units	SR-PSA	SR-MR	ATR-MR	WGS-MR
Capital cost	$	253014	236594	237291	235746
Main components	$	214159	214159	214159	214159
Air compressor	$	-	-	5240	-
Hydrogen compressor	$	20780	25424	25424	25424
PSA	$	18075	-	-	-
Membranes(1)	$	-	1655	2352	807
O&M (annual)(2)	$	12651	12062	12359	12020
Insurance (annual)(3)	$	7590	7237	7415	7212
Annual uniform capital cost(4)	$	41177	39260	40227	39122
Total annual	$	61418	58559	60001	58353
Total cost	$/kg$_{H2}$	3.50	3.45	3.58	3.61
Capital recovery	$/kg$_{H2}$	2.23	2.12	2.18	2.12
Production cost	$/kg$_{H2}$	1.27	1.33	1.41	1.50

(1): Considering membrane cost at 5382 $\$/m^2$ (500 $\$/ft^3$)

(2): 5 % of capital cost

(3): 3 % of capital cost

(4): Interest rate 10 %. Lifetime 10 years

Criteria for membrane unit design

As indicated above, the criteria used for membrane reactor design was that of limiting flux at the exit of the reactor to a value between 0.25 and 0.30 of the average flux. The application of this condition allowed for the sizing of the required membrane area. Figure 7 shows the profiles of hydrogen flux across the membrane resulting from the calculations in all three MR configurations. The figure shows the flux profile from the inlet up to the total membrane area (as also indicated in Table 3). As the figure shows, the WGS-MR gives the most efficient use of the membrane surface followed by the SR-MR. This effect is a consequence of the higher hydrogen content due to the shift reaction.

The adopted design criteria is not general, and it should be modified accordingly depending on the application or on the configuration, such as the use or not of a sweep gas on the permeate side (not used in this work). In order to validate the use of the design criteria adopted, a sensitivity analysis on the membrane area design was performed. The results of this analysis are summarized in Table 5. The table shows the performance results of the SR-MR system for +/-10 % of membrane area with respect to the designed configuration shown above in Table 3, plus a configuration that gives the minimum membrane area while still respecting all other design parameters in Table 2. The results show that the direct effect of membrane area on production costs is not too strong, when a +/-10 % change in membrane area leads to a +/-0.8 % in production costs.

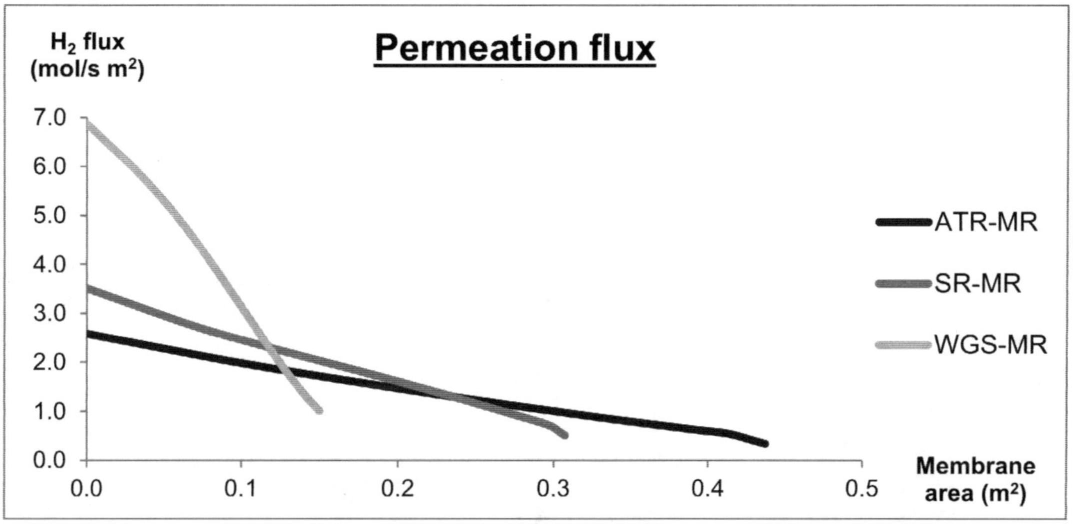

Figure 7: Permeation flux profile for all three membrane reactors

Table 5: Sensitivity to membrane design criteria

	units	SR-MR optimized	SR-MR + 10 % membr.	SR-MR - 10 % membr.	SR-MR min membr.
NG input	$kW_{(HHV)}$	203.0	202.4	204.6	211.9
main	%	73.7	71.9	78.3	100.0
burner	%	26.3	28.1	21.7	0.0
Total electric consumption	kW	18.0	17.9	18.1	18.8
Membrane area	m^2	0.31	0.34	0.28	0.23
Thermal efficiency	% $H2/NG_{(HHV)}$	87.3%	87.5%	86.5%	83.6%
Energy efficiency	%	80.2%	80.4%	79.5%	76.8%
Production cost	$\$/kg_{H2}$	1.33	1.32	1.34	1.39
NG cost	$\$/kg_{H2}$	0.92	0.91	0.92	0.96
Electricity cost	$\$/kg_{H2}$	0.41	0.41	0.41	0.43

Regarding the cost of the membrane, as already addressed, this work considered 5382 $\$/m^2$ as in the DOE target for 2015 [8]. The assumption on the cost of palladium based perm-selective membranes is a delicate matter, because it is not only associated to the requirement of noble metals but to elaborate preparation techniques. Recent studies reported production costs of 10764 $\$/m^2$ or lower [9,17,18], however uncertainty remains while the technology is not commercial. In order to consider the effect of membrane cost a sensitivity analysis was conducted showing the results in terms of total cost of hydrogen in Figure 8. At 23681 $\$/m^2$ the total cost of hydrogen for the SR-MR solution levels up to that of SR-PSA, then at 43056 $\$/m^2$ the capital costs between these two systems becomes equal leaving the difference to the cost of production.

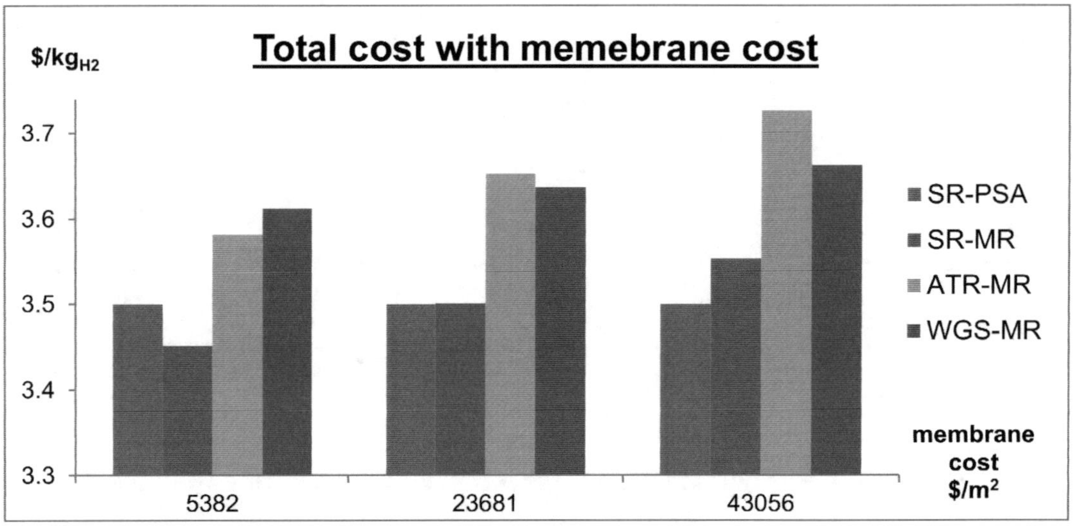

Figure 8: Change in total cost of hydrogen production with membrane price

Conclusions

This work performed a comparative analysis of membrane based configuration for distributed production of hydrogen for filling stations using membrane reactor technology with a conventional system based on steam reformer and PSA and. Within the membrane solution, three different layouts were evaluated in the analysis, discussing results obtained for steam reforming membrane reactor, autothermal membrane reactor and water gas shift membrane reactor.

Results show how the SR-MR and ATR-MR configuration can perform at higher thermal efficiency than the conventional system. This can make them very attractive with elevated costs of natural gas. The SR-MR and ATR-MR configurations also show higher energy efficiency (80.2 and 81.0 % respectively) than the SR-PSA system (75.6 %), however due to the higher fraction of electric consumption, the hydrogen production cost equal to 1.33 and 1.41 $\$/kg_{H2}$ for SR-MR and ATR-MR respectively, remain higher than the 1.27 $\$/kg_{H2}$ of the reference case.

The study reflects that with a proper design of the membrane reactor the capital cost of the system could actually be reduced compared to the traditional solution, even considering a specific cost of the perm-selective membranes much higher than the expected threshold of 5382 $\$/m^2$. This reduction in the investment cost takes place thanks to the exclusion of the PSA unit for purification of hydrogen. As a result, the calculation of the total cost of hydrogen considering 10 years lifetime of the system places the SR-MR solution as more competitive than the SR-PSA with 3.45 $\$/kg_{H2}$ rather than 3.50 $\$/kg_{H2}$. The ATR-MR and WGS-MR solutions result in slightly higher costs of 3.58 $\$/kg_{H2}$ and 3.61 $\$/kg_{H2}$ respectively. The estimations show that up to 23681 $\$/m^2$ the total cost of hydrogen for the SR-PSA and the SR-MR systems would be levelled.

The use of membrane reactor technology presents promising outlook in the field of distributed production of hydrogen. Although the solution is both technically and economically feasible, the design of the membrane reactor unit requires careful attention and expertise.

References

[1] Scandinavian Hydrogen Highway Partnership, Online:http://www.scandinavianhydrogen.org/h2moves. (2012).

[2] HyER- European Association for Hydrogen and fuel cells and Electro-mobility in European Regions, Online:www.hyer.eu/category/about-hyer. (n.d.).

[3] HyRaMP, [online]:www.hy-ramp.eu/category/about-hyramp. (2011).

[4] R. Dittmeyer, Membrane reactors for hydrogenation and dehydrogenation processes based on supported palladium, Journal of Molecular Catalysis A: Chemical. 173 (2001) 135-184.

[5] Y. Chen, Y. Wang, H. Xu, G. Xiong, Efficient production of hydrogen from natural gas steam reforming in palladium membrane reactor, Applied Catalysis B: Environmental. 81 (2008) 283-294.

[6] J. Tong, Y. Matsumura, H. Suda, K. Haraya, Experimental Study of Steam Reforming of Methane in a Thin (6 μM) Pd-Based Membrane Reactor, Industrial & Engineering Chemistry Research. 44 (2005) 1454-1465.

[7] S.-K.K. Ryi, N. Xu, A. Li, C.J. Lim, J.R. Grace, Electroless Pd membrane deposition on alumina modified porous Hastelloy substrate with EDTA-free bath, International Journal of Hydrogen Energy. 35 (2010) 2328-2335.

[8] U.S. DoE, DOE - Hydrogen and Fuel Cell Program, Online:www.hydrogen.energy.gov. (2011).

[9] A. Damle, A. Manheim, K. Randolph, High Performance Palladium-Based Membrane for Hydrogen Separation and Purification - II.C.3 FY 2011 Annual Progress Report, 2011.

[10] Fuel Cells 2000, Fuel Cell Vehicles, Online:www.fuelcells.org/info/charts/carchart.pdf. (2012).

[11] C.S. Hsu, P.R. Robinson, Practical Advances in Petroleum Processing, Springer, 2006.

[12] A. Kohl, R. Nielsen, Gas Purification, 5th ed., Gulf Professional Publishing, 1997.

[13] D.Y. Peng, D.B. Robinson, A new two constant equation of state, Ind Eng Chem Fundm. 15 (1976) 59-64.

[14] U.S. DoE, U.S. Natural Gas Prices, Online:www.eia.gov/dnav/ng/ng_pri_sum_dcu_nus_m.htm. (2012).

[15] U.S. DoE, EIA - Electricity Data, Online:www.eia.gov/electricity/monthly/epm_table_grapher.cfm?t=epmt_5_3. (2012).

[16] Directed Technologies Inc., Cost and Performance comparison of Stationary Hydrogen Fueling Appliances, 2002.

[17] J. Schwartz, Advanced Hydrogen Transport Membranes for Coal Gasification, 2011.

[18] S.C. Emerson, Experimental Demonstration of Advanced Palladium Membrane Separators for Central High-Purity Hydrogen Production, in: Annual Merit Review Proceedings Hydrogen Production and Delivery, 2010.

(562i) Experimental and Simulation Studies of Effects of Operating Conditions and Membrane Properties On the Performance of Zeolite Membrane Reactors for High Temperature Water Gas Shift Reaction

Seok-Jhin Kim, **Shaowei Yang** and Junhang Dong, Chemical Engineering, University of Cincinnati, Cincinnati, OH

Experimental and Simulation Studies of Effects of Operating Conditions and Membrane Properties on the Performance of Zeolite Membrane Reactors for High Temperature Water Gas Shift Reaction

Seok-Jhin Kim, Shaowei Yang, Junhang Dong

Department of Chemical & Materials Engineering, University of Cincinnati, Cincinnati, OH 45220, USA

ABSTRACT

Modified microporous MFI-type zeolite membranes have been studied as high temperature water gas shift (WGS) membrane reactors (MR) using a cerium-doped ferrite (Fe/Ce) catalyst. The effects of the reaction conditions and the membrane separation properties on the CO-conversion (c_{CO}) in the MR were investigated experimentally and by simulations using a simple one-dimensional model which was validated by the experimental data. The experimental results have demonstrated that the zeolite MRs with moderate H_2 separation performance are capable of overcoming the equilibrium CO-conversion ($c_{CO,e}$) existing in traditional packed-bed reactors at operating temperatures of >500°C where CO reaction rate is rapid enough to effectively minimize the permeation of the unreacted CO. The model calculations have shown that near-completion CO-conversion (c_{CO}>99.5%) may be achieved under realistic operating temperature, pressure, space velocity and catalyst load even for membranes with moderate H_2 selectivity, e.g. ($a_{H2/CO2}$<50) and H_2 permeance (<3×10^{-7} mol/s·m^2·Pa).

(607a) Pore Scale Models for Imbibition of CO2 Analogue Fluids in Etched Micro-Model Junctions Using Micro-Fluidic Experiments and Direct Flow Calculations

Emily Chapman[1], **Jianhui Yang**[1], John Crawshaw[2] and Edo Boek[1], (1)Imperial College London, London, United Kingdom, (2)Imperial College London, London

In this paper, we investigate pore scale models for the storage of CO_2 in subterranean rock formations. Various mechanisms for storage and trapping have been proposed, including structural, solution, mineralisation and the immobilisation of CO_2 by capillary trapping. Currently, the capillary trapping mechanism seems to be a suitable option, as it is a fast way to render the injected supercritical CO2 immobile. In addition, this trapping mechanism does not rely entirely upon the cap rock maintaining its integrity on geologic time scales. One important parameter for the success of capillary trapping is the residual saturation after CO_2 injection in the formation. The saturation is determined by the wetting properties of the rock formation. Often, and particularly in carbonate reservoirs, the rock may be wetting to CO_2, especially when injected under supercritical conditions. In this case, the mechanism of entry of CO_2 into the pores of the rock matrix is determined by spontaneous imbibition. To understand this problem at the pore scale, it is essential to have a clear view of the mechanism of capillary filling in pore junctions. In the 1980s, Lenormand and co-workers carried out pioneering work on displacement mechanisms of fluids in etched networks [1]. They observed that for drainage, the mechanism of displacement is simply determined by the Young-Laplace law, stating that the capillary pressure is proportional to surface tension and the cosine of the contact angle, and inversely proportional to the radius of the capillary. In addition, Lenormand *et al.* claimed that capillary filling rules for imbibition are also determined by the Young-Laplace law. The rules for capillary filling for both drainage and imbibition are currently used in pore network models. However, in recent years, several spurious results were obtained for spontaneous imbibition in pore network models. For this reason, we decided to revisit and investigate the validity of the capillary filling rules for the case of spontaneous imbibition, using a combination of micro-fluidic experiments and direct Navier-Stokes CFD flow calculations, using the lattice-Boltzmann (LB) method. The micro-fluidic cells consisted of single junctions with unequal throats, etched into PMMA. First, we validated the case of drainage by injecting DI water into the oil-wet PMMA models. Indeed we observed that, when the pore body has been filled, the subsequent displacement shows drainage into the largest capillary throat first, in agreement with the Young-Laplace law. Then we carried out a set of imbibition experiments. This was done by imbibing decane, which has shown to be an excellent analogue fluid at ambient conditions for CO_2 under supercritical conditions, into the junction. Each time, we observed that the fluid does not imbibe into the smallest throat, as predicted by the Young-Laplace law. Instead, we observed that the capillary filling rules for displacement are determined by the local geometry of the junction. To validate the experimental results, we carried out LB flow calculations in an exact representation of the 2D experimental geometry. Indeed we observed that the flow calculations reproduce the experimentally observed displacements in detail. This means that the capillary filling rules in pore network models (PNMs) will have to be overhauled. In this paper, we will describe the experimental and modeling results in detail. Also, we will discuss the implications for CO2 injection at the Darcy scale.

This work was carried out as part of the activities of the Qatar Carbonates & Carbon Storage Research Centre (QCCSRC). We gratefully acknowledge the funding of QCCSRC provided jointly by Qatar Petroleum, Shell, and the Qatar Science and Technology Park.

[1] G. Lenormand, C. Zarcone and A. Sarr, *J. Fluid Mech.* **135 , 337-353** (1983).

(607b) Selection of Domain Size for Lattice Boltzmann Simulation to Calculate Flow Properties of Porous Media

Md Rakibul Sarker[1], Shameem Siddiqui[1], Dustin McIntyre[2], Grant Bromhal[2] and Mohamed Soliman[1], (1)Texas Tech University, Lubbock, TX, (2)National Energy Technology Laboratory, Morgantown, WV

Understanding fluid flow through porous media is important in many fields of science and engineering. Some of these fields are ground water flow, production of petroleum fluids, CO_2 sequestration, air filtration, fuel cells, breathing etc. This paper describes a three dimensional lattice Botlzmann simulation study conducted on high resolution micro-CT images of porous media obtained from an actual reservoir rock in order to determine permeability and to evaluate the heterogeneity of the rock. A memory efficient computer code that uses the sparse matrix algorithm, along with the parallel processing technique using the Message Passing Interface (MPI) was used in the simulation.

The Lattice Boltzmann method (LBM) is commonly used to simulate fluid flow through porous media because of its ease of implementation in computer programming. The discretization required for LBM is inherent and readily available for the segmented micro-CT images in the form of pixels. LBM is an alternative way to solve the Navier-Stokes equation that describes the motion of fluids. LBM considers a definite number of discrete nodes or positions in the domain in which fluid can only move from one node to the other along the particular set of velocity directions. In the D3Q19 stencil system of LBM, 19 velocity directions are considered in the three dimensional domain. For LBM implementation, for each node in space needs to hold 19 velocity values along with other variables. To keep the account of all these variables at every time step and to update their values for the next time step, a large amount of computational memory is required. This computational memory requirement limits the maximum domain size possible to simulate in a computer with a finite amount of memory. For handling large domain sizes, computer memory requirements for the simulation can be met by using a distributed memory system. The question is how big the domain should be to simulate fluid flow to approximate the core level results from the micro-CT images. This paper proposes some new insights that help select the domain size for the LBM simulation of micro-CT data.

In this study different properties of the porous media such as porosity, permeability, specific surface area and surface to volume ratio were calculated on different domain sizes starting from 64-cube up to 520-cube of the porous media obtained from the micro-CT images. The micro-CT images were obtained from the CT scan of a core size sample which was 1.5 inches in diameter and almost three inches in length. Only a portion of the rock sample was scanned to produce images with resolution of 2.32 microns in x, y and z directions. Variation of different properties was plotted against their domain size to see its effect on the properties. The porosity and permeability were measured experimentally on the same core plug with established laboratory core analysis methods. The permeability values for the smaller domain sizes fluctuated due to the microscopic domain effect and stabilized to match the experimentally measured permeability for the domain size beyond the 500-cube. The computer code written for this simulation was first verified by reproducing the analytical results from known flow models such as, Poiseuille flow through narrow slits, flow through circular conduit, etc. Porosity and surface area were calculated by counting the appropriate pixels of the images. The ratio of porosity and specific surface area was plotted against the domain size and found to be asymptotic at the domain size larger than the 500-cube. So, the optimum domain size to simulate fluid flow through porous media for calculating permeability should be free from microscopic domain effects should be selected where the ratio of porosity and specific surface area values become asymptotic. One of the practical applications of the lattice Boltzmann model and the methodology developed is the ability to predict the flow and storage capacities of hydrocarbon reservoirs using small cuttings generated through the drilling of wells without resorting to the expensive and risky process of coring.

(607c) A Lattice Boltzmann Model with Realistic Equations of State for Modeling Pore-Scale Effects On Capillary Trapping of CO2 During Geological Carbon Sequestration

Marcel Schaap, Soil, Water and Environmental Science, University of Arizona, Tucson, AZ and Dorthe Wildenschild, School of Chemical , Biological, and Environmental Engineering (CBEE), Oregon State University, Corvallis, OR

With large proven global coal reserves, it is highly likely that combustion of coal and coal-derived products will remain part of the global energy future. Compared to other fossil fuels, coal-based power yields the highest CO_2 output per energy unit, indicating the critical need for some form of carbon mitigation regardless of future global economic developments and associated emission scenarios. Currently feasible and relatively mature carbon capture and storage (CCS) technologies would prevent about 90% of the fossil carbon from reaching the atmosphere. Geological carbon sequestration, however, also carries significant risks such as leakage of the injected gas back into the vadose zone and atmosphere through abandoned wells or underground cracks and fault systems (geological or man made by, for example fracking). Despite very broad research efforts by industry, government, and academic groups, there are still substantial uncertainties related to the effectiveness of the trapping, dissolution, and precipitation processes controlling the permanent storage of CO_2.

Typical sequestration scenarios involve the injection of supercritical CO_2 into the saline aquifer while displacing brine in what is equivalent to a water drainage process. Due to density differences between supercritical CO_2 and reservoir brine (applicable under most injection scenarios), the CO_2 will tend to buoyantly migrate upward until an impermeable barrier is encountered, possibly following complex flow patterns along high permeability channels. Once injection stops, buoyant forces will continue to propel the s.c. CO_2 upwards, while re-invading brine from below reoccupies the pore space in a water imbibition process, trapping a large part of the s.c. CO_2 with capillary forces. The s.c. CO_2 will likely remain in the capillary trapped state for tens to hundreds of years until it finally dissolves in the brine as a negatively buoyant phase. Depending on the weathering kinetics of the host rock, the ultimate fate of the CO_2 is a carbonate mineral precipitate – a process that may take thousands of years to complete.

To date, capillary trapping (otherwise known as residual trapping or relative permeability hysteresis trapping) is a somewhat under-explored mechanism in geologic sequestration of CO_2. Central to capillary trapping is its pore-scale nature: pore-scale geometry, fluid velocities as well as brine- CO_2 interfacial physics will strongly determine the magnitude of capillary trapping. Because of its stable immobilization characteristics, capillary trapping has several advantages over structural trapping, including not being affected by potentially compromised reservoir caprock; facilitating enhanced dissolution of gaseous CO_2 into the brine; and allowing gaseous CO_2 to be distributed over a larger reservoir volume, thus enhancing the potential for mineral weathering and carbonate precipitation. To date, relatively few experiments and models have addressed capillary trapping mechanisms in detail and at relevant reservoir conditions. Compared to most current large-scale reservoir studies, our research takes several steps back in scale to observe and model trapping at the pore-scale and to learn how these findings translate into continuum scale properties that can subsequently support improved modeling of sequestration at large spatio-temporal scales.

Two parallel, but strongly interacting, approaches are followed in our research. The Oregon State University (OSU) group, under leadership of Dr. Dorthe Wildenschild conducts X-ray micro-tomography measurements of the capillary trapping process using proxy fluids to explore the effect of varying interfacial tension, viscosity and flow rate on the resulting trapped amount of CO_2. Simultaneously, the OSU group is building an x-ray tomography compatible core holder and experimental setup where we can address similar questions under reservoir conditions, i.e., with s.c. CO_2. We are using both synchrotron-based and conventional x-ray computed microtomography to quantify trapped CO_2 amount, distribution, and morphology as a function of the three properties (viscosity, interfacial tension, and flow rate).

The group at the University of Arizona under leadership of Dr. Marcel Schaap is developing functionality into lattice Boltzmann models that allow for the physical conditions found in the experiments as well as relevant pressure and temperatures found in candidate geological reservoirs. Particularly important here is to define a model that can handle the correct density, and viscosity ratios between brine and s.c CO2, while it is also critical that pressure, temperature and brine concentration-dependent interfacial tension is handled with accuracy. The current focus is to test several van der Waals-type Equations of State (EOS, the relation between pressure, density and temperature) for the pure water-vapor and the CO_2 system near their respective critical points. We show that a wide range of conditions can be simulated and demonstrate that liquid and vapor densities and surface tension agree with published experimental data. We also show that modifications to the parameterization of the equation of state are necessary to obtain a better quantitative match for the water-vapor and CO_2 systems as well as the brine – super critical CO_2 system.

Currently, work is underway to verify the Lattice Boltzmann models using micro-tomography results obtained by the OSU group. We are also working on implementing mixed wettability into our models, to allow for the presence of residual hydrocarbons in aquifers. The models, combined with the experiments should subsequently allow us which types of porous media, and which flow conditions lead to better trapping. Ultimately, but not covered in the current presentation, we will upscale the experimental and modeling results to verify existing transport equations (e.g. Lands model) and, where necessary propose new transport equations for use in reservoir-scale modeling studies.

The research in this presentation is supported in part by DOE-BES geosciences grants DE-FG02-11ER16278 and DE-FG02-11ER16277

(607d) Insights From Image-Based Modeling of Fluid Flow in Bulk Proppant Packs and Propped Fractures Under Varying Confining Stress

Yijie Shen[1], Paula Sanematsu[2], Seth M. Bradley[2], Clinton S. Willson[2], Ali Takbiri Borujeni[3], Karsten E Thompson[3] and Mayank Tyagi[3], (1)Cain Department of Chemical Engineering, Louisiana State Univeristy, Baton Rouge, LA, (2)Department of Civil and Environmental Engineering, Louisiana State Univeristy, Baton Rouge, LA, (3)Craft & Hawkins Department of Petroleum Engineering, Louisiana State Univeristy, Baton Rouge, LA

Ceramic proppant was imaged using x-ray computed microtomography (microCT) in three different geometries: bulk proppant, multilayer proppant pack between Berea walls, and single-layer proppant between shale walls. In all cases, the proppant packs were exposed to a range of confining stresses that varied between zero and 20,000 psi. The resulting microCT images were segmented, analyzed for structural and porosity changes, and then used for image-based flow modeling of low- and moderate-Reynolds number flow.

The images show expected changes as stress is increased: rearrangement of the packing structure, corresponding reduction in porosity, and some embedding at rock walls. At the largest stresses, individual proppant particles began to break apart.

Numerical simulation was performed using finite element and lattice Boltzmann methods. Results from both methods were used to compute permeability and Forchheimer coefficient (Beta factor) for the structures. Simulation results show reasonable agreement with experimental (vendor-reported) conductivity values (both permeability and beta). However, our results show that conductivity is less sensitive to loading than has been reported. Another somewhat surprising result is that fracture conductivity for the single-layer proppants confined between shale is similar to what would be predicted from bulk-proppant permeability, despite the significantly different flow geometry in the monolayer fracture. Possible reasons for this result as well as insights derived from flow visualization are presented in the paper.

Recent Trends of Streamline Simulation Technology Applications in Oil Industry

A. A. Al-Najem, S. Siddiqui, M. Soliman, Texas Tech University

This paper was prepared for a presentation at the 2012 AIChe Annual Meeting in Pittsburgh, PA October 28 - November 2, 2012.

Abstract

Streamline and streamtube methods have been used in fluid flow computations for many years. Early applications for hydrocarbon reservoir simulation were first reported by Fay and Pratts in the 1950s. Streamline-based flow simulation has made significant advances in the last 15 years. Today's simulators are fully three-dimensional and fully compressible and they account for gravity as well as complex well controls. Most recent advances also allow for compositional and thermal displacements.

In this paper, we present a comprehensive review of the diverse applications of streamline simulation technology in petroleum industry. This paper offers a comprehensive review of the major development in this area. This work involved the review of more than 200 technical papers and gives a chronological advancement of streamline simulation technology from 1996 till now. In view of the fact that this state of-the-art technology has been employed for a wide range of applications, we have defined three major application areas that symbolize the relevance and validity of streamline simulation in addressing petroleum engineering issues. These are history matching, reservoir management and upscaling, ranking and characterization of fine-grid geological models.

Streamline simulation has undergone several phases within its short stretch in the petroleum industry. Initially, the main focus was on the speed advantage and less on fluid flow physics. Next, the focus was shifted to extend its applicability to more complex issues such as compositional and thermal simulations, which require the inclusion of more physics, and potentially reducing the advantage of computational time. Recently, the focus has shifted towards the application of streamline technologies to areas where it can complement finite difference simulation such as revealing important information about drainage areas, flood optimization and improvement of sweep efficiency, quantifying uncertainties, etc.

Introduction of Streamlines Simulation

Streamlines are integrated curves that are locally tangential to a defined velocity field at a given instant in time (Datta-Gupta 2007 and Thiele *et al.* 2010) as illustrated in Figure 1. Modeling fluid flow and transport using streamlines dates back to the study of well pattern and total recovery by Muskat and Wyckoff in 1934. Streamline-based flow simulation has made significant advances in the last 15 years. A great historical overview of the earlier streamlines work was presented by Batycky (1997), Datta-Gupta and King (1998), Thiele (2001), Moreno *et al.* (2004), Datta-Gupta (2007).

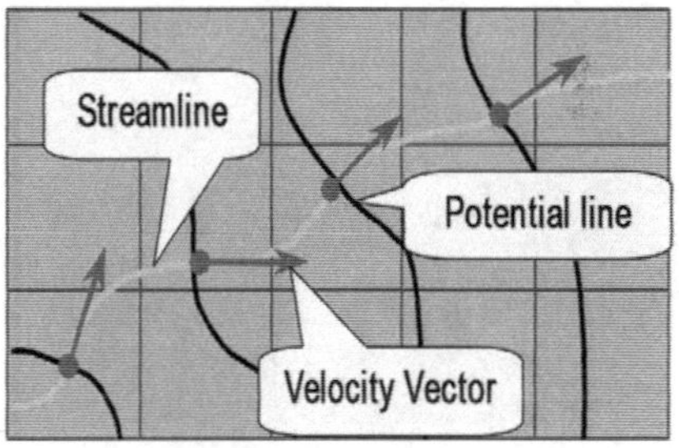

Figure 1: Streamline Tracing and Velocity Vector Mapping in a Saturation Grid (Ibrahim *et al.* 2007)

Streamline simulation is an IMPES type reservoir simulation which solves pressure equation implicitly and then saturation/conservation equations explicitly. Thus, streamline simulators operate on the principle of de-coupling the pressure equation from the saturation equation. This simplification helps in decomposing a heterogeneous 3D domain into a number of 1D streamlines where all fluid calculations are done.

Figure 2 is a generalized flow diagram for streamline simulation. Starting with some input data such as grid geometry, rock and fluid properties, well locations, injection rates and boundary conditions, the 3D grid is initialized. Then, using finite difference approximations, the pressure distribution is derived to generate instantaneous velocity vectors perpendicular to the computed pressure contours. The velocity field is then used to trace the streamlines. The specified wells (i.e. source or sink) and boundary conditions govern the initiation and termination of all traced streamlines which once established make up the second grid system to help solve the fluid flow equation along the streamlines. The second grid system works under a local timestep controlled by fluid movement computations, making it a time-variant grid. The solution obtained from the second grid is then mapped back onto the original 3D Cartesian grid to account for fluid phase distribution and saturations. For next cycle, the pressure is solved again and the streamlines are redrawn. The process continuous tills the end of simulation.

This dual grid approach is the underlying difference between streamline simulators and conventional finite difference simulators.

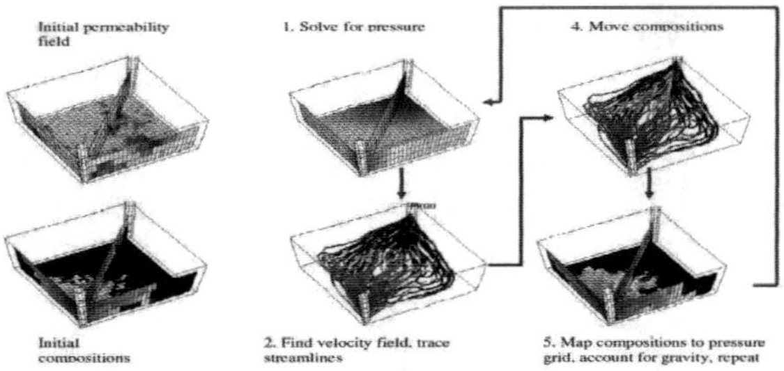

Figure 2: Generalized flow diagram for streamline simulation (Gerritsen M. 2008)

What we have just explained involves a lot of mathematical calculations on both grid systems in order to solve pressure and transport equations. We refer the reader to the following references where the mathematical formulations have been thoroughly explained and covered. These include but not limited to (Batycky *et al.* 1997, Ingebrigsten *et al.* 1999, Doi and Suzuki 2000, Lolomari *et al.* 2000, Gautier *et al.* 2001, Jessen and Orr Jr. 2002, Di Donato and Blunt, 2003, Moreno *et al.* 2004, Gerristen *et al.* 2005, Mallison *et al.* 2006, Cheng *et al.* 2006).

There are several advantages of employing streamlines simulation for modeling fluid flow. One of the main advantages that attracted researchers at first place to streamline technology was its computational speed. Due to the 1D nature, transport calculations are not constrained by grid instability leading to larger time steps. Batycky (1997) reported that streamline technique was faster than FD simulation by factors of 1 to 3 orders of magnitude. The second main advantage was the minimization of numerical artifacts (artificial diffusion and grid orientation effects) as a result of the effective decoupling of streamline grid from the underlying geological grid. Another advantage is the visualization potential. For example, visualizing the source-sink relationships based on streamline density help obtain quantitative flow indicator.

Like finite difference (FD) simulation, streamline simulation has its limitations. The most important limitations are mapping between coordinates and fluid flow complex physics. Streamline simulation contains two separate grids; an underlying physical grid where the pressures and the velocities are calculated, and the streamline time-of-flight grid where the fluid transportation is calculated. A re-generation of streamlines is done at each pressure update. This means that the saturations from the old set of streamlines must be mapped back to a new set of streamlines. Streamlines transport saturations rather than conserved volumes, which are only implicitly defined in the time-of-flight coordinate. Because of the re-sampling of implicit volume from time-of-flight coordinates to physical coordinates, potential mass balance errors and, to some extent, numerical dispersion may be introduced.

The second major limitation of streamline simulation results from its main advantage defined above: computational speed. When dealing with complex physics like high compressibility, capillary effects and phase behavior, the computational speed decreases. This is due to the need for more frequent re-sampling of streamlines, which means more frequent solving of the pressure equation.

1. Petroleum Engineering Applications of Streamline Simulation

State of-the-art streamline technology has been employed for a wide range of applications. In this section, we defined three major application areas that symbolize the relevance and validity of streamline simulation in addressing reservoir engineering concerns. These are history matching, reservoir management and upscaling, ranking and characterization of fine-grid geological models. The following sections highlight the work that has been done under each category and its sub-categories.

1.1. History Matching

History matching is an inverse problem that is known to be highly nonlinear and ill-posed, producing a set of non-unique solutions (Subbey *et al.* 2003, Maucec *et al.* 2007, Stephen *et al.* 2007, Kazemi and Stephen 2010). Obtaining a reliable history match typically requires numerous flow simulations. One major benefit of streamline simulation is that due to its efficiency, many more flow simulations can be evaluated in a given time compared to standard FD simulators. As a consequence, many authors have employed streamline simulation in history matching algorithms which require hundreds or thousands of flow simulations (for example, Subbey *et al.* 2003). Beyond the speed of the flow simulation, there are many additional benefits that streamline simulation provide to history matching, many of which have been already described in other reviews, for example Datta-Gupta 2007, Stenerud and Lie (2007)and Thiele *et al.* 2010. In this section, we review many of the diverse applications of streamline simulation to history matching, ranging from the simple use of streamline-delineated drainage zones in traditional history matching practices, to sophisticated use of streamline information for data integration.

1.1.1. Delineation of Drainage Zones

One significant benefit of streamline simulation applied to history matching is that the streamlines identify instantaneous flow directions and drainage regions of the producing wells, which are often very complex and varying throughout the lifetime of a reservoir. When a producing well does not match its historical data at a particular time, the streamlines can associate the mismatch with the petrophysical properties in a particular drainage zone. Several authors have employed this association in various history matching methods, described below.

Emanuel and Milliken (1998) described a history matching methodology that uses flow paths described by 3D streamlines to assist in changing reservoir parameters. Two examples involving more than 60 wells were described, which showed its power and utility. In both cases, very satisfactory well-by-well history matches were obtained after just a few iterations.

Using dynamic streamline-derived drainage zones, Hoffman and Caers (2003) proposed a regional probability perturbation method (RPM) for modifying geostatistical realizations in a geologically consistent manner. The methodology is an extension of the global perturbation method (GPM) and it relies on a region-wise perturbation of the probability model. Tests using a two-facies (sand and shale), fractured reservoir model with three fault blocks showed that the RPM converged faster and it showed better matches for pressure and water-cuts than GPM.

To facilitate the choice of regions-of-interest to history match, Maschio and Schiozer (2004) used streamline simulation to determine flow paths among producers and injectors or between an aquifer and a producer. They combined this with other two methods namely, the global history match and the regions-of-interest history match. The proposed approach was applied to an offshore field with 22 years of production history. The results showed the use of the regions of main flow path determined by streamlines provided better results.

Devegowda *et al.* (2007) examined two techniques for covariance localization using streamline-specific information to mitigate small sample sizes and non-linear dynamics in non-Gaussian settings problems associated with the use of Ensemble Kalman Filter (EnKF). The first approach utilizes streamline trajectories to delineate regions that will have largest influence on solution. The second approach uses streamline-based sensitivities, which will be discussed below. The proposed methods of covariance localization are shown to exhibit superior performance vis-à-vis the traditional form of the EnKF demonstrated with synthetic examples and real field cases.

Kazemi and Stephen (2010) investigated a history matching technique by focusing on sub-volumes of the parameter space and used streamlines to help choose where changes are required. The streamline-guided approach gave a 70% improvement in history matching from base model and around 40% reduction of misfit in prediction. In a more sophisticated example, Watanabe and Datta-Guptta (2011) defined streamline zones per fluid phase order to improve the traditional distance-based localizations used in Ensemble Kalman Filter (ENKF). The discontinuities in water and gas phase streamlines were used to define flow relevant covariance localization for water-cut and GOR data assimilation during EnKF updating. The authors compared both methods using two cases and results showed that phase streamlines based localization outperforms traditional distance based localization in terms of capturing low permeability barriers affecting gas flow and high permeability streaks contributing to aquifer supports.

1.1.2. Direct Methods

Streamline-based history matching techniques can go beyond the ability to delineate drainage zones. One set of examples are described in this subsection as direct methods. Direct methods were developed first by Wang and Kovscek (2000), and employ the fractional flow curves at the producing wells and relate them to changes effective permeability (and/or porosity) values for history matching. This approach was benchmarked with 2-D incompressible cases where the effect of gravity was not important and illustrated the computational efficiency, generality, and robustness of the proposed procedure.

The approach was extended by Agarwal and Blunt (2001) and Agarwal and Blunt (2003), who developed a history matching method for water cut data using a commercial, compressible three-phase streamline simulator with gravity. Data from a North Sea field was used to test this full-physics technique, which gave a reasonable history match and a good prediction of future performance. Lately, Batycky *et al.* (2007) applied a similar method to a large waterflood.

de Hann *et al.* (2001) and Kretz *et al.* (2004) proposed similar approaches for history matching of time-lapse seismic. de Hann *et al.* (2001) used time lapse 3D seismic saturation contours to determine the velocity of the flow between saturation contour pairs. By applying Darcy's Law to this velocity field, a realistic permeability map could be inverted. The authors tested this technique on five randomly created cases and it allowed identifying different zones of high and low permeabilities. Kretz *et al.* (2004) modified a given set of flow properties (e.g. porosity, permeability, cell volume, etc.) for history matching observed fluid fronts obtained from 4D seismic. The approach was tested using several 2D cases. The matching of fluid fronts needed few iterative steps leading to a more active use of 4D seismic in reservoir management.

1.1.3. Optimization Methods

Gradient-based optimization methods require the calculation of sensitivity coefficients. By drawing a correspondence between streamline modeling and ray tracing in seismology, Vasco *et al.* (1998) developed a method for analytically calculating the sensitivity coefficients of the time-of-flight with respect to the grid-block petrophysical properties. The time-of-flight can then be related to the breakthrough of injected fluids or tracer. The fact that the sensitivity coefficients are calculated analytically greatly enhances the efficiency of the optimization procedure. Because the sensitivity calculations involve evaluation of 1D integrals along streamlines, it scales very well with respect to model size (Qassab *et al.* 2003, Cheng *et al.* 2006). This breakthrough development has led to a significant body of research and applications on history matching, some of which are described below.

He *et al.* (2001) developed a 'generalized travel time' inversion method for production data integration that minimizes a 'travel time shift' derived through maximizing a cross-correlation between observed and computed production responses at each well. The travel time sensitivities were computed analytically using a single streamline simulation that accounts for gravity and changing field conditions and the minimization is relatively insensitive to initial model. Starting with well logs and seismic data, the Goldsmith San Andres Unit (GSAU) in West Texas field example water-cut history was integrated in less than 2 hours using a simple PC.

Gautier *et al.* (2001) extended the previous work of Vasco *et al.* to compute sensitivity coefficients of pressure, saturation and production data in two-phase flow and non-unit mobility ratios for varying boundary conditions. The results showed that objective function gradient was non-monotonic when sensitivity coefficients for saturation only were considered. In contrast, when only pressure is considered, the objective function becomes smooth and converges fast.

Qassab *et al.* (2003) applied the streamline-based production data integration method proposed by He *et al.* (2001) to a giant Saudi Arabian carbonate reservoir, to condition a multimillion-cell geologic model in order to match 30 years of historical production responses. For this field, production data integration was carried out in less than 6 hours on a PC and geologic model derived was found to be consistent with field observations.

Vasco *et al.* (2003) developed a trajectory-based streamline approach to invert time-lapse seismic amplitude and travel time to image reservoir flow properties, such as permeability, by computing and matching time-of-flight (TOF). The application to actual time-lapse data from 40 years of historic production of the Bay Marchand field in the Gulf of Mexico indicated that the method was robust even in the presence of noise.

By defining a generalized travel time inversion (GTTI), Cheng *et al.* (2004) proposed an approach to history match FD models using streamline trajectories, time-of-flight and parameter sensitivities in an inversion algorithm to update reservoir model during FD simulation. The approach was applied to a synthetic example and two field examples with multimillion cell geologic models and with 30 years of production history. The entire history matching of the field example took 9 hours on a PC.

Al-Harbi (2004) extended a streamline model to treat fracture and matrix as separate continua connected through a transfer function. In addition, the author computed reservoir properties versus production response sensitivities in Naturally Fractured Reservoir (NFR) using the GTTI. These derived sensitivities were applied in conjunction with a dual porosity FD simulator to help incorporating the compressibility effects and the complex physics. Al-Huthali and Datta-Gupta (2004) and Al-Huthali and Datta-Gupta (2004) developed a streamline-based technique to describe fluid transport in NFR through a dual-media approach. This technique was tested on 5 and 9 spot patterns and results showed close agreement with recovery and saturation profiles with a marked reduction in numerical dispersion and grid orientation effects. Using the double-continua formulation developed by Al-Huthali and Datta-Gupta (2004) and flow paths produced by streamline simulator, Lino (2006) developed a simulation model for tracer response analysis in NFR. The model was validated using heterogeneous permeability distributions and a multiple injection/production scheme including mega-fractures. The saturation fronts and tracer responses were less smeared than those obtained from a commercial FD simulator, in addition to improved depiction of channeling flow.

Vargas-Guzmán *et al.* (2009) used streamline-based generalized travel time inversion (GTTI) to efficiently and rapidly condition a one million cells prior geological model with measured dynamic data and interpret spatial distributions of high permeability zones. The permeability modifications statistical analysis revealed that marcopores and fractures are the main causes of high permeability zones. Additionally, the existence of localized high permeability fracture zones was associated with specific rocks and high curvature regions.

Stenerud *et al.* (2007) combined multi-scale streamline method with the 'generalized travel-time inversion' method for history matching high-resolution geologic models. The approach helped to derive a fast quasi-linear method, demanding a few iterations, and enforce basis functions to be adaptively updated only in areas with relatively large sensitivities to production response. This approach was tested using a highly detailed 3D geo-model consisting of more than 1 million simulation cells and 69 producers and history matching of 7 years period was accomplished in less than 40 minutes on an ordinary PC.

Using streamline based water-cut (WC) and GOR sensitivities; Oyerinde *et al.* (2007) developed a three-phase flow history matching method that involves simultaneous GTTI inversion of WC, GOR, and flowing BHP data. This approach was examined using a highly faulted, West African reservoir with over 30 years of production history. The results of water-cut and GOR match illustrated the effectiveness of this method.

Devegowda *et al.* (2007) used streamline based sensitivities for covariance localization applied to EnKF.

Efendiev *et al.* (2008) used streamlines based sensitivities generated through a single streamline simulation to provide analytical approximations to a two stage Markov Chain Monte Carlo (MCMC) approach used for sampling of permeability field. The numerical tests showed several fold increase in the acceptance rate and significant reduction in computational time required for MCMC sampling.

To depict the movement and trapping of CO2 in the aquifer, Rey *et al.* (2010) developed a streamline tracing algorithm using component fluxes generated by FD compositional simulator. The authors also used a compositional streamlines simulator to determine the sensitivity of the time-lapse seismic attributes specifically, interpreted saturation differences to changes in reservoir properties such as permeability and porosity. These sensitivities are then used in an inverse modeling algorithm to calibrate the geologic model to time-lapse seismic data.

1.1.4. Data Integration Methods

A separate approach which is distinctive from the optimization methods described above fall under the category of data integration methods. These methods use the same streamline-specific information which is use for the direct methods, but instead of applying the history matching modifications directly to the grid block properties, the information is incorporated into the geomodeling process. As such, these methods are termed "geologically-consistent".

The first example of this approach is described by Caers *et al.* (2002), who adapted the method of Wang and Kovscek (2000) to modify the geological model while at the same time maintaining the spatial correlation structure of the grid block properties. Similar approaches were proposed by Le Ravalec-Dupin and Fenwick (2002) and Caers (2003). These authors developed a procedure that combines a streamline-based history matching method and a geostatistical parameterization technique to estimate the distributions of porosity and permeability in heterogeneous petroleum reservoirs. A special feature of the suggested methodology is that a single fluid flow simulation can be run for a group of reservoir model perturbations. This property makes the entire history match process computationally efficient.

Gross *et al.* (2004) and Fenwick *et al.* (2005) employed a combination of geostatistical tools to relate production data to petrophysical properties at multiple scales. The methodology was applied for history matching a giant Saudi Arabian carbonate oil reservoir with over 500 producers and with over 50 years of historical data. The permeability corrections provided by the streamlines helped achieving noticeable history matching improvements.

Gross (2006) developed a statistical technique to establish consistency across field data such as facies type, permeability, porosity, and net-to-gross by obeying interdependence revealed using streamline simulators. The technique was tested using a 10,000 grid block, 2D model with 16 producers and 9 injectors. The results showed satisfactory match of water-cut and pressure without destroying geological consistency.

1.2. Reservoir Management

Managing a water and miscible flooding requires an understanding of how injectors displace fluid to producers (Grinestaff 1999). A key element to a successful flood is good sweep efficiency, which can be significantly impacted by reservoir heterogeneity resulting in areas of water cycling and poor sweep as flood matures (Izgec *et al.* 2010).Traditionally, most flood management has been restricted to static allocations or sensitivity studies centered on finite difference simulation (Thiele and Batycky 2003, Batycky *et al.* 2005, Ghori *et al.* 2007, Izgec *et al.* 2010). Static allocations, which use reservoir properties and neighboring well distances, have been used by engineers to quantify injector to producer relationship. However, such approach did not permit the evaluation of changes in injection profiles, water influx and cross flow over time (Grinestaff 1999). Likewise, detailed finite difference simulation can provide some answers related to injectors/producers connectivity but computational times and data extraction makes well level decisions on a daily basis too expensive (Grinestaff 1999).

As an alternative, streamline simulation is an ideal tool to help optimize large floods because it provides efficient and fast means to capture detailed fluid movements resulting from conformance control (vertical and/or areal by mechanical means), infill drilling, well conversion and pattern re-alignment (Grinestaff 1999, Thiele and Batycky 2003, Izgec *et al.* 2010). Thus, total flow rates and phase rates between well pairs can be calculated. This important information can help engineers identify areas of extreme cycling, patterns with poor sweep, or local voidage imbalances (Thiele and Batycky 2003, Batycky *et al.* 2005, Singh *et al.* 2010).

Following is a literature review of the work in which streamlines have been applied to optimize flood management and hence allow engineers to focus on understanding reservoir and well behavior rather than reacting to it (Grinestaff 1999). In this section, four areas have been identified namely sweep efficiency, rate optimization, well placement and enhanced oil recovery. The review period has observed intense efforts by researchers especially during the last three years to fully and widely implement streamlines technology in addressing reservoir management issues. A very interesting observation resulting from exploiting streamlines in reservoir management is the introduction of several streamline based reservoir engineering indicators as a result of the powerful aspects of streamlines visualization. Some of those are time-of-flight to the producer (TTP), Predominant Streamline Fields (PSF), Weak Streamline Fields (WSF), Strength Index of Streamline Field (SISF), depletion capacity maps, flow-based well allocation factors, streamline-based flood efficiency maps and variations of TOF distribution maps.

1.2.1. Sweep Efficiency

Chakravarty *et al.* (2000) conducted a study of the high permeability, high recovery (over 70%) Saladin (Barrow Group) reservoir (offshore Western Australia) using 3D Streamline Simulation to investigate any remaining potential prior to abandonment. A very good history match was achieved in a short time and results indicated two distinct areas of significant bypassed oil and drilling in one of recommended target areas has yielded significant recovery of bypassed oil.

Grinestaff and Caffrey (2000) used a high-resolution 3D streamline model to study regional and individual wells performance of the Northwest Fault Block (NWFB) area of Prudhoe Bay, Alaska. Streamline modeling provided accurate results of vertical and area wide sweep efficiencies and water cycling. Injection was hence reduced by 30-40%, resulting in increased production and cost savings.

By examining the time-of-flight to the producer (TTP), Sharif and MacDonald (2001) used streamline simulation to evaluate flow behavior and production performance of 105 scenarios of a 4.2 million blocks geologic model. By observing evolution of drained volume as TTP increases, the results showed that wells drained most of swept volume for TTP<1,000, after which increase in drained volume is not significant. Additionally, between TTP of 1,000 and 10,000, presence of pockets of additional drained volume was identified.

Thiele and Batycky (2003) described an automated approach to optimize injection and production well efficiencies in a waterflood using streamline-based flow simulation. By knowing injection efficiencies or well allocation factors (WAF) across the field for each injector, water can be reallocated from low-efficiency to high efficiency wells.

Liu (2010) introduced Predominant Streamline Fields (PSF) and Weak Streamline Fields (WSF) by statistically analyzing the Strength Index of Streamline Field (SISF), which is a function of both allocation factor and swept volume of specific well pair at a given time step. The results of testing this on the Gudong field, China, indicated that PSF behavior was consistent with a lot of other history data, which helped EOR and infill drilling decisions in this field.

Using a streamline tracing algorithm, Al-Zawawi *et al.* (2011) enhanced a FD simulator, to generate streamlines. The authors presented two cases in which number of injectors was reduced without impacting overall injection efficiency in the first case while injection efficiency was improved by optimizing well location in the second case.

To assess the impact of scale precipitation deep in the reservoir on scale management economics, Akuanyionwu and Wahid (2011) used a commercial streamline simulator to model in-situ scale precipitation in the Green field (North Sea) and tracked scaling dynamic concentration ions present at the high-risk producer area. Because streamlines showed a drop in scaling ion concentration (sulfate) in some areas as the water moves toward the producers, frequencies of estimated scale squeeze treatment were highly reduced and thus, eliminated the need of building a sulfate reducing plant.

1.2.2. Rate Optimization

Grinestaff (1999) demonstrated the ability of streamline simulation, dynamic injection pattern allocations to proactively manage waterflood patterns through time. For a major field simulated using this modeling tool, over 15% of total flood throughput was removed by lowering injection rates leading to a 75% reduction in total fluid rate with no loss of oil production.

Guimarães *et al.* (2005) developed an automated optimization methodology combining three tools: traditional simulation, streamline simulation and quality map. By applying efficiency of injector well (Ef) function obtained from streamline simulation, water injection and allocation of new wells were more reliably controlled. The proposed methodology was applied to an offshore field with water injection and results showed that efficient control of fluid flow allowed the best production strategy.

Thiele and Batycky (2006) employed streamline information to develop the concept of Injector Efficiency (IE). They developed a heuristic optimization procedure which rebalances injection volumes from inefficient to efficient injectors, thus improving oil production. A similar procedure was proposed by Ghori *et al.* (2007) who utilized streamline simulation to optimize and set injection targets for 160 water injectors supporting 520 oil producers in a Saudi Arabian field. Using predefined independent segments established from streamline patterns, rates in efficient injectors were increased and rates of inefficient injectors were decreased. This proposed method increased the injectors' efficiencies by 10% and overall efficiency of each segment by 5%-7%.

Rodriguez *et al.* (2008) used streamlines technology in conjunction with IPR target per segment, reservoir pressure distribution and maximum water injection plant facilities to effectively manage water injection in a giant field. The new approach improved injection efficiency for all segments simultaneously in a single run.

To give a better measure of an injector's true effectiveness, Batycky *et al.* (2008) developed a streamline based flood surveillance method using flow-based well allocation factors (WAF). By comparing results of this approach against full history-matched flow simulation models of several field cases, the authors confirmed that only first order flow effects such as well locations, historical rates, major geological features are required to build such model.

Using a penalized misfit function for optimization, Taware *et al.* (2010) proposed a streamline based approach for computing sensitivity of injection and production rates in smart wells, given multiple geologic models, to maximize oil recovery and NPV. The results of 2D and 3D cases showed that as the acceleration component was increased, the proposed arrival time optimization approached the NPV optimization.

Alhuthali *et al.* (2010) computed the sensitivity of arrival times with respect to well rates using streamlines. The approach was tested on the Brugge field and a super-giant Middle Eastern field containing vertical and horizontal wells and completed with ICVS. The results showed increased cumulative oil production while substantially decreasing associated water production.

Izgec *et al.* (2010) and Izgec *et al.* (2011) used streamline simulations to derive Hydrocarbon F-Φ Curve and Hydrocarbon Lorenz Coefficient (LC-HC) and used them for optimum waterflood management. The authors tested this approach using both synthetic data and field data. The speed, flexibility to start optimizing at any arbitrary time regardless of history in addition to ability to handle large problems illustrated the power of this method.

Bostan *et al.* (2011) utilized streamline simulations to optimize well allocation factor of an undersaturated oil field in Iran with an area of 10 km^2. Based on streamlines information, using same amount of water, injection efficiency was improved by about 7% in 3 years.

Su *et al.* (2011) used two streamline optimization algorithms to optimize 60 years waterflooding of the Burgan Field in Kuwait. The first optimizes injection efficiency based on total production rates and the second is utilizes remaining oil ranking, in which streamline bundles with more remaining oil get more injection. Both methods were very effective and oil gains were about the same for the first 25 years. However, the authors believed that the second method would result in higher ultimate recovery based on the comparison between some wells that were not completely shut-in when used remaining oil ranking algorithm.

By equalizing average time-of-flight amongst producing wells using analytically computed weighting factors of injection and production rates, Park and Datta-Gupta (2011) proposed a waterflood rate optimization workflow using streamline-based flood efficiency maps. The 60,000-grid block Brugge field model and a CO2 pilot project in the Goldsmith San Andrea Unit (GSAU) were tested. The results not only showed optimized flood rates but confirmed the practicality of evaluating injection efficiency using the variations of TOF distribution maps.

1.2.3. Well Placement

To select new wells best locations, understand drainage volume and speed up process, Dehdari *et al.* (2008) employed two tools: compositional simulation using different property maps, and streamlines simulation for optimizing one of Iranian oil fields. This methodology helped in defining several scenarios to maximize oil production and the streamline technique improved recovery by 22 MM bbls compared to the conventional methods.

Singh *et al.* (2010) used a streamline model with Singular Value Decomposition technique (SVD) to estimate flood front movement in the mature Assam oil field. The authors demonstrated the calculation of breakthrough time; streamline distribution and areal sweep efficiencies, sensitivity of solution to number of boundary points and number of image wells for four cases. Such information resulted in identifying fresh areas for infill drilling. A comparison of this technique with FD technique showed a remarkable similarity.

Sayyafzadeh *et al.* (2010) used streamline density information to understand fluid movement within the reservoir and to develop a best practice for future developments such as infill drilling or converting dead producers into injectors. Following this approach, the simulation results showed that the ultimate oil recovery increased by 70%, which compares favorably with the EOR processes by requiring less investment and operating cost.

Sehbi *et al.* (2011) optimized number of hydraulic fracture stages in tight gas reservoirs by examining the evolution of gas streamlines derived drainage volumes as a function of number of fracture stages. The application to the Cotton Valley formation showed that fractures in the middle have considerably less streamline density compared to outer fractures in an indication of the shielding of the inner fractures by the outer fractures. Ten hydraulic fracture stages with a fracture half-length of 500 ft was found to be the optimal completion strategy for the horizontal well studied.

Kang *et al.* (2011) optimized well placement in a naturally fractured tight gas reservoirs by examining the evolution of the gas streamlines derived well drainage volumes. The application to a tight gas field in the Rocky Mountain region showed that presence of natural fractures tends to enhance well drainage volumes and accelerate production depending upon distribution and orientation of fractures. Additionally, defining undrained volumes helped introducing a "depletion capacity" map for optimization of infill locations.

1.2.4. Enhanced Oil Recovery

Maure *et al.* (2001) utilized streamline simulation to infer inter-well connectivity and allocation factors in Victor Claro reservoir layers C2-D2, La Ventana field, northern Cuyo Basin in the Mendoza Province of Argentina to understand the microbial enhanced oil recovery (MEOR) mechanisms and improve bio-treatments. The results showed that the original oil production decline rates improved and the two combined mechanisms helped explain the changes.

Seto *et al.* (2003) developed a streamline compositional simulator to model enhanced condensate recovery by gas injection. The authors compared streamline and FD results for 2D and 3D cases and the results showed that numerical dispersion associated with compositional FD simulation resulted in optimistic recovery predictions while compositional streamline simulation reduced these effects, thus producing more realistic predictions. In addition, speed-up factors of two to three orders of magnitude were observed.

Li (2004) extended UTCOMP streamline simulation (Strator) to model four-phase flow for a nine-component injection of the 50-50 combination of Prudhoe Bay Gas (PBG) and Natural Gas Liquid (NGL) into a reservoir containing a 12-component viscous oil (Shrader). Results indicated that Strator was many times faster than FD simulator. In addition, the viscous fingering was clearly observed during streamline simulation while displacement front was very blurry during FD simulation due to high numerical dispersions.

Bhambri and Mohanty (2005) and Bhambri and Mohanty (2005) developed a four-phase streamline module that works with an existing FD simulator to study gas or WAG injections in medium viscosity reservoirs. The authors tested this module using 2D and 3D homogenous quarter of five-spot models with vertical wells. Results showed that four phases exist near gas-oil displacement front and sweep efficiency increases with WAG ratio.

Bhambri (2007) developed black oil and compositional streamline modules with parallel option to study large scale waterflooding, gas injections, and WAG processes in low temperature reservoirs. The comparison with FD revealed close agreement with a speed up factor of 12 for a 10,000 grid cell 2D waterflood simulation. However, water saturation maps showed some instabilities and oscillations for very big time steps (100 days) and its resolution is affected by number of streamlines launched.

By using a data reduction algorithm to reduce number of calls to the Rieman solver, Juanes and Lie (2005) and Juanes and Lie (2008) developed a framework that employ streamline method, a front tracking algorithm and the analytical Riemann solver to efficiently simulate first-contact miscible (FCM) gas injection processes. The Authors tested this framework using the 10th SPE comparative solution project and demonstrated that WAG scheme recovery efficiency was higher than waterflooding or gas injection alone.

To delineate the advantages and limitations of streamline simulations for high mobility ratio waterfloods, Osako *et al.* (2009) used streamline simulation to predict the performance of mechanistic and field-scale heavy oil waterflood models for oil-water viscosity ratio range of 10 to 1000. Besides showing similar results as FD, streamline showed less grid orientation effects and computational advantage for large-scale field model. However, for viscosity ratios of ten or higher, SL generated instabilities and the results were sensitive to pressure update.

In the domain of polymer floods, in addition to the work of Thiele *et al.* (2008), and AlSofi *et al.* (2009), Clemens *et al.* (2010) used streamline based well pattern to efficiently manage polymer injection projects. Based upon comparison of the cumulative oil produced as a function of polymer injected,, the authors introduced a new metric: the polymer injection efficiency as a function of time for each pattern. The simulation results from the Romanian field tested showed an anticipated recovery increase of 5.3% of OOIP compared to the base waterflood case.

Nielsen *et al.* (2010) developed a two-phase microbial enhanced oil recovery (MEOR) compositional streamline simulator with gravity effects to study oil displacement by water containing bacteria, substrate, and surfactant. This streamline simulator was found to produce similar results as FD simulator for 1D and 2D displacements with increased and faster oil recovery.

1.3. Upscaling, Ranking, and Characterizing Fine-Grid Geological Models

Recent enhancements in software, hardware, integration expertise, data collection and interpretation have improved the industry's ability to build detailed 3D large multi-million cell geologic models of the subsurface (Idrobo *et al.* 2000). Due to their computational time and costs, upscaling is needed to efficiently and reasonably translate detailed grids to coarser grids without compromising the anticipated reservoir performance (Chawathé and Taggart 2001). Equally, there exists a need to rank a large number of generated reservoir model realizations to quantify the potential sources of errors or uncertainties. For both tasks, streamlines method is especially appealing. The streamline time-of-flight provides a powerful means of visualization. By comparing the time-of-flight from the fine scale detailed geologic model with that from an upscaled model, the validity of upscaling techniques are examined in addition to level of upscaling (Datta-Gupta 2007). Additionally, a unique feature of streamlines method is its ability to efficiently compute the sensitivity of the production data to reservoir parameters such as porosity and permeability providing a great potential for considering a large number of plausible reservoir descriptions (Efendiev *et al.* 2008).

Following is a literature review of the work in which streamlines has been applied for bridging the gap between fine scale detailed geological models and flow simulation models through providing a judicious upscaling technique and appropriate ranking criteria to select a subset of equi-probable reservoir realizations.

1.3.1. Upscaling

Portella and Hewett (1998) developed an upscaling method that generates a 2D or 3D coarse grid, based on streamtubes. It also calculates upscaled relative permeability curves, or pseudo-functions using semi-analytical streamtube method. The results from streamtube coarse grid were in very good agreement with the fine grid solution.

By defining impermeable media based on increased tortuosity of streamline flow path, Doi and Suzuki (2000) applied a streamline method for upscaling vertical permeability of a Middle Eastern heterogeneous limestone reservoir. The final effective permeability showed that the streamline method was effective as it was able to match the simulated formation pressure with the historical data.

Chawathe and Taggart (2001) used 3D streamline simulation for vertically and areally upscaling a 0.6 million-cell sector of a gas condensate field on the West coast of Africa in order to better understand flux distributions. A high-permeability streak seen in the static geological model was clearly identified in addition to discovery of approximately three more relatively prolific layers.

By combining the methods of minimizing variation and flux, manual judgment and streamline simulation, Nair and Al-Maraghi (2006) applied a hybrid upscaling approach to the 157-layer geological model of a carbonate reservoir in Kuwait to generate a vertically upscaled model with 42 layers. The streamline simulation results showed that flow paths are directly related to the overall heterogeneity and water arrival time was closely related to the presence (or removal) of high permeability pathways. The streamline model results were comparable with the fine-scale model.

1.3.2. Uncertainty Quantification and Ranking of Geological Models

To rigorously accounts for interaction between flow field and underlying heterogeneity, Idrobo *et al.* (2000) proposed a connectivity criterion based on streamline time-of-flight and used it to rank geostatistical realizations. For the test example on North Robertson Unit field, West Texas, 50 permeability realizations were ranked based on volumetric sweep efficiency and the percent recovery and the swept volumes at different times were clearly consistent with the ranking.

By accounting for gravity, fluid contacts, changing streamlines, fractional flow effects, choice of well locations, rates, boundary conditions, and patterns, Gilman *et al.* (2002) described a practical process for dynamically ranking various geologic realizations for a mature field with 60 years of historical production using uniform well spacing and streamline simulation. The streamline simulation gave a 50% variation in future oil recovery using a uniformly spaced, 40-acre 5 spot pattern.

Kovscek and Wang (2005) employed a streamline simulator as a proxy for uncertainty quantification of CO2 sequestration. Streamline simulation allowed for a selection of a representative subset of equi-probable reservoir models that encompass uncertainty with respect to true reservoir geology. In a similar approach, Acosta (2010) used a streamline simulator to rank 20 equi-probable geostatistical realizations for a model with 1.8 million active cells in the Ceuta-Tomoporo field in West Venezuela by finding the least cumulative errors of the historical and simulated oil productions and the water-cuts.

Scheidt and Caers (2009) used streamline simulation as a distance measure of dissimilarity between reservoir models. Using clustering and distance measures, their approach identifies a small subset of reservoir models for standard flow simulation. The uncertainty in future production was captured accurately when simulating fewer than 10% of the total number of reservoir models.

Odai and Ogbe (2011) utilized average breakthrough time (ABT) derived from a 3D streamline simulator for ranking equi-probable realizations. Tests indicated that high ABT very reasonably correlated with cumulative waterflood recovery.

1.3.3. Characterization of Reservoir Heterogeneity

To quantify the uncertainties of small-scale geological features, Hastings *et al.* (2001) proposed a semi-analytical streamline method based on a new solution for the Buckley-Leverett flow in heterogeneous media with varying relative permeabilities. The method assumed that larger scale heterogeneities (faults, channel bodies etc.) define the streamlines geometry but small scale properties (e.g. cross-bedding) governed the flow along them which was validated using a fluvial reservoir cross-section. The ability to model contrasts in relative permeability helped accurately reproduces the main features of the production curve with smoother profile compared to a commercial streamline simulator.

Pranter *et al.* (2005) used streamline simulations to assess the variability of petrophysical properties on fluid flow of the dolomite outcrops model of the Mississippian Madison formation at Sheep Canyon, Wyoming. Results indicated that as the component of the long-range periodicity increases, a corresponding increase in breakthrough time and sweep efficiency occurs. Additionally, heterogeneity caused by lateral petrophysical cyclicity should be incorporated for hole effect magnitudes that are greater than 10% of the petrophysical variance. To quantify intermediate-scale heterogeneity of fluvial reservoirs, Pranter *et al.* (2007) conducted single phase, 2D and 3D streamline simulations on the fluvial point-bar outcrop model, Williams Fork Formation, Colorado to compare relative effects of lithology, grain size, and petrophysical heterogeneity on breakthrough time (BTT), reservoir volume swept at BTT, and sweep efficiency. For 2D models, results showed that lithologic heterogeneity has a greater effect on recovery efficiency than petrophysical heterogeneities. However, the 3D models showed that lithologic distribution has a more significant effect on BTT and sweep efficiency compared to perophysical distribution if well patterns oriented perpendicular to paleoflow direction.

Using streamline TOF and volumetric flow rate information, Shook and Mitchell (2009) developed a new method for estimating heterogeneity in Earth models. This method was tested using 450 models that were constructed using a wide range of Dykstra-Parsons coefficient and correlation lengths, and two different well patterns. This study showed that the Lorenz coefficient – as determined from dynamic data – is the single best measure of heterogeneity.

Summary

In this paper, we presented a comprehensive review of the evolution and advancement of streamline simulation technology. This paper offered a general overview of most of the material available in the literature on the subject. This work included the review of more than 200 technical papers and gave a chronological advancement of streamline simulation technology from 1996 to 2011. Firstly, three major areas are identified. These are development of streamline simulators, enhancements to current streamline simulators and applications. In view of the fact that this state of-the-art technology has been employed for a wide range of applications, we defined three major application areas that symbolize the relevance and validity of streamline simulation in addressing reservoir engineering concerns. These are history matching, reservoir management and upscaling, ranking and characterization of fine-grid geological models.

Streamline simulation has undergone several phases within its short stretch in the petroleum industry. Initially, the main focus was on the speed advantage and less on fluid flow physics. The focus then shifted to extend its applicability to more complex issues such as compositional and thermal simulations, which require the inclusion of more physics, and potentially reducing the advantage of computational time. Recently, the focus has shifted towards application of streamline technologies to areas where it can complement finite difference simulation such as revealing important information about drainage areas, flood optimization and improvement of sweep efficiency, quantifying uncertainties, etc.

References

Acosta, A. 2010. Hierarchical Pixel-Based Geostatistical Modeling of Estuarine Reservoir Through Streamline Simulation in Ceuta-Tomoporo Field, West Venezuela. Paper SPE 132762 presented at the Trinidad and Tobago Energy Resources Conference, Port of Spain, Trinidad , 27-30 June. http://dx.doi.org/10.2118/132762-MS.

Agarwal, B. and Blunt, M.J. 2001. Full-Physics, Streamline-Based Method for History Matching Performance Data of a North Sea Field. Paper SPE 66388 presented at the SPE Reservoir Simulation Symposium, Houston, Texas, 11-14 February. http://dx.doi.org/10.2118/66388-MS.

Agarwal, B. and Blunt, M. 2003. A Streamline-Based Method for Assisted History Matching Applied to an Arabian Gulf Field. Paper SPE 84462 presented at the SPE Annual Technical Conference and Exhibition, Denver, Colorado, 5-8 October. http://dx.doi.org/10.2118/84462-MS.

Al-Harbi, M., Cheng, H., He, Z. and Datta-Gupta, A. 2004. Streamline-Based Production Data Integration in Naturally Fractured Reservoirs. Paper SPE 89914 presented at the SPE Annual Technical Conference and Exhibition, Houston, Texas, 26-29 September. http://dx.doi.org/10.2118/89914-PA.

Al-Huthali, A.H., Datta-Gupta, A. 2004. Streamline Simulation of Water Injection in Naturally Fractured Reservoirs. Paper SPE 89443 presented at the SPE/DOE Symposium on Improved Oil Recovery, Tulsa, Oklahoma, 17-21 April. http://dx.doi.org/10.2118/89443-MS.

Al-Huthali, A.H., Datta-Gupta, A., Yuen, B. and Fontanilla, J.P. 2010. Field Applications of Waterflood Optimization via Optimal Rate Control With Smart Wells. Paper SPE 118948 presented at the SPE Reservoir Simulation Symposium, The Woodlands, Texas, 2-4 February. http://dx.doi.org/10.2118/118948-MS.

AlSofi, A.M., LaForce, T.C., and Blunt, M.J. 2009. Sweep Impairment Due to Polymers Shear Thinning. Paper SPE 120321 presented at the SPE Middle East Oil and Gas Show and Conference, Bahrain, Bahrain, 15-18 March. http://dx.doi.org/10.2118/120321-MS.

Al-Zawawi, A.S., Hayder, E.M., Baddourah, M.A., Mohd Ghazali, B.A. and Hidayat, W. 2011. Using Streamline and Reservoir Simulation to Improve Water Flood Management. Paper SPE 141794 presented at the SPE Middle East Oil and Gas Show and Conference, Manama, Bahrain, 25-28 September. http://dx.doi.org/10.2118/141794-MS.

Batycky, R.P. 1997. A Three Dimensional Two Phase Field Scale Streamline Simulation. PhD dissertation. Stanford University, Stanford, California (January 1997).

Batycky, R.P., Thiele, M.R., Baker, R.O. and Chung, S.H. 2005. Revisiting Reservoir Flood-Surveillance Methods Using Streamlines. Paper SPE 95402 presented at the SPE Annual Technical Conference and Exhibition, Dallas, Texas, 9-12 October. http://dx.doi.org/10.2118/95402-MS.

Bhambri, P. and K.K. Mohanty. 2005. Streamline Simulation of Four-Phase WAG Processes. Paper SPE 96940 presented at the SPE Annual Technical Conference and Exhibition, Dallas, Texas, 9-12 October. http://dx.doi.org/10.2118/96940-MS.

Bhambri, P. 2007. A Three dimensional Four Phase Compositional

Streamline Simulator with Paralell Implementation. PhD Dissertation. University of Houston, Houston, Texas (May 2007).

Bostan, M., Kharrat, R. and Ghorbani, D. 2011. Implementing a Novel Method for Injection Efficiency Optimization in Water Flooding Process: Case Study. Paper SPE 142133 presented at the SPE Production and Operations Symposium, Oklahoma City, Oklahoma, USA, 27-29 March. http://dx.doi.org/10.2118/142133-MS.

Chakravarty, A., Liu, D.B., Meddaugh, W.S. 2000. Application of 3D Streamline Methodology in the Saladin Reservoir and Other Studies Paper SPE 63154 presented at the SPE Annual Technical Conference and Exhibition, Dallas, Texas, 1-4 October. http://dx.doi.org/10.2118/63154-MS.

Chawathe, A. and Taggart, I. 2001. Insights into Upscaling Using 3D Streamlines. Paper SPE 66379 presented at the SPE Reservoir Simulation Symposium, Houston, Texas, 11-14 February. http://dx.doi.org/10.2118/66379-MS.

Cheng, H., Kharghoria, A., Zhong, H. and Datta-Gupta, A. 2004. Fast History Matching of Finite-Difference Models Using Streamline-Derived Sensitivities. Paper SPE 89447 presented at the SPE/DOE Symposium on Improved Oil Recovery, Tulsa, Oklahoma, 17-21 April. http://dx.doi.org/10.2118/89447-PA.

Cheng, H., Oyerinde, D., Datta-Gupta, A. and Milliken, W. 2006. Compressible Streamlines and Three-Phase History Matching. Paper SPE 99465 presented at the SPE/DOE Symposium on Improved Oil Recovery, Tulsa, Oklahoma, USA, 22-26 April. http://dx.doi.org/10.2118/99465-MS.

Clemens, T., Abdev, J. and Thiele, M. 2010. Improved Polymer-Flood Management Using Streamlines. Paper SPE 132774 presented at the SPE Annual Technical Conference and Exhibition, Florence, Italy, 19-22 September. http://dx.doi.org/10.2118/132774-MS.\

Datta-Gupta, A. and King, M.K. 2007. Streamline simulation: theory and practice. Textbook Series, SPE, Richardson, Texas.

Datta-Gupta A. 2000. Streamline simulation: a technology update. J. Pet Tech 52 (12): 68-73, 84. SPE 65604. http://dx.doi.org/10.2118/65604-MS.

de Haan, S.C., Oldenziel, T., Berentsen, C.W.J. and van Kruijsdijk, C.P.J.W. 2001. Water-Cut Forecasting and Permeability Inversion from Time-Lapse Seismic using a Stream-Line Technique. Paper OTC 13165 presented at the Offshore Technology Conference, Houston, Texas, 30 April-3 May. http://dx.doi.org/10.4043/13165-MS.

Dehdari, V., Aminshahidy, B. and Tabatabaei-nejad, A. 2008. Well Spacing and Recovery Optimization of One of Iranian Oil Fields by Using Streamline and Reservoir Simulation. Paper SPE 112985 presented at the SPE Western Regional and Pacific Section AAPG Joint Meeting, Bakersfield, California, USA, 29 March-2 April. http://dx.doi.org/10.2118/112985-MS.

Devegowda, D. Arroyo-Negrete, E., Datta-Gupta, A. and Douma, S.G. 2007. Efficient and Robust Reservoir Model Updating Using Ensemble Kalman Filter with Sensitivity-Based Covariance Localization. Paper SPE 106144 presented at the SPE Reservoir Simulation Symposium, Houston, Texas, U.S.A. 26-28 February. http://dx.doi.org/10.2118/106144-MS.

Di Donato, G., Huang, W. and Blunt, M. 2003. Streamline-Based Dual Porosity Simulation of Fractured Reservoirs. Paper SPE 84036 presented at the SPE Annual Technical Conference and Exhibition, Denver, Colorado, 5-8 October. http://dx.doi.org/10.2118/84036-MS.

Doi, M. and Suzuki, S. 2000. Formulation of the Re-Development Scheme: Reservoir Modeling for the Highly Heterogeneous Reservoir to Assess IOR. Paper SPE 59440 presented at the SPE Asia Pacific Conference on Integrated Modeling for Asset Management, Yokohama, Japan, 25-26 April. http://dx.doi.org/10.2118/59440-MS.

Efendiev, Y., Datta-Gupta, A., Ma, X. and Mallick, B. 2008. Modified Markov Chain Monte Carlo Method for Dynamic Data Integration Using Streamline Approach. Math Geoscience 40: 213-232.

Emanuel, A.S. and Milliken, W.J. 1998. History Matching Finite Difference Models with 3D Streamlines. Paper SPE 49000 presented at the SPE Annual Technical Conference and Exhibition, New Orleans, Louisiana, 27-30 September. http://dx.doi.org/10.2118/49000-MS.

Fenwick, D., Thiele, M., Agil, M., Hussain, A., Humam, F. and Caers, J. 2005. Reconciling prior geologic information with production data using streamlines: application to a giant Middle-Eastern oil field. Paper SPE presented at the SPE Annual Technical Conference and Exhibition, Dallas, Texas, 9-12 October. http://dx.doi.org/10.2118/95940-MS.

Gautier, Y., Nœtinger, B. and Roggero, F. 2001. History Matching Using a Streamline-Based Approach and Exact Sensitivity Coefficients. Paper SPE 71626 presented at the SPE Annual Technical Conference and Exhibition, New Orleans, Louisiana 30 September-3 October. http://dx.doi.org/10.2118/71626-MS.

Ghori, S.G., Jilani, S.Z., Al-Huthali, A.H., Krinis, D. and Kumar, A. 2007. Improving injector efficiencies using streamline simulation: a case study in a giant Middle East field. Paper SPE 105393 presented at the SPE Middle East Oil and Gas Show and Conference, Kingdom of Bahrain, 11-14 March. http://dx.doi.org/10.2118/105393-MS.

Gilman, J.R., Meng, H-Z, Uland, M.J., Dzurman, P.J. and Cosic, S. 2002. Statistical Ranking of Stochastic Geomodels Using Streamline Simulation: A Field Application. Paper SPE 77374 presented at the SPE Annual Technical Conference and Exhibition, San Antonio, Texas, 29 September-2 October. http://dx.doi.org/10.2118/77374-MS.

Grinestaff, G.H. 1999. Waterflood Pattern Allocations: Quantifying the Injector to Producer Relationship with Streamline Simulation. Paper SPE 54616 presented at the SPE Western Regional Meeting, Anchorage, Alaska, 26-27 May. http://dx.doi.org/10.2118/54616-MS.

Grinestaff, G.H. and Caffrey, D. J. 2000. Waterflood Management: A Case Study of the Northwest Fault Block Area of Prudhoe Bay, Alaska, Using Streamline Simulation and Traditional Waterflood Analysis. Paper SPE 63152 presented at the SPE Annual Technical Conference and Exhibition, Dallas, Texas, 1-4 October. http://dx.doi.org/10.2118/63152-MS.

Gross, H. 2006. History-Matching Multiple Attributes with Streamline sand Statistics. Paper SPE 106520 presented at the SPE Annual Technical Conference and Exhibition, San Antonio, Texas, USA, 24-27 September. http://dx.doi.org/10.2118/106520-STU.

Gross, H., Thiele, M.R., Alexa, M.J., Caers, J.K. and Kovscek, A.R. 2004. Streamline-Based History Matching Using Geostatistical Constraints: Application to a Giant, Mature Carbonate Reservoir. Paper SPE 90069 presented at the SPE Annual Technical Conference and Exhibition, Houston, Texas, 26-29 September. http://dx.doi.org/10.2118/90069-MS.

Guimarães, M.S., Schiozer, D.J. and Maschio, C. 2005. Use of Streamlines and Quality Map in the Optimization of Production Strategy of Mature Oil Fields. Paper SPE 94746 presented at the SPE Latin American and Caribbean Petroleum Engineering Conference, Rio de Janeiro, Brazil, 20-23 June. http://dx.doi.org/10.2118/94746-MS.

Hastings, J.J., Muggeridge, A.H. and Blunt, M.J. 2001. A New Streamline Method for Evaluating Uncertainty in Small-Scale, Two-Phase Flow Properties. Paper SPE 66349 presented at the SPE Reservoir Simulation Symposium, Houston, Texas, 11-14 February. http://dx.doi.org/10.2118/66349-MS.

He, Z., Datta-Gupta, A. and Yoon, S. 2001. Streamline-based Production

Data Integration Under Changing Field Conditions. Paper SPE 71333 presented at the SPE Annual Technical Conference and Exhibition, New Orleans, Louisiana, 30 September-3 October. http://dx.doi.org/10.2118/71333-MS.

Hoffman, B.T., Caers, J. 2003. Geostatistical History Matching Using a Regional Probability Perturbation Method. Paper SPE 84409 presented at the SPE Annual Technical Conference and Exhibition, Denver, Colorado, 5-8 October. http://dx.doi.org/10.2118/84409-MS.

Idrobo, E.A., Choudhary, M.K. and Datta-Gupta, A. 2000. Swept Volume Calculations and Ranking of Geostatistical Reservoir Models Using Streamline Simulation. Paper SPE 62557 presented at the SPE/AAPG Western Regional Meeting, Long Beach, California, 19-22 June. http://dx.doi.org/10.2118/62557-MS.

Ingebrigtsen, L., Bratvedt, F., Berge, J. 1999. A Streamline Based Approach to Solution of Three-Phase Flow. Paper SPE 51904 presented at the SPE Reservoir Simulation Symposium, Houston, Texas, 14-17 February. http://dx.doi.org/10.2118/51904-MS.

Izgec, O., Sayarpour, M. and Shook, G.M. 2010. Optimizing Volumetric Sweep Efficiency in Waterfloods by Integrating Streamlines, Design of Experiments, and Hydrocarbon F-{capital Phi} Curves. Paper SPE 132609 presented at the SPE Western Regional Meeting, Anaheim, California, USA, 27-29 May. http://dx.doi.org/10.2118/132609-MS.

Jessen, K., and Orr Jr., F.M. 2002. Compositional Streamline Simulation. Paper SPE 77379 presented at the SPE Annual Technical Conference and Exhibition, San Antonio, Texas, 29 September-2 October. http://dx.doi.org/10.2118/77379-MS.

Juanes, R., Lie, K.-A. 2005. A Front-Tracking Method for Efficient Simulation of Miscible Gas Injection Processes. Paper SPE 93298 presented at the SPE Reservoir Simulation Symposium, The Woodlands, Texas, 31 January-2 February. http://dx.doi.org/10.2118/93298-MS.

Kang, S., Datta-Gupta, A. and Lee, J. 2011. Impact of Natural Fractures in Drainage Volume Calculations and Optimal Well Placement in Tight Gas Reservoirs. Paper SPE 144338 presented at the SPE Annual Technical Conference and Exhibition, Denver, Colorado, USA, 30 October-2 November. http://dx.doi.org/10.2118/144338-MS.

Kazemi, A. and Stephen, K.D. 2010. Optimal Parameter Updating in Assisted History Matching of the Nelson Field using Streamlines as a Guide. Paper SPE 131540 presented at the SPE EUROPEC/EAGE Annual Conference and Exhibition, Barcelona, Spain, 14-17 June. http://dx.doi.org/10.2118/131540-MS.

Kovscek, A.R. and Wang, Y. 2005. Geologic storage of carbon dioxide and enhanced oil recovery. I. Uncertainty quantification employing a streamline based proxy for reservoir flow simulation. *Energy Conversion and Management* 46:1920-1940.

Kretz, V., Valles, B. and Sonneland, L. 2004. Fluid Front History Matching Using 4D Seismic and Streamline Simulation. Paper SPE 90136 presented at the SPE Annual Technical Conference and Exhibition, Houston, Texas, 26-29 September. http://dx.doi.org/10.2118/90136-MS.

Le Ravalec-Dupin, M. and Fenwick, D.H. 2002. A Combined Geostatistical and Streamline-Based History Matching Procedure. Paper SPE 77378 presented at the SPE Annual Technical Conference and Exhibition, San Antonio, Texas, 29 September-2 October. http://dx.doi.org/10.2118/77378-MS.

Li, D. 2004. Four-Phase Streamline-Based Compositional Simulation of Gas Injections for a Viscous Oil. Paper SPE 88790 presented at the Abu Dhabi International Conference and Exhibition, Abu Dhabi, United Arab Emirates, 10-13 October. http://dx.doi.org/10.2118/88790-MS.

Lino, A. Arihara, N., and Okatsu, K. 2006. Tracer-Flow Simulation for Characterization of Naturally Fractured Reservoirs by Streamline Method. Paper SPE 100183 presented at the SPE Europec/EAGE Annual Conference and Exhibition, Vienna, Austria, 12-15 June. http://dx.doi.org/10.2118/100183-MS.

Liu, X., Wang, Z.P. and Wang, Y. 2010. Medium-High Permeability, Mature Reservoir Management Study: An Innovative Methodology and Case Study. Paper SPE 130134 presented at the International Oil and Gas Conference and Exhibition in China, Beijing, China, 8-10 June. http://dx.doi.org/10.2118/130134-MS.

Lolomari, T., Bratvedt, K., Crane, M., Milliken, W.J., Tyrie, J.J. 2000. The Use of Streamline Simulation in Reservoir Management: Methodology and Case Studies. Paper SPE 63157 presented at the SPE Annual Technical Conference and Exhibition, Dallas, Texas, 1-4 October. http://dx.doi.org/10.2118/63157-MS.

Mallison, B.T., Gerritsen, M.G. and Matringe, S.F. 2006. Improved Mappings for Streamline-Based Simulation. *SPE Journal* 11(3): 294-302. Paper SPE 89352-PA. http://dx.doi.org/10.2118/89352-PA.

Maschio, C. and Schiozer, D. J. 2004. Integration of Streamline Simulation and Automatic History Matching. Paper SPE 89850 presented at the SPE Annual Technical Conference and Exhibition, Houston, Texas, 26-29 September. http://dx.doi.org/10.2118/89850-MS.

Maučec, M., Douma, S., Hohl, D., Leguijt, J., Jimenez, E.A. and Datta-Gupta, A. 2007. Streamline-Based History Matching and Uncertainty--Markov-Chain Monte Carlo Study of an Offshore Turbidite Oil Field. Paper SPE 109943 presented at the SPE Annual Technical Conference and Exhibition, Anaheim, California, USA, 11-14 November 2007. http://dx.doi.org/10.2118/109943-MS.

Maure, A., Dietrich, F., Gómez, U., Vallesi, J, and Irusta, M. 2001. Paper SPE 69652 presented at the SPE Latin American and Caribbean Petroleum Engineering Conference, Buenos Aires, Argentina 25-28 March. http://dx.doi.org/10.2118/69652-MS.

Moreno, J., Kazemi, H., and Gilman, J.R. 2004. Streamline Simulation of Countercurrent Water-Oil and Gas-Oil Flow in Naturally Fractured Dual-Porosity Reservoirs. Paper SPE 89880 presented at the SPE Annual Technical Conference and Exhibition, Houston, Texas, 26-29 September. http://dx.doi.org/10.2118/89880-MS.

Nair, C.V.G. and Al-Maraghi, E. 2006. A Practical Approach in Building Upscaled Simulation Model for a Large Middle East Carbonate Reservoir Having Long Production History. Paper SPE 100270 presented at the SPE Europec/EAGE Annual Conference and Exhibition, Vienna, Austria, 12-15 June. http://dx.doi.org/10.2118/100270-MS.

Nielsen, S.M., Jessen, K., Shapiro, A.A., and Stenby, E.H. 2010. Microbial Enhanced Oil Recovery: 3D Simulation with Gravity Effects. Paper SPE 131048 presented at the SPE EUROPEC/EAGE Annual Conference and Exhibition, Barcelona, Spain, 14-17 June. http://dx.doi.org/10.2118/131048-MS.

Obi, E.-O. I., and M. J. Blunt (2006), Streamline-based simulation of carbon dioxide storage in a North Sea aquifer, *Water Resour. Res.*, 42, W03414, http://dx.doi.org/10.1029/2004WR003347.

Odai, L. and Ogbe, D.O. 2011. An Approach for Ranking Realizations to Characterize Reservoirs for Fluid Flow Simulation. Paper SPE 150738 presented at the Nigeria Annual International Conference and Exhibition, Abuja, Nigeria, 30 July – 3 August. http://dx.doi.org/10.2118/150738-MS.

Osako, M.K., Hoang, V. and Balasubramanian, G. 2009. Evaluation of Streamline Simulation Application to Heavy Oil Waterflood. Paper SPE 122922 presented at the Latin American and Caribbean Petroleum Engineering Conference, Cartagena de Indias, Colombia, 31 May-3 June. http://dx.doi.org/10.2118/122922-MS.

Oyerinde, A., Datta-Gupta, A. and Milliken, W. 2007. Experiences with Streamline-based Three Phase History Matching. Paper SPE 109964 presented at the SPE Annual Technical Conference and Exhibition, Anaheim, California, 11-14 November. http://dx.doi.org/10.2118/109964-MS.

Park, H. and Datta-Gupta, A. 2011. Reservoir Management Using Streamline-based Flood Efficiency Maps and Application to Rate Optimization. Paper SPE 144580 presented at the SPE Western North American Regional Meeting , Anchorage, Alaska, USA, 7-11 May. http://dx.doi.org/10.2118/144580-MS.

Portella, R.C.M. and Hewett, T.A. 1998. Upscaling, Gridding, and Simulation Using Streamtubes. Paper SPE 49071 presented at the SPE Annual Technical Conference and Exhibition, New Orleans, Louisiana, 27-30 September. http://dx.doi.org/10.2118/49071-MS.

Pranter, M.J., Hirstius, C.B. and Budd, D.A. 2005. Scales of Lateral Petrophyscial Heterogeneity in Dolomite Lithofacies as Determined from Outcrop analogs: Implications of 3-D Reservoir Modeling. *AAPG Bulletin* 89(5): 645 -662.

Pranter, M.J., Ellison, A.I., Cole, R.D. and Patterson, P.E. 2007. Analysis and Modeling of Intermediate Scale Reservoir Heterogeneity based on a Fluvial Point-bar Outcrop Analog, Williams Fork Formation, Piceance Basin, Colorado. *AAPG Bulletin* 91(7): 1025 -1051.

Qassab, H., Khalifa, M., Pavlas, R., Afaleg, N., Ali, H., Kharghoria, A., He, Z., Lee, S.H., Datta-Gupta, A. 2003. Streamline-based Production Data Integration Under Realistic Field Conditions: Experience in a Giant Middle-Eastern Reservoir. Paper SPE 84079 presented at the SPE Annual Technical Conference and Exhibition, Denver, Colorado, 5-8 October. http://dx.doi.org/10.2118/84079-MS.

Rey, A., Taware, S.V. and Datta-Gupta, A. 2010. Use of Time-Lapse Seismic Data for Heterogeneity Characterization during CO2 Sequestration in Saline Aquifers. Paper SPE 139519 presented at the SPE International Conference on CO2 Capture, Storage, and Utilization, New Orleans, Louisiana, USA, 10-12 November. http://dx.doi.org/10.2118/139519-MS.

Rodriguez, A., Dawood, N.J., Soremi, Y. and Taiban, A. 2008. Practical Aspects of Streamline Application to Water Injection Management of a Giant Carbonate Reservoir. Paper SPE 112899 presented at the SPE North Africa Technical Conference & Exhibition, Marrakech, Morocco, 12-14 March. http://dx.doi.org/10.2118/112899-MS.

Sayyafzadeh, M., Pourafshary, P. and Rashidi, F. 2010. Increasing Ultimate Oil Recovery by infill drilling and converting Weak Production Wells to Injection wells using streamline simulation. Paper SPE 132125 presented at the International Oil and Gas Conference and Exhibition in China, Beijing, China, 8-10 June. http://dx.doi.org/10.2118/132125-MS.

Scheidt, C. and Caers, J. 2009. Uncertainty Quantification in Reservoir Performance Using Distances and Kernel Methods - Application to a West Africa Deepwater Turbidite Reservoir. SPE Journal 14 (4): 680-692. SPE-118740-PA. http://dx.doi.org/10.2118/118740-PA.

Sehbi, B., Kang, S., Datta-Gupta, A. and Lee, J. 2011. Optimizing Fracture Stages and Well Completions in Tight Gas Reservoirs Using Drainage Volume calculations. Paper SPE 144365 presented at the SPE North American Unconventional Gas Conference and Exhibition, Woodland, Texas,14-16 June. http://dx.doi.org/10.2118/144365-MS.

Seto, C.J., Jessen, K. and Orr Jr., F.M. 2003. Compositional Streamline Simulation of Field Scale Condensate Vaporization by Gas Injection. Paper SPE 79690 presented at the SPE Reservoir Simulation Symposium, Houston, Texas, 3-5 February. http://dx.doi.org/10.2118/79690-MS.

Sharif, A. and MacDonald, A. 2001. Evaluating the Uncertainty of 3D Geologic Reservoir Models. *World Oil* 222(5):44-48.

Shook, G.M. and Mitchell, K.M. 2009. A Robust Measure of Heterogeneity for Ranking Earth Models: The F-PHI Curve and Dynamic Lorenz Coefficient. Paper SPE 124625 presented at the SPE Annual Technical Conference and Exhibition, New Orleans, Louisiana, 4-7 October. http://dx.doi.org/10.2118/124625-MS.

Singh, A.K., Jain, V.K. and Marathe, R.V. 2010. Flood Front Estimation with Streamline Modeling: A Case Study of a Mature Field in Assam, India. Paper SPE 128993 presented at the SPE Oil and Gas India Conference and Exhibition, Mumbai, India 20-22 January. http://dx.doi.org/10.2118/128993-MS.

Stenerud, V.R., Kippe, V., Datta-Gupta, A. and Lie, K.-A. 2007. Adaptive Multiscale Streamline Simulation and Inversion for High-Resolution Geomodels. Paper SPE 106228 presented at the SPE Reservoir Simulation Symposium, Houston, Texas, U.S.A., 26-28 February. http://dx.doi.org/10.2118/106228-PA.

Stephen, K.D., Shams, A. and MacBeth, C. 2007. Faster Seismic History Matching in a UKCS Reservoir. Paper SPE 107147 presented at the EUROPEC/EAGE Conference and Exhibition, London, U.K.11-14 June. http://dx.doi.org/10.2118/107147-MS.

Su, H., Ali, F.M., Orjuela, J.A. and Gheorghiu, S. 2011. Advanced Reservoir Mangaement of Greater Burgan Field. Paper SPE 148198 presented at the SPE Reservoir Characterization and Simulation Conference and Exhibition, Abu Dhabi, UAE, 9-11 October. http://dx.doi.org/10.2118/148198-MS.

Subbey, S., Christie, M., Sambridge, M. 2003. A Strategy for Rapid Quantification of Uncertainty in Reservoir Performance Prediction. Paper SPE 79678 presented at the SPE Reservoir Simulation Symposium, Houston, Texas, 3-5 February. http://dx.doi.org/10.2118/79678-MS.

Taware, S., Sharma, M., Alhuthali, A.H. and Datta-Gupta, A. 2010. Optimal Water Flood Management Under Geological Uncertainty Using Accelerated Production Strategy. Paper SPE 133882 presented at the SPE Annual Technical Conference and Exhibition, Florence, Italy, 19-22 September. http://dx.doi.org/10.2118/133882-MS.

Thiele, M.R. Streamline Simulation. 2001. 6th International Forum on Reservoir Simulation. Schloss Fuschl, Austria September 3-7.

Thiele, M.R. and Batycky, R.P. 2006. Using streamline-derived injection efficiencies for improved waterflood management. SPE Res Eval & Eng 9 (2): 187-196. SPE 84080. http://dx.odi:10.2118/84080-PA.

Thiele, M.R., Batycky, R.P., Pöllitzer, S. and Clemens, T. 2008. Polymer-Flood Modeling Using Streamlines. Paper SPE 115545 presented at the SPE Annual Technical Conference and Exhibition, Denver, Colorado, USA, 21-24 September. http://dx.doi.org/10.2118/115545-MS.

Thiele, M.R., Batycky, R.P. and Fenwick, D.H. 2010. Streamline simulation for modern reservoir-engineering workflows. J. Pet Tech 62 (1): 64-70. SPE 118608. http://dx.odi:10.2118/118608-MS.

Thiele, M.R. and Batycky, R.P. 2003. Water Injection Optimization Using a Streamline-Based Workflow. Paper SPE 84080 presented at the SPE Annual Technical Conference and Exhibition, Denver, Colorado, 5-8 October. http://dx.doi.org/10.2118/84080-PA.

Vargas-Guzmán, J.A., Al-Gaoud, A., Datta-Gupta, A., Jimenez, E.A. and Oyeriende, D. 2009. Identification of High Permeability Zones from Dynamic Data Using StreamlineSimulation and Inverse Modeling of Geology. *Journal of Petroleum Science and Engineering* 69:283-291.

Vasco, D.W., Yoon, S. and Datta-Gupta, A. 1998. Integrating Dynamic Data into High-Resolution Reservoir Models Using Streamline-Based Analytic Sensitivity Coefficients. Paper SPE 49002 presented at the SPE Annual Technical Conference and Exhibition, New Orleans, Louisiana, 27-30 September. http://dx.doi.org/10.2118/49002-MS.

Vasco, D.W., Datta-Gupta, A., He, Z., Behrens, R., Rickett, J. and

Condon, P. 2003. Reconciling Time-Lapse Seismic and Production Data Using Streamline Models: The Bay Marchand Field, Gulf of Mexico. Paper SPE 84568 presented at the SPE Annual Technical Conference and Exhibition, Denver, Colorado, 5-8 October. http://dx.doi.org/10.2118/84568-MS.

Wang, Y. and Kovscek, A.R. 2000. A Streamline Approach for History-Matching Production Data. Paper SPE 59370 presented at the SPE/DOE Improved Oil Recovery Symposium, Tulsa, Oklahoma, 3-5 April. http://dx.doi.org/10.2118/59370-MS.

Watanabe, S. and Datta-Guptta, A. 2011. Use of Phase Streamlines for Covariance Localization in Ensmble Kalman Filter for Three-Phase History Matching. Paper SPE 144579 presented at the SPE Western North American Region Meeting, Anchorage, Alasaka, USA, 7-11 May. http://dx.doi.org/10.2118/144579-MS.

Hydrogeochemical modelling in coastal carbonate aquifer: porosity and permeability evolution

Rachida Bouhlila and Ezzeddine Laabidi

National Engineering School of Tunis, BP N37, Le Belvédère, 1002, Tunis-Tunis EL Manar University. Tunisia

Tel : +216 70 860 396 ; Fax : +216 70 860 396

bouhlila.rachida@enit.rnu.tn, laabidimhe@yahoo.fr

Abstract

Hydrogeological properties that govern subsurface flow are porosity, permeability, water-water and water-solid interaction, pore and particle size distributions. Porosity and permeabilities can be modified due to dissolution and precipitation of minerals. In several hydrogeochemical situations, there is an interaction between the solid matrix and the carbonate species according to salts dissolution and precipitation reactions (evaporate deposition in Sebkha anc Chotts, calcite dissolution in seawater-freshwater mixing zone). The dissolution of such rocks can easily induce a development of the porosity and permeability as a result of the mixing processes in the coastal carbonate aquifers (Romanov and Dreybrodt, 2006). The modeling of such problem requires a set of highly nonlinearly coupled equations. GEODENS model, presented in this work, can solve these equations by a finite element procedure. GEODENS is a FORTRAN code (Bouhlila and Laabidi, 2008) for modeling partly or fully saturated density dependent flow and multispecies reactive solute transport in porous media under both local chemical equilibrium and kinetic conditions. It can handle geochemical reactions such as mineral dissolution-precipitation processes. Its main purpose is to represent the physicochemical processes in the subsurface system. The code considers flow, transport and geochemical reactions in porous media. GEODENS comprises two modules, (i) the density dependent flow and multispecies transport module and (ii) the geochemical module. The mathematical formulation of the first module leads to a nonlinear and strongly coupled equations. In order to solve such equations, a finite element method called has been developed with a consistent numerical scheme of gravity terms to calculate Darcy velocities (Voss, C. I., 1984). The second module focuses on salts and brine geochemistry, using the Pitzer model (Pitzer, K. S. et al., 1984). This geochemical module allows the calculation of ions and solvent activities as well as the density of the solution. Reactions of salts, including dissolution- precipitation processes are controlled by diffusion and are represented by a first order kinetic law. The hydrogeochemical model presented in this paper integrates the two modules described above. The code iteratively calculates the quantities of the different salts that may precipitate or dissolve in a solution when the system is displaced from its equilibrium. This may occur due to evaporation or by mixing with a different solution. The system's hydrodynamic parameters may also change due to mineral dissolution-precipitation reactions. It is hoped that the code will be a useful tool for understanding hydrological and geochemical processes in arid regions. Through the examples simulated in this work we have shown firstly that our model is in good agreement with the previous experiment work (Singurindy and Berkowitz, 2004) and numerical one (Romanov and Dreybrodt, 2006). The model tested and validated is used to simulate the effect of the calcite dissolution on the penetration length of the seawater intrusion, and in order to provide general results we simulate these processes using Henry problem geometry with the same hydrodynamic parameters and boundary conditions.

1. Introduction

Hydrogeological properties that govern subsurface flow are porosity, permeability, water-water and water-solid interaction, pore and particle size distributions. Porosity and permeabilities can be modified due to dissolution and precipitation of minerals, In several hydrogeochemical situations, there is an interaction between the solid matrix and the carbonate species according to salts dissolution and precipitation reactions; in the following we quote three situations involving these processes.

- During the evaporation of the brines evolving in the carbonate cycle, the first salts to be deposited are calcite and dolomite. They often constitute the first evaporate deposition of Sebkhas, Chotts and other Playas (Perthuisot, 1975).
- In order to limit the greenhouse effects on world climate change, the sequestration of CO_2 in deep geological formation is an increasingly evoked solution. It is clear that the injection of this gas in the carbonate rocks can induces dissolution processes. This will, in the long term, compromise the reliability of this solution.
- In the limestone coastal aquifers, seawater intrusion can be enhanced as a result of the calcite dissolution in the freshwater–seawater mixing zone. The quantification of these various processes is necessary for predicting their impacts on such environments. The coupled hydrogeochemical modelling of this phenomenon in porous media constitutes a tool to analyze and predict these processes.

Calcite dissolution in limestone coastal aquifers is a phenomenon well known by the geochemists (Rezaei et al., 2005; Romanov and Dreybrodt, 2006: Singurindy and Berkowitz, 2004; Sanford and Konikow, 1989 ;...). The dissolution of such rocks can easily induce a development of the porosity and permeability as a result of the groundwater salinity processes in the coastal fringe of these aquifers (Romanov and Dreybrodt, 2006). The physicochemical processes taking place in the mixing zone of such aquifers can be seen as a density dependent flow and multispecies reactive transport problem (Bouhlila and Laabidi, 2008). Henry problem (Henry, 1960), relating to a semi-analytical resolution of the flow and transport equations in a vertical section of a coastal aquifer, is widely studied (Dierch, and Kolditz ,2002; Ataie-Ashtiani et al., 1999; Bear and Kapuler, 1981; Benson et al., 1998; Croucher and O'Sullivan, 1995; Gotovac et al., 2003; Kolditz et al., 1998; Sanz and Voss, 2006). It is also often used to check the models developed to simulate density dependent flow and coupled transport processes in porous media (Bouhlila, 1999). The seawater intrusion is often measured by the solutes penetration distance from the shore line. Obviously, this distance is affected by the aquifer hydrodynamic parameters and by the freshwater flux (Abarca et al., 2006). In this paper, we test our results obtained by the GEODENS code (Bouhlila and Laabidi, 2008) by comparing them to the experimental results, concerning the calcite dissolution in 2D flow cell, these results have been compared to the experimental work of Singurindy and Berkowitz (2004), and to the theoretical work of Phillips (1991). Finally, we simulate a fully coupled 2D problem concerning the seawater intrusion with a calcite solid matrix.

2. Geochemistry of carbonate environment

Carbonates are minerals characterized by the CO_3^{2-} ion. The most abundant ones are calcite ($CaCO_3$), aragonite ($CaCO_3$), and dolomite (Ca, Mg)$(CO_3)_2$. Dolomite is a frequent mineral in various sedimentary rocks such as dolomitic limestones, evaporates, etc. as in the metamorphic rocks, which are derived from them (Morel, 1994). The carbonated rocks are sedimentary ones formed at least by 50% of carbonate (calcite, dolomite, and aragonite).

If we consider that the solution is closed, equilibrium in a simple carbonate solution is described by the six following species: $CO_2(aq)$, H_2CO_3, HCO_3^-, CO_3^{2-}, H^+ and OH^-, bounded by four independent equilibrium aqueous reactions. According to the chemical components method (Morel, 1994), only two components are necessary to characterize the geochemistry of such solutions. Thus, in the coupled hydrogeochemical modelling of these processes, we will consider, in addition to the six major ions of sea water (Na^+, K^+, Ca^{2+}, Mg^{2+},

Cl^-, SO_4^{2-}), the two species H^+ and HCO_3^-. We also retain all the 19 evaporitic salts that could be generated in such solutions (Laabidi, 2007). With the eight components concentrations for each time step, we calculate the ions and water activities and we test the saturation index for each salt. According to the kinetic law those with an index greater than one will precipitate and those with an index less than one and present in the solid matrix will be dissolved. Even if this procedure is rather intuitive, it works perfectly when the number of salts to react is small.

3. Governing equations

GEODENS model is a two-dimensional coupled model for flow, transport and geochemical reactions for multi-components/multi-species system in porous media. The mainly objective is to solve the physicochemical processes in the subsurface system. The microscopic porous media is now an environment of flow, transport and precipitation-dissolution reactions of such brine in this volume. GEODENS is composed by two basic modules: flow and reactive transport. In the flow module, the Galerkine finite element method is used to discretize the modified flow equation in order to simulate the density-dependent flow. In the reactive chemical transport, solute transport is coupled with chemical equilibrium or kinetic to treat the interaction among chemical species as they are transported by subsurface flow. The equations governing the movement of fluid via a variably saturated porous media subject to variable density conditions can be derived from mass and momentum conservation principles. For this study only two-dimensional flow is considered and in the following we present the equations that govern the density dependent flow and multi-species reactive model in porous media (Bear, 1988).

3.1. Flow equation

The conservation of the liquid phase is expressed as follows:

$$\frac{\partial}{\partial x_i}(\rho q_i)-\frac{\partial(\rho\phi)}{\partial t}+Q_m=0 \quad , \tag{1}$$

where ϕ [L^3L^{-3}] is the porosity, ρ [ML^{-3}] is the fluid density, Q_m [$ML^{-3}T^{-1}$] is the source/sink term mass flux per unit of volume, and *t* [T] is time.

The Darcy flux q_i [LT^{-1}] is given by:

$$q_i=-\frac{k_{ij}}{\mu}\left(\frac{\partial p}{\partial x_j}-\rho g\right), \tag{2}$$

where k_{ij} [L^2] is the intrinsic permeability, μ [$ML^{-1}T^{-1}$] is the fluid dynamic viscosity, *p* [$ML^{-1}T^{-2}$] is the fluid pressure, and *g* [LT^{-2}] is the gravitational acceleration.

3.2. Solute transport equation

GEODENS model describes transient advective dispersive and reactive transfer of specie α (α =1, N_α), by the following equation

$$L\left(C^{\alpha}\right)+Q_m C^{\alpha}+\phi f^{\alpha}=\frac{\partial\left(\rho\phi C^{\alpha}\right)}{\partial t}. \tag{3}$$

C^{α} [MM^{-3}] is the concentration of the solute α, and N_α is the total number of solutes, and $L\left(C^{\alpha}\right)$ represents the transfer operator applied to C^{α} including the convection and the diffusion-dispersion D_{ij} term and is expressed as:

$$L\left(C^{\alpha}\right)=\frac{\partial}{\partial x_i}\left\{\rho\phi D_{ij}\frac{\partial C^{\alpha}}{\partial x_j}-\rho q_i C^{\alpha}\right\}. \tag{4}$$

The f^{α} term represents the geochemical mass flux of the solute α. If N_r is the number of reactions that concern the solute α, f^{α} is the sum of each reaction contributions into this mass flux (Bouhlila, 1999), and it is written as follow:

$$f^{\alpha}=\sum_{r=1}^{N_r}\rho\left(\frac{\partial C^{\alpha}}{\partial t}\right)_r \tag{5}$$

3.3. Impact on porosity and permeability

The physicochemical processes are written at macroscopic scale in GEODENS. The pore geometry and volume are continuously modified by the dissolution-precipitation reactions of salts. The porosity in any nodes of the mesh and at every time step Δt is updated as following:

$$\phi(t+\Delta t)=\phi(t)+\sum_{js=1}^{N_s} r_{js}V_{js}\Delta t. \tag{6}$$

V_{js} is the molar volume ($m^3\ mol^{-1}$) of the salt j, r_{js} is the rate ($mol\ m^{-3}\ s^{-1}$) of the mineral precipitation (-) or dissolution (+), and N_s is the total number of salts.

Sanford and Konikow (1989) proposed an empirical model for the sedimentary rocks to represent the permeability variations due to the dissolution of calcite in a coastal aquifer giving the following formula:

$$Log(\phi)=ak+b. \tag{7}$$

According to these authors and for sedimentary rocks, *a* is about 20 and *b* is deduced from the initial known values of *k* and ϕ.

3.4. Kinetic of calcite dissolution

The undersaturation of calcite in the mixture zone induces dissolution of the carbonate rocks. The dissolution reaction (essentially kinetic), increases the permeability in the cavities that induces an increase of saltwater penetration. Many authors studied the kinetic of calcite dissolution in low-temperature aqueous systems. We can find in Sanford and Konikow (1989), and in Freedman and Ibaraki (2001) a detailed description of this process. In fact, evaporates dissolution and precipitation reactions are rather limited by ions diffusion processes into the aqueous solutions. Then, they can be written as first order reactions. In GEODENS model, the kinetic of salts dissolution or precipitation reactions are represented by the geochemical flux, f^{α}, and is written as follows:

$$\phi f^{\alpha} = \sum_{r=1}^{N_r} \rho \phi k_{cr} \left(1 - \Omega_r\right), \qquad (8)$$

where Ω_r [L] is the saturation index of the salt *r,* and k_{cr} [T^{-1}] is the kinetic coefficient of the reaction.

4. Model Validation: porosity evolution in a 2D flow cell

Recently, Singurindy and Berkowitz (2004) conducted a laboratory analysis of the carbonate dissolution-precipitation process in coastal environments. These laboratory experiments focus on the porosity changes due to the precipitation or the dissolution of the carbonate rocks during the mixing process.

The basic domain is shown in Fig. 1. The domain is 16 cm×16 cm×1 cm; two reservoirs supply the domains, one containing a freshwater solution and the other containing the NaCl water solution. The two waters were saturated with the calcite and under atmospheric pressure (P_{CO2}=-3.5atm) for the precipitation experiment, and under closed system condition (P_{CO2}=1atm) for the dissolution experiment.

Table 1 Chemical composition of boundary solution used in the calculations.

Solution	Na^+	K^+	H^+	Mg^{2+}	Ca^{2+}	Cl^-	CO_3^{2-}	SO_4^{2-}	HCO_3^{2-}
Saltwater-(PCO2=1atm)	42.251	.000	1.26x10-6	.000	0.633	42.437	2.29x10-4	.000	1.495
Freshwater-(PCO2=1atm)	.000	.000	1.12x10-6	.000	0.331	.000	7.75x10-5	.000	1.075

Units for (total) concentrations are g Kg^{-1}

The first reservoir (freshwater solution) supplies the domain through the upper boundary under fixed head boundary (h=1cm). The second reservoir (NaCl water solution) supplies the domain through the right hand side under the same fixed head value. In Table 1, we present the chemical composition of boundary solutions used in the calculations. The two remaining boundaries present the outlets of the domain under fixed head boundary equal to zero (h=0m).

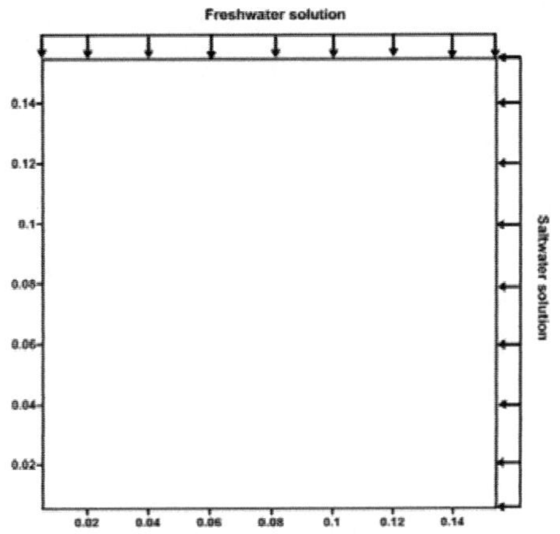

Figure 1 Geometry and boundaries conditions of the modeled domain used in the Laboratory experiments.

A grid of 441 elements was used to simulate this laboratory experience (dissolution experiment). Fig. 2 shows the concentration isolines of the chloride (Cl^{-}) in the modelling domain after the mixing of the two solutions.

An analytical solution to calculate the porosity change in the mixing zone was suggested by Phillips (1991).

$$\rho s \frac{\partial \phi}{\partial t} = -\phi D \left(\frac{\partial^2 c}{\partial s^2} \right) (\nabla s)^2 \ , \qquad (9)$$

where ϕ is the porosity, *D* is the coefficient of dispersion, ∇s is the salinity gradient and $\frac{\partial^2 c}{\partial s^2}$ is the second derivate of the equilibrium concentration c of the solution.

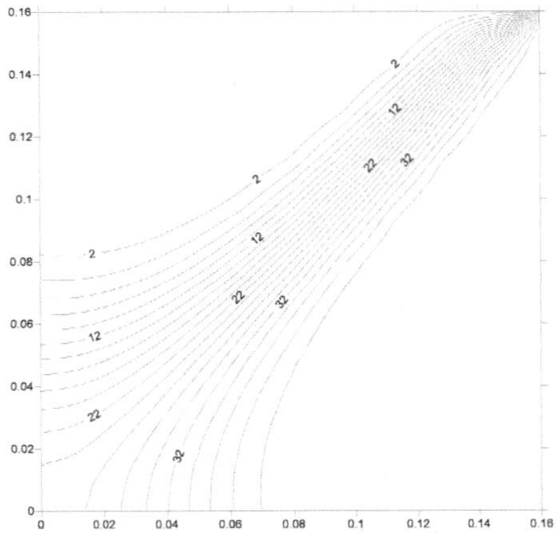

Figure 2 Spatial distribution of the Cl^{-} concentration (g/kg) in the modeling domain.

In order to validate our approach we used GEODENS to simulate the flow, transport and geochemical reactions in the geometry used by Singurindy and Berkowitz (2004) in their laboratory experiment. Fig. 3 shows the spatial distribution of initial change of porosity. We have found the same structure and location of the porosity change as in the simulation of Romanov and Dreybrodtn(2006) and of Singurindy and Berkowitz (2004). The laboratory experience shows a porosity change along the domain diagonal (the mixing zone) as in the present work.

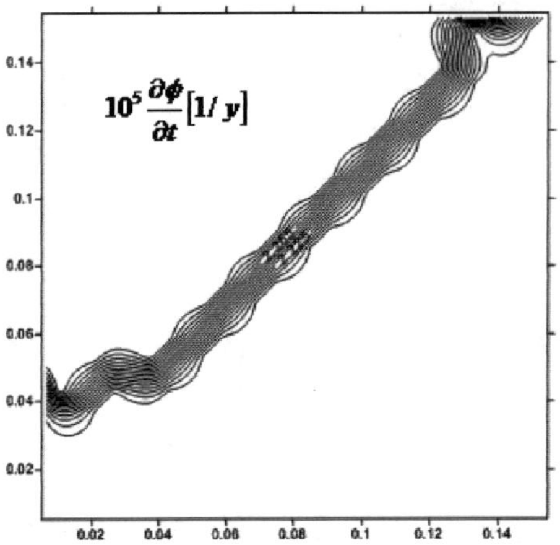

Figure 3 Spatial distribution of the modeled porosity change.

The change of porosity has been determined as 0.240×10^{-4} year^{-1} in the laboratory experience of Singurindy and Berkowitz (2004), in this work an average of 0.224×10^{-4} year^{-1} were determined, this value is reached after many simulations using different K_c values as presented in Table 2. The results show that the simulation with a kinetic coefficient value equal to 1.2×10^{-10} moles/s is in good agreement with the experience. In Table 2, we present the dissolved quantity of calcite and the change of the porosity evaluated analytically, experimentally and numerically (this study).

Table 2 Theoretical, Experimental and Numerical (this study) Estimates of Dissolution Rates and Porosity Change.

	kc (moles/s)	dΦ/dt (year^{-1})	Q_{CaCO3} (kg/m^3/d)
Theory	-	3.60E-05	2.70E-04
Experiment	-	2.40E-05	1.80E-04
This study (numerical)	2.00E-09	3.73E-04	5.53E-03
	1.00E-09	1.86E-04	2.76E-03
	2.00E-10	3.74E-05	5.53E-04
	1.20E-10	**2.24E-05**	**3.32E-04**
	1.00E-10	1.86E-05	2.76E-04

5. Henry Reactive problem: Simulation and discussion

We use in this study the same geometry and parameters as in Henry problem. The Henry saltwater intrusion problem concerns a vertical slice through an isotropic, homogeneous, confined aquifer. For the flow boundary, a freshwater constant flux is applied from the inland boundary and a hydrostatic pressure is distributed on the sea boundary. For the transport boundary, it is not easy to represent the sea side. In this work we introduce a mixed boundary condition in the seaside to avoid difficulties in the finite element solution near the outflow region of the aquifer (Segol et al., 1975). Segol et al (1975) replaced the original homogeneous Dirichlet boundary by a mixed Dirichlet-Neumann boundary. The upper 20 cm of the new boundary was a zero concentration flux exit boundary, and the lower 80 cm remained a constant concentration Dirichlet boundary.

Henry (1960) solved the governing equations in a nondimensional form and presented a semianalytical solution method for the steady state distribution of salt concentration and stream function. The Henry problem has been widely used as a test case of density dependent groundwater flow models benchmarked (Voss and Souza, 2006; Simpson and Clement, 2003; E.Abarca et al. 2006). The correspondent geometrical and hydrodynamic parameters are listed in Table 3.

Table 3 Physical parameters of the Henry's problem.

parameters	ϕ	ρ_{sw} (kg m^{-3})	Q_m (kg s^{-1})	K (m^2)	g (ms^{-2})	α_L, α_T (m)	d_0 (m^{2}s^{-1})	D (m)	L (m)
values	0.35	1025	0.066	1.02×10^{-2}	9.8	.0	6.6×10^{-6}	1.0	2.0

ϕ = Porosity, α_L = Longitudinal dispersivity coefficient

ρ_{SW} = Density of saltwater, α_T = Transverse dispersivity coefficient

Q_m = Recharge rate, K = Hydraulic conductivity

d_0 = Coefficient of molecular diffusion, D = Elevation, L = Length

Seawater intrusion studies are usually concerned with the depth of inland penetration of saltwater which characterizes the size of the contaminated zone. The penetration of the seawater intrusion wedge (denoted as L_p) is defined as the distance between the seaside boundary and the point where the 50% mixing isoline intersects the aquifer bottom (E.Abarca et al. 2006).

In the original Henry problem, a pseudo-permanent regime is reached after 100 minutes (Voss, 1984; Bouhlila, 1999). If the solid matrix is calcareous, the matrix dissolution processes are perpetual and no permanent regime is reached, provided that calcite remains present in the system. In order to test the effect of the dissolution of the calcite on the penetration of the seawater intrusion, we simulate the Henry problem with a kinetic coefficient equal to 1.0×10^{-5} moles/s. Fig. 4 shows that during the mixing of the saltwater and freshwater, continuous calcite dissolutions occur. The quantity of the dissolved calcite is extremely related to the rate of calcite dissolution and to the order of the mixing (fraction of the saltwater).

In summary the penetration length increases and is mainly controlled by the horizontal permeability, the amount of calcite dissolved in the mixing zone increases with time (Fig. 4B), inducing an increase in porosity (Fig. 4C) due to the pore volume released by calcite dissolution phenomenon. The new permeability result of the porosity evolution is shown in Fig. 4D. This new hydrodynamic distribution controls the variation of the velocity field and the shape of the concentration repartition as shown in Fig. 4A.

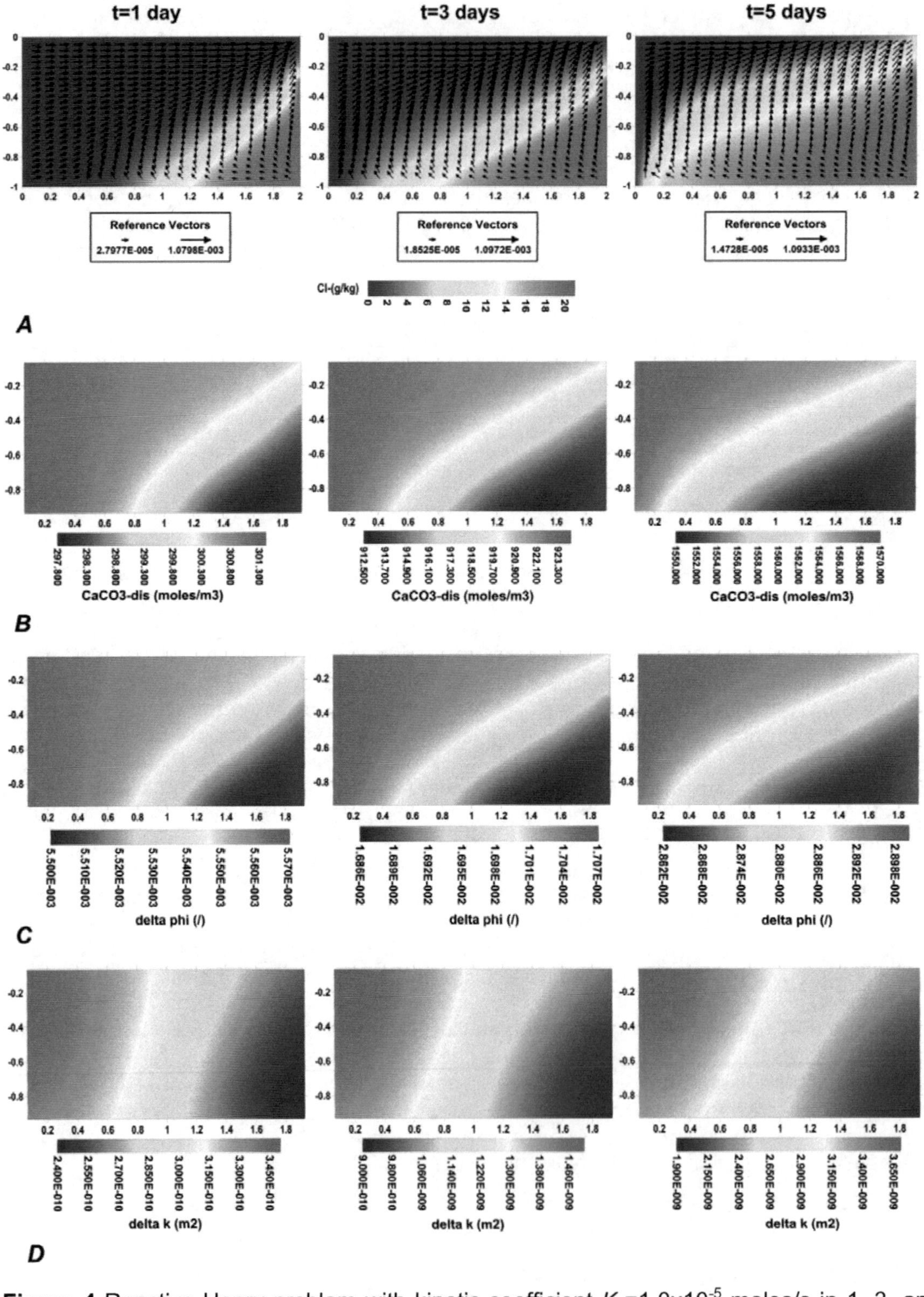

Figure 4 Reactive Henry problem with kinetic coefficient K_c=1.0x10^{-5} moles/s in 1, 3, and 5 days. Spatial distribution of the Cl^- concentration and velocity field (A), quantity of dissolved calcite (B), porosity change (C), and hydraulic conductivity change (D)

Maximal values of the porosity change occur at the freshwater side and in the transition zone, the maximal value reached after 5 days is about 2.898×10^{-2}. Regarding to the

theoretical work of Phillips (1991), this is reasonable because in these regions, overlaps between hydraulic gradients and the second derivative are maximal.
Romanov, and Dreybrodt (2006) described the initial porosity changes by assuming that the hydraulic properties of the aquifer do not change as porosity evolves. Actually, changes in porosity cause an increase in permeability. The maximal value of the permeability change reached after 5 days is about 3.650×10^{-9} m^2.

To conclude, the dissolution induces an increase in the porosity which is related to the permeability. This would enhance further seawater flow and mixing in the simulated section. Chemical dissolution in such environment is another characteristic of real aquifers which the Henry problem does not account for. Calcite dissolution may affect the horizontal permeability which affects both seawater penetrations and the flux of saltwater that enters the aquifer through the seaside boundary.

6. Conclusion

Dissolution of limestone in the saltwater–freshwater mixing zone in coastal carbonate aquifers is described by a density dependent flow and multi species reactive transport model. This requires a set of highly nonlinearly coupled reaction–diffusion–advection equations. GEODENS model, presented in this work, can solve these equations by a finite element procedure.
The seawater intrusion phenomenon is a real risk on the subsurface water resources in several areas of the world. This risk can be increased by the solid matrix dissolution processes in the saltwater–freshwater mixing zone in the case of limestone aquifer. We tried in this work to present a relative complete modelling scheme of this phenomenon in order to quantify this risk. Through the examples simulated in this work we have shown firstly that our model is in good agreement with the previous experiment work (Singurindy and Berkowitz, 2004) and numerical one (Romanov and Dreybrodt, 2006). We have also shown that the changes of porosity are created in narrow regions at the freshwater side of the mixing zone.
We discuss the behavior of the solution in terms of two parameters: velocity field which have a direct effect on the flux from the seaside, and the penetration length. These two parameters are a direct results of the geochemical phenomenon during the mixing process. Furthermore the protocol written in GEODENS can easily describe this relation. At each time step, the geochemical fluxes are calculated with the previous time step results according to an explicit procedure. The porosity is calculated from the pore volume released by calcite dissolution in every node and each time step. The new porosity value is used to update the permeability using an empirical equation presented in Sanford and Konikow (1989). The new hydrodynamic distribution is the new inputs for the next time step to solve the problem in every nodes of the mesh.
To conclude, the model presented in this work is a numerical tool to predict and quantify the effect of the calcite dissolution below a limestone costal aquifer. This model seems to be in good agreement with the majority of the previous works (numerical, theoretical and experimental).

References

Abarca E., Carrera J., Sanchez-Vila X., Dentz M., 2006. Anisotropic dispersive Henry problem. Advances in Water Resources. doi:10.1016/j.advwatres.2006.08.005.

Ataie-Ashtiani B., Volker R.E., Lockington D.A., 1999. Tidal effects on sea water intrusion in unconfined aquifers. Journal of Hydrology 216, 17-31.

Back, W., B. B. Hanshaw, T. E. Pyle, L. N. Plummer, and A. E. Weidie, 1979. Geochemical significance of groundwater discharge and carbonate solution to the formation of Caleta Xel Ha, Quintana Roo, Mexico. Water Resources Research, 15, 1521-1525.

Bear J., 1988. Dynamic of Fluids in Porous Media. Dover, New York, USA.

Bear J., Kapuler I.,1981. A numerical solution for the movement of an interface in a layered coastal aquifer. Journal of hydrology, 50, 273-298.

Benson DA., Carey AE., Wheatcraft SW., 1998. Numerical advective flux in highly variable velocity fields exemplified by saltwater intrusion. Journal of Contaminant Hydrology, 34:207-33.

Bouhlila R., 1999. Ecoulements, Transports et Réactions Géochimiques Couplés Dans Le Milieu Poreux. Thèse de doctorat d'état es sciences. ENIT

Bouhlila R., Laabidi E., 2008. Impacts of calcite dissolution on seawater intrusion processes on coastal aquifers: Density dependent flow and multi species reactive transport. IAHS Publications, 320, 2008.

Croucher A.E., O'Sullivan M.J., 1995. The Henry Problem for seawater intrusion. Water Resources Research, 31(7):1809-14.

De Marsily G., 1981. Hydrogéologie quantitative. Ed. Masson, 212 p.

Dierch, H. J., Kolditz, O.,2002. Variable-density flow and transport in porous media. Advances in Water Resources, 25, 899-944.

Freedman V., Ibaraki M., 2002. Effects of chemical reactions on density-dependent fluid flow: on the numerical formulation and the development of instabilities. Advances in Water Resources 25, 439-453

Gledhill D.K., Morse, J.W., 2006. Calcite dissolution kinetics in Na–Ca–Mg–Cl brines. Geochimica et Cosmochimica Acta 70, 5802-5813.

Gotovac H., Andricevic R., Gotovac B., Kozulic V., Vranjes M., 2003. An improved collocation method for solving the Henry problem. Journal of Contaminant Hydrology 64, 129-149.

Guo, W., Langevin, C.D., 2002. User's Guide to SEAWAT: A Computer Program for Simulation of Three-dimensional Variable-density Ground-water Flow, US Geological Survey Techniques of Water-resources Investigations Book 6, USA.

Henry, H. R., 1960. Salt intrusion into coastal aquifers . General Assembly of Helsinki 25-7-6-8, 196. International Association of Scientific Hydrology, 52, 478–548.

Kolditz O., Ratke R., Dierschb H. G., Zielke W., 1998. Coupled groundwater flow and transport: 1. Verification of variable density flow and transport models. Advances in Water Resources, Vol. 21, No. 1. pp. 21-46.

Laabidi E., 2007. Ecoulement saturé – non saturé et transport multi espèces réactifs dans les milieux poreux. Extension de MEFDENS aux ions carbonatés. Mémoire de mastère MHE- ENIT, Université Tunis El Manar, Tunisia.

Morel F., 1994. Chimie des eaux naturelles et des interfaces dans l'environnement. Ed. Masson, 87 p.

Morse, J.W., Berner, R.A., 1972. Dissolution kinetics of calcium carbonate in seawater: II. A kinetic origin for lysocline. American Journal of Science, 272, 840-851.

Morse, J.W., Arvidson R.S., 2002. The dissolution kinetics of major sedimentary carbonate minerals. Earth-Science Reviews 58, 51-84.

Parkhurst, D.L., Apello, C.A.J., 1999 (Version 2). User's Guide to PHREEQC - A Computer Program for Speciation, Reaction-path, 1D-transport, and Inverse Geochemical Calculations, Technical Report 99-4259, US Geological Survey, USA.

Perthuisot, 1975. La Sebkha El Melah de Zarzis. Genèse et évolution d'un bassin salin paralique. Thèse de L'Ecole Normale Supérieure de Paris, France.

Phillips, O. M., 1991. Flow and Reactions in Permeable Rocks. Cambridge University Press, New York.

Plummer, L.N., Parkhurst, D.L., Wigley, T.M.L., 1979. Critical review of the kinetics of calcite dissolution and precipitation. In: Jenne, E. (Ed.), Chemical Modeling-Speciation, Sorption, Solubility and Kinetics in Aqueous Systems. American Chemical Society, Washington, DC, pp. 537-573.

Pitzer, K. S., Peiper, J. C., Busey, R. H., 1984. Thermodynamic properties of aqueous sodium chloride solutions. Journal of Physical and Chemical Reference Data 13, 1-102.

Price, R. M., and J. S. Herman, 1991. Geochemical investigation of saltwater intrusion into a coastal carbonate aquifer: Mallorca, Spain. Geological Society of America, v. 103; no. 10; p. 1270-1279.

Rezaei, M., Sanz, E., Raeisi, E., Ayora, C., Vazquez-Sune, E., Carrera, J., 2005. Reactive transport modeling of calcite dissolution in the fresh-saltwater mixing zone. Journal of Hydrology 311, 282-298.

Romanov, D., Dreybrodt, W., 2006. Evolution of porosity in the saltwater-freshwater mixing zone of coastal carbonate aquifers: An alternative modelling approach. Journal of Hydrology, 329, 661-673.

Sanford, W. E., Konikow, L. F., 1989. Simulation of calcite dissolution and porosity changes in saltwater mixing zones in coastal aquifers. Water Resources Research, 25, 655-667.

Sanz E., Voss C.I., 2006. Inverse modeling for seawater intrusion in coastal aquifers: Insights about parameter sensitivities, variances, correlations and estimation procedures derived from the Henry problem. Advances in Water Resources 29, 439-457.

Segol, G., Pinder, G. F., Gray, W. G., 1975. A Galerkin-finite element technique for calculating the transient position of the saltwater front. Water Resources Research. 11, 343-347.

Simpson M.J., Clement T.P., 2003. Theoretical analysis of the worthiness of Henry and Elder problems as benchmarks of density-dependent groundwater flow models Advances in Water Resources 26, 17-31.

Singurindy O.,Berkowitz B., 2004. Carbonate dissolution and precipitation in coastal environments: Laboratory analysis and theoretical consideration. Water Resources Research, VOL. 40, W04401, doi:10.1029/2003WR002651.

Voss, C. I., 1984. SUTRA: a finite element simulation model for saturated–unsaturated fluid density dependent groundwater flow with energy transport of chemically-reactive simple-species solute transport. US Geological Survey Water-Resources Investigations 84-4369.

(623b) Structure Sensitivity of Methanol Synthesis From Carbon Dioxide On Copper-Based Catalysts

Sittichai Natesakhawat[1,2], Victor Abdelsayed[3,4], Jonathan W. Lekse[1], John P. Baltrus[1], Paul R. Ohodnicki Jr.[1], Bret H. Howard[1], Xingyi Deng[1,3] and Christopher Matranga[1], (1)U.S. Department of Energy, National Energy Technology Laboratory, Pittsburgh, PA, (2)Department of Chemical and Petroleum Engineering, University of Pittsburgh, Pittsburgh, PA, (3)URS, South Park, PA, (4)U.S. Department of Energy, National Energy Technology Laboratory, Morgantown, WV

Thermocatalytic conversion of carbon dioxide to methanol has recently gained considerable attention because it is regarded as an optimal solution to mitigate CO_2 emissions while generating revenues to offset carbon management costs. Currently, commercial synthesis of methanol is accomplished by hydrogenation of CO-rich syngas over $Cu/ZnO/Al_2O_3$ catalysts. However, the low activity and stability of existing formulations create major barriers towards using them in realistic CO_2-rich conditions. Despite extensive studies of methanol synthesis over Cu-based catalysts, the nature of active sites and structure sensitivity of methanol synthesis on Cu surfaces are not fully understood. Thus, a better understanding of active sites and structure-activity relationships is essential for the rational design of new catalysts that would be suitable for CO_2 conversion applications.

We have investigated coprecipitated Cu-based catalysts possessing different metal oxide additives, Cu crystallite sizes, and varying degrees of support crystallinity. Using a variety of ex situ and in situ characterization techniques, we have verified that Cu^0 species are active sites for methanol synthesis from a stoichiometric CO_2/H_2 mixture. The possibility of Cu^+ species acting as active sites is excluded because these species are not observed on the surface of the working catalysts. The reaction is structure-sensitive with smaller Cu particles demonstrating higher turnover frequency. An explanation for an activity enhancement upon promotion will be presented. Possible reaction pathways for CO_2 hydrogenation to methanol will also be discussed.

Fluorine-modified Cu/Zn/Al/Zr catalysts via hydrotalcite-like precursors for the CO_2 hydrogenation to methanol

Peng Gao[1,2], Feng Li[1], Fukui Xiao[1], Ning Zhao[1], Wei Wei[1,] and Yuhan Sun[1,3,]**

[1]*State Key Laboratory of Coal Conversion, Institute of Coal Chemistry, Chinese Academy of Sciences, Taiyuan, PR China*

[2]*University of the Chinese Academy of Sciences, Beijing 100049, PR China*

[3]*Low Carbon Energy Conversion Technology Center, Shanghai Advanced Research Institute, Chinese Academy of Sciences, Shanghai, PR China*

(Corresponding author. Address: Institute of Coal Chemistry, Chinese Academy of Sciences, South Taoyuan Road 27#, 030001, P. R. China. Tel: +86-0351-4049612, Fax: +86-0351-4041153. E-mail: weiwei@sxicc.ac.cn, yhsun@sxicc.ac.cn)*

Introduction

It is well known that carbon dioxide emissions have induced global warming. As a cheap, nontoxic and abundant C1 feedstock, chemical utilization of CO_2 is a challenge and important topic. Methanol is a starting material for several important chemicals and can also be used as a fuel additive. It can also be converted to high-octane gasoline, aromatics, ethylene, as well as other useful petrochemicals.[1] The Cu/ZnO catalyst is well-known for high activity and selectivity for CO_2 hydrogenation to methanol. Support such as Al_2O_3 can further increase the catalytic performance on methanol synthesis. Furthermore, the promoter such as Zr is known to affect the copper dispersion and surface basicity, which in turn affects the activity and the adsorption of CO_2.[2, 3] Apart from catalyst compositions, the nature of precursor plays an important role for the catalytic performance. Composite catalysts could be obtained by the controlled thermal decomposition of hydrotalcite-like compounds (HTlcs) with the general formula $[M^{2+}{}_{1-x}M^{3+}{}_{x}(OH)_2]^{x+}\,(A^{n-})_{x/n}{\bullet}mH_2O$, where M^{2+} and M^{3+} are the divalent cation and trivalent cation respectively [4, 5]. The as-obtained composite oxides possessed homogeneous dispersion of M^{2+} and M^{3+} at an atomic level, high stability against sintering, synergetic effects between the elements and strong basic properties [6-8]. Therefore, HTlcs with various metal cations are among the most investigated catalyst precursors for the remarkable properties of the final catalysts.

In order to better understand and design the highly active catalysts for the CO_2 hydrogenation to methanol, the reaction mechanism has been widely investigated by many researchers. For the Cu/ZrO_2 or $Cu/ZnO/ZrO_2$ catalyst, a bifunctional (dual-site) mechanism of CO_2 hydrogenation is currently accepted. It is assumed that the adsorption and dissociation of hydrogen occur on the Cu sites, while the adsorption of CO_2 takes place over the ZrO_2 or ZnO and ZrO_2 sites. The atomic hydrogen transports from the surface of Cu onto the surface of ZrO_2 on which formate species form via spillover and undergo stepwise hydrogenation to methylenebisoxy, methoxide, and finally methanol [9-12]. Therefore, a high exposed Cu surface area and appropriate adsorption amount and adsorption strength of CO_2 are favorable for the hydrogenation of CO_2. As well known, the fluorine ion was a basic anion and promoted many base-catalytic reactions. It could be expected that the introduction of fluoride into Cu/Zn/Al/Zr catalyst via hydrotalcite-like (htl) precursor will affect the surface basicity of catalyst, which in turn affects the adsorption of CO_2. The fluorine anions can be introduced into the interlayer region of HTlcs based on the so called "memory effect" [13, 14]. The reconstruction of the htl structure by the memory effect took place by the rehydration of the derived mixed metal oxide formed by the calcination of their htl precursor, using liquid water or different aqueous solutions containing the ddesired anions.

In the present work, the fluorine-modified Cu/Zn/Al/Zr mixed oxide was prepared by the reconstruction of the htl structure from Cu/Zn/Al/Zr mixed oxide using aqueous solution containing F^- and were used as catalysts for the methanol synthesis from CO_2 hydrogenation. The effect of the introduction of fluorine by the reconstruction method on the physicochemical properties of precursor materials and final catalysts were discussed in detail.

Experimental

Preparation of Catalysts

The $Cu_2Zn_1Al_{0.7}Zr_{0.3}$ HTlcs intercalated with CO^{3-} (HTs-CO_3) were synthesized by coprecipitation method, and corresponding calcined HTlcs (CHTs-CO_3) were obtained. The reconstructed forms of $Cu_2Zn_1Al_{0.7}Zr_{0.3}$ HTlcs with fluorine anions in the interlayer region (HTs-F) were prepared using the "memory effect" according to the literature. Briefly, the as-prepared mixed metal oxide (CHTs-CO_3) were treated separately with an aqueous solution of NaF under N_2 atmosphere for 48 h. The obtained HTs-F was further calcined in an oven at 500 °C for 4 h. Then the fluorine-modified mixed oxides (CHTs-F) were obtained.

Characterization of Catalysts

Powder X-ray diffraction (XRD) patterns were recorded on a Panalytical X'Pert Pro X-ray diffractometer with Cu Kα radiation, in the step mode (0.0167°, 10 s) in the range $5° < 2\theta < 75°$.

Thermal decomposition of the catalyst precursors were studied by thermogravimetric (TG) method by using NETZSCH TG209F1 Thermal Analyzer. Measurements were performed in the temperature range of 303-1023 K with linear temperature program $\beta = 10$ K min^{-1} in continuous flow of synthetic air (30 mL min^{-1}).

The dispersion of Cu (D_{Cu}) and exposed Cu surface area (S_{Cu}) were determined by dissociative N_2O adsorption and carried out on Micromeritics AutoChem 2920 instrument using the procedure described by Yuan et al [15].

Temperature program reduction (TPR) was carried out in a U-tube quartz reactor. The samples (50 mg) were purged with Ar (30 mL min^{-1}) at 423 K to remove physically adsorbed water and then reduced in a flow of 5 vol.%H_2 + Ar (30 mL min^{-1}) at a heating rate of 5 K min^{-1} up to 873 K. TCD was used to monitor the consumption of H_2.

The basicity of the catalyst was measured by CO_2 temperature-programmed desorption (CO_2-TPD). The catalyst was first reduced at 603 K in H_2 flow of 30 mL min^{-1} for 2 h. After cooling to room temperature, the catalyst was saturated with pure CO_2 (30 mL min^{-1}) at 323 K for 60 min, and then flushed with Ar flow (40 mL min^{-1}) to remove any physisorbed molecules. Afterward, the TPD experiment was started with a heating rate of 10 K min^{-1} under Ar flow (40 mL min^{-1}), and the desorbed CO_2 was detected by an AMETEK mass spectrometer.

Evaluation of Catalysts

Activity measurements in the hydrogenation of CO_2 were carried out in a continuous-flow, high-pressure, fixed-bed reactor. Catalyst (1.5 mL, 40–60 mesh) diluted with 1.5 mL quartz sand (60 mesh) was placed in a stainless steel tube reactor. Prior to reaction, the catalyst was reduced in pure H_2 at a flow-rate of 80 mL min^{-1} under atmospheric pressure. The reduction temperature was programmed to increase from room temperature to 603 K. The reactor was then cooled to room temperature. After reduction, the activities of the catalyst samples in CO_2 hydrogenation process were determined under reaction conditions of 503-543 K, 5.0 MPa, $n(H_2):n(CO_2) = 3:1$, GHSV=4000 h^{-1}.

Results and discussion

Textural properties of the prepared materials

The XRD patterns of the precursor materials shown in Fig. 1 are typical for hydrotalcite-like compounds. The XRD pattern of the reconstructed HTs-F sample, which was subject to a possible dissolution–recrystallization process, displayed significantly higher and sharper basal peaks. The apparently increase of the crystallinity upon reconstruction could be explained by a partial recrystallization of the amorphous phase existing in the mixed oxides forms [16]. Generally, the crystallographic parameters a and c were calculated by employing least-squares refinement assuming a hexagonal crystal system for samples. In comparison with HT–CO_3 HTlcs (c=22.76 Å,

a=3.086 Å), HTs-F showed an increase in c (22.86 Å) with slight decrease in a (3.082 Å). The increase in c value could be ascribed to the less electrostatic force of F^- with respect to CO_3^{2-} anions. Thermal decomposition of these precursors results in the formation of poorly crystallized CuO phase (Fig. 1).

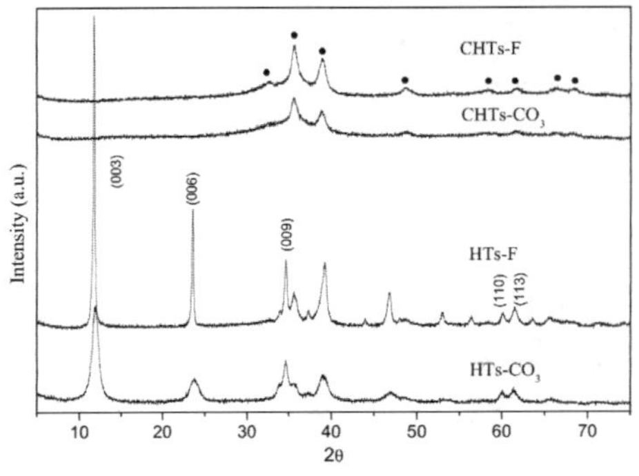

Fig. 1. XRD patterns of the prepared samples. (●) CuO

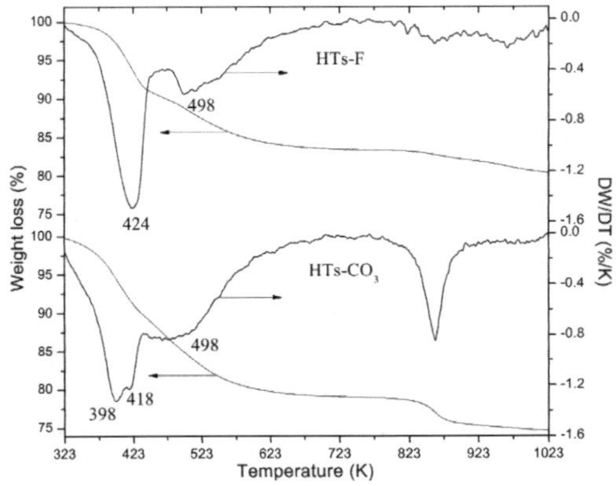

Fig. 2. Thermogravimetry and differential thermogravimetry (TG-DTG) profiles of precursors in air.

Fig. 2 presents TG-DTG profiles of the Cu/Zn/Al/Zr HTlcs during heating in air. In general, the thermal decomposition (weight loss) of the HTlcs consists of three steps. The first two-stage peak occurs at 373-473 K corresponding to elimination of physically adsorbed and interlayer water molecules. The second step of the weight loss at 473-623 K corresponds to removal of hydroxyl groups from the htl network and the loss of anion, resulting in the collapse of the layered structure [17, 18]. A progressive weight loss was observed for HTs-CO_3 sample between 773 and 973 K, which can be attributed to the decomposition of Cu oxocarbonates formed during the former steps of the thermal decomposition of the htl structure [19, 20]. However, this weight loss for HTs-F was much lower compared with HTs-CO_3, which indicated that a little CO^{3-} was present in the interlayer region of HTs-F. In addition, the peak temperatures for the first step shifted toward higher temperature with the introduction of fluorine, which might be attributed to the increase in the strength of hydrogen bonds between water molecules and interlayer F^- ions [21].

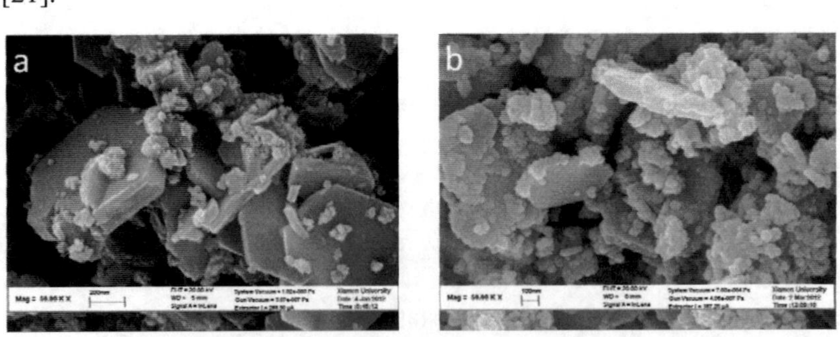

Fig. 3. SEM images of prepared fluorine-modified samples (a) HTs-F and (b) CHTs-F.

The reconstructed HTlcs possessed relatively close-packed structure compared to the parent HTlcs [16]. Obviously, such compact aggregates can be observed from the SEM micrograph of the precursor HTs-F in Fig. 3. Some defects appeared on the edges of the plates and small particles can also be found on the surface of plates. The crystals calcined after 773 K were destroyed and broken into small pieces, though some plates maintained the similar morphology of the original precursor during thermal treatment (Fig.3). In addition, the round shaped particles seem to have grown on the destroyed plates.

Table 1 Physicochemical properties of the calcined products.

Sample	BET surface area	Pore volume	Cu surface area [a]	Cu dispersion [a]
	($m^2 g^{-1}$)	($cm^3 g^{-1}$)	($m^2 g^{-1}$)	(%)
CHTs-F	38	0.15	29.1	11.52
CHTs-CO_3	69	0.31	41.5	14.37

[a] Calculated from N_2O dissociative adsorption.

After calcination at 773 K, both specific surface area and pore volume of CHTs-F sample are much lower than those of CHTs-CO_3 (Table 1). The drop of the specific surface area can be attributed to the formation of defects on the edge of plates and compact aggregates for the precursor during reconstruction [16, 22]. Moreover, the exposed copper surface area (S_{Cu}) and dispersion of copper (D_{Cu}) were measured by reactive N_2O adsorption technique (see Table 1). The decrease of the S_{Cu} and D_{Cu} in CHTs-F sample might be related to the formation of aggregates of individual particles upon the reconstruction of precursor.

The reducibility of Cu/Zn/Al/Zr catalysts

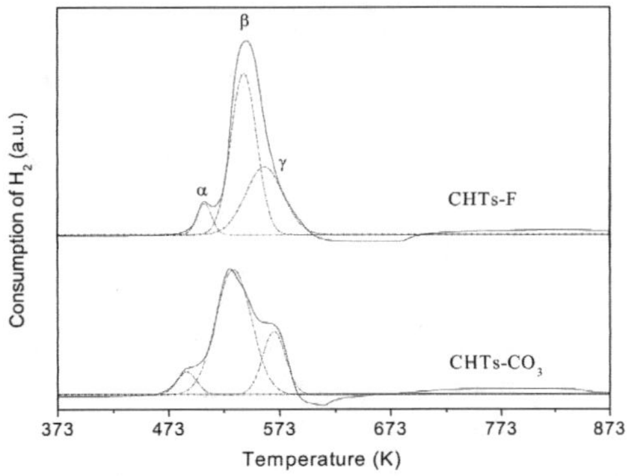

Fig. 4. TPR profiles of the catalysts.

Table 2 Temperature of reduction peaks and their contributions to the TPR pattern over Cu/Zn/Al/Zr catalysts.

Sample	TPR peak position [temperature(K)] and concentration (%)[a]		
	Peak α	Peak β	Peak γ
CHTs-F	505(6.5)	541 (57.8)	559 (35.7)
CHTs-CO_3	489 (31.2)	531 (52.0)	568 (16.8)

[a] Values in parentheses are the contribution (%) of each species.

In order to investigate the reduction behavior of the Cu/Zn/Al/Zr catalysts, TPR measurements were carried out. As shown in Fig. 4, the reduction profiles of the two samples exhibit a broad reduction profile together with shoulders in the temperature range 473–623 K. In order to get more insight into the TPR results the profiles were deconvoluted into several Gaussian peaks. The peak positions and their contributions derived from deconvolution are summarized in Table 2.

The peak β and peak γ might be ascribed to the reduction of two different types of CuO phase, e.g. the low temperature peak (peak β) is due to the reduction of dispersed CuO, whereas the peak appearing at high temperature (peak γ) is attributed to the reduction of bulk CuO [23]. There existed another peak at lower temperature (peak α) which can be assigned to the reduction of highly dispersed CuO. It can seen that the positions of peak α and peak β for CHTs-F are higher compared with the CHTs-CO_3 sample, while the change of peak γ position is reverse. As shown in Table 2, the summation of the contribution of peak α and peak β to the TPR pattern decreases with the introduction of fluorine. The result indicated that the amount of easily reducible well-dispersed copper oxide for CHTs-F is lower than that for CHTs-CO_3, which is similar to the result of D_{Cu} determined by means of N_2O chemisorption.

The surface basicity of Cu/Zn/Al/Zr catalysts

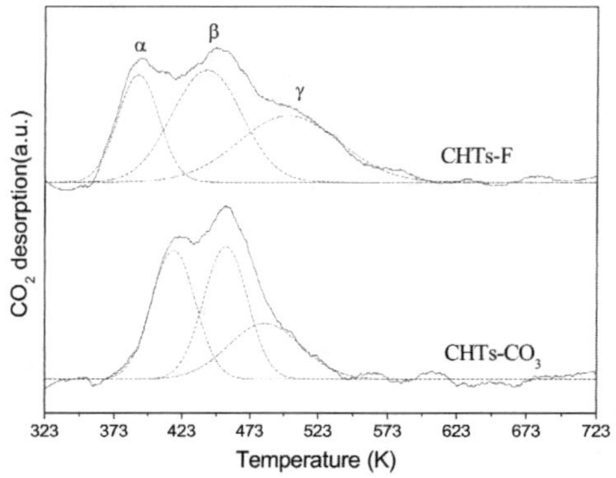

Fig. 5. CO_2-TPD curves of pre-reduced catalysts.

Table 4 The basicity and distribution of basic site over reduced Cu/Zn/Al/Zr catalysts with different Zr content.

Sample	Number of total basic sites	Density of total basic sites	Number of basic sites (μmol g^{-1}) and contribution[a]		
	(μmol g^{-1})	(μmol m^{-2})	Site α	Site β	Site γ
CHTs-F	257.4	5.720	59.8 (23.2)	106.7 (41.5)	90.9 (35.3)
CHTs-CO_3	234.0	3.382	85.8 (36.7)	88.9 (38.0)	59.3 (25.3)

[a] The value in the parenthesis is the contribution of single basic site to the number of total basic site.

Fig. 5 shows the CO_2 desorption profiles obtained for pre-reduced catalysts. All profiles are able to be deconvoluted into three Gaussian peaks, which could be assigned to the weak (α peak), moderate (β peak) and strong (γ peak) basic sites, respectively. The basicity of all samples is listed in Table 4. The density of total basic sites was also calculated by dividing the number of total basic sites by the BET specific surface areas. It can be seen from Table 4 that the enhancement of both the number and the density of total basic sites was significant with the introduction of fluorine. The increase can also be observed for the number of basic sites β and γ, which can be explained as follows:

For CHTs-CO_3, the weak basic sites were related to the OH^- groups, the moderate basic sites were ascribed to the metal-oxygen pairs and the strong basic sites were associated with the coordinatively unsaturated O^- ions [24]. In addition, upon the introduction of fluorine into HTs-F, the metal-fluorine pairs and the coordinatively unsaturated F^- ions were formed after calcination [14]. Similar to the mechanisms of basic sites that appeared in CHTs-CO_3, the weak, moderate and strong basic sites in CHTs-F might be attributed to the OH^- groups, metal-oxygen and metal-fluorine pairs, and coordinatively unsaturated O^- and F^- ions, respectively. Therefore, the formation of metal-fluorine pairs and the coordinatively unsaturated F^- ions can increase the amounts of the moderate and strong basic sites.

Moreover, the contributions of single basic site to the total basic sites were also listed in Table 4. It can be seen that the proportion of strong basic site γ to the total basic site increased remarkably with introduction of fluorine, while the change of the contribution of basic site β to the total basic sites was pretty small.

Catalytic performance

Table 5 The catalytic performance for CO_2 hydrogenation to methanol over the catalysts.

Sample	Temperature	CO_2 Conversion	TOF×10^3	Selectivity, (C-mol%)			CH_3OH Yield
	(°C)	(%)	(s^{-1})	CO	HC	MeOH	(g ml^{-1} h^{-1})
CHTs-F	230	14.3	5.60	35.0	1.1	63.9	0.12
	250	20.8	8.24	48.1	0.7	51.2	0.15
	270	25.8	10.00	54.7	0.5	44.8	0.17
$CHTs-CO_3$	230	15.5	5.46	45.1	1.1	53.9	0.11
	250	22.5	7.93	52.2	0.7	47.1	0.14
	270	26.6	9.22	56.6	0.6	42.9	0.16

Reaction conditions: P = 5.0 Mpa, GHSV = 4000 h^{-1}, $H_2:CO_2$ (atomic) = 3:1.

The catalytic performance of the two Cu/Zn/Al/Zr catalysts in the methanol synthesis from CO_2 hydrogenation is summarized in Table 5. Methanol, carbon monoxide and water were the main products under the reaction conditions. Trace amount of methane and higher hydrocarbons were also detected. As illustrated in Table 5, the conversion of CO_2 for CHTs-F was slightly lower than that for $CHTs-CO_3$ at 503-543 K. According to the bifunctional mechanism of CO_2 hydrogenation, the exposed copper surface area (S_{Cu}) is an important parameter for the catalytic performance of copper-based catalysts. Some researchers reported that the variation trend of CO_2 conversion of Cu-based catalysts is consistent with the change of S_{Cu} [23, 25-27]. In our study, the decrease of S_{Cu} was pronounced with introduction of fluorine, however, the conversion of CO_2 for CHTs-F was slightly lower than that for $CHTs-CO_3$. Moreover, in order to compare the catalytic efficiency produced by unit site of copper atom on the surface of the catalysts, the turnover frequency (TOF), which represents the number of CO_2-molecule hydrogenated on unit site of exposed copper atom per second (s^{-1}), had been calculated from the exposed copper surface area for various catalysts. Results from this study (see Table 5) showed that, the TOF for CHTs-F was higher than that for $CHTs-CO_3$. These results reveal that the activity of a catalyst is not only related to S_{Cu} but also to other causes. Behrens et al. reported that the high activity for CO_2 hydrogenation to methanol was generated by the presence of steps at the Cu surface which can be stabilized by bulk defects at the surface [28]. "Many defects" might form in the lattice of CuO with the introduction of fluorine into Cu/Zn/Al/Zr mixed oxides, which largely contributed to the activity, and similar results were proposed by other researchers [14, 29]. Therefore, the change of the activity is small with the introduction of fluorine, though the S_{Cu} is much lower, and the higher TOF for CHTs-F is easy to understand by considering this value calculated on the basis of the full exposed Cu surface area.

It is noteworthy that the CH_3OH selectivity for CHTs-F is markedly higher than that for $CHTs-CO_3$ sample. Based on the bifunctional mechanism of CO_2 hydrogenation, Guo et al.[10] reported that both methanol and CO are produced dominantly via the formate intermediate. Compared to the formate species adsorbed on moderate basic sites, the one adsorbed on strong basic sites prefers to be hydrogenated into methanol further to form methanol rather than dissociate to form CO. In our research, it is noticed that the methanol selectivity is affected by the distribution of basic sites. The proportion of strong basic site γ to the total basic site for CHTs-F was much higher than that for $CHTs-CO_3$, while the change of the contribution of basic site β to the total basic sites was pretty small according to the above analysis of CO_2-TPD. Therefore, the introduction of fluorine into Cu/Zn/Al/Zr mixed oxides sharply improved the strong basic sites, and then significantly increased the CH_3OH selectivity in the methanol synthesis from CO_2 hydrogenation.

Conclusions

The fluorine-modified Cu/Zn/Al/Zr mixed oxides were prepared from the reconstructed forms of HTlcs with fluorine anions in the interlayer region. For the HTs-F precursor, the defects on the edge of plates and compact aggregates were formed during reconstruction. With the introduction of fluorine, the exposed copper surface area decreased remarkably, and the density of total basic sites and the proportion of strong basic site to the total basic site increased significantly. The CHTs-F catalyst showed much higher CH_3OH selectivity than the corresponding pure CHTs-CO_3 catalyst in the methanol synthesis from CO_2 hydrogenation due to the much higher proportion of strong basic site. In addition, the CO_2 conversion is not only related to the exposed copper surface area, and the conversion of CO_2 decreased slightly with the introduction of fluorine, though copper surface area was much lower.

Acknowledgements

This work was supported by the Natural Science Foundation of China (20906099 and 50976116); Ministry of Science and Technology of the People's Republic of China (2009BWZ003); "Strategic Priority Research Program-Climate Change: Carbon Budget and Related Issues" of the Chinese Academy of Sciences, Grant No. XDA05010109, XDA05010110, XDA05010204.

Reference

1 G.A. Olah, A. Geoppert, G.K.S. Prakash, Beyond Oil and Gas: *The Methanol Economy, 1st ed.*, Wiley-VCH, Weinheim, 2006, pp. 173–187, 239–245.
2 X. An, J.L. Li, Y.Z. Zuo, Q. Zhang, D.Z. Wang, J.F. Wang, *Catal. Lett.* 118 (2007) 264.
3 D. Tichit, N. Das, B. Coq, R. Durand, *Chem. Mater.* 14 (2002) 1530.
4 Y.N. Jiao, W.G. Hou, *Colloid Surf., A* 296 (2007) 62.
5 L.H. Zhang, F. Li, D.G. Evans, X. Duan, *Mater. Chem. Phys.* 87 (2004) 402.
6 L.H. Zhang, C. Zheng, F. Li, D.G. Evans, X. Duan, *J. Mater. Sci.* 43 (2008) 237.
7 A. Alejandre, F. Medina, P. Salagre, X. Correig, J.E. Sueiras, *Chem. Mater.* 11 (1999) 939.
8 G. Busca, U. Costantino, F. Marmottini, T. Montanari, P. Patrono, F. Pinzari, G. Ramis, *Appl. Catal. A: Gen.* 310 (2006) 70.
9 Q.J. Hong, Z.P. Liu, *Surf. Sci.* 604 (2010) 1869.
10 X. Guo, D. Mao, G. Lu, S. Wang, G. Wu, *J. Mol. Catal. A: Chem.* 345 (2011) 60.
11 I.A. Fisher, H.C. Woo, A.T. Bell, *Catal. Lett.* 44 (1997) 11.
12 X.M. Liu, G.Q. Lu, Z.F. Yan, J. Beltramini, *Ind. Eng. Chem. Res.* 42 (2003) 6518.
13 G.D. Wu, X.L. Wang, W. Wei, Y.H. Sun, *Appl. Catal. A: Gen.* 377 (2010) 107.
14 G.D. Wu, X.L. Wang, B. Chen, J.P. Li, N. Zhao, W. Wei, Y.H. Sun, *Appl. Catal. A: Gen.* 329 (2007) 106.
15 Z.L. Yuan, L.N. Wang, J.H. Wang, S.X. Xia, P. Chen, Z.Y. Hou, X.M. Zheng, *Appl. Catal. B: Environ.* 101 (2011) 431.
16 E. Angelescu, O.D. Pavel, R. Birjega, M. Florea, R. Zavoianu, *Appl. Catal. A: Gen.* 341 (2008) 50.
17 J. Cheng, J.J. Yu, X.P. Wang, L.D. Li, J.J. Li, Z.P. Hao, *Energy Fuels* 22 (2008) 2131.
18 Z.P. Xu, H.C. Zeng, *J. Phys. Chem. B* 104 (2000) 10206.
19 R. Trujillano, M.J. Holgado, F. Pigazo, V. Rives, *Phys. B (Amsterdam, Neth.)* 373 (2006) 267.
20 M. Behrens, I. Kasatkin, S. Kuhl, G. Weinberg, *Chem. Mater.* 22 (2010) 386.
21 E. Lima, M.D. Martinez-Ortiz, R.I.G. Reyes, M. Vera, *Inorg. Chem.* 51 (2012) 7774.
22 K. Takehira, T. Kawabata, S. Shishido, K. Murakami, T. Ohi, D. Shoro, M. Honda, K. Takaki, *J. Catal.*231 (2005) 92.
23 X.M. Guo, D.S. Mao, G.Z. Lu, S. Wang, G.S. Wu, *J. Catal.*271 (2010) 178.
24 Y.X. Liu, K.P. Sun, H.W. Ma, X.L. Xu, X.L. Wang, *Catal. Commun.* 11 (2010) 880.
25 J. Sloczynski, R. Grabowski, P. Olszewski, A. Kozlowska, J. Stoch, M. Lachowska, J. Skrzypek, *Appl. Catal. A: Gen.* 310 (2006) 127.
26 M. Saito, T. Fujitani, M. Takeuchi, T. Watanabe, *Appl. Catal. A: Gen.* 138 (1996) 311.
27 G.C. Chinchen, P.J. Denny, D.G. Parker, M.S. Spencer, D.A. Whan, *Appl. Catal. A: Gen.* 30 (1987) 333.
28 M. Behrens, F. Studt, I. Kasatkin, S. Kuhl, M. Havecker, F. Abild-Pedersen, S. Zander, F. Girgsdies, P. Kurr, B.L. Kniep, M. Tovar, R.W. Fischer, J.K. Norskov, R. Schlogl, *Science* 336 (2012) 893.
29 R.J. Chimentao, S. Abello, F. Medina, J. Llorca, J.E. Sueiras, Y. Cesteros, P. Salagre, *J. Catal.* 252 (2007) 249.

DEHYDROGENATION OF ETHYLBENZENE WITH CO_2 OVER V_2O_5/Al_2O_3-ZrO_2 CATALYST

Shujiao Zhang, Taiyuan University of Technology, Taiyuan, PR China
Xiaohong Li, Taiyuan University of Technology, Taiyuan, PR China
Jie Feng, Taiyuan University of Technology, Taiyuan, PR China
Wenying Li[1]*, Taiyuan University of Technology, Taiyuan, PR China*
Hongxia Fan, Taiyuan University of Technology, Taiyuan, PR China
Qian Wang, Taiyuan University of Technology, Taiyuan, PR China

Abstract

V-containing catalysts supported on alumina, modified with different amounts of zirconium, were prepared by impregnation method. Dehydrogenation of ethylbenzene with CO_2 was runing over these catalysts in a fixed-bed-down flow stainless steel reactor for 8 hours. Results found that a small amount of zirconium into the support leads to a significant increase of the catalytic activity, but the addition of ZrO_2 is not the more the better. Based on the same loading of VO_x as the active component, the optimal addition amount of ZrO_2 is 10 wt% in Al_2O_3-ZrO_2 mixed oxide. X-ray photoelectron spectroscopy, thermogravimetric analysis and Raman spectroscopy were used for investigating the reasons for the catalytic activity decline of the catalysts. Partial reduction of vanadium oxides and a few active sites localized by the coke on the catalyst surface were the dominating.

Keywords

ethylbenzene, dehydrogenation, V_2O_5/Al_2O_3-ZrO_2 catalysts, carbon dioxide.

Introduction

Styrene is now commercially produced by the large-scale ethylbenzene dehydrogenation process with steam, which has inherent drawbacks, such as thermodynamically limited, energy consuming, higher reaction temperature (550-650 °C) and low ethylbenzene pressure [1, 2]. Since 1937, dehydrogenation of ethylbenzene to produce styrene was developed and applied into industry. Both material and energy consumption of this technique has reached to the limit for 70 years development. It is necessary to develop a new technology to improve ethylbenzene conversion and enhance thermal efficiency. An alternative method of using oxygen as an oxidant is found to overcome above-mentioned defects. However, this process has a number of safety issues and suffers from a

[1] Corresponding author. Tel. & Fax: +86 351 6018453.
E-mail addresses: zhangshujiao@tyut.edu.cn (ZhangSJ), lixiaohong@tyut.edu.cn (LiXH), fengjie@tyut.edu.cn (FengJ),ying@tyut.edu.cn (LiWY).

significant loss of styrene selectivity by producing substantial oxidation products [3, 4]. As a result, it does not make into reality so far. As a mild oxidant, CO_2 is considered to be a proper substance to replace O_2. Compared with H_2O, dehydrogenation of ethylbenzene to styrene with CO_2 can not only reduce energy consumption, but also provide both economic and environmental benefits, which conforms with the trend of the green chemistry [5, 6]. Meanwhile, CO_2 utilization provides several advantages in catalysis [7, 8]. It can accelerate the reaction rate, enhance produce selectivity, alleviate thermodynamic equilibrium limitations, suppress total reaction, prolong catalyst lifetime and prevent hot spots on the catalyst surface. However, the present commercial catalyst is ineffective in the presence of CO_2. The indispensable requirement is seeking for a suitable catalyst system.

Recently, several attempts are carried out in the process of ethylbenzene dehydrogenation to produce styrene with CO_2 (EDSC). It is well known that the type of the support has an important influence on the catalyst properties. In the course of exploring, Al_2O_3 is considered to be a proper support [9], and widely used when FeO_x-species [10-13] or VO_x-species [1, 14-16] as the active component. However, among these catalysts, the catalytic activity is much below to the thermodynamic equilibrium conversion. The catalytic property could be promoted by improving the support. Recently, ZrO_2 has received considerable attention in catalysis and material science [1, 5, 17, 18]. Its bifunctional properties of acid and base, reducing and oxidizing ability, and high thermal stability, are suit for the process of EDSC. In general, ZrO_2 has low surface area, more expensive price than many traditional oxide materials, what makes it unsuitable for industrial applications. In order to modulate the catalytic property better, mixing the pure oxides is a beneficial way to display each advantages of Al_2O_3 and ZrO_2.

In the present investigation, catalytic activity of individual oxides and Al_2O_3-ZrO_2 mixed oxides as supports in the process of EDSC has been evaluated. Based on the same loading of active component, the optimal amount of ZrO_2 in Al_2O_3-ZrO_2 support is also explored. The reason for the catalyst deactivation has also been delineated by X-ray photoelectron spectroscopy (XPS), Thermo-gravimetric (TG) analysis and Raman spectroscopy.

Experimental

Catalyst preparation description

Al_2O_3-ZrO_2 mixed oxide (ZrO_2 was 10-90 wt%) was prepared by precipitation method. The process was as follows: aluminum nitrate and zirconyl (IV) nitrate hydrate were first dissolved in de-ionized H_2O and stirred till the solution became homogeneous, then added aqueous ammonium solution (25%, diluted to 10%) until pH value of the solution was in the scope of 7 to 8. The white precipitate was separated by vacuum filtration and washed with de-ionized H_2O several times till nitrate ion was completely removed, then dried at 120 °C for 12 hours in an oven and finally torrefied at 550 °C for 6 hours in a muffle furnace. After that, VAZ (V_2O_5/Al_2O_3-ZrO_2) catalyst was prepared by impregnation of Al_2O_3-ZrO_2 mixed oxide with an oxalic acid solution of ammonium metavanadate. The impregnated samples were dried at 120 °C for 12 hours and torrefied in air at 550 °C for 6 hours. This catalyst hereinafter named V4AZx, 4 means the content of VOx in mmol unit per 1 g Al_2O_3-ZrO_2 carrier, x stands for ZrO_2 mass percentage in Al_2O_3-ZrO_2 carrier.

Catalyst characterization

The surface area and pore size distribution measurements were tested by nitrogen physical adsorption at 77 K using a Beijing JWGB JW-BK1212W instrument. The specific surface area of samples was analyzed by the BET method and pore size distributions were calculated on the basis of BJH method.

X-Ray power diffraction patterns of the samples were recorded on a Dandongfangyuan DX-2700 diffractometer with Ni filtered Cu Kα radiation operated at 40 kV and 30 mA in the 2θ range of 10-80 ° with a scan speed of 8 °/min.

Temperature-programmed reduction with hydrogen (H_2-TPR, 10 vol.% H_2 in argon) of the catalysts was performed from 60 to 800 °C with heating rate of 10 °C/min in a Micrometrics Autochem II 2920 apparatus, which equipped with a TCD detector for monitoring of the H_2 consumption.

X-ray photoelectron spectroscopy analysis was made on a VG (ESCALAB 250) spectrometer by using Al Kα radiation as the excitation source. Charging of catalyst samples was corrected by setting the binding energy of adventitious carbon (C 1s) at 284.6 eV.

Thermo-gravimetric analysis was detected by Germany NETZSCH (STA409C) instrument. Air flow rate was 100 ml/min. From the room temperature to 900 °C with the heating rate was 10 °C/min.

Raman spectra of selected spent catalysts were obtained using Renishaw inVia Raman Microscopy. A 514 nm laser was used as the exciting source on the sample surface and a power of 1 mW. The range of the wavenumbers varied from 100 to 4000 cm^{-1}.

Catalytic activity evaluation system

The catalytic tests were carried out in a fixed-bed-down flow stainless steel reactor (i.d. 8 mm and length 560 mm) using 0.40 g of catalyst at 550 °C under atmospheric pressure. The feed rate of ethylbenzene was 0.01 $mL{\cdot}min^{-1}$ and CO_2 was 18.3 $mL{\cdot}min^{-1}$ (the molar ratio of CO_2 and ethylbenzene was 10). Gaseous (H_2, CO, CH_4 and CO_2) and liquid (benzene, toluene, ethylbenzene and styrene) products were analyzed simultaneously by on-line gas chromatography (GC, Tianmei Instrument, 7890 II, China) equipped with TCD and FID detectors.

Results and discussion

Catalytic activity of VAZ catalyst in the EDSC process

The catalytic activity and product selectivity results (1 hour on stream) for ethylbenzene dehydrogenation to styrene with CO_2 (EDSC) over different catalysts (single and AZ50 support) are presented in Figure 1. Based on the similar styrene selectivity (≥97%), V4AZ50 (38%) catalytic activity is obvious higher than V4AZ0 (23%) and V4AZ100 (13%). This proves that compared with the single catalyst, supported catalyst has absolute dominance. It may be due to the interaction in the mixed oxide support. It is well known that mixed oxide support has some characteristics which is impossible for the single support, such as [19]: (i) improvement of the dispersion of the active compound or creation the new active site; (ii) modification of the reducibility of the oxide precursors through change of the interaction between the active phase and support and (iii) reduction the deactivation owing to the coke formation. Moreover, in other reaction system, Al_2O_3-ZrO_2 supported catalysts also exhibit better catalytic properties than the catalysts supported on pure Al_2O_3 or ZrO_2 [20, 21]. Therefore, the research on the optimal ratio of ZrO_2 in the mixed oxide support is meaningful for the catalytic activity in the EDSC process.

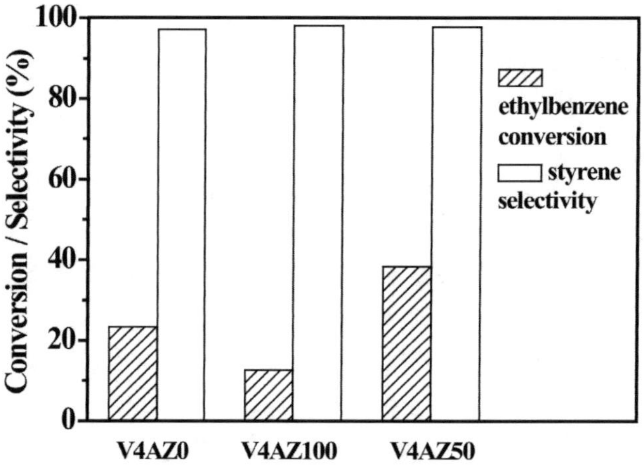

Figure 1 ethylbenzene conversion and styrene selectivity with different catalysts.

The catalytic activity (1 hour on stream) with different ZrO_2 content is listed in Figure 2. From the graph, it can be seen that all catalysts yield very high selectivity towards styrene (≥96%), but the ethylbenzene conversion differs greatly. It is obvious to see that V4AZ10 has the highest catalytic activity. In other words, the proportion of ZrO_2 content with 10 wt% in Al_2O_3-ZrO_2 mixed oxide is the most suitable. Table 1 displays the effect of catalyst stability according to different ZrO_2 mass percentage in Al_2O_3-ZrO_2 mixed oxide. As the reaction proceeds, catalytic activity drops by different degrees. Meanwhile, high styrene selectivity keeps above 96% during the reaction. Since ethylbenzene conversion declines, selectivity to styrene over all V_2O_5/Al_2O_3-ZrO_2 catalysts are slightly increased relative to their initial styrene selectivity (Fig. 2), so it does not list in Table 1. The addition of ZrO_2 into V_2O_5/Al_2O_3 catalyst increases the initial styrene yield significantly, but has almost no performance to its stability.

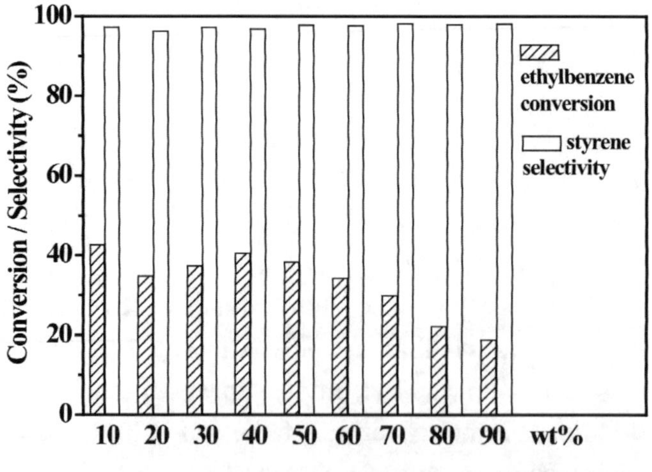

Figure 2 Effect of different ZrO_2 content on the activity of V4AZ catalyst.

Table 1 Variation of catalytic activity and selectivity with time-on-stream

Time	Ethylbenzene conversion (%)								
(hour)	V4AZ10	V4AZ20	V4AZ30	V4AZ40	V4AZ50	V4AZ60	V4AZ70	V4AZ80	V4AZ90
1	42.6	34.8	37.3	40.4	38.3	34.2	29.9	22.1	18.8
2	41.2	34.0	33.2	33.1	30.2	28.3	25.2	18.7	16.7
3	38.2	29.9	29.7	28.0	24.9	21.2	19.3	17.6	13.4
4	32.2	24.2	23.4	21.5	20.6	19.2	15.3	17.8	12.2
5	32.5	19.2	16.4	18.5	18.7	17.9	17.8	16.2	11.1
6	31.4	17.1	16.5	16.4	17.5	17.7	17.0	15.7	10.6
7	29.2	12.3	14.6	15.2	16.9	16.7	165	15.1	9.6
8	27.6	13.1	13.9	15.0	17.2	16.7	16.1	13.8	9.4

Investigation on textural properties of VAZ catalyst

The specific surface area and pore volume of single (Al_2O_3 and ZrO_2) and mixed (AZ10-AZ90) supports together with fresh and used catalysts (V4AZ10-V4AZ90) is listed in Table 2. The specific surface area between Al_2O_3 and ZrO_2 differs greatly, but all composite supports exhibit reasonable large specific surface area in the range 161-218 m^2/g, which means that it may increase the contact between reactants and catalyst, as well as disperse more active VO_x-species or distribute active VO_x-species more uniformly on its surface. It fits in with what we need. With loading the active component, the specific surface area and pore volume of all samples has decreased. It seems to be that the deposited active oxides penetrate into pores of the support, which causes the narrow of its pore diameter and the blockage of some microspores, or the solid-state reaction between the supporting oxides and the dispersed active oxide. After using for 8 hours, the specific surface area and pore volume of all samples lower again. The reason may be interpreted for the carbon deposition on the surface of the catalyst. This would explain the difference of color for catalyst from pale yellow to black after the reaction. It will be further explored in the following section.

Table 2 Nitrogen adsorption characteristics of various supports and catalysts

Samples	S_{BET} (m^2/g)	Pore volume (cm^3/g)	Samples	S_{BET} (m^2/g)	Pore volume (cm^3/g)	Samples	S_{BET} (m^2/g)	Pore volume (cm^3/g)
Al_2O_3	197	0.29	V4AZ0-fresh	108	0.18	V4AZ0-used	63	0.10
ZrO_2	64	0.25	V4AZ100-fresh	69	0.21	V4AZ100-used	63	0.19
AZ10	185	0.39	V4AZ10-fresh	114	0.25	V4AZ10-used	74	0.15
AZ20	165	0.26	V4AZ20-fresh	133	0.26	V4AZ20-used	61	0.11
AZ30	198	0.32	V4AZ30-fresh	115	0.23	V4AZ30-used	71	0.13
AZ40	161	0.28	V4AZ40-fresh	119	0.25	V4AZ40-used	74	0.13
AZ50	185	0.39	V4AZ50-fresh	101	0.22	V4AZ50-used	80	0.15
AZ60	167	0.32	V4AZ60-fresh	110	0.23	V4AZ60-used	64	0.11
AZ70	218	0.37	V4AZ70-fresh	80	0.20	V4AZ70-used	67	0.12

AZ80	186	0.24	V4AZ80-fresh	95	0.16	V4AZ80-used	45	0.06
AZ90	200	0.22	V4AZ90-fresh	96	0.12	V4AZ90-used	16	0.03

Crystal phase modification of VAZ catalyst by the addition of ZrO_2

According to Figure 3, with exception of pure ZrO_2 in monocline phase to a great extent, all samples are in X-ray amorphous state. For single and mixed supports, when ZrO_2 proportion is less than 30 wt% in Al_2O_3-ZrO_2 mixed oxide, its crystal phase tends to pure Al_2O_3. With ZrO_2 additive increasing, the mixed supports transfer to show ZrO_2 crystal phase gradually (not included here). On the whole, Al_2O_3-ZrO_2 mixed oxides display amorphous state. For the catalyst V4AZ10, without displaying active component characteristic peaks, only weak or broad lines of ZrO_2 are found in the XRD profile. The catalysts spectra with other ZrO_2 contents are similar to V4AZ10. It is illustrated that the active component distributed homogeneously on the support or the crystallites formed are less than XRD technique's detection. After using for 8 hours, samples' XRD pattern almost keep constant.

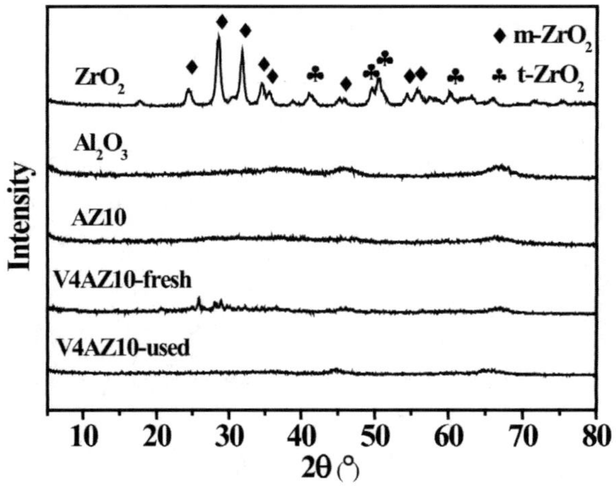

Figure 3 XRD patterns of single and mixed support with V4AZ10 catalyst.
Notation: m-ZrO_2, monocline ZrO_2 phase; t- ZrO_2, tetragonal ZrO_2 phase.

Analysis of interaction between the support and the active component in VAZ catalyst by TPR

To assess the reducibility of the dispersed compounds and interaction between support and the active component, temperature programming reduction of H_2 (H_2-TPR) is performed. The H_2-TPR profiles of various samples calcined at 550 °C are displayed in Figure 4. It can be seen that the support of AZ10 does not have the reducibility. This clearly shows that the AZ10 mixed oxide is quite stable at the reaction temperature. After impregnating the active component, samples' TPR patterns reveal different information about the relative reducibly. The V4AZ0 exhibits two reduction peaks at 454 and 556 °C respectively, meanwhile, V4AZ100 only has one reduction peak at 467 °C. Compared to them, V4AZ10 has the characteristics of these two. In general [22, 23], pure V_2O_5 is reduced by hydrogen in three characteristic steps. These three steps of the reduction process of V_2O_5 is described as $V_2O_5 \rightarrow V_6O_{13} \rightarrow VO_2 \rightarrow V_2O_3$. However, the TPR profiles of the vanadium oxides contained catalysts exhibits peaks at lower temperatures (Fig. 4) compared to the pure V_2O_5, indicating higher reducibility of all catalysts due to the influence of the AZ support. The V4AZ10 reduction profile exhibits a broad peak near 460 °C can be attributed to the reduction of surface vanadium oxide [24]. The other peak at 580 °C is assigned to the reduction of bulk-like species, which may exist in a highly dispersed state on the support surface [25, 26], they may also be due to the strong interaction between active compound and support.

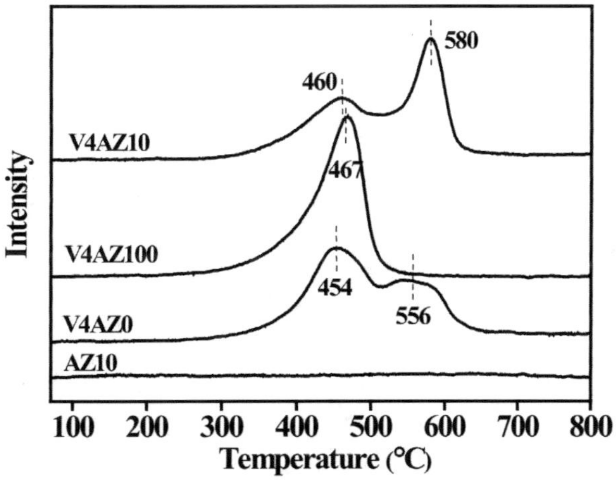

Figure 4 H_2-TPR profiles of AZ10, V4AZ0, V4AZ10 and V4AZ100.

Effect of vanadium valence state changes on VAZ catalytic activity by XPS analysis

To investigate the difference of the surface composition in V4AZ10 between fresh and used, XPS measurement is performed and the corresponding binding energy values of each element is listed in Table 3. Based on the standard spectrum database of X-ray photoelectron spectroscopy, the binding energy of Al_{2p} in pure Al_2O_3 and Zr_{3d} in pure ZrO_2 is 74.7 eV (CAS 10043-01-3) and 182.9 eV (CAS 14644-61-2), respectively. From Table 3, it can be seen that in the fresh V4AZ10, the binding energy of Al_{2p} and Zr_{3d} is 74.2 and 182.2 eV, which changes little comparing with the pure oxide. This may be due to existing as the large number element in the fresh V4AZ10, whereas the slight variation indicates the interaction between Al_2O_3 and ZrO_2. It may interpret the variation of catalyst activity with single oxide support (Fig. 1).

Table 3 XPS analysis of V4AZ10 between fresh and used

Samples	V_{2p} (eV)		O_{1s} (eV)		Al_{2p} (eV)		Zr_{3d} (eV)	
	BE	FWHM	BE	FWHM	BE	FWHM	BE	FWHM
V4AZ10-fresh	517.9	1.77	531.2	2.42	74.2	1.57	182.2	1.67
V4AZ10-used	517.1	1.60	530.5	2.12	74.1	1.47	182.3	1.25

Notation: BE, binding energy; FWHM, full width at half maximum.

For the fresh V4AZ10 catalyst, its binding energy for V_{2p} is mostly corresponded to V^{5+}, indicating that the fresh catalyst possesses the highest oxidation state of V. Meanwhile, the peak position and the full width at half maximum of V_{2p} for the used V4AZ10 are decreased relative to the fresh one, from 517.9 to 517.1 eV and 1.88 to 1.73 eV, respectively. This phenomenon reveals the reduction of surface vanadium species during the reaction. To explain the change of V_{2p} before and after the reaction, it is fitted for V^{5+}, V^{4+} and V^{3+} components using the deconvolution method. The result is displayed in Figure 5. It points out that a less half of vanadium species changes into the low valence. As we all know, the high-valent vanadium species is the active center in the EDSC reaction. The declination in the amount of V^{5+} may be one of the reasons for the catalytic activity decreasing.

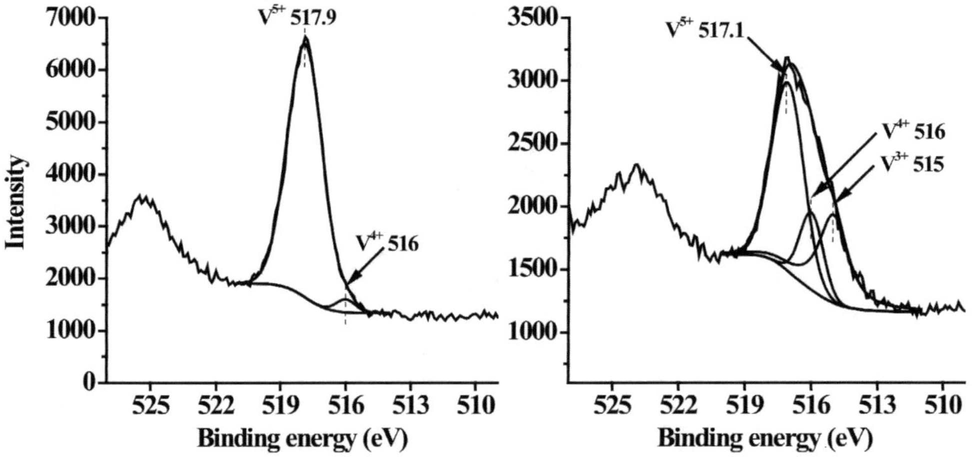

Figure 5 V_{2p} XPS profiles in V4AZ10 before and after the reaction

Analysis of carbon deposition over VAZ catalyst after the EDSC process

As the reaction proceeds, catalytic activity decreases significantly (Table 1). Besides the reduction of active center, coke formation may be the other reason. In the used V4AZ10 after 8 hours on stream, the amount of carbon deposit is measured about 4.2 wt%, which can be seen in Figure 6. It will be an interpretation for the specific surface area and pore volume of V4AZ10 decreasing after the reaction. In addition, during the re-oxidation test, this used catalyst treats in an air stream at 550 °C for 30 min, then changes gas circuit to its reaction condition. It is found that its activity is recovered after re-oxidation.

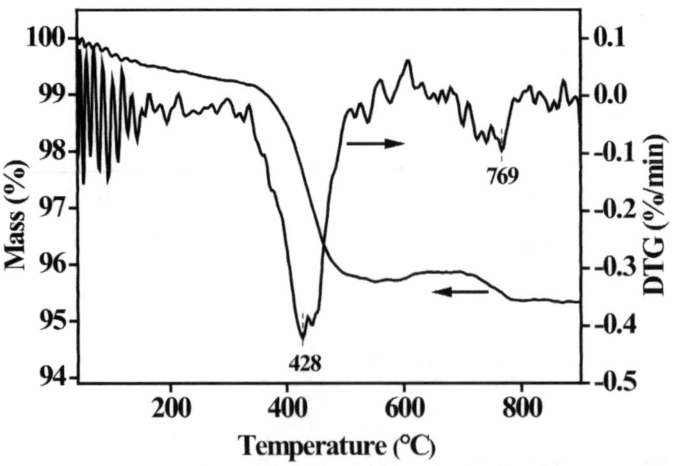

Figures 6 TG-DTG profile of used V4AZ10 catalyst.

The Raman spectra of various samples after used for 8 hours are presented in Figure 7. In this graph, it can be seen that all of the three peak positions are the same for each catalyst. It is known that

[27, 28] D and G bands are belonged to disorder-induced planes of carbon and graphite, respectively, are attributed to the bands at 1351 cm^{-1} and 1577 cm^{-1}. The spectra of spent catalysts has the similar intensity between the D and G bands reveal the presence of disordered carbon species (1339.3 cm^{-1}) as well as graphite (1599.9 cm^{-1}). This result indicates the nature of the carbon and reveals the disordered and graphite carbon deposition from the process of EDSC, in agreement with TG analysis of spent V4AZ10 catalyst. The G-band suggests the occurrence of a hexagonal lattice of graphite carbon [29, 30]. Graphitic species could be formed from CO decomposition, which have been observed in reforming reactions [31]. The third Raman band appears at 2865.9 cm^{-1}. Li [32] has used Ultraviolet Raman (UV-Raman) spectroscopy to characterize coke formation in ZSM-5 and USY zeolites under propene at temperatures from 300 K to 773 K. One of the band in the 3000 cm^{-1} region is assigned to the stretching vibrations of C-H bonds of adsorbed olefin, it may be symmetric and asymmetric stretching vibrations of CH group [33]. Besides these, this band will be assigned to vibration of CH_2 aliphatic group from styrene, which indicates that this monomer may be the coke precursor [10].

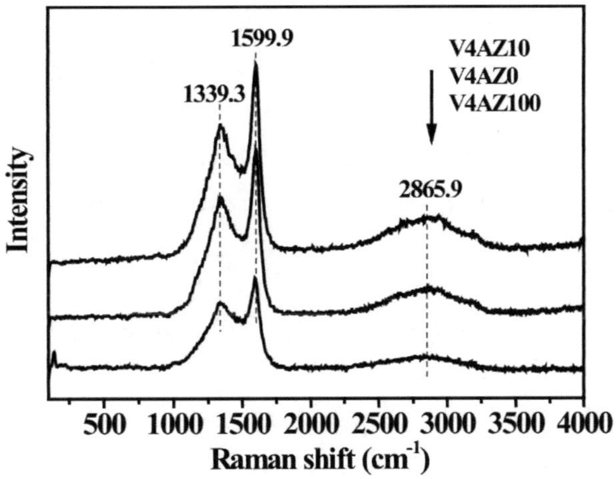

Figure 7 Raman spectra of V4AZ10, V4AZ0 and V4AZ100 samples.

Conclusions

Comparing with V4AZ0 and V4AZ100, V4AZ10 displays a significant increase of its catalytic activity and selectivity in the process of ethylbenzene dehydrogenation to styrene with CO_2 (EDSC), which means that Al_2O_3-ZrO_2 mixed oxide support is proper for this reaction. Based on the analysis of XPS, TG and Raman, causes for deactivation may be two aspects: (1) partial reduction of vanadium oxidation from high valence to the low. (2) carbon deposition covers a few active center on catalyst surface.

Acknowledgements

The authors gratefully acknowledge the financial support from NNSF of China (No. 20846002), Program for Changjiang Scholars (2009) in Ministry of Education, P. R. China, Specialized Research

Fund for the Doctoral Program of Higher Education (No. 20091402120013) and Shanxi Scholarship Council of China.

References

1. Rao, K. N.; Reddy, B. M.; Abhishek, B.; Seo, Y.-H.; Jiang, N. and Park, S.-E. (2009), "Effect of Ceria on the Structure and Catalytic Activity of V_2O_5/TiO_2-ZrO_2 for Oxidehydrogenation of Ethylbenzene to Styrene Utilizing CO_2 as Soft Oxidant", *Applied Catalysis B: Environmental*, July, pp. 649-656.
2. Cai, W.; Li, H. and Zhang, Y. (2004), "Selective Oxidation of Ethylbenzene to Styrene with Carbon Dioxide", *Progress in Chemistry***,** May, pp. 406-413.
3. Cavani, F. and Trifirò, F. (1995), "Alternative Processes for the Production of Styrene", *Applied Catalysis A: General*, September, pp. 219-239.
4. Hong, D.-Y.; Vislovskiy, V. P.; Hwang, Y. K.; Jhung, S. H. and Chang, J.-S. (2008), "Dehydrogenation of Ethylbenzene with Carbon Dioxide over MgO-modified Al_2O_3-supported V-Sb Oxide Catalysts", *Catalysis Today*, November, pp. 140-145.
5. Reddy, B. M.; Han, D.-S.; Jiang, N. and Park, S.-E. (2008), "Dehydrogenation of Ethylbenzene to Styrene with Carbon Dioxide Over ZrO_2-based Composite Oxide Catalysts", *Catalysis Surveys from Asia*, January, pp. 56-69.
6. Zhang, S.; Li, X.; Feng, J.; Li, W.; Wang, Q. and Fan, H. (2012), "Ethylbenzene Dehydrogenation with CO_2 over V_2O_5-Fe_2O_3/Al_2O_3-TiO_2 Catalyst", *Advanced Materials Research*, July, pp. 301-305.
7. Burri, D. R.; Choi, K. M.; Han, D.-S.; Koo, J.-B. and Park, S.-E. (2006), "CO_2 Utilization as an Oxidant in the Dehydrogenation of Ethylbenzene to Styrene over MnO_2-ZrO_2 Catalysts", *Catalysis Today*, April, pp. 242-247.
8. Hong, D.-Y.; P.Vislovskiy, V.; Park, S.-E.; Park, M.-S.; S.Yoo, J. and Chang, J.-S. (2005), "Dehydrogenation of Ethylbenzene with Carbon Dioxide as Soft Oxidant over Supported Vanadium-Antimony Oxide Catalyst", *Bulletin of the Korean Chemical Society*, May, pp. 1743-1748.
9. Sato, S.; Ohhara, M.; Sodesawa, T. and Nozaki, F. (1988), "Combination of Ethylbenzene Dehydrogenation and Carbon Dioxide Shift-reaction over a Sodium Oxide/Alumina Catalyst", *Applied Catalysis*, October, pp. 207-215.
10. Nogueira, I. M.; Sabadia, G. Q.; Moreira, A. A.; Filho, J. M. and Oliveira, A. C. (2011), "Investigation of the Deactivation of Iron Nanocomposites by Coking in the Dehydrogenation of Ethylbenzene", *Journal of Molecular Catalysis A: Chemical*, September, pp. 81-92.
11. Tope, B. B.; Balasamy, R. J.; Khurshid, A.; Atanda, L. A.; Yahiro, H.; Shishido, T.; Takehira, K. and Al-Khattaf, S. S. (2011), "Catalytic Mechanism of the Dehydrogenation of Ethylbenzene over Fe-Co/Mg(Al)O Derived from Hydrotalcites", *Applied Catalysis A: General*, August. pp. 118-126.
12. Mimura, N. and Saito, M. (2000), "Dehydrogenation of Ethylbenzene to Styrene over Fe_2O_3/Al_2O_3 Catalysts in the Presence of Carbon Dioxide", *Catalysis Today*, January, pp.173-178.
13. Saito, M.; Kimura, H.; Mimura, N.; Wu, J. and Murata, K. (2003), "Dehydrogenation of Ethylbenzene in the Presence of CO_2 over an Alumina-supported Iron Oxide Catalyst", *Applied*

Catalysis A: General, June, pp. 71-77.

14. Sun, A.; Qin, Z.; Chen, S. and Wang, J. (2004), "Ethylbenzene Dehydrogenation in the Presence of Carbon Dioxide over Alumina Supported Catalysts", *Catalysis Today*, July, pp. 273-279.
15. Reddy, B. M.; Lee, S.-C.; Han, D.-S. and Park, S.-E. (2009), "Utilization of Carbon Dioxide as Soft Oxidant for Oxydehydrogenation of Ethylbenzene to Styrene over V_2O_5-CeO_2/TiO_2-ZrO_2 Catalyst", *Applied Catalysis B: Environmental*, September, pp. 230-238.
16. Vislovskiy, V. P.; Chang, J.-S.; Park, M.-S. and Park, S.-E. (2002), "Ethylbenzene into Styrene with Carbon Dioxide over Modified Vanadia-alumina Catalysts", *Catalysis Communications*, April, pp. 227-231.
17. Burri, D. R.; Choi, K.-M.; Han, S.-C.; Burri, A. and Park, S.-E. (2007), "Selective Conversion of Ethylbenzene into Styrene over K_2O/TiO_2-ZrO_2 Catalysts: Unified Effects of K_2O and CO_2", *Journal of Molecular Catalysis A: Chemical*, December, pp. 58-63.
18. Park, J.-N.; Noh, J.; Chang, J.-S. and Park, S.-E. (2000), "Ethylbenzene to Styrene in the Presence of Carbon Dioxide over Zirconia", *Catalysis Letters*, January, pp. 75-78.
19. Damyanova, S.; Petrov, L.; Centeno, M. A. and Grange, P. (2002), "Characterization of Molybdenum Hydrodesulfurization Catalysts Supported on ZrO_2-Al_2O_3 and ZrO_2-SiO_2 Carriers", *Applied Catalysis A: General*, August. pp. 271-284.
20. Chary, K. V. R.; Praveen Kumar, C.; Venkat Ramana Rao, P. and Venkat Rao, V. (2004), "Dispersion and Reactivity of V_2O_5 Catalysts Supported on Al_2O_3-ZrO_2", *Catalysis Communications*, July, pp. 479-484.
21. Murakami, Y.; Niwa, M.; Hattori, T.; Osawa, S.-i.; Igushi, I. and Ando, H. (1977), "Reaction Mechanism of Ammoxidation of Toluene : I. Kinetic Studies of Reaction Steps by Pulse and Flow Techniques", *Journal of Catalysis*, July, pp. 83-91.
22. Bosch, H.; Kip, B. J.; Ommen, J. G. v. and Gellings, P. J. (1984), "Factors Influencing the Temperature-programmed Reduction Profiles of Vanadium Pentoxide", *Journal of the Chemical Society, Faraday Trans. 1*, November, pp. 2479-2488.
23. Boningari, T.; Koirala, R. and Smirniotis, P. G. (2012), "Low-temperature Selective Catalytic Reduction of NO with NH_3 over V/ZrO_2 Prepared by Flame-assisted Spray Pyrolysis: Structural and Catalytic Properties", *Applied Catalysis B: Environmental*, October, pp. 255-264.
24. Lewandowska, A. E. and Bañares, M. A. (2006), "In Situ TPR/TPO-Raman Studies of Dispersed and Nano-scaled Mixed V-Nb Oxides on Alumina", *Catalysis Today*, August, pp. 323-331.
25. Reddy, E. P. and Varma, R. S. (2004), "Preparation, Characterization, and Activity of Al_2O_3-supported V_2O_5 Catalysts", *Journal of Catalysis*, July, pp. 93-101.
26. Shiju, N. R.; Anilkumar, M.; Mirajkar, S. P.; Gopinath, C. S.; Rao, B. S. and Satyanarayana, C. V. (2005), "Oxidative Dehydrogenation of Ethylbenzene over Vanadia-alumina Catalysts in the Presence of Nitrous Oxide: Structure-activity Relationship", *Journal of Catalysis*, January, pp. 484-492.
27. Pinheiro, A. L.; Pinheiro, A. N.; Valentini, A.; Filho, J. M.; Sousa, F. F. d.; Sousa, J. R. d.; Rocha, M. d. G. C.; Bargiela, P. and Oliveira, A. C. (2009), "Analysis of Coke Deposition and Study of the Structural Features of MAl_2O_4 Catalysts for the Dry Reforming of Methane", *Catalysis Communications*, August, pp. 11-14.
28. Pinheiro, A. N.; Valentini, A.; Sasaki, J. M. and Oliveira, A. C. (2009), "Highly Stable Dealuminated Zeolite Support for the Production of Hydrogen by Dry Reforming of Methane",

Applied Catalysis A: General, December, pp. 156-168.

29. Batista, A. H. d. M.; Ramos, F. S. O.; Braga, T. P.; Lima, C. L.; de Sousa, F. F.; Barros, E. B. D.; Filho, J. M.; Oliveira, A. S. d.; Sousa, J. R. d.; Valentini, A. and Oliveira, A. C. (2010), "Mesoporous MAl_2O_4 (M=Cu, Ni, Fe or Mg) Spinels: Characterisation and Application in the Catalytic Dehydrogenation of Ethylbenzene in the Presence of CO_2", *Applied Catalysis A: General*, April, pp. 148-157.
30. Batista, A. H. d. M.; Sousa, F. F. d.; Honorato, S. B.; Ayala, A. P.; Filho, J. M.; Sousa, F. W. d.; Pinheiro, A. N.; Araujo, J. C. S. d.; Nascimento, R. F.; Valentini, A. and Oliveira, A. C. (2010), "Ethylbenzene to chemicals: Catalytic Conversion of Ethylbenzene into Styrene over Metal-containing MCM-41", *Journal of Molecular Catalysis A: Chemical*, September, pp.86-98.
31. San-José-Alonso, D.; Juan-Juan, J.; Illán-Gómez, M. J. and Román-Martínez, M. C. (2009), "Ni, Co and Bimetallic Ni-Co Catalysts for the Dry Reforming of Methane", *Applied Catalysis A: General*, September, pp. 54-59.
32. Li, C. and Stair, P. C. (1997), "Ultraviolet Raman Spectroscopy Characterization of Coke Formation in Zeolites", *Catalysis Today*, January, pp. 353-360.
33. Li, J.; Xiong, G.; Feng, Z.; Liu, Z.; Xin, Q. and Li, C. (2000), "Coke Formation During the Methanol Conversion to Olefins in Zeolites Studied by UV Raman Spectroscopy", *Microporous and Mesoporous Materials*, March, pp. 275-280.

(623e) A Comparative Study of Activated CO2 Adsorption and Protonation On the γ-Al2O3, β-Ga2O3, In2O3 Surfaces

Jingyun Ye[1,2], Changjun Liu[3] and Qingfeng Ge[1], (1)Chemistry and Biochemistry, Southern Illinois University Carbondale, Carbondale, IL, (2)Chemical Engineering and Technology , Tianjin University, Tianjin, China, (3)Chemical Engineering and Technology, Tianjin University, Tianjin, China

A Comparative Study of Activated CO_2 Adsorption and Protonation on the γ-Al_2O_3, β-Ga_2O_3, In_2O_3 Surfaces

Jingyun Ye,[1,2] Changjun Liu[1] and Qingfeng Ge[2,*]

[1] *School of Chemical Engineering and Technology, Tianjin University, Tianjin, 300072, China,*

[2] *Department of Chemistry and Biochemistry, Southern Illinois University, Carbondale, IL,62901, USA.*

Emails: yejingyun@siu.edu(Ye); ughg_cjl@yahoo.com(Liu); qge@chem.siu.edu(Ge).

Increased consumption of fossil fuels due to economic development and global population growth has resulted in a significant increase in anthropogenic CO_2. Carbon capture and storage have been suggested as the ultimate solution to mitigate the global CO_2 emission. On the other hand, catalytically recycling CO_2 into liquid fuels and valuable chemicals will help to reduce the net CO_2 emissions into the atmosphere, and thereby alleviate the greenhouse effect caused by CO_2. These processes include CO_2 reforming of methane to produce synthesis gas ($CO + H_2$)[1,2] and CO_2 hydrogenation to produce methanol.[3,4] Adsorption of CO_2 on the catalysts is the first step in catalytic CO_2 conversion and plays an important role in the subsequent conversion of CO_2. Herein, we first studied and compared the activated adsorption of CO_2 from the gas phase and the physisorbed precursor initial states on the γ-Al_2O_3, β-Ga_2O_3, In_2O_3 surfaces. Our results show that activation barriers and adsorption energies to form chemisorbed CO_2 from the gas phase or physisorbed precursor state are similar by the vertical and parallel CO_2 adsorption models. The stability of the activated CO_2 on these oxide surfaces follows: $In_2O_3 > \gamma$-$Al_2O_3 > \beta$-Ga_2O_3. We also examined the effect of water on CO_2 adsorption and activation on these oxide surfaces as water always exists, either in the reaction environment or as a reaction product. For co-adsorbed CO_2 and H_2O, we found them to be attractive on γ-Al_2O_3 but repulsive on β-Ga_2O_3 and In_2O_3, as reflected in the overall adsorption energy. We also find that CO_2 protonation on γ-Al_2O_3 is more facile than on β-Ga_2O_3 and In_2O_3. We anticipate that the different chemistries of H_2O on these oxide surfaces have important implication to the reactivity and selectivity of CO_2 hydrogenation reaction on the catalysts involving these oxides either as support or promoter.

1. Song, C. *Catalysis Today* **2006**, *115*, 2.

2. Ruckenstein, E.; Wang, H. Y. *Journal of Catalysis* **2002**, *205*, 289.

3. Arena, F.; Barbera, K.; Italiano, G. *J. Catal.* **2007**, *249*, 185.

4. Zhao, Y.; Yang, Y.; Mims, C. A.; Peden, C. H.; Li, J.; Mei, D. *J. Catal.* **2011**, *281*, 199.

(623f) C-H Activating Carboxylation of Terminal Alkynes with CO2 and Its Application in Synthesis

Yugen Zhang[1], Jackie Y. Ying[2], Dingyi Yu[3] and Siti Riduan[3], (1)green synthesis, IBN, singapore, Singapore, (2)Institute of Bioengineering and Nanotechnology, Singapore, Singapore, (3)IBN, singapore, Singapore

C-H Activating Carboxylation of Terminal Alkynes with CO_2 and Its Application in Synthesis

Yugen Zhang, Dingyi Yu, Siti Nurhanna Riduan, and Jackie Y. Ying

Institute of Bioengineering and Nanotechnology

31 Biopolis Way, The Nanos, Singapore 138669, Singapore

E-mail: ygzhang@ibn.a-star.edu.sg

Carbon dioxide (CO_2) is a non-toxic, non-combustible, non-flammable gas. The use of carbon dioxide as a renewable and environmentally friendly source of carbon in organic synthesis is a highly attractive approach, but its real world application remains a great challenge. The major obstacles for commercialization of most current protocols are their poor catalytic performance and harsh, severe reaction conditions. It is important to develop new reactions and new protocols for CO_2 transformations at mild conditions and in a cost-efficient manner. This presentation will discuss a new way for transforming CO_2 to carboxylic acid through C-H bond activation of terminal alkynes (Scheme 1). Various propiolic acids have been synthesized in good to excellent yields under ambient conditions without consumption of any organometallic or organic reagent additives. Different approaches have been developed for this transformation. A copper-catalyzed and copper-N-heterocyclic carbene co-catalyzed C-H activating carboxylation of terminal alkynes with CO_2 has led to the formation of propiolic acids at ambient conditions. A transition metal-free system and a recyclable heterogeneous catalyst system have also been developed for this reaction. Furthermore, the application of this reaction in the stereoselective coupling of terminal alkynes and thiols under carbon dioxide atmosphere will be presented (Scheme 2).

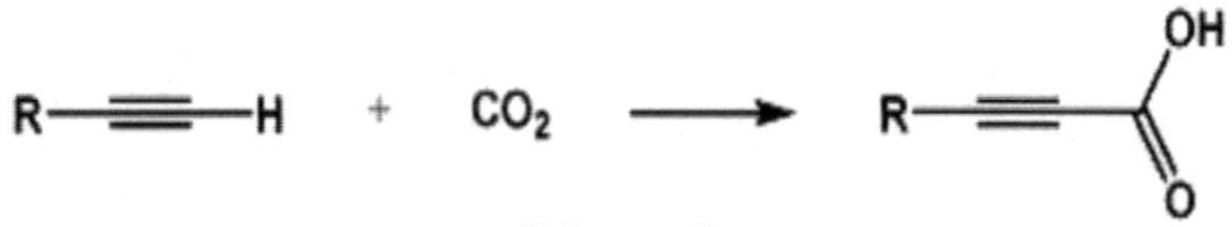

Scheme 1

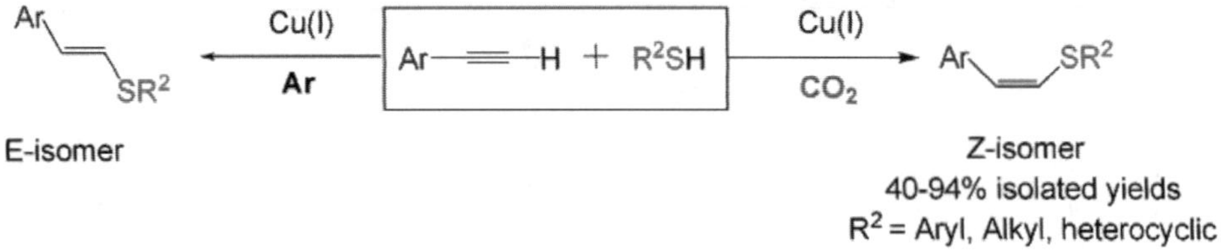

Scheme 2

[1] D. Yu and Y. Zhang, *PNAS*, **2010**, *107*, 20184.

[2] D. Yu and Y. Zhang, *Green Chem.*, **2011**, *13*, 1275.

[3] D. Yu, M. Tan and Y. Zhang, *Adv. Synth. Catal.*, **2012**, *354*, 969.

[4] S. N. Riduan, J. Y. Ying and Y. Zhang, *Org. Lett.* **2012**, *14*, 1780.

[5] S. N. Riduan and Y. G. Zhang, *Dalton Trans.* **2010**, *39*, 3347.

American Institute of Chemical Engineers Annual Meeting, Symposium on 'Chemical Production from CO_2,' TDB03, Pittsburgh, Pennsylvania, USA, October 28 November 2, 2012.

(685a) Photochemical Reduction of CO_2 Using Delafossite Oxides

Jonathan W. Lekse[1], James, P. Lewis[2], M. Kylee Underwood[3] and Christopher Matranga[1], (1)U.S. Department of Energy, National Energy Technology Laboratory, Pittsburgh, PA, (2)Department of Physics, West Virginia University, Morgantown, WV, (3)Physics, West Virginia University, Morgantown, WV

The photochemical reduction of CO_2 in the presence of H_2O to form CO, CH_4 and other light gases with industrial value is an interesting approach for dealing with CO_2 emissions. This approach can generate tangible revenue to help offset carbon capture and storage costs by generating a product stream with industrial demand from a CO_2 feedstock. Delafossite materials of the general stoichiometry ABO_2 are a new class of photocatalysts being considered for this application. Recent theoretical calculations have indicated that B-site alloying in these systems breaks the inversion symmetry of the crystal giving rise to symmetry forbidden optical transitions across the band structure of the material. B-site alloying can also be used to modulate the delafossite band structure to create new, low energy band gaps, as well as to align band edge positions for the photochemical redox reactions needed specifically for CO_2 applications. The photochemical activity of $CuAlO_2$, $CuAl_{0.9}Fe_{0.1}O_2$, $CuGaO_2$ and $CuGa_{1-x}Fe_xO_2$ (x=0.05, 0.1, 0.15, 0.2) for the reduction of CO_2 will be presented. The photoreduction of CO_2 in the presence of H_2O vapor using $CuAlO_2$, $CuAl_{0.9}Fe_{0.1}O_2$, $CuGaO_2$ and $CuGa_{1-x}Fe_xO_2$ produces CO with little evidence for other products such as H_2 or hydrocarbons. The observed optical spectra will be compared to the theoretical band structure in these systems to better understand how B-site alloying can be used to control the optical activity of this material for photochemical applications. Additionally, the calculated band edge positions will be compared to CO_2 reduction and H_2O oxidation potentials to better understand the reaction products and yields observed experimentally.

(685b) Two-Step Solar Thermochemical Cycle for Splitting CO2 Via Ceria Redox Reactions - Experimental Investigation with a 3.8 Kw Solar Reactor

Philipp Furler[1], Jonathan Scheffe[1] and Aldo Steinfeld[1,2], (1)Department of Mechanical and Process Engineering, ETH Zurich, Zurich, Switzerland, (2)Solar Technology Laboratory, Paul Scherrer Institute, Villigen, Switzerland

Two-step solar thermochemical cycle for splitting CO_2 via ceria redox reactions – Experimental investigation with a 3.8 kW solar reactor

Philipp Furler [1], Jonathan Scheffe [1], Aldo Steinfeld [1,2]

[1] Department of Mechanical and Process Engineering, ETH Zurich
Zurich 8092, Switzerland

[2] Solar Technology Laboratory, Paul Scherrer Institute
Villigen PSI 5232, Switzerland

Abstract

Solar thermochemical approaches for CO_2 splitting inherently operate at high temperatures and utilize the entire solar spectrum. As such, they provide an attractive path to solar fuels production with high energy conversion efficiencies without the use of precious metal catalysts. In contrast to direct thermolysis of CO_2, two-step thermochemical cycles further bypass the CO/O_2 separation problem. Cerium oxide (ceria) has emerged as a highly attractive redox intermediate because of its favorable thermodynamics and kinetics at moderate temperatures. Reduction proceeds via the formation of oxygen vacancies and the release of gaseous O_2, resulting in the subsequent change in stoichiometry (δ). Oxidation is capable of proceeding with CO_2, thereby releasing CO and re-incorporating oxygen into the lattice. The two-step CO_2splitting solar thermochemical cycle based on oxygen-deficient ceria is represented by:

High-T reduction: $CeO_2 \rightarrow CeO_2\text{-}\delta + \delta/2\ O_2$ (1)
Low-T oxidation with CO_2: $CeO_2\text{-}\delta + \delta\ CO_2 \rightarrow CeO_2 + \delta\ CO$ (2)

We report on recent experimental studies for splitting CO_2 using concentrated solar energy. A novel, macro-porous CeO_2 structure was developed and tested in a 3.8 kW solar reactor prototype which was previously described in detail.[1, 2] Experiments were conducted at ETH's High-Flux Solar Simulator under conditions that closely approximates the heat transfer characteristics of highly concentrating solar systems such as solar towers and parabolic dishes. The reactor engineering design, experimental setup, and the novel CeO_2structure are described in detail and measured product compositions and solar-to-fuel energy conversion efficiencies are presented.

1. W. C. Chueh, C. Falter, M. Abbott, D. Scipio, P. Furler, S. M. Haile and A. Steinfeld, Science, 2010, 330, 1797-1801.
2. P. Furler, J. R. Scheffe and A. Steinfeld, Energy & Environmental Science, 2012, 5, 6098-6103.

(685c) Applicability of Perovskite-Based Oxides As Redox Materials for Thermochemical CO2 Dissociation

Jonathan Scheffe[1], David Weibel[1] and Aldo Steinfeld[1,2], (1)Department of Mechanical and Process Engineering, ETH Zurich, Zurich, Switzerland, (2)Solar Technology Laboratory, Paul Scherrer Institute, Villigen, Switzerland

Two-step thermochemical cycles utilizing metal oxide based intermediates are capable of dissociating CO_2 to CO using concentrated solar energy as the source of high-temperature process heat. The first step consists of a high-temperature, endothermic reduction of a metal oxide and the subsequent evolution of oxygen. The second step consists of a lower temperature exothermic oxidation reaction, in which CO_2 is reduced to CO and oxygen is re-incorporated into the metal oxide. Because the metal oxide is recycled, the net reaction is simply the dissociation of CO_2, whose products can be further processed into liquid hydrocarbons via Fischer-Tropsch and other catalytic processes. In contrast to the direct thermolysis, CO and O_2 are derived in different steps, thereby eliminating the need for high-temperature separation.

Ceria-based ($CeO_{2-\delta}$) materials are attractive as metal oxide intermediates because of their high-temperature morphological and crystallographic phase stability [1]. Rapid kinetics are also achievable compared to other systems because of relatively large ambipolar diffusion coefficients. There are, however, many other nonstoichiometric oxides which have yet to be evaluated as redox splitting materials but exhibit many of the advantages of ceria-based materials. One such class of materials is known as perovskites, which have the crystallographic form $ABO_{3-\delta}$. A and B can represent a wide range of cations and have a large influence the degree of reduction as a function of both O_2 partial pressure (pO_2) and temperature.

A large quantity of thermodynamic data exists for many perovskites which describe the degree of nonstoichiometry (δ) as a function of pO_2 and temperature. Based on this data we have identified a number of materials which may be suitable as metal oxide redox intermediates. In short, we have modeled oxygen nonstoichiometry and extrapolated the data to high temperatures Subsequently, relevant thermodynamic parameters were computed and equilibrium CO concentrations determined as a function of reduction conditions (T, pO_2) and ensuing oxidation temperature. We have coupled thermodynamic modeling with experimental demonstration (thermogravimetric analysis) of select materials in order to further identify their suitability as redox materials. Data presented here will focus primarily on thermodynamic modeling and experimental screening of various perovskite-based oxides in relation to thermochemical CO_2 splitting. A comparison of these systems with CeO_2-based materials from a thermodynamic and kinetic perspective will be highlighted.

References

1. Chueh, W.C. and S.M. Haile, *A thermochemical study of ceria: exploiting an old material for new modes of energy conversion and CO_2 mitigation.* Philosophical Transactions of the Royal Society a-Mathematical Physical and Engineering Sciences, 2010. **368**(1923): p. 3269-3294.

Bench-Scale Electrochemical Production of Synthesis Gas

Eric J. Dufek, Tedd E. Lister, and Michael E. McIlwain
Interfacial Chemistry Department, Idaho National Laboratory, Idaho Falls, ID
tedd.lister@inl.gov

Introduction

This paper describes the development of bench-scale systems for the electrochemical reduction of CO_2 to synthesis gas (syngas). Syngas is a significant building block chemical currently derived from fossil energy. Electrochemical reduction of CO_2 is a process that could integrate intermittent alternative energy sources into the electric grid by using excess electricity to convert CO_2 and H_2O to a useful product. Electrochemical reduction of CO_2 has been demonstrated in lab experiments for many years where reduced carbon has been produced in high yields using moderate conditions. The Idaho National Laboratory (INL) effort has focused on developing these systems into working bench-scale and eventually pilot-scale processes [1-4]. The challenge is to perform the electrolysis with the minimum energy input and produce a quality product stream at high throughput. The eventual goal of this project was to develop systems that, if viable, could be advanced to larger scale operations.

Background on CO_2 reduction

Previous work in the electrochemical reduction of CO_2 has largely focused on study of the cathodic half-cell reaction. As a whole, these reports have provided a wealth of information on the conditions necessary to produce desired products including CO [5-6] and simple hydrocarbons such as methane [7-8], ethylene [8-9], carboxylic acids [10] and alcohols [11]. A key factor in determining the product distribution has been the cathode material [5,12-14]. Metals that strongly adsorb CO (Pt group metals) and hydrogen tend to favor hydrogen evolution. Metals with weak CO adsorption and moderate hydrogen overvoltage (Ag, Au, Zn and Cu) tend to form CO, although Cu electrodes primarily form hydrocarbons. In poorly catalytic metals (Hg, Cd, Pb, In, Sn) formic acid is the primary product as C-O bonds are not broken without surface adsorption. To date, hydrocarbon generation on Cu has been studied with the most detail, however the broad product distribution, including the gases (CH_4, C_2H_4) and formic acid and its high sensitivity to impurities makes it a less than ideal cathode choice [15].

A limited number of studies have reported on complete systems for CO_2 reduction which are capable of being readily scaled-up [6,16-17]. A major technical challenge in the electrochemical reduction of CO_2 is the mass-transport limitation associated with low CO_2 solubility in aqueous solutions. At planar metal electrodes this results in relatively low CO_2 reduction current densities. This can be alleviated through the incorporation of a gas diffusion electrode (GDE) in the electrochemical cell [6,9,18-19]. While issues arise with the use of Cu, Ag has previously been shown to be highly effective at the production of CO [5-6,13,20].

The production of syngas mixtures has been investigated using conventional electrodes and has shown great promise at low current densities. As an example, in 1994 Hori at al. reported faradaic efficiencies values of 87.1% for CO and 10.2% for H_2 at 5 mA cm^{-2} (-1.14 V vs NHE) on Au foil (and similar values for Ag) [13]. Yano et al. reported the use of a Ag mesh GDE for producing a syngas mixture but did not report current density [21]. Hori et al produced a syngas mixture using a Ag coated anion exchange membrane at 100 mA cm^{-2} at -3 V vs SHE [22]. At a cathode voltage of -1.3 V the current density was 20 mA cm^{-2} producing 92.1% CO.

Due to issues with CO_2 solubility in aqueous solutions at RT it is desirable to perform CO_2 electrolysis in a closed pressurized system. This has been demonstrated in simple electrolysis cells contained within a pressure vessel up to 50 atm of CO_2. By introducing CO_2 through a GDE a current density of 300 mA cm^{-2} at -1.22 V (vs Ag/AgCl) with 86% Faradaic efficiency for CO was observed [23]. This amounts to a 60X increase in current density over ambient pressure on a foil cathode.

Experimental

Ambient pressure system

A commercially available filter-press type flow cell (Model AB, ElectroCell Systems) was modified in-house to incorporate Ag GDEs (Silflon, Gaskatel). A Ag-based GDE served as the cathode and a O_2-evolving dimensionally stable anode (DSA) was employed. An advantage to the use of commercial electrodes, especially the Ag GDE is that similar electrodes systems are widely used in the chloro-alkali electrolysis industry as depolarizing electrodes [24]. Even after several years of use these electrodes show minimal degradation in performance [25]. Nafion® 424 cation exchange membranes separated the cell compartments. Each cell component was sealed via 1/16" thick EPDM rubber. The anode and cathode active geometric areas were approximately 10 cm^2. A Ag/AgCl (3 M NaCl) reference electrode (Bioanalytical Systems) connected via salt bridge was used to measure cathode voltage (E_c). High purity CO_2 (Norco) was introduced to the back of the GDE using a manifold/current collector constructed of Type 304 stainless steel. The manifold had evenly spaced holes in a recessed area behind the active part of the electrode to promote even distribution of CO_2 to the back surface of the GDE. CO_2 gas was passed through a hydrocarbon trap (Restek). A mass-flow controller (1479A, MKS Inc.) metered CO_2 to the GDE.

Both catholyte and anolyte solutions were cycled through the cell at 30-40 mL min^{-1} using a peristaltic pump and C-flex tubing (Masterflex-Cole Parmer). Flow of CO_2 was initiated prior to introducing solutions into the cell. Solutions reservoirs were immersed in a water bath to control cell temperature. The temperature was monitored using a thermocouple at the cathode outlet. A LabVIEW (National Instruments)-based data acquisition system was used to capture the cell voltage (E_{cell}) and E_c. Polarization of the cell at a constant current was accomplished through the use of a DC power supply (30V, 3A, BK Precision).

Pressurized electrolysis system

A custom, pressurized electrolysis cell was acquired from Giner, Inc. This was incorporated into the system shown in Figure 1. The cell consists of 316 stainless steel endplates/current collectors. The anode and cathode compartments use polysulfone electrolyte flow channels separated by a Nafion® 115 membrane. The cathode had a flow frame to introduce CO_2 (welding grade, Norco) to the backside of the GDE. Electrical contact to the GDE was accomplished using a porous Ag sheet (800 μm pores, Sterlitech) and 316 stainless steel mesh in contact with the endplate. The Ag sheet was used to minimize the likelihood of solution contact with the current collector. An Ir-based DSA supported on Ni mesh served as the anode, prepared in-house using an established method [26]. The discs were overlayed and placed in the electrolysis cell. The geometric area for both the anode and cathode was 8.4 cm^2. During operation electrolyte was continuously recirculated from the gas/liquid separators through the cell and back using a Model 81946 high pressure pump (Micropump) equipped with a variable speed controller.

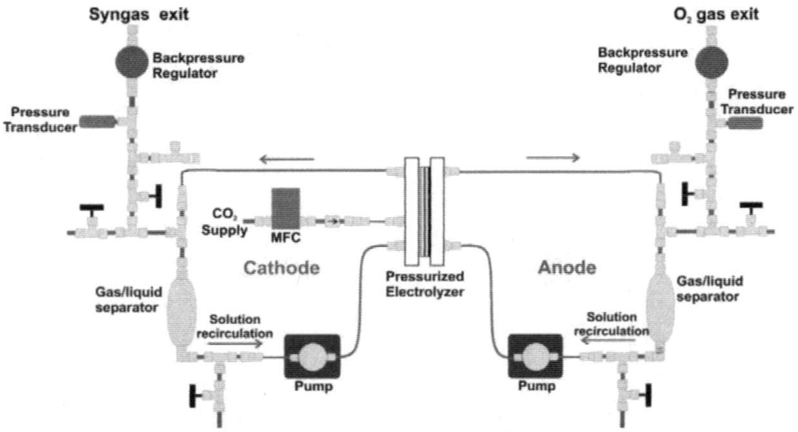

Figure 1: Schematic of the pressurized electrolysis system.

Experiments were performed at constant pressure and temperature with a fixed volume of catholyte (160 mL using either 0.5 M K_2SO_4 or 0.5 M K_2SO_4:1.0 M $KHCO_3$) and anolyte (120 mL, 2.5 M KOH). Electrolyte solutions were continuously cycled through the cell at rates of 100 mL min^{-1}. The cathode side of the cell was pressurized using CO_2 while the anode was pressurized with Ar (for pressure balance). Prior to each experiment the cell was pressurized and allowed to equilibrate for 1.25 h prior to the application of current. The temperature within the cell was monitored using a thermocouple imbedded in the cathode endplate. Constant current was applied using a 35 V, 10 A DC power supply (BK Precision). CO_2 was introduced to the backside of the GDE using a calibrated Aalborg CO_2 mass flow controller (MFC) with the pressure of the CO_2 source maintained at 6 atm above the cathode pressure.

Product analysis

Head space analysis of the product gas for both the cathode and anode was accomplished using an injection loop on a gas chromatograph (GC) (Hewlett Packard 5890) equipped with thermal conductivity detectors. Prior to introduction to the injection loop, the product gas was passed through an ice cooled water trap to condense water vapor. For H_2, O_2 and CO analysis Ar carrier gas and an open tubular column (Molecular Sieve 5A, Hewlett Packard) was used. CO_2 was determined using a CP PoraPLOT U capillary column (Varian, Inc.) using He carrier gas. Formic acid analysis was performed on the cathode solutions (after acidification) using GC (Model 6890, Hewlett Packard) equipped with a Nukol column (Supelco) and He carrier gas. An Optiflow 520 (Supelco) digital bubble flow meter was used to determine the flow of product gas prior to each GC measurement.

Results and Discussion

Part 1: Ambient pressure CO_2 reduction

Baseline polarization data

The initial focus of work was to demonstrate syngas production using an ambient pressure electrolysis cell [1]. Baseline polarization measurements are presented in Figure 2. Increasing the temperature has a profound effect on both E_{cell} and E_c. At 70 mA cm^{-2} E_{cell} drops by 1.57 V from 4.45 V at 18°C to 2.88 V at 70°C. In conjunction with the decrease in E_{cell} a drop in E_c from –2.19 V to -1.87 V for the same temperatures is observed. A small portion of the E_{cell} drop is due to the E_c while the remainder associated with the anode and reduced impedance for the electrolyte and membrane. A key implication of these changes is that the current window where E_{cell} remains below 4 V increases by close to 100 mA cm^{-2} at 70°C. The product gas changes significantly with current density, as shown in Figure 2C. At low current density a CO rich gas is observed which switches to H_2 rich at higher current densities. The current density where the cross-over occurs is actually observed to increase with temperature, with the data at 35°C having the highest crossover current.

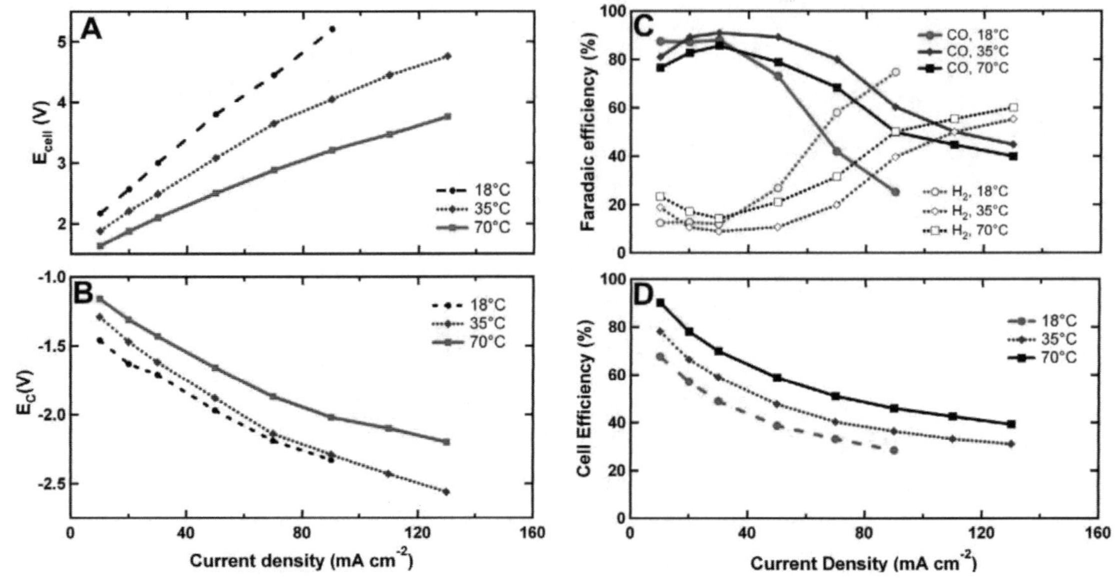

Figure 2: Cell performance at 18, 35, and 70°C, CO_2 flow 20 mL min^{-1}
A) E_{cell} and B) E_c, C) Faradaic efficiency and D) electrical efficiency.

Using the FE and the E_{cell} the overall cell efficiency can be determined [6]. The calculated values are shown in Figure 2D where cell efficiency decreases with increasing current density. This trend closely follows previously reported values for full cell analysis of CO_2 reduction [6]. The key is that by increasing the cell temperature the cell efficiency increases for all the currents investigated. This behavior is directly related to the decreases seen in the E_{cell} values. It is recognized that this method of calculating the cell efficiency ignores the energy required in heating the cell. Commercial-scale cells are designed to use current-resistance losses in the cell to heat the cell to improve energy efficiency.

Control of syngas mixtures

Optimal production of chemicals from syngas requires the correct H_2:CO ratio. Ideally for methanol production a ratio of 2:1 is desired while dimethyl ether (DME), a possible replacement for diesel fuel, production using a mixed catalyst performs best with a ratio between 1 and 1.5:1 [27-28]. From Figure 2, the H_2:CO ratio varies from 0.28 to 1.5:1 when moving from 10 mA cm^{-2} to 130 mA cm^{-2} at 70°C. Another means to increase the H_2:CO ratio is to reduce the flow of CO_2 to the cathode. Figure 3 contains the data from a set of experiments where the CO_2 flow rate was varied between 10 and 30 mL min^{-1}. As can be seen, limiting the amount of CO_2 available to reach the electrode surface results in a increase in the H_2:CO ratio. While the ratio increases there is little if any observed change in either E_{cell} or E_c.

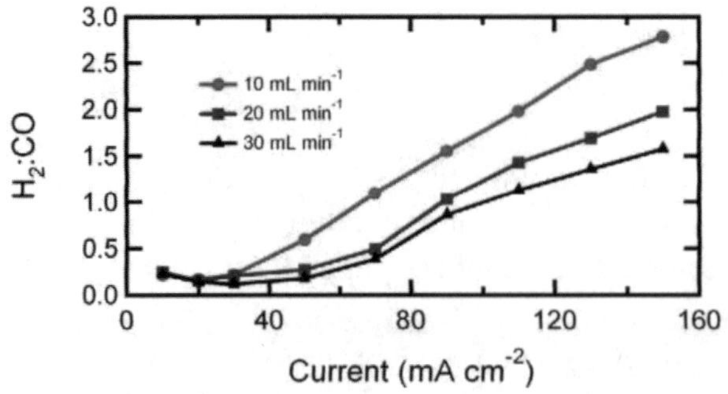

Figure 3: CO_2 flow rate dependence, 0.8 M K_2SO_4 catholyte, 70°C.

Production of syngas mixtures

In Figure 4a-b the cell performance for a run where the current density was held at 117.8 mA cm^{-2} with a CO_2 flow rate of 10 mL min^{-1} are shown. These conditions are roughly what would be necessary to produce a syngas H_2:CO ratio of 2. For the first 30 min of cell operation after the initial application of current both E_{cell} and E_c gradually decrease before stabilizing for the remainder of the 4 hour run. At the very end of the run a slight increase in E_{cell} is most likely associated with the consumption of OH^- in the anode solution (no additions were made). In Figure 4c-d the current density was held at 95 mA cm^{-2} with a CO_2 flow rate of 30 mL min^{-1} to mimic the conditions necessary to produce syngas for DME production. In a manner similar to Figure 4a following an initial decrease in the H_2:CO ratio the system stabilized for the remainder of the experiment. Due to the decreased current the electrolyte life for the lower H_2:CO ratio is increased because less OH^- is consumed at the anode per unit time.

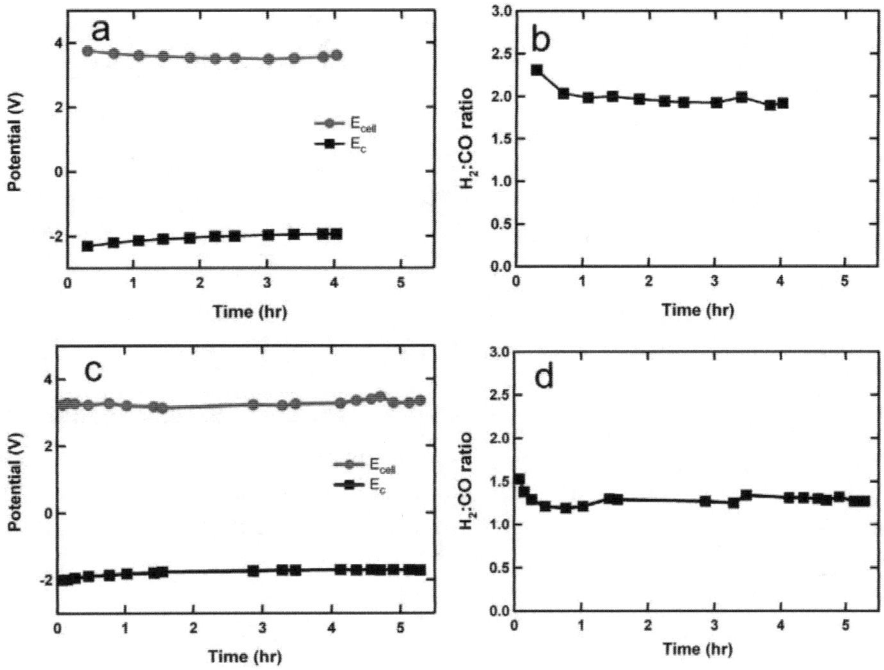

Figure 4: Generation of syngas at 70°C in 0.8 M K_2SO_4 catholyte. a-b) Methanol mixture: CO_2 flow 10 mL min^{-1} and 117.8 mA cm^{-2} c-d) DME mixture: CO_2 flow of 30 mL min^{-1} and 95 mA cm^{-2}.

Part 2: Pressurized system for continuous reduction of CO_2

Baseline data for pressurized system

To realize greater throughput, a pressurized system was built and operated [4].Three significant advantages are obtained by performing the reduction at elevated pressure: 1) increased solubility of CO_2, 2) downstream handoff to catalytic processes and 3) potential to operate above the boiling point. Previous studies of pressurized systems for CO_2 reduction have primarily focused on cell configurations for small batch conversion. Such designs are ill-suited for large scale electrolysis applications. The present system is capable of extended operation and is amenable to future scale up.

For baseline comparison, the pressurized system was compared to the ambient pressure system tested in this lab using similar conditions [1-2]. The highest current density for CO_2 reduction investigated at ambient pressure using Silflon® GDEs prior to this report was 225 mA cm^{-2}, with the majority of the current going to the production of H_2 (2:1 H_2:CO). The pressure system displays similar behavior to the previous report at ambient pressure (Figure 5). However, significant increases in the FE* of CO generation are observed as the pressure is increased to 18.5 atm (current efficiency is reported as Faradaic efficiency (FE*) with the realization that a lag exists between the electrode processes and sampling due to the pressurized nature of the system). The increase in FE* equates to a ~5 fold increase in CO generation and closely mirrors prior work that found pressures above 10 atm enabled conversion of CO_2 to CO at efficiencies above 80% at 300 mA cm^{-2} [5].

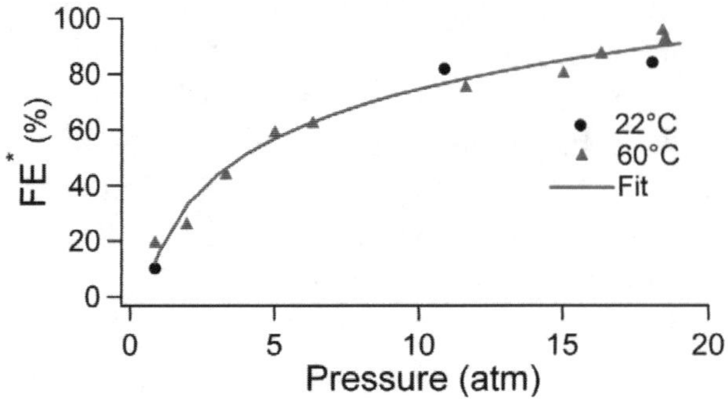

Figure 5: Impact of pressure on CO FE* at 225 mA cm^{-2} at 21°C and 60°C. CO_2 flow rate 85 mL min^{-1} through the GDE, catholyte 0.5 M K_2SO_4.

During the operation of commercial-scale electrolysis cells, efficiencies within the cell (overpotential, electrolyte impedance) result in the generation of heat leading to elevated operating temperatures. To mimic this condition the pressurized cell was heated prior to operation. Elevating the temperature had little impact on selectivity for CO generation (Figure 5) but did lead to a decrease in E_c from 4.25 to 3.52 V at 18.5 atm and 225 mA cm^{-2}. Similar decreases have been observed for ambient pressure reduction of CO_2, where regardless of current density the decrease in cathode potential was ~0.3 V.[1] Given the same cathode material it is expected that the extent of minimization of cathode potential for the present cell is similar.

Figure 6 shows the product distribution and E_c as the current was increased at 60°C. Current densities which can be accessed while generating a predominantly CO stream range up to 350 mA cm^{-2} while maintaining E_c below 4.6 V. Hara and Sakata's work suggest CO remains the predominant product to current densities far in excess of 350 mA cm^{-2} [5], but due to the increases in E_c, current densities in this range were not investigated.

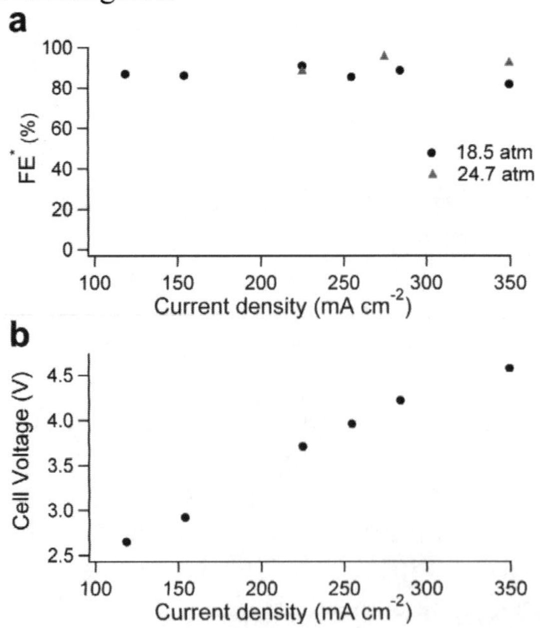

Figure 6: a) CO FE* at 60°C, CO_2 flow rate 85 mL min^{-1}, K_2SO_4 catholyte, b) E_c for CO_2 reduction using Silflon® GDEs at 18.5 atm.

Operation at both 60 and 90°C (Figure 7) was investigated to further assess benefits of increased temperature. Temperatures above 90°C, although achievable, were not investigated due to the possibility of system degradation. At 18.5 atm and 225 mA cm^{-2} a gradual decline in CO efficiency occurred at both temperatures with the E_c decreasing to below 3 V at 90°C. The final efficiencies for CO generation were 90% and 82% at 60 and 90°C respectively. When the pressure was increased to 24.7 atm and the current density was increased to 275 mA cm^{-2} a similar decline in CO efficiencies was observed with the CO efficiency remaining above 78% for the duration of the experiment at 90°C (87% at 60°C).

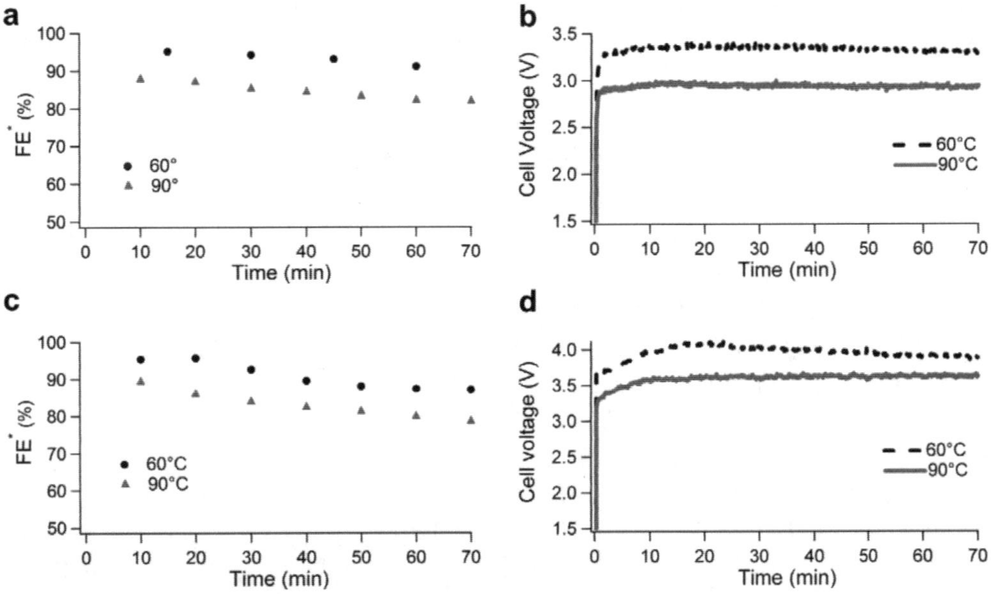

Figure 7: Comparison of cell performance at 60 and 90°C at a-b) 18.5 atm and 225 mA cm^{-2} (0.5 M K_2SO_4 and 1.0 M $KHCO_3$ catholyte), CO_2 flow rate 150 mL min^{-1}, c-d) 24.6 atm and 275 mA cm^{-2}

At just under 3 V and 225 mA cm^{-2}, the data at 18.5 atm indicates that larger quantities of CO can be generated at significantly lower E_c when the reduction of CO_2 occurs at elevated pressure instead of under ambient pressure [1]. The energy efficiency (ε) of a cell can be determined from the faradaic efficiencies of the various products generated (in this case FE*), the E_{cell} and the enthalpies associated with the electrode processes ($\Delta rH°_{H2}$ = 286 kJ mol^{-1}, $\Delta rH°_{CO}$ = 283 kJ mol^{-1}) [6]. At 60ºC and 225 mA cm^{-2} the E_{cell} drops from 4.01 to 3.71 V when the pressure is increased from ambient to 18.5 atm. This corresponds to an increase in ε from 37% to 40%. To achieve similar ε at ambient pressure the current density is limited to 125 mA cm^{-2} with less CO generation [1]. The bulk of the E_{cell} decrease is attributed to a Nernstian shift in the reduction potential for CO_2 as the pressure is increased and to a minimization of electrode blockage by bubbles. Elevation of the temperature to 90ºC results in additional cell efficiency increases with the ε being 50% with only a marginal decrease in the quantity of CO produced. Future investigations at more elevated temperatures are expected to generate additional increases in ε. Preliminary investigation over multiple, independent experiments at a CO_2 flow rate of 85 mL min^{-1} (18.5 atm, 60º C, 225 mA cm^{-2}, using the same GDE) found the system to display high reproducibility with no indication of a downward trend in CO FE* between runs. The standard deviations for these repeated studies remained below 2.7%.

Technical challenges and potential applications

While much was accomplished on this project there is much more to be done to forward the electrochemical reduction of CO_2. The main technical obstacle is the kinetic limitation for CO_2 reduction. This fact points to catalyst development to enhance kinetics while maintaining the desired product output. Another solution would be to increase temperatures to increase kinetics. There was a small kinetic effect found in the INL work at 70°C. For water electrolysis the limitation in temperature is system material stability. Due to the low electrical efficiency one must consider applications that are not as financially tied to the cost of electric power. One application could be where significant renewable electrical energy exists with limited ability to tie to the grid. Remote wind farms could be used to power a remote fuel or chemical production facility. The locally generated product could be used locally for farming or mining operations. Another situation could be envisioned for production of commodity chemicals which use syngas normally derived from natural gas or coal. The profit margin for chemicals is larger than fuels production thus providing some ability to absorb the reduced efficiency. Finding a suitable anodic reaction that performs a useful process in place of O_2 evolution would also improve viability.

References

1) E. J. Dufek, T. E. Lister, M. E. McIlwain, J. Appl. Electrochem. **41** (2011) 623.
2) E. J. Dufek, T. E. Lister, M. E. McIlwain, J. Electrochem. Soc. **158** (2011) B1384.
3) E. J. Dufek, T. E. Lister, M. E. McIlwain, Electrochem. Solid St. **15** (2012) B48.
4) E. J. Dufek, T. E. Lister, M. E. McIlwain, J. Electrochem. Soc. **159** (2012) F514.
5) K. Hara, A. Kudo, T. Sakata, J. Electroanal. Chem. **391** (1995) 141.
6) C. Delacourt, P.L. Ridgway, J.B. Kerr et al, J. Electrochem. Soc. **155** (2008) B42.
7) K. Hara and T. Sakata, J. Electrochem. Soc. **144** (1997) 539.
8) K. Hara, A. Tsuneto, A. Kudo et al, J. Electroanal. Chem. **434** (1997) 239.
9) R.L. Cook, R.C. Macduff, A.F. Sammells, J. Electrochem. Soc. **137** (1990) 607.
10) D.T. Whipple, E.C. Finke, P.J. Kenis, Electrochem. Solid St. **13** (2010) B109.
11) K.W. Frese, S. Leach, J. Electrochem. Soc. **132** (1985) 259.
12) M. Azuma, K. Hashimoto, M. Hiramoto et al, J. Electroanal. Chem. **260** (1989) 441.
13) Y. Hori, H. Wakebe, T. Tsukamoto et al, Electrochim. Acta **39** (1994)1833.
14) M. Azuma, K. Hashimoto, M. Hiramoto et al, J. Electrochem. Soc. **137** (1990) 1772.
15) M. Gattrell, N. Gupta, A. Co, J. Electroanal. Chem. **594** (2006) 1.
16) H. Li, C. Oloman, J. Appl. Electrochem. **35** (2005) 955.
17) T. Yamamoto, D.A. Tryk, A. Fujishima et al, Electrochim. Acta **47** (2002) 3327.
18) M.N. Mahmood, D. Masheder, C.J. Harty, J. Appl. Electrochem. **17** (1987) 1223.
19) T. Yamamoto, D.A. Tryk, K. Hashimoto et al, J. Electrochem. Soc. **147** (2000) 3393.
20) N. Hoshi, M. Kato, Y. Hori, J. Electroanal. Chem. **440** (1997) 283.
21) H. Yano, F. Shirai, M. Nakayama et al, J. Electroanl. Chem. **533** (2002) 113.
22) Y. Hori, H. Ito, K. Okano et al, Electrochim. Acta **48** (2003) 2651.
23) K. Hara, T. Sakata, *B. Chem. Soc. Jpn.*, **70**, 571 (1997).
24) I. Moussallem, J. Jorissen, U. Kung et al, J. Appl. Electrochem. **38** (2008) 1177.
25) K.H. Tetzlaff, R. Walz, C.A. Gossen, J. Power Sources 50 (1994) 311.
26) C. Comninellis, G. P. Vercesi, J. Appl. Electrochem. **21** (1991) 136.
27) G. Olah, A. Goeppert, G.K.S. Prakash et al, *Beyond Oil and Gas: The Methanol Economy*, 2006 Wiley, Weinheim.
28) G.R. Moradi, R. Ghanei, F. Yaripour, Int. J. Chem. React. Eng. **5** (2007)1542.

(685e) Electrode and Catalyst Design for Electrochemical Reduction of CO2

Molly Jhong, Department of Chemical & Biomolecular Engineering, University of Illinois at Urbana-Champaign, Urbana, IL, Sichao Ma, Department of Chemical & Biomolecular Engineering, University of Illinois at Urbana Champaign, Urbana, IL and Paul J. A. Kenis, Chemical and Biomolecular Engineering, University of Illinois at Urbana-Champaign, Urbana, IL

Carbon dioxide is believed to be a major contributor to climate change due to its accumulation in the atmosphere as a greenhouse gas. To significantly reduce carbon emissions several strategies are being explored, including carbon capture and sequestration (CCS) and increasing the use of carbon-neutral energy sources such as wind, solar, and nuclear.[1] Electrochemical reduction of CO_2 to useful fuels or small organic molecules is more promising than sequestration as it produces value-added products and provides a means of storing power from intermittent sources such as wind and solar in a convenient and high energy density form. While the electrochemical conversion of CO_2 has great potential, significant technological advances with regard to electrode structure and catalyst design are still needed for this process to become economically viable.

Many early efforts have focused on exploring different catalysts to make various products such as formic acid and syngas. More recently, several researchers have proposed various reactor designs to electrochemically convert CO_2 into formic acid or syngas. Previously, we developed an electrochemical flow reactor for CO_2 reduction in which the anode and cathode are separated by a flowing liquid electrolyte. Operating this reactor at acidic pH, we succeeded in converting CO_2 to formic acid with high Faradaic and energetic efficiency of 89 and 45%.[2] Furthermore, we studied efficient conversion of CO_2 to CO, and found ionic liquid electrolyte, EMIM BF_4, significantly reduced the overpotential to <0.2 V as opposed to ~0.8 V in aqueous electrolytes. The ionic liquid acts as co-catalyst, stabilizing the unstable intermediate CO_2^- radical.[3] While these results are encouraging, to date few efforts have focused on the effect of electrolyte composition and electrode structure. Better understanding of the relationship between electrode structure and performance is needed (i) to enhance catalyst utilization, (ii) to reduce catalyst loading, and (iii) to improve electrode durability. To address these challenges, we demonstrate a systematic characterization of the structure-activity relationships of complex electrodes as a function of component materials choice and synthesis (e.g., catalyst design), of electrode fabrication and processing methods (e.g., catalyst deposition and the choice of the gas diffusion layer), and of electrochemical environment (e.g., electrolyte composition).

With respect to catalyst design, we observed CO_2 reduction is highly dependent on the morphology of the catalyst layer. The particle size of the catalyst has a tremendous influence on performance. Moreover, unsupported catalysts are most commonly used for CO_2 reduction; however, the catalyst utilization is very low (less than 10%) due to agglomeration. To overcome this issue, we have developed supported catalyst via loading catalyst particles on highly electrically conductive supports in uniform fashion. This approach provides an extensive three-phase boundary where effective transport of electrons, protons and CO_2 can take place, resulting in significant performance enhancement.

Regarding electrode fabrication and processing, we investigated how physical properties of the gas diffusion layer such as porosity and hydrophobicity influence electrode performance in terms of effective gas transport of both reactant (CO_2) and product (CO and H_2) gases. Also, to improve electrode performance and durability we have developed a fully automated air-brushing deposition technique that significantly enhances performance up to 40%.

Regarding electrochemical environment, we investigated the influence of electrolyte composition on the electrochemical reduction of CO_2 to CO. In specific, we studied the effect of cation and anion size on partial current densities of the desired CO and undesired H_2 products. For example, we have observed that large cations enhance the

CO_2 reduction reaction to CO while suppressing hydrogen evolution reaction, resulting in high Faradaic yield toward CO production.

1. C. Oloman and H. Li, *Chemsuschem*, **1**, 385 (2008).

2. D. T. Whipple, E. C. Finke and P. J. A. Kenis, *Electrochem Solid St*, **13**, D109 (2010).

3. B. A. Rosen, A. Salehi-Khojin, M. R. Thorson, W. Zhu, D. T. Whipple, P. J. A. Kenis and R. I. Masel, *Science*, **334**, 643 (2011).

(685f) CO2 - Ionic Liquids Synergies: Reactive Capture and Conversion of CO2 to Chemicals

Valerie Eta[1], **Pasi Virtanen**[1], Ikenna Anugwom[1], Päivi Mäki-Arvela[1] and Jyri-Pekka Mikkola[1,2], (1)Åbo Akademi University, Turku, Finland, (2)Umeå University, Umeå, Sweden

CO_2 - Ionic Liquids synergies: Reactive capture and conversion of CO_2 to chemicals

Valerie Eta[a], Pasi Virtanen[a], Ikenna Anugwom[a], Päivi Mäki-Arvela[a], Jyri-Pekka Mikkola[a,b]

[a]*Laboratory of Industrial Chemistry and Reaction Engineering, Process Chemistry Centre, Åbo Akademi University, Biskopsgatan 8, 20500 Turku/Åbo, Finland.*

[b]*Technical Chemistry, Department of Chemistry, Chemical – Biological Center, Umeå University, SE-90187 Umeå.*

Novel technologies based on the utilization of abundant and non-toxic carbon dioxide (CO_2) for the synthesis of materials and chemicals may significantly contribute towards the development of sustainable industrial practices. Naturally, myriad substances are synthesized from CO_2 and consumed by humans. The conversion of CO_2 to fuels and chemicals is an attractive method of utilizing CO_2 from the viewpoint of green chemistry and sustainable development. The use of CO_2 in the synthesis of chemicals such as dimethyl carbonate (DMC) or switchable ionic liquids (SILs) constitutes an environmentally attractive alternative to hazardous and toxic reagent, and resulting in waste minimisation. DMC exhibits versatile physical and chemical properties, such as low toxicity and excellent biodegradability and thus, is an environmentally benign building block for other chemicals.

In the present study, i) DMC was synthesised from CO_2 and methanol, and ii) CO_2 was used as feedstock for the synthesis of switchable ionic liquids (SILs) where the entire –COO– moiety is maintained.

The direct synthesis of DMC from methanol and CO_2 is characterized by low yields due to the reaction equilibrium and the thermodynamic limitations (ΔG^o_{298K}= +26.3 kJ/mol). Alkoxide ionic liquids possessing alkylimidazolium and benzalalkonium cations were prepared, characterised and tested together with ZrO_2-MgO catalyst for the synthesis of DMC from methanol and CO_2. By using the novel ionic liquid as additives, *ca* 12 % conversion of methanol, and 90 % selectivity to DMC was obtained at 120 °C and 7.5 MPa [1]. The water abstracting potential of the ionic liquids influenced the conversion of methanol and the selectivity to DMC. The alkoxide ionic liquids were recovered and reused in DMC synthesis without loss in activity and selectivity.

Switchable ionic liquids are compounds which can be switched from molecular compounds to ionic liquid by bubbling acid gas, such as CO_2, through the liquid. The formed ionic liquid can be switched back to molecular building blocks - and CO_2 will be released - by bubbling an inert gas, such as N_2, through the liquid or applying vacuum. Novel SILs containing 1,8-diazabicyclo-[5.4.0]-undec-7-ene (DBU), glycerol and CO_2 were synthesized, characterized and tested for the selective fractionation of lignocellulosic materials. The structures of the synthesized SILs were confirmed by NMR and FTIR. The results showed that carbonate containing ionic liquid was thermally stable up to 120 °C and was capable of selectively fractionate birch chips at 100 °C in a batch reactor (Fig. 1). FTIR spectra indicated that DBU was protonated via formation of N–H and C=NH+ bands. Furthermore, deprotonation of glycerol was confirmed by FTIR. NMR spectra of the SIL showed the formation of glyceryl carbonate [2]. As such, the challenges associated with the use of ionic liquid, such as the high costs and the risk assessment of the complete life cycle of ionic liquids could be minimized by using cheap and less toxic SILs.

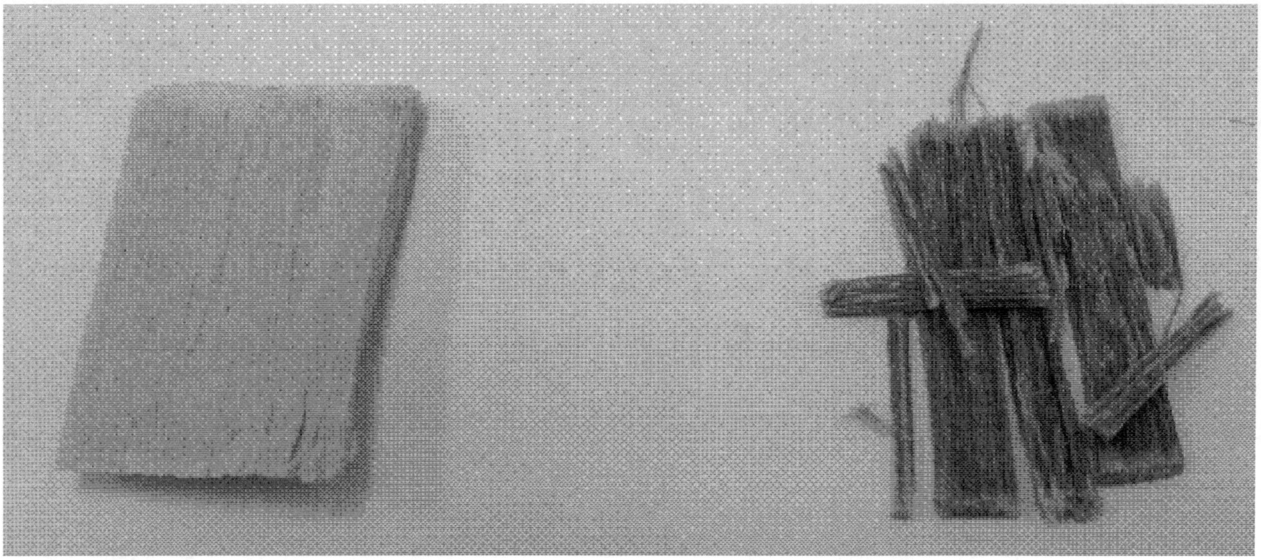

Figure 1. Photograph of untreated (left) and SIL treated birch chip. Treatment conditions: T = 100 °C, t = 120 h.

References

[1] Eta V., Mäki-Arvela P., Salminen E., Salmi T., Murzin D., Mikkola J-P.; Catal Lett (2011) 141:1254–1261

[2] Anugwom I., Mäki-Arvela P., Virtanen P., Willför S., Damlin P., Hedenström M., Mikkola J-P.; **Affiliation Information**

1. Laboratory of Analytical Chemistry,Åbo Akademi University,Process Chemistry Centre, Åbo-Turku, Finland

[3] ***Holzforschung***, (2012) in press, DOI: 10.1515/hf-2011-0226

(688a) Numerical Simulation of Reservoir-Wellbore Fluid Flow Simulations At Geologic CO2 Sequestration Sites: Coupled or De-Coupled?

Rajesh Pawar[1], Haruko Wainwright[2], Robert Dilmore[3], Jens Birkholzer[2], Jim Houseworth[2] and Grant Bromhal[4], (1)Computational Hydrology Group, Los Alamos National Laboratory, Los Alamos, NM, (2)LBL, Berkeley, CA, (3)National Energy Technology Laboratory, U.S. Department of Energy, Pittsburgh, PA, (4)National Energy Technology Laboratory, Morgantown, WV

Leakage of CO_2 and brine through pre-existing wellbores is considered to be one of the main potential risks at geologic CO_2 sequestration sites. To assess impacts of this potential weakness, it is critical to accurately estimate leakage rates through failed wellbores. Currently, there is very limited field data available on CO_2 leakage rates through wells. In lieu of field data, numerical simulations that accurately capture the nature of multi-phase fluid flow through wellbores can be used to provide estimates of the leakage rates for a variety of wellbore leakage parameters. A few earlier studies have reported results of numerical simulations of leakage of CO_2 and brine through wellbores from CO_2 storage reservoirs. There are different types of approaches used for modeling wellbore flow:

- Models that include only the wellbore flow and not the sequestration reservoir. These models assume constant pressure and/or saturation boundary conditions at the bottom hole.
- Models that include coupled wellbore and sequestration reservoir flow. These models solve the equations for fluid flow in the wellbore and reservoir implicitly (either semi-analytical or numerical approaches).
- System models that include coupled wellbore and sequestration reservoir flow. Unlike the models in the second bullet above, the coupling in these models is not implicit.

System models are being increasingly used for quantitative risk assessment (QRA) of geologic CO_2 sequestration operations. Typical QRAs require a probabilistic treatment of quantities such as wellbore leakage rates and need a large number of Monte-Carlo realizations of system performance to predict the likelihoods of different types of events. Most system models uses the following approach: 1) divide the full system (e.g., the sequestration reservoir to the surface) into component sub-systems, 2) use some type of approximation such as a reduced order models to capture the behavior of each sub-system, and 3) combine those component models into a large, but efficient, model of the full system. This approach assumes that the component models can be combined into the full systems model without any feedback. It is quite computationally efficient when the reduced order models are developed to produce very rapid results. Typically, significant effort is required to develop appropriate reduced order models that can accurately capture the underlying physics of a system component.

One concern for carbon storage systems, however, is the conditions under which the assumption that the separate systems model components produce negligible feedback between each other is valid. For wellbore leakage, the multi-phase fluid flow through or around the well is dependent on the evolution of pressure and saturation in the reservoir which is contact with wellbore. The reservoir pressures and saturations in turn will be affected by the amount of leakage through wellbore and the geologic properties of reservoir. In this study, we are testing the validity of the no feedback approximation of coupled wellbore-reservoir fluid flow by a model comparison study. We utilize both Finite Element Heat and Mass (FEHM), Los Alamos National Laboratory's (LANL) fluid flow simulator, and TOUGH2-ECOM, Lawrence Berkeley National Laboratory's (LBNL) fluid flow simulator. These simulators were used to perform detailed numerical studies of leakage rates through wellbores in a coupled wellbore-reservoir model. We also performed simulations of fluid migration through wellbores decoupled from the reservoir and without feedback, using some semi-analytical models for wellbore leakage, as well as LANL's CO_2-PENS system model for geologic CO_2 sequestration sites. Appropriate wellbore flow model boundary conditions were investigated. Results of these model calculations are compared to determine the validity of an approximation of no feedback in the system models.

(688b) Investigating the Role of Coupled Thermal-Hydrological-Mechanical (THM) Processes During Geological Sequestration of CO2

Sharad Kelkar[1], Rajesh Pawar[2], George A. Zyvoloski[3], Philip Stauffer[4] and Hari Viswanathan[4], (1)Earth and Environmental Sciences (EES-6), Los Alamos National Laboratory (LANL), Los Alamos, NM, (2)Computational Hydrology Group, Los Alamos National Laboratory, Los Alamos, NM, (3)Earth & Environmental Sciences Division, Los Alamos National Laboratory, Los Alamos, NM, (4)EES, LANL, Los Alamos, NM

Geologic sequestration of CO_2 is one of the technologies being considered for mitigating impact of anthropogenic emissions of CO_2. During injection of large quantities of CO_2 into geological formations, the coupled fluid flow, thermal and geomechanical effects are expected to be important. Mechanical failure (fracturing) of the caprock, or to reactivation of existing geological faults due to high injection pressures are some of the potential risks in sequestration reservoirs. Developing effective reservoir management strategies to minimize these risks will require accurate models for the fracture-stress interaction due to variations in temperature and fluid pressure. The equations describing each of the THM processes are complex, involving material properties with highly nonlinear behavior – for example the permeability of fractured rocks can vary by many orders of magnitude over the pressure and temperature ranges of interest. The time spans of interest range from minutes to tens of years, and the spatial scales range from centimeters to kilometers. The need to accommodate the wide range scales also requires the simulator to handle a wide range of pressure and temperature gradients. To address these issues, a fully coupled nonlinear Thermo-Hydro-Mechanical (THM) model is being developed as part of the Finite Element Heat and Mass Transfer (FEHM) code at the Los Alamos National Laboratory. FEHM is a porous flow simulator that uses continuum mechanics approach with control volume (CV) and finite element (FE) discretization. It can be used to simulate non-isothermal, multiphase, multi-component fluid flow, heat transfer and mechanical processes using either implicit or explicit solution approaches. We have implemented a novel technique for mapping between CV formulation for fluid-flow equations and FE formulation for stress equations to handle stress-dependant variations in permeability. This approach allows for implicit solution of the coupled THM balance equations as well as effective representation of stress-dependent permeability changes. In this presentation, we will describe development of the model and its numerical implementation. We will demonstrate application of the approach using an example of a faulted reservoir. Our numerical studies are focused on understanding how faults/fault gouges behave during large-scale CO_2 injection including coupled fault displacements, potential permeability enhancement due to shear as well as tensile failures and CO_2 migration.

(688c) Reservoir Model Optimization Using the Seismic Response to CO2 Injection and Stochastic Inversion

Abelardo Ramirez, Lawrence Livermore National Laboratory, Livermore, CA

Reservoir Model Optimization Using the Seismic Response to CO_2 Injection and Stochastic Inversion

Abelardo L. Ramirez[1], Kathy Dyer[1], Donald J. White[2], Yue Hao[1], James Johnson[3]

1-Lawrence Livermore National Laboratory

2-Geological Survey of Canada

3-Schlumberger-Doll Research Center

Abstract

4D reflection seismic data were used to map CO_2 migration during Phase 1 of the Weyburn-Midale Project, while an extensive fluid sampling program documented the geochemical evolution triggered by CO_2-brine-oil-mineral interactions. Our goal is to use these data to optimize the reservoir model and thereby improve site characterization and dependent predictions of long-term CO_2 storage in the Weyburn-Midale reservoir.

We have developed a stochastic inversion tool that identifies porosity/permeability models that optimize agreement between the observed and predicted seismic response. The tool explicitly integrates reservoir flow modeling, facies-based geostatistical methods, and a novel stochastic inversion technique to identify optimal porosity/permeability models. Reservoir model optimization is accomplished through stepwise refinement of its permeability magnitude, anisotropy, and heterogeneity.

Our algorithm starts by proposing one reservoir model realization. Then, using a multi-phase/multi-component flow simulator, we predict reservoir conditions associated with CO_2/H_2O injection, HC/H_2O withdrawal, and CO_2 migration. Using these conditions, we predict P-wave velocities as a function of fluid-phase saturations, pressure, and porosity using Gassmann's equation. The velocity stack is used together with calculated bulk densities to generate an intra-reservoir reflectivity series. These series are convolved with an estimate of the source wavelet to predict time-series waveforms that can be directly compared to observed seismic response. The likelihood for each model is computed and used to decide whether the proposed reservoir model produces acceptable agreement between predicted and observed 4D seismic reflection data. Then, a new reservoir model is proposed and the process is repeated until the process converges. We will present the results from a synthetic data study to illustrate the method's performance.

This work has been performed under the auspices of the U.S. Department of Energy by Lawrence Livermore National Laboratory under Contract DE-AC52-07NA27344.

(688d) Assessing the Role of Thermal Stresses in Caprock Integrity During CO2 Storage

Gennady Gor, Department of Civil and Environmental Engineering, Princeton University, Princeton, NJ and Jean-Herve Prevost, Civil and Environmental Engineering, Princeton University, Princeton, NJ

Deep saline aquifers provide unbeatable capacities for long term CO_2 storage, provided CO_2 does not leak. Upon CO_2 injection the pressure in the aquifer increases and induces stresses in the caprock. If CO_2 is injected at the temperature different from the ambient temperature in the aquifer, there could appear thermal stresses also. We study the effect of CO_2 injection temperature on the caprock integrity. Using coupled thermo-poromechanical multi-phase simulations [1], we reveal that thermal expansion of the rock significantly enhances the stresses developing as a result of the fluid pressure buildup. Calculations show that when CO_2 is injected at a temperature below the ambient value, the stresses in the caprock above the horizontal injection well become tensile and even exceed the tensile strength after several years of continuous injection [2]. These stresses lead to the caprock fracturing. Once the fractures form, they start to propagate because of the high pressure in the reservoir and may serve as pathways for CO_2 leakage. Based on the results of our simulations using the method from [3] we estimate the rate of fracture propagation. We show that this rate is strongly affected by the value of the permeability in the caprock. Finally, our results show that injection of CO_2 at temperature close to the ambient value in the aquifer significantly reduces the risk of caprock fracturing and, therefore, possible leakage.

1. Preisig, M. and Prevost, J.H. Coupled multi-phase thermo-poromechanical effects. Case study: CO_2 injection at In Salah, Algeria, *International Journal of Greenhouse Gas Control*, 5, 1055-1064 (2011).

2. Gor, G. Y. and Prevost, J.H. Effects of thermal stresses on caprock integrity during CO_2 storage, *International Journal of Greenhouse Gas Control* (submitted)

3. Gor, G.Y., Stone, H.A., Prevost, J.H. Fracture propagation driven by fluid outflow from a low-permeability reservoir, *Journal of the Mechanics and Physics of Solids* (submitted) http://arxiv.org/pdf/1203.4543

Cement Degradation Under Conditions Relevant to Geological Carbon Sequestration

Li Li, Leopold Brunet, Zuleima Karpyn

Department of Energy and Mineral Engineering, Pennsylvania State University, University Park, PA, 16802

Brian Strazisar, and Barbara Kutchko

National Energy Technology Laboratory, Pittsburgh, PA, 15236-0940

Abstract

Assessing the potential risks of CO_2 leakage from deep subsurface is one of the major challenges for geological carbon sequestration. Injected CO_2 can dissolve into in situ brine to produce acidity for cement wellbore degradation. This degradation can potentially change the properties of the cement and decrease its structural integrity. Wellbore integrity assessment is essential for predicting the risk of leakage because degraded cement can open pathways for CO_2 leakage. In this work, we develop a reactive transport model based on experimental observations to obtain quantitative and mechanistic understanding of the coupled reactions and diffusion processes involved in cement degradation. This work also aims to quantify the effect of initial cement properties, in particular porosity and the abundance of portlandite, on the rate and extent of cement degradation.

The results show that the leaching of calcium ions during the acid attack of the cement leads to precipitation of calcite near the cement brine interface. This precipitation yields a localized reduction in porosity, which reduces the diffusion process and the penetration of the degradation inside the cement sample. The calcite precipitation is largely affected by the initial porosity and portlandite content. Increasing content of portlandite allows a thicker calcite precipitation thus decreasing the extent of degradation. Increasing initial porosity leads to faster diffusion of the acidic brine and less porosity reduction by the calcite precipitation. As a result, it increases the rate and extent of cement degradation. The ratio of the initial portlandite content and the porosity is a critical indicator of the severity of cement degradation. The results allow us to better understand cement behavior when exposed to CO_2-saturated brine under geological sequestration conditions and to predict the rate and extent of cement degradation with given cement properties.

1. Introduction

A promising option for carbon mitigation would be the use of Carbon Capture and Storage (CCS) (Bachu 2003; Solomon and (eds.). 2007; Friedlingstein, Houghton et al. 2010). However , there are major concerns regarding the possibility and consequences of CO_2 leakage associated with CCS (White, Strazisar et al. 2003; Benson and Cole 2008; Jenkins, Cook et al. 2012). One identified potential leakage pathway is through abandoned wells (Gasda, Bachu et al. 2004; Nordbotten and Celia 2006). Acidic CO_2-saturated brine can react with in-situ cement, potentially altering its structural integrity and physical properties and increasing the likelihood of leakage. Thus it is important to understand and quantify CO_2- brine-cement interactions in order to identify the key factors that lead to cement alteration, and the reaction mechanisms involved (Kutchko, Strazisar et al. 2007; Duguid and Scherer 2010; Huet, Tasoti et al. 2011).

The first objective of this work is to develop a reactive transport model that couples reaction thermodynamics and kinetics with transport processes using constraints from experimental observations of Kutchko et al. under diffusion-controlled conditions (Kutchko, Strazisar et al. 2007; Kutchko, Strazisar et al. 2008). The second objective is to use the model to systematically understand and quantify the role of cement properties, in particular initial porosity and portlandite content, in determining the rates and extent of degradation under conditions relevant to carbon sequestration. We focus on diffusion-controlled conditions because flow velocities are typically very slow and diffusion controlled conditions prevail in the deep subsurface (Matteo and Scherer).

2. Methodology

Imaging and long-term alteration data from Kutchko et al. were used as constraints to develop a reactive transport model that characterizes the cement/ CO_2/brine interaction processes (Kutchko, Strazisar et al. 2007; Kutchko, Strazisar et al. 2008). We then used the model to predict the extent of alteration under diffusion-controlled conditions with an array of cement properties. A total of 81 cases were investigated in this work.

Reactive transport modeling

The reactive transport code CrunchFlow (Gaus 2010) was used to simulate the chemical reaction and transport process involved in cement- CO_2-brine interaction. Reactive transport models solve mass, energy and momentum conservation equations. The expression of the mass conservation equation is as follows (modified from (Steefel and Lasaga 1994)):

$$\frac{\partial(\phi C_i)}{\partial t} = \nabla \cdot (\phi D_i \nabla C_i) \pm \sum_{reactions} v_{ir} \cdot R_{ir} \quad \text{(Eq. 1)}$$

Where ϕ is the porosity, C_i is the concentration of the species i, D_i is the effective diffusion coefficient of the species i, R_{ir} is the reaction rate of the rth reaction involving the species and v_{ir} its stoichiometric number.

The left-hand side term is the mass accumulation rate of species i while the right-hand side takes into account all the process leading to mass change of the species i. The first right-hand side term accounts for diffusion, while the second accounts for the mass change due to chemical reactions involving species i.

Effective diffusion coefficients in porous media are typically based on Archie's law. The form used in CrunchFlow is as follows:

$$D_e = \phi^m \cdot D_w \quad \text{(Eq. 2)}$$

Where D_e is the effective diffusion coefficient in porous media (cm^2/s), D_w is the diffusion coefficient in water, ϕ is the porosity, and m is the cementation factor. This cementation factor accounts for the tortuosity of porous media and its effect on diffusion (Gaus 2010).

The main reaction process involved is as follows: CO_2 will dissolve into the brine lowering its pH. C-S-H and portlandite will dissolve at low pH leaching Ca^{2+} and increasing the pH. The dissolution of C-S-H leads to the formation of silica gel (amorphous silica). The rise in Ca^{2+} concentration could potentially induce a carbonation process (i.e. the formation of calcite). Other fast aqueous reactions will also occur and are included in the model.

2.1. Description of the system

A reactive diffusion model is developed using data from Kutchko et al. as constraints (Kutchko, Strazisar et al. 2007; Kutchko, Strazisar et al. 2008). Class H cement is represented as a simplified paste composed of C-S-H and portlandite. The initial portlandite content is fixed at 16.4%wt as reported by Kutchko et al. this leads to a content of 63.7%wt for the C-S-H. An initial porosity of 25% is chosen. A 1%wt NaCl brine is used with a CO_2 concentration of 0.5mol/L based on the temperature and pressure condition reported by Kutchko et al.(Duan and Sun 2003). The brine pH is calculated to be 2.9. (Kutchko, Strazisar et al. 2007) The initial pH of the cement core is fixed at 13 (Glasser 1997).

The simulation was conducted in one dimension for a 10-millimeter domain composed of 1,000 cells with a resolution of 10μm. The first 6 mm represents the cement core and the rest of the domain represents CO_2 saturated brine. Thermodynamic parameters were obtained from the EQ 3/6 database (Wolery, Jackson et al. 1990) and thermoddem (Marty, Tournassat et al. 2009). Reaction kinetic parameters, including the rate constants, and pH dependence, were used as fitting parameters in order to obtain a fitting of the Kutchko SEM-BSE imaging data. The rate constants obtained for the best fit were similar to those reported in literature (Marty, Tournassat et al. 2009). A cementation factor of 3.95 is used. This is consistent with the observations of Huet et al. that porosity reduction during cement degradation leads to major changes in diffusivity values (Huet, Tasoti et al. 2011).

3. Results and discussion

The main purpose of this base case simulation is to develop a reactive transport model that reproduces the cement alteration SEM-BSE images in the experiments of Kutchko et al.(Kutchko, Strazisar et al. 2007). In the experiment, a class H cement core was exposed to CO_2 saturated brine for 9 days. Figure 1-A shows an SEM-BSE image from this experiment after 9 days. Three distinct alteration zones appear to have been formed (zone I, II and III in figure 1-A). Zone I is a degraded zone sitting near the cement-brine interface, zone II appears to be a highly carbonated layer and zone III consists of seemingly unaltered cement.

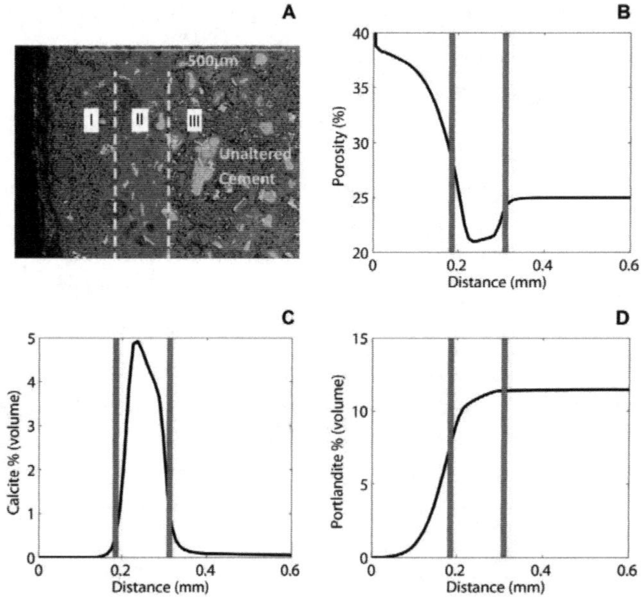

Figure 1 – Laboratory observation and simulation output at day 9 of exposure. Only the cement core is shown, with the cement-brine interface at 0 mm. (A) SEM-BSE image of cement after 9 days of exposure to CO_2 saturated brine at 50°C and 30.3 MPa (Kutchko, Strazisar et al. 2007); (B)-(D) Predicted spatial profiles of properties at day 9 for (B) porosity, (C) calcite volume fraction, (D) portlandite volume fraction.

The model reproduce the zoning seen in figure 1-B,C,D. As expected zone I is degraded significantly with porosity values increasing to as high as 40% at the interface and a near total depletion of the portlandite figure 1-B,D. No calcite precipitation is observed in this zone while calcite concentrations as high as 5% are observed in zone II with a very localized distribution (almost no calcite is observed in zone III) figure I-C. This carbonation results in a decrease of 4% in porosity in zone II, which will lead to a significant decrease in effective diffusion coefficient, reducing further diffusion of acidic brine toward the cement inner core.

4. Conclusion

A reactive transport model for the simulation of the coupled process of diffusion and reactions involved in cement exposure to CO_2 saturated brine was developed. This model reproduced the results observed in experimental work and thus confirms the proposed mechanism for cement interaction with CO_2 saturated brine. The dissolution of the Ca(II) containing species provide a source of Ca(II) allowing precipitation of calcite, which leads to the reduction of the porosity. This will reduces the effective diffusion coefficient and further penetration of acidic brine. It shows that over the long term, the extent of cement alteration will be dependent on the formation and characteristics of the calcite zone. In addition, this work focuses on diffusion-controlled condition. If flow does occur, the system might see a much deeper alteration of the cement , as it have been shown in some studies [31].

Acknowledgement

This work was supported by the Carbon Sequestration Program of the U. S. DOE National Energy Technology Laboratory (NETL) and U. S. DOE National Energy Technology Laboratory—Regional University Alliance (RUA) Partnership.

Reference

Bachu, S. (2003). "Screening and ranking of sedimentary basins for sequestration of CO_2 in geological media in response to climate change." Environmental Geology **44**(3): 277-289.

Benson, S. M. and D. R. Cole (2008). " CO_2 Sequestration in Deep Sedimentary Formations." Elements **4**(5): 325-331.

Duan, Z. and R. Sun (2003). "An improved model calculating {CO\scriptsize{2}} solubility in pure water and aqueous {NaCl} solutions from 273 to 533 {K} and from 0 to 2000 bar." Chemical Geology **193**(3-4): 257-271.

Duguid, A. and G. W. Scherer (2010). "Degradation of oilwell cement due to exposure to carbonated brine." International journal of greenhouse gas control **4**(3): 546-560.

Friedlingstein, P., R. A. Houghton, et al. (2010). "Update on CO_2 emissions." Nature Geosci **3**(12): 811-812.

Gasda, S. E., S. Bachu, et al. (2004). "Spatial characterization of the location of potentially leaky wells penetrating a deep saline aquifer in a mature sedimentary basin." Environmental Geology **46**(6-7): 707-720.

Gaus, I. (2010). "Role and impact of CO_2-rock interactions during CO_2 storage in sedimentary rocks." International journal of greenhouse gas control **4**(1): 73-89.

Glasser, F. P. (1997). "Fundamental aspects of cement solidification and stabilisation." Journal of Hazardous Materials **52**(2–3): 151-170.

Huet, B., V. Tasoti, et al. (2011). "A review of Portland cement carbonation mechanisms in CO_2 rich environment." Energy Procedia **4**(0): 5275-5282.

Jenkins, C. R., P. J. Cook, et al. (2012). "Safe storage and effective monitoring of CO_2 in depleted gas fields." Proceedings of the National Academy of Sciences **109**(2): E35-E41.

Kutchko, B. G., B. R. Strazisar, et al. (2007). "Degradation of Well Cement by CO_2 under Geologic Sequestration Conditions." Environmental Science & Technology **41**(13): 4787-4792.

Kutchko, B. G., B. R. Strazisar, et al. (2008). "Rate of CO_2 Attack on Hydrated Class H Well Cement under Geologic Sequestration Conditions." Environmental Science & Technology **42**(16): 6237-6242.

Marty, N. C. M., C. Tournassat, et al. (2009). "Influence of reaction kinetics and mesh refinement on the numerical modelling of concrete/clay interactions." Journal of Hydrology **364**(1-2): 58-72.

Matteo, E. N. and G. W. Scherer "Experimental study of the diffusion-controlled acid degradation of Class H Portland cement." International journal of greenhouse gas control(0).

Nordbotten, J. M. and M. A. Celia (2006). "An improved analytical solution for interface upconing around a well." Water Resources Research **42**(8).

Solomon, S., D. Qin, M. Manning, Z. Chen, M. Marquis, K.B. Averyt, M. Tignor and and H. L. M. (eds.). (2007). "Contribution of Working Group I to the Fourth Assessment Report of the Intergovernmental Panel on Climate Change."

Steefel, C. I. and A. C. Lasaga (1994). "A coupled model for transport of multiple chemical species and kinetic precipitation/dissolution reactions with application to reactive flow in single phase hydrothermal systems." Am. J. Sci. **294**: 529-592.

Taylor, H. F. W. (1997). Cement Chemistry, Thomas Teldford.

White, M. C., B. R. Strazisar, et al. (2003). "Separation and capture of CO_2 from large stationary sources and sequestration in geological formations-coalbeds and deep saline aquifers." Journal of the Air & Waste Management Association (1995) **53**(6): 645.

Wolery, T. J., K. J. Jackson, et al. (1990). "CURRENT STATUS OF THE EQ3/6 SOFTWARE PACKAGE FOR GEOCHEMICAL MODELING." Acs Symposium Series **416**: 104-116.

Laboratory and Numerical Simulation based Investigation of Two-Phase Brine-CO2 Displacement Processes Using Cores from Three CO2-Sequestration Candidate Formations

Shameem Siddiqui, Talal Gamadi, Otman Algadi, Abdulrahman A. Alquraishi*[§] *and Dustin McIntyre*[+]

Texas Tech University, Lubbock, Texas 79409-3111, United States
§King Abdulaziz City of Science and Technology, Riyadh, S. Arabia
+National Energy Technology Laboratory, Morgantown, West Virginia, United States

Abstract

Capturing carbon dioxide (CO2) from power plants and other sources of its emissions and injecting it into deep geologic formations for permanent storage purposes has become an important area of research recently. Some of the good formation candidates for CO2 sequestration are deep saline aquifers because they have a high storage capacity and because they are available in almost all sedimentary basins of the world. The effectiveness of the sequestration process in saline aquifers depends highly on the two-phase CO2-brine relative permeability behavior and the latter can only be measured in the laboratory. Although relative permeability data for hydrocarbon bearing formations are generally available, very little is known about the relative permeability of CO2-brine two phase systems, especially for the cases involving actual reservoir rocks. In this paper, we present relative permeability data generated on three saline formations that are being explored as potential candidates for CO2 capture and sequestration. These are - the Knox sandstone, the St. Peter sandstone, and the Sylvania formation. The latter is divided into an upper sandstone formation and a lower dolomite formation, and both are investigated separately. Three 1.5 inch diameter by 2 inch long core plugs from each formation, drilled in the direction of flow, were used to form composite cores in order to conduct drainage CO2-brine relative permeability measurements using the unsteady state method at simulated reservoir conditions. In order to understand the various phenomena taking place during multiphase displacement processes, all of these experiments were conducted inside an X-ray CT (computerized tomography) scanner. Pixel-by-pixel image subtraction was used to calculate individual saturations of brine and CO2 at various stages of displacement. CT-based heterogeneity measurements were used to compare the heterogeneity of different plugs and to assess the effect of heterogeneity on displacement. Additionally, the CT-based image data were used to see if capillary end effects, gravity segregation and bypassing occurred during each test.

The results indicate low CO2 end point relative permeability (k_{rg}) in the range of 0.036 to 0.34 for the three formations. These low values can be attributed to rock heterogeneity as indicated by the saturation distribution determined from the scanner images. Such low relative permeability values would tend to decrease injectivity and hence increase the displacement efficiency. This was reflected on the residual brine saturations which were in the range of 0.25 to 0.45 indicating good storage potential for the investigated formations. This paper also includes some lab-scale numerical simulation runs which were conducted for history matching of the laboratory data and to have better confidence on the relative permeability data. The three-dimensional grid (50x50x120) was developed using the actual porosity data generated from the CT-scanning of the core plugs at 1.25 mm axial resolution. The porosity data were converted to permeability values using the Kozeny-Carman equation and the average permeability derived this way for each core plug was compared against the value measured individually using the brine permeability based data. For numerical simulation the laboratory conditions used for the relative permeability tests were used and ECLIPSE simulator with laboratory units was used to describe the cores in three dimensions. The results showed a good agreement between the lab-derived and simulator generated brine recovery curves and pressure drop between the inlet and outlet ends giving confidence in the process.

Introduction

Carbon capture and sequestration (CCS) in deep geological formations has lately emerged as a promising technology to mitigate greenhouse gases emissions in short to medium terms. Sedimentary basins worldwide are capable of holding around 320 gigatons of CO2[1]. CO2 can be sequestered in different potential geological storage formations such as depleted oil or gas reservoirs, coal beds and saline aquifers. Among these, the saline aquifers have the largest sequestration capacity. CO2 sequestration in saline aquifers in sedimentary basins can be obtained through four main mechanisms[2]: CO2 dissolution in the formation water (solubility trapping), geochemical reactions with aquifer fluids and rocks (mineral trapping), structural trapping (gravity and capillary effects) and hydrodynamic trapping when the aquifer prevents the CO2 to seep out due to low density difference. Because of the depth of the typical saline aquifers CO2 is injected as a supercritical fluid, which by itself behaves in a complex way. CO2 injected in deep saline aquifers will migrate through the porous medium in the form of a plume around the injection well that will try to rise to the top of the formation and expand below the cap rock laterally due gravity segregation. The process of lateral expansion of CO2 as an immobile phase is controlled by the residual saturation trapping mechanism due to CO2-brine relative permeability and capillary pressure, measurements of which are required by numerical simulators that are used to predict the fate of the overall CO2 injection process. Therefore, Understanding multi-phase flow and trapping is crucial in order to effectively utilize the large storage capacity of the saline aquifers. The main disadvantage of using saline aquifers for CO2 storage is the lack of information compared to the oil and gas reservoirs. In this paper we discuss the results of some multi-phase (supercritical CO2-brine) flow tests that were conducted on samples collected from three different saline formations. These formations are - the St. Peter, the Knox and the Sylvania formation, all of which are potential candidates for CCS in the US Midwest region. Unsteady state drainage relative permeability tests measurements were conducted at simulated reservoir conditions of pressure and temperature in order to determine the ultimate distribution of the CO2 plume and the storage capacity of the investigated formations.

Experimental Setup, Materials and Procedures

The experiments described in this paper were conducted with a coreflooding unit (Figure 1) having X-ray CT visualization capability. It mainly consists of a 10,000 psi low X-ray attenuation Hassler type core holder capable of handling up to a foot long core sample of 1.5 inch diameter; two dual-piston positive displacement pumps; and two one-liter floating piston transfer vessels filled with brine and supercritical CO2, respectively. The pressure drop across the core was measured using differential pressure transducers of different ranges mounted in parallel. Furthermore, two line pressure transducers were used to measure the inlet and outlet pressures. A set of 1/8 inch high-pressure steel and PEEK tubing and several two and three-way valves were used to direct and control the flow. The back-pressure was maintained on the system by passing the effluents through a dome type back-pressure regulator (BPR). Produced fluids were passed through a separator where gas was directed to gas flow meters and water was collected in graduated tubes placed in a fraction collector. Heating tapes and temperature controllers were used to heat the system to the desired temperature. Confining pressure was applied using distilled water injected by a hand pump. A vacuum pump was used to evacuate the core sample and the flow lines after each test. The X-ray CT scanner used was a Universal Systems HD300 based on a multislice CT scanner (8 slices per rotation, 1.25 mm apart) operating at 140 kV and 7 mA.

The operating parameters such as flow rate; injection time; cumulative volumes injected; inlet, outlet and differential pressures; and the system temperature were recorded continuously using a data acquisition system. The tests were conducted in the horizontal displacement mode, to simulate the primary direction of flow that would occur in most reservoir applications.

The plug samples used for the relative permeability tests were obtained from whole cores from three wells. These are, Blan Well No.1 (Hancock County, Kentucky, covering 5114 - 5115 ft, from the Knox group, Gunter sandstone formation); Briggs Well No. 1-12 (Arenac County, Michigan, covering 10811.0 - 10842.5 ft, from the Middle Ordovician St. Peter sandstone formation) and Gregory Well No. 2 (Midland County, Michigan, covering 4721 - 4726 ft, an upper sandstone facies and 4795 - 4801 ft from a lower dolomite facies of the Sylvania formation.

Three core plugs from each of these formations were used to make a core in each of the four unsteady-state relative permeability experiments. Composite cores are generally preferred in special core analysis tests as they represent a broader variation of reservoir properties than a single plug. Table 1 presents the core plug depths and measured physical properties in addition to the overall porosity and permeability of the three-plug composite cores at simulated reservoir conditions. The average porosities of the composite cores were measured using both the vacuum saturation technique and the CT scanning technique[3] and both sets generally agree well with each other. The permeability of the composite cores was measured by flowing brine at different flow rates with the core at simulated reservoir conditions and then applying Darcy's law.

The aqueous phase consisted of distilled water containing 10% by weight Sodium Iodide (NaI), except the test for the St. Peter samples for which 10% by weight Potassium Iodide (KI) was used. NaI and KI were used because of their high attenuation during the scanning process. The use of KI as opposed to NaI as a tracer (dopant) is generally preferred if there is some clay mineral present in the cores, which was the case of the St.

Peter sandstone. The displacing phase used was high purity supercritical CO2. The physical properties of these fluids at corresponding experimental conditions are listed in Table 2. Brine properties of density and viscosity were measured using glass pycnometer and Cannon-Fenske capillary viscometer. Supercritical CO2 properties were obtained from the National Institute for Standards and Technology (NIST) Web-book.

The relative permeability measurements were conducted using the unsteady state method which consisted of injecting several pore volumes of supercritical CO2 into the brine saturated core at the volumetric flow rate of 2 cm^3/min until the brine production ceased. The effluent volumes and the pressure drops across the core samples were recorded continuously and the relative permeability curves were constructed utilizing the Johnson, Bossler and Naumann[4] (JBN) method.

Experimental Results

Relative permeability tests with each of the four three-plug composite cores were done as close to the same manner as possible which included placing the core in the core holder, applying confining pressure and reservoir temperature, pulling vacuum on the core, saturating with brine, flowing brine at different rates to measure the absolute permeability, injecting CO2 while monitoring and recording effluent volumes and differential pressures across the core. The entire core was scanned from end to end multiple times at the same locations to record CT numbers, which were converted to saturations using the image subtraction technique described in the literature by Akin and Kovscek[5]. Although this technique does not require any assumption about uniform saturation[3] throughout the core at the end of the injection of a particular phase, it requires the core plugs to be cleaned (preferably in a Soxhlet extractor) after the coreflooding test, placed inside the coreholder and scanned at the same locations to establish a second baseline required for calculating saturations. A slight positional accuracy may sometimes result in obtaining inaccurate values of saturation.

Figures 2 through 5 show the calculated in-situ saturation distributions at various stages of injection of supercritical CO2 inside the cores that were initially saturated with brine. Each of these figures contains two additional graphs on the second y-axis and they are the CT-derived porosity and the coefficient of variation, C_v. The C_v, which is the standard deviation divided by the mean of CT attenuation (or CT number in Hounsfield Units) is calculated for each slice based on dry scan data, is a good classifier of the intra-slice heterogeneity[6]. A low value of C_v indicates low heterogeneity and vice versa. Among the four composite cores, the Sylvania (upper) samples showed the highest overall C_v values.

Figures 2 through 5 also show some saturation spikes at the interface between the first and the second and the second and the third plugs, resulting in local saturation disturbances. For the tests involving the plugs from the St. Peter and the Knox formations the interface was left empty whereas for the tests involving the plugs from the upper and the lower facies of the Sylvania formation a filter paper each was placed in the interface between the core plugs which resulted in slight improvement in local saturation values near the interface. The practice of using filter paper is controversial as the filter papers can retain significant amount of a particular fluid during coreflooding causing material balance anomalies. On the other hand natural fractures are common in most reservoirs and what can be seen in the interface of cores during coreflooding may also be experienced during field applications.

Figures 6 through 9 represent the unsteady-state relative permeability curves generated for the four composite core samples using the JBN technique. The red dots represent the oil relative permeability values and the black ones, the brine relative permeability. The solid lines are based on the curve-fitting of the respective data using the Corey[7] equation (sometimes called Brooks and Corey method). The Corey equation involves first calculating Sw*, the normalized saturation of the wetting phase as follows:

$$S_w^{*} = \frac{S_w - S_{wir}}{1 - S_{wir}} \quad \text{.........} \quad (1)$$

Here, S_w is the water (or brine) saturation and S_{wir} is the irreducible water saturation. By introducing the Corey exponents N_w and N_g for water and CO2 gas, respectively, and by including the relative permeability end points when the other phase is at the residual saturation, a generalized form of Corey equations can be written for the brine relative permeability, k_{rw}, and the gas relative permeability, k_{rg}:

$$k_{rw} = k_{rw@S_{gr}} \times (S_w^{*})^{N_w} \quad \text{.........} \quad (2)$$

$$k_{rg} = k_{rg@S_{wir}} \times (1 - S_w^{*})^{N_g} \quad \text{.........} \quad (3)$$

Table 3 lists all the end point saturation values and the Corey exponents for the four composite coreflooding tests. The irreducible water saturation values for the two Sylvania cores are the lowest and hence they have superior potential for storing CO2. In this work no attempt was made to saturate the brine prior to the injection of CO2 as it is physically impossible to do so in an aquifer. The in-situ saturation values represent the actual distribution of the brine and the CO2 phases within the core plugs.

Numerical Simulation

A commercial simulator (ECLIPSE) was used for history matching of the laboratory-derived data for one of the experiments. Slice-by-slice porosity data from one of the experiments (Sylvania formation, upper facies) were used to populate cells of a 50x50x120 grid system representing the core using the laboratory scale. The data were used to mimic the recovery and pressure history. As a first step three Kozeny-Carman type equations were developed for the three core plugs to generate permeability values in each grid block using an approach similar to that of Perrin et al.[9]. The equations are given below:

$$k = \frac{1.25}{\left(0.0316\phi^{0.7593}\right)^2} \frac{\phi^3}{(1-\phi)^2} \quad \text{(Plug 1) ....} \quad (4)$$

$$k = \frac{0.0721}{\left(0.0316\phi^{0.7593}\right)^2} \frac{\phi^3}{(1-\phi)^2} \quad \text{(Plug 2) ....} \quad (5)$$

$$k = \frac{0.06}{\left(0.0316\phi^{0.7593}\right)^2} \frac{\phi^3}{(1-\phi)^2} \quad \text{(Plug 3) ....} \quad (6)$$

The average permeability derived this way for each core plug was compared against the value measured individually using the brine permeability based data which were 40.0 md, 2.33 md and 1.85 md, respectively. The pressure vs. time plot for simulation and experimental data is shown in Figure 10 and the pore volume of brine produced vs. pore volumes of supercritical CO2 injected is shown in Figure 11. The two sets of data are in good agreement. Some differences exist as the simulator cannot mimic the exact distribution of fluids at the inlet and outlet faces. However, this approach shows tremendous potential for future use in simulating actual CO2 sequestration processes.

Conclusions

Unsteady-state drainage relative permeability of supercritical CO2-brine fluid pair were investigated for samples representing three potential saline aquifers for CCS purposes. The following lists the results obtained:

1. In general, low CO2 relative permeability and high irreducible water saturation are observed for the formations investigated except for the reasonably low residual saturation of the two Sylvania intervals indicating their good potential storage capacity.
2. Complete fluid displacement is primarily affected by core heterogeneity, fluid gravity segregation and rock internal capillary effects. The flooding front images indicate no gravity segregation, hence, capillary end effect and heterogeneity are considered to be the main factors behind the low Kr_{CO2} and high S_{wir} values for the St. Peter and Knox sandstones.
3. In-situ saturation data show some local saturation anomaly due to the capillary end effects at the interface between core plugs. This anomaly was partly eliminated by the use of filter papers. At the field scale vertical fractures can potentially present similar phenomena.
4. Numerical simulation with both lab scale and field scale data appears to be the next logical step in characterizing the sequestration potentials of these reservoirs.

Acknowledgments

The authors like to acknowledge the financial support of Texas Tech University, Indiana and Kentucky Geological Surveys and Western Michigan University. Special thanks go to John Rupp of Indiana Geological Survey; Dave Harris of Kentucky Geological Survey; Dave Barnes and Duane Hampton of Western Michigan University for providing the core samples. Special thanks go to Mr. Joe McInerney for his assistance in ordering parts and setting up various items for this research.

References

1. Koide H.G., Tazaki, Y., Noguchi, Y., Iijirna, M., Ito, K. and Shindo, Y. (1993), "Carbon Dioxide Injection into Useless Aquifers and Recovery of Natural Gas Dissolved in Fossil Water," Energy Convers Mgmt, 24(9-11), pp. 921-4.

2. Shukla, R., Ranjith, P., Haque, A. and Choi, X. (2010), "A Review of Studies on CO2 Sequestration and Caprock Integrity," Fuel 89, pp. 2651-2664.
3. Siddiqui, S., Funk, J. and Khamees, A. (2000), "Static and Dynamic Measurements of Reservoir Heterogeneities in Carbonate Reservoirs," SCA 2000-06, SCA Symposium, Abu Dhabi, UAE, October 18 - 22, 2000.
4. Johnson, E.F., Bossler, D.P. and Naumann, V.O. (1959), "Calculation of Relative Permeability from Displacement Experiments," Pet. Trans. AIME, 216, pp. 370-372.
5. Akin, S. and Kovscek, A.R. (2003), "Computed Tomography in Petroleum Engineering Research," in Akin S. and Kovscek, A.R. (eds.) Applications of X-ray Computed Tomography in Geosciences, Vol. 215, pp. 23-38, Geological Society, London.
6. Siddiqui, S., Okasha, T.M., Funk, J.J. and Al-Harbi, A.M. (2006), "Improvements in the Selection Criteria for Special-Core-Analysis Samples," SPE Reservoir Evaluation and Engineering Journal, 9(6), pp. 647-653.
7. Corey, A. T. (1954), "The Interrelation Between Gas and Oil Relative Permeabilities," Producers Monthly, November, 1954, pp. 38-41.
8. Perrin, J-C, Krause, M. Kuo, C-W, Miljkovic, L., Charoba, E. and Benson, S.M. (2009), "Core-Scale Experimental Study of Relative Permeability Properties of CO2 and Brine in Reservoir Rocks," Energy Procedia 1, pp. 3515-3522.

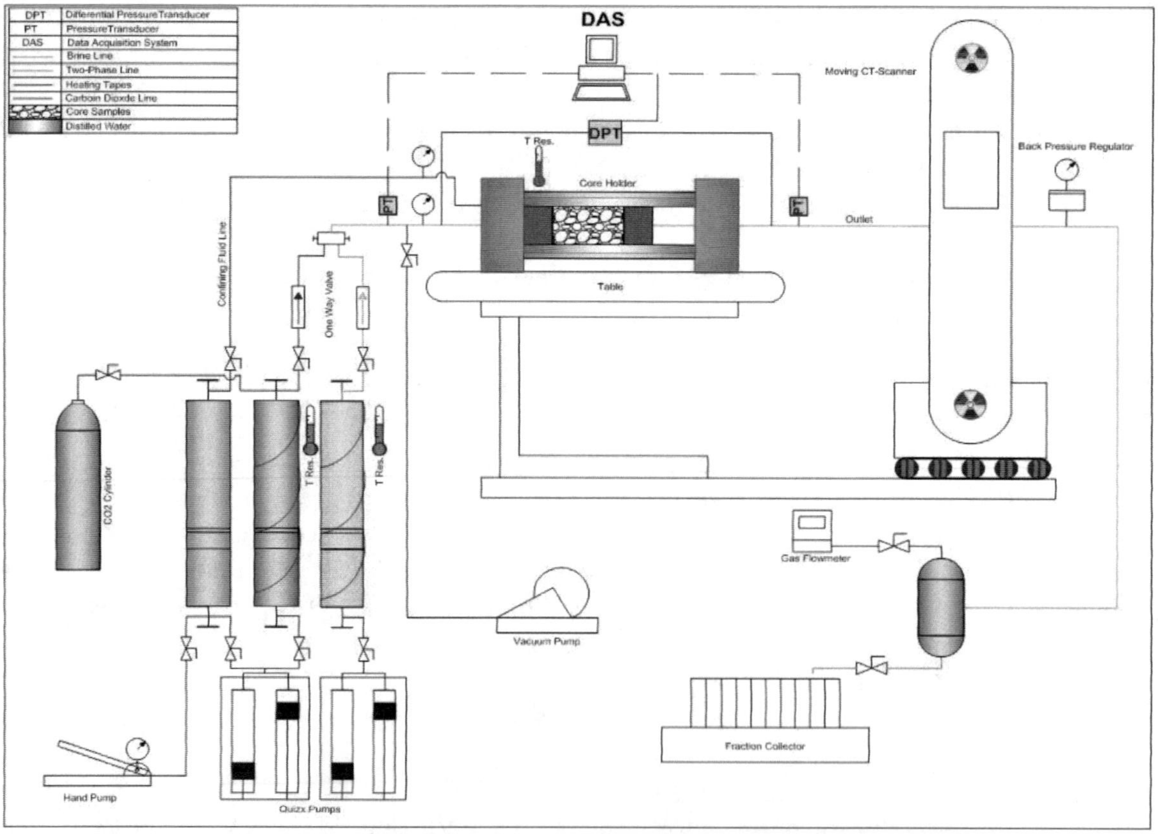

Figure 1: Coreflooding unit used to conduct the relative permeability tests.

Table 1. Core Plug Data

Formation	Plug #	Depth (ft)	Length (cm)	Plug ϕ* (%)	K (md)	Core ϕ (%)	Core K (md)	Test Temp. (°F)	Back Pressure (psi)
St. Peter SST	4	10,832.5	5.14	12.0	193.7	10.2	5.5	180	2500
	5	10,839.0	5.05	3.4	685.1				
	6	10,842.5	5.14	11.8	4.9				
Knox SST	1	5114.0	6.25	11.5	273.6	12.2	250.0	120	2000
	2	5114.2	6.26	11.5	270.6				
	3	5114.3	5.27	11.1	236.0				
Upper Sylvania SST	1	4722.0	5.05	10.0	40.7	9.8	4.0	120	
	3	4723.1	5.21	9.0	2.3				
	4	4724.4	5.18	7.6	1.9				
Lower Sylvania Dolomite	1	4797.1	5.21	12.0	65.8	14.6	26.1	120	
	2	4795.7	5.13	15.5	26.0				
	3	4799.2	5.10	11.5	20.5				

*The plug porosity values are averaged from CT scan images of saturated and dry cores.

Table 2. Fluid Data

Fluids	Density (g/cm^3)	Viscosity (cp)
10% NaI brine @ 120 °F	1.066	0.6750
10% KI brine @ 180 °F	1.042	0.5585
Supercritical CO2 @ 120 °F	0.685	0.0543
Supercritical CO2 @ 180 °F	0.374	0.0293

Table 3: Comparison between steady- and unsteady-state measurements of relative permeability for the Briggs 1-12 composite core

Formation	S_{wir}	k_{rw} at S_{wmax}	k_{rCO2} at S_{wir}	Corey Exponent N_w	Corey Exponent N_{CO2}
St. Peter SST	0.4310	1.0	0.0626	4.0	1.0
Knox SST	0.3398	1.0	0.1295	2.3	1.5
Upper Sylvania SST	0.2509	1.0	0.1857	1.8	6.5
Lower Sylvania Dolomite	0.2897	1.0	0.1404	1.6	1.7

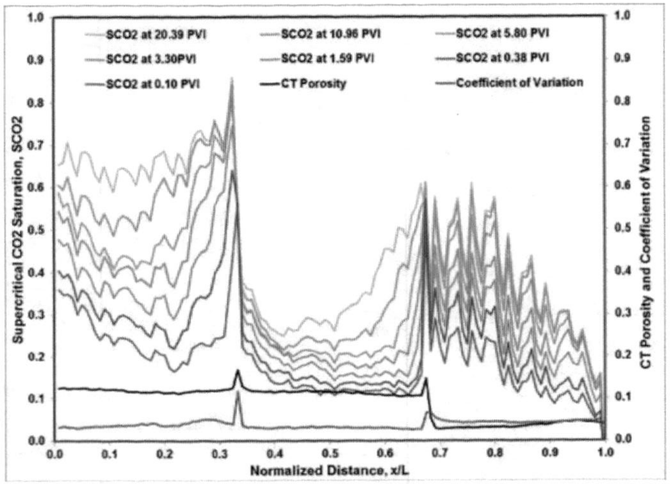

Figure 2: CT-derived saturation distribution in the St. Peter composite core during supercritical CO2 injection.

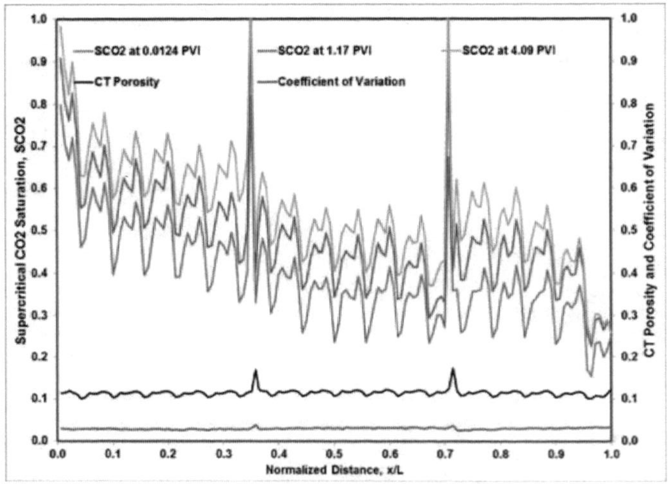

Figure 3: CT-derived saturation distribution in the Knox composite core during supercritical CO2 injection.

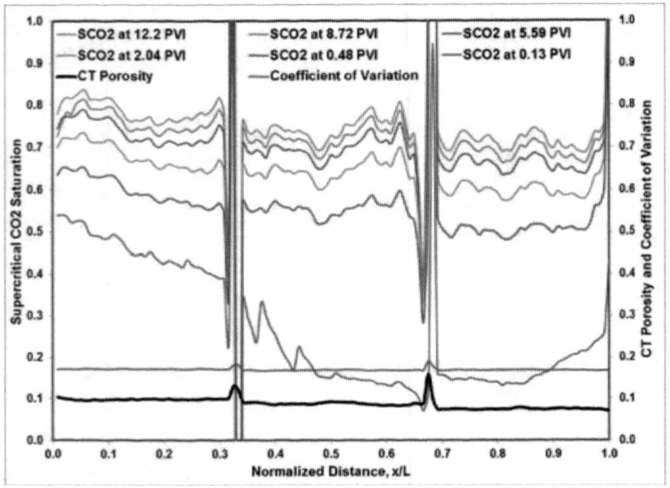

Figure 4: CT-derived saturation distribution in the Sylvania (upper) composite core during supercritical CO2 injection.

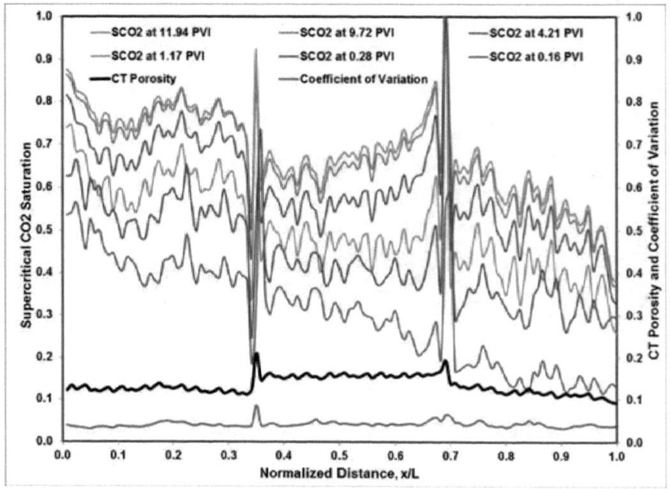

Figure 5: CT-derived saturation distribution in the Sylvania (lower) composite core during supercritical CO2 injection.

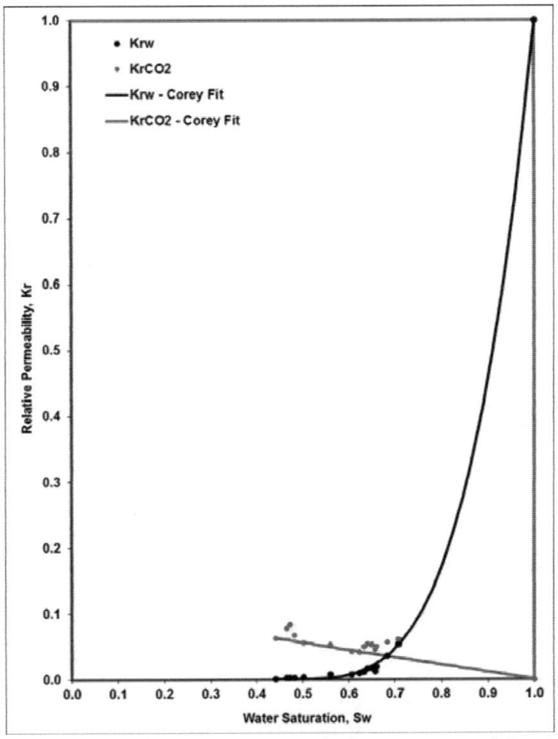

Figure 6: Unsteady-state brine-supercritical CO2 relative permeability for the St. Peter composite core sample.

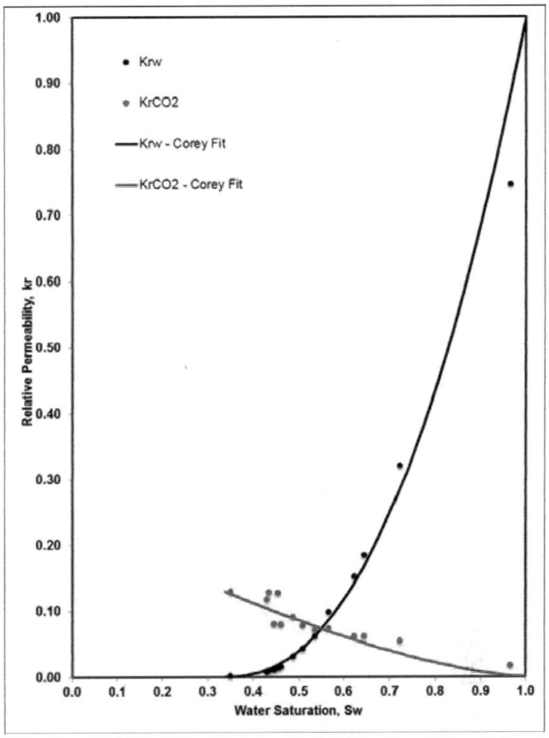

Figure 7: Unsteady-state brine-supercritical CO2 relative permeability for the Knox composite core sample.

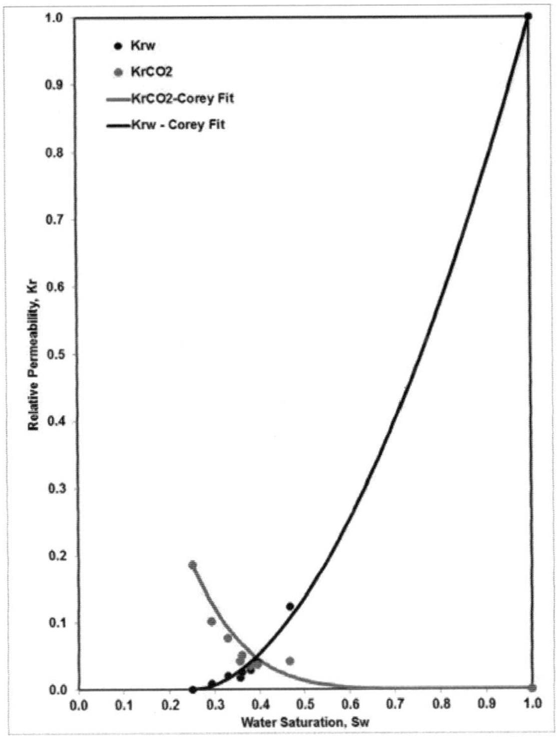

Figure 8: Unsteady-state brine-supercritical CO2 relative permeability for the Sylvania (upper) composite core sample.

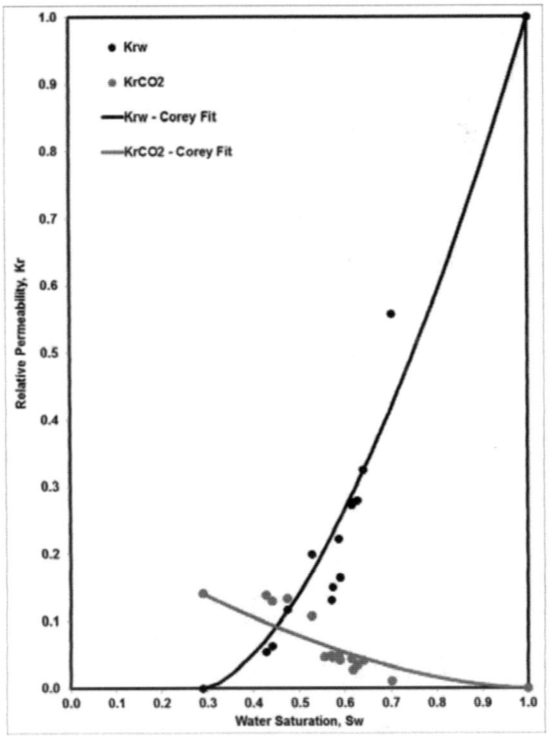

Figure 9: Unsteady-state brine-supercritical CO2 relative permeability for the Sylvania (lower) composite core sample.

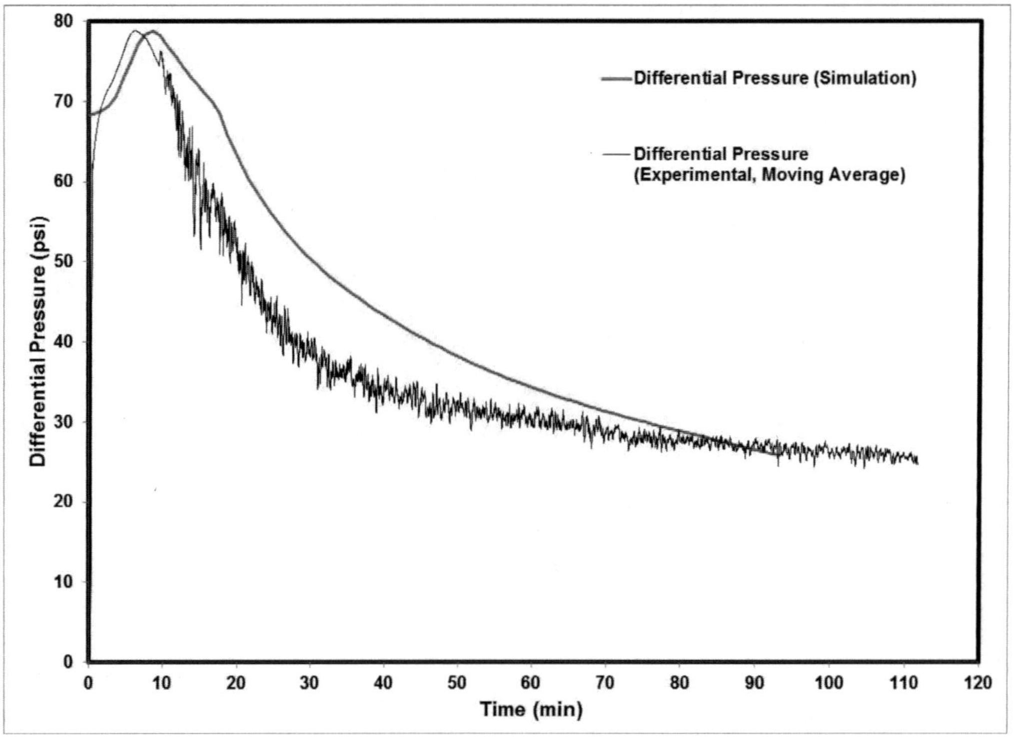

Figure 10: Differential pressure vs. time, Lab and Simulation.

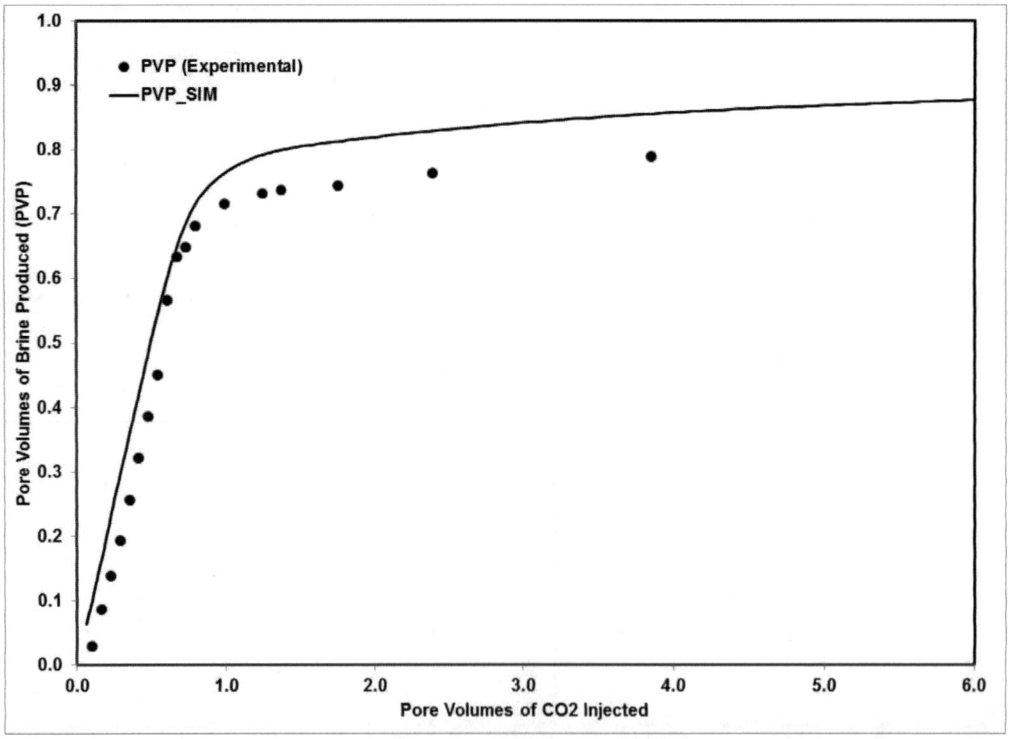

Figure 11: Pore volumes of brine recovered vs. pore volumes of CO2 injected, Lab and Simulation.

(688g) Modification of Fracture Hydraulic Properties by CO2-Acidified Brine Flow

Hang Deng, Brian R. Ellis and Catherine Peters, Department of Civil and Environmental Engineering, Princeton University, Princeton, NJ

Geological storage of CO_2 is a promising option for carbon mitigation, but its application faces challenges such as the possibility of CO_2 leakage. Quantification of the leakage risks requires characterization of the porosity and permeability of the porous media, and prediction of dynamic evolution of these properties caused by rock-CO_2-brine interactions. Fractures in porous media are of special interest as they can be created, activated and propagated during CO_2 injection (e.g. In Salah CO_2 storage site). Moreover, they serve as preferential-flow pathways, and allow constant flushing of the solutes, which may lead to enhanced rock- CO_2-brine reactions.

In this study, we aim to quantify the changes in fracture permeability and flow velocity fields caused by exposing a fractured carbonate caprock sample to CO_2-acidified brine flow. Before and after a flow-through experiment, X-ray computed micro-tomography scans were taken. The images were segmented and processed to reconstruct the initial and final fracture geometries. 1-D empirical equation, 2-D steady state flow model, and 3-D computational fluid dynamics (CFD) simulation were applied on the reconstructed fracture geometries to estimate the intrinsic permeabilities prior to and after the experiment. The capabilities of the 1-D empirical equation, 2-D steady state flow model and 3-D CFD simulation in quantitatively capturing the permeability evolution are also examined. Furthermore, comparing the detailed pre- and post- fluid velocity field provides important insights on the direction of local geochemical reactions. The simulation results serve to deepen our understandings of the modification of fracture hydraulic properties by CO_2-acidified brine flow, and hence to improve our predictive capability on evolution of caprock integrity and leakage risks.

(772a) New Experimental Data and Physico-Chemical Model for the Solubility of Carbon Dioxide in Aqueous Solutions of Monoethanolamine

Michael Wagner, Inga von Harbou, Jaeik Kim, Gerd Maurer and Hans Hasse, Laboratory of Engineering Thermodynamics, University of Kaiserslautern, Kaiserslautern, Germany

The most widely used solvent for Post Combustion Carbon Capture (PCC) is a 30 mass % aqueous solution of monoethanolamine (MEA). That solvent is generally accepted as a reference to which new PCC solvents are to be compared. [1, 2] Quite a large amount of experimental data is available for the vapor-liquid equilibrium (and related physico-chemical properties) such of such mixtures and one might assume that there is little need for new data. However, a closer look reveals that literature data for the solubility of carbon dioxide in a 30 mass % aqueous solution of MEA scatter widely. Even when obvious outliers are removed, the remaining uncertainty is so large that results of process simulations can be widely tuned simply by selecting certain data sets to which models are fitted.

Therefore in the present work, a comprehensive study on the solubility of carbon dioxide in aqueous solutions of MEA was carried out using two different experimental techniques: at low partial pressures of carbon dioxide (from 5 to 80 kPa) a headspace gas chromatographic technique (HS-GC) was used whereas at high pressures (from 0.4 to 8.5 MPa) a synthetic method was applied. Both apparatuses have been used in our laboratory for more than twenty years in many studies and are known to yield very reliable and accurate data. Aqueous MEA solutions with 15 and 30 mass % MEA were investigated at 313 K, 353 K and 393 K. The investigated pressure range corresponds to a molar ratio of carbon dioxide to MEA in the liquid from 0.1 to 1.3.

A physico-chemical model was developed to describe the new experimental data. Such modeling is challenging as it has to take simultaneously into account the complex chemical reactions and the strong physical non-ideality of the studied reactive electrolyte solutions. Five reactions are included: the autoprotolysis of water (I), the formation of bicarbonate (II) and carbonate (III), the protonation of MEA (IV) and the formation of carbamate (V). The extended Pitzer model is used to describe the liquid phase non-ideality. The equilibrium constants of reactions (I) – (V) were taken from the literature. The results are checked using NMR spectroscopic literature data on the speciation. [3, 4, 5]

Only a small number of parameters for interactions between the dominant species, are fitted. The model gives a good description of the entire experimental gas solubility data set as well as a good agreement with literature data for the liquid phase speciation (from NMR investigations).

Besides presenting the new experimental and modeling results, we will also briefly review of the previous approaches both regarding the experiments and the modeling and we will discuss the important impact of the physico-chemical data on the process simulation.

[1] Mangalapally, H. P.; Hasse, H., *Chem. Eng. Res. Des.* **2011**, *89*, 1216-1228.

[2] Wang,M.; Lawal, A.; Stephenson, P.; Sidders, J.; Ramshaw, C.,*Chem. Eng. Res. Des.* **2011**, *89*, 1609-1624.

[3] Hilliard, M. D. A, Ph.D. thesis, University of Texas at Austin, 2008.

[4] Böttinger, W.; Maiwald, M.; Hasse, H., *Fluid Phase Equilib.* **2008**, *263*, 131-143.

[5] Jakobsen, J. P.; Krane, J.; Svendsen, H. F., *Ind. Eng. Chem. Res.* **2005**, *44*, 9894-9903.

(772b) Mass Transfer Evaluation of CO2 Absorption in a Wetted Wall Column

Reynolds A. Frimpong, Joseph E. Remias, James K. Neathery and Kunlei Liu, Center for Applied Energy Research, University of Kentucky, Lexington, KY

The chemical absorption of CO_2 in solvents is considered as one of the mature process options for reducing CO_2 emissions in power plants. Solvents with fast mass transfer rates and high CO_2 carrying capacity are therefore desired and continue to be explored to reduce equipment size and its associated capital cost. The wetted wall column provides a relatively facile assessment of solvent mass transfer characteristics and was used in this study to analyze the performance of 30 wt% monoethanolamine (MEA) for CO_2 absorption at utility flue gas concentrations that will pertain over the height of an absorber in post-combustion CO_2 capture process.

In this work, different CO_2 gas concentrations were absorbed by MEA at various carbon loadings and the flux measured. The overall mass transfer of CO_2 into MEA was evaluated by considering the relative contributions of the gas and liquid side resistances. The applicability of the pseudo-first order approximation and the instantaneous reaction were assessed for the different conditions to determine the appropriate reaction regimes. It was determined that at lean solvent conditions with low driving force (representative of conditions at the top of an absorber), mass transfer was well modeled by the pseudo-first order approximation.

(772c) Kinetic Study of Carbon Dioxide Absorption by Aqueous Solutions of N-Methyldiethanolamine and Piperazine

Alberto Servia[1,2], Nicolas Laloue[1], Julien Grandjean[1], Sabine Rode[2] and Christine Roizard[2], (1)IFP Energies nouvelles, Solaize, France, (2)Laboratoire Réactions et Génie de Procédés (LRGP) - ENSIC, Nancy, France

The CO_2 post-combustion capture process consists on the chemical absorption of this acid gas into an aqueous solution of alkanolamines. To optimise both investment and operating costs the solvent must present high reaction rates with CO_2 and low reaction enthalpies. Chakravarty et al., 1985 observed that adding small quantities of a primary or a secondary amine to an aqueous solution of a tertiary amine considerably enhances the CO_2 absorption rate without significantly increasing the reaction enthalpy.

1. Introduction

The kinetics of CO_2 absorption with aqueous blends of N-MethylDiEthanolAmine (MDEA) and PiperaZine (PZ) has been widely studied (Zhang et al., 2001, Bishnoi and Rochelle, 2002, Edali et al., 2010, Samanta and Bandyopadhyay, 2011). Nevertheless, some discrepancies remain concerning the reaction mechanisms proposed in the literature. For instance, the synergy between both amines is still not completely understood and proposed kinetics models cannot fully explain the experimental data at very high loadings. The piperazine carbamate and dicarbamate hydrolysis were not investigated at these conditions neither the eventual participation of other bases in the interaction between PZ and piperazine derivatives with CO_2. Finally, the models used in the literature to deduce the kinetics of CO_2 absorption by aqueous blends of PZ and MDEA from experiments present some limitations. They all assume a constant CO_2 partial pressure in the gas phase. This hypothesis can lead to noticeable errors if high CO_2 quantities are absorbed.

The objective of this work is to investigate the kinetics of CO_2 absorption by aqueous blends of MDEA and PZ. Experiments are performed in a wetted wall column under various operating conditions and simulated using a rigorous reactor model.

2. Experimental setup

The experiments are conducted in a wetted wall column at temperatures ranging from 24 to 60 °C on unloaded and loaded aqueous solutions of MDEA and PZ. The solutions loading is up to 0,3 mol_{CO2}/mol_{amine} while the MDEA and PZ concentrations range between 2 – 4 M and 0,2 - 1 M, respectively.

The wetted wall column is a suitable equipment to obtain kinetic data of gas-liquid systems presenting high reaction rates due to the high values of liquid mass transfer coefficients (k_L) associated to this device. Within the reactor, the gas phase flows counter-currently with the liquid that overflows from the inside of a cylinder to form a thin liquid film. The gas phase, composed of CO_2 and Nitrogen (N_2), is water-saturated before being in contact with the liquid in the reactor to prevent from water mass transfer in the reaction zone. The experimental flux is calculated using the variation of the CO_2 gas concentration at both the inlet and the outlet of the reactor, measured by an in-line infra-red spectrometer.

The gas phase hydrodynamics was characterized to determine the gas phase mass transfer coefficient (k_G) by performing experiments of CO_2 absorption in aqueous solutions of MonoEthanolAmine (MEA). Experiments conducted at different solvent concentrations enable simultaneous determination of the k_G and the gas-liquid contact area that was found to be very close to the geometric area.

3. Model description

A 2D stationary model has been developed using COMSOL software to predict the absorption flux of CO_2 in aqueous blends of PZ and MDEA in the wetted wall column. This model couples hydrodynamics, mass transfer, chemical reactions and gas-liquid equilibrium. One of the model originalities is the description of the CO_2 partial pressure evolution within the gas phase, instead of considering a constant CO_2 partial pressure given by the logarithmic average between the reactor inlet and outlet.

The concentration profile of the chemical species within the liquid phase is obtained by the simultaneous resolution of the mass balance for each compound (equation 1) and the electroneutrality condition

$0 = D_i\left(\frac{\partial^2 C_i}{\partial r^2} + \frac{1}{r}\frac{\partial C_i}{\partial r} + \frac{\partial^2 C_i}{\partial z^2}\right) - v(r)\frac{\partial C_i}{\partial z} + \sum_{j=1}^{n} v_{ij} R_j$	Equation 1

where r and z represents radial and axial coordinates respectively

The liquid phase velocity profile is determined by the Navier-Stokes equation for an incompressible fluid associated with specific boundary conditions. Thus, the liquid phase velocity is 0 at the liquid-solid surface (r=0) and it presents a maximum at the gas liquid-interface (r=th).

The reaction rates R_j are determined by five reversible chemical reactions occurring in the liquid phase:

$CO_2 + OH^- \xleftrightarrow{k_1, k_{-1}} HCO_3^-$	Equation 2
$CO_2 + PZ + H_2O \xleftrightarrow{k_2, k_{-2}} PZCOO^- + H_3O^+$	Equation 3
$CO_2 + PZ + MDEA \xleftrightarrow{k_3, k_{-3}} PZCOO^- + MDEAH^+$	Equation 4
$CO_2 + PZCOO^- + H_2O \xleftrightarrow{k_4, k_{-4}} PZ(COO)_2^{2-} + H_3O^+$	Equation 5
$CO_2 + PZCOO^- + MDEA \xleftrightarrow{k_5, k_{-5}} PZ(COO)_2^{2-} + MDEAH^+$	Equation 6

The five instantaneous chemical reactions are modeled by algebraic equations involving their equilibrium constants.

$OH^- + H_3O^+ \xleftrightarrow{K_1} 2H_2O$	Equation 7
$MDEA + H_3O^+ \xleftrightarrow{K_2} MDEAH^+ + H_2O$	Equation 8

$PZ + H_3O^+ \xleftrightarrow{K_3} PZH^+ + H_2O$	Equation 9
$PZCOO^- + H_3O^+ \xleftrightarrow{K_4} H^+PZCOO^- + H_2O$	Equation 10
$CO_3^{2-} + H_3O^+ \xleftrightarrow{K_5} HCO_3^- + H_2O$	Equation 11

A partial differential-algebraic system composed by 12 equations is solved to determine the concentration profiles $C_i(r,z)$ in the liquid phase. The used boundary conditions are given as follows.

$C_i(r,h) = C_i^{eq}$	Equation 12
$D_i \frac{\partial C_i(r,0)}{\partial z} = 0$	Equation 13
$D_i \frac{\partial C_i(0,z)}{\partial r} = 0$	Equation 14
$D_i \frac{\partial C_i(th,z)}{\partial r} = 0 \quad i \neq CO_2$	Equation 15
$D_{CO_2} \frac{\partial C_{CO_2}(th,z)}{\partial r} = k_G \left(P_{CO_2} - HC_{CO_2}\right)$	Equation 16

where *h* represents the reactor height, *H* the Henry constant and P_{CO2} the CO_2 partial pressure in the gas phase.

The concentration of each species at equilibrium conditions are provided by a thermodynamic model that accounts for the deviations from the ideal behavior through an electrolyte NRTL approach. It has been validated by comparison with gas-liquid equilibrium experimental data found in the literature. The calculated concentrations are used to determine the equilibrium constants in equations 7 to 11. The CO_2 Henry constant typically used in the literature is based on a N_2O analogy. In this model, the CO_2 Henry constant is determined by the ratio between P_{CO2} and the molecular CO_2 concentration at equilibrium provided by the thermodynamic model.

As CO_2 is absorbed, P_{CO2} presents a decreasing profile within the reactor. Consequently, the proposed model takes into account the evolution of the CO_2 partial pressure through a plug-flow model:

$$0=\frac{\partial(v_G C_{CO_2})}{\partial z}+k_G a(P_{CO_2}-HC_{CO_2})$$	Equation 17

where a represents the ratio between the gas-liquid contact area (A) and the reactor volume.

Moreover, the CO_2 transfer from the gas phase into the liquid phase engenders a variation of the gas velocity that is modeled through a mass balance on the N_2.

$$v_G=\frac{Q_G^{in}(1-y_{CO_2}^{in})P}{A(P-C_{CO_2}^{G}RT)}$$	Equation 18

where $Q_G{}^{in}$ and $y_{CO2}{}^{in}$ represent the gas volume flow and the CO_2 molar fraction at the reactor inlet, respectively.

4. Preliminary results

This model has been validated by comparison with experimental data obtained on unloaded aqueous solutions of PZ ranging from 0,2 to 1 M and at temperatures between 24 and 60 °C. The reaction mechanism proposed by Bishnoi and Rochelle, 2002 and their kinetic constants associated to equations 3 and 5 were used to model the kinetics of the CO_2/PZ system. Equations 4 and 6 were not considered since the aqueous solutions did not contain MDEA.

Figure 1 presents the simulated normalized concentration profiles of PZ as a function of the radial coordinate for various PZ concentrations for experiments conducted at an inlet P_{CO2} of 1050 Pa. The results show a depletion of the PZ concentration at the gas-liquid interface up to 10%. This implies that the interpretation of the experimental data using a simple model considering two mass transfer resistances and assuming the pseudo-first order regime for PZ remains possible but is not fully rigorous. Actually, this type of approach is only suitable when very low driving forces are involved since it does not account for the existing $PZCOO^-$ concentration present at the gas-liquid interface. It appears that a numerical approach is needed for a rigorous representation of the CO_2 absorption experiments conducted in the wetted wall column for various operating conditions, even in unloaded solutions. A numerical approach is essential in the case of loaded solutions to account for $PZCOO^-$ and other species present in the liquid.

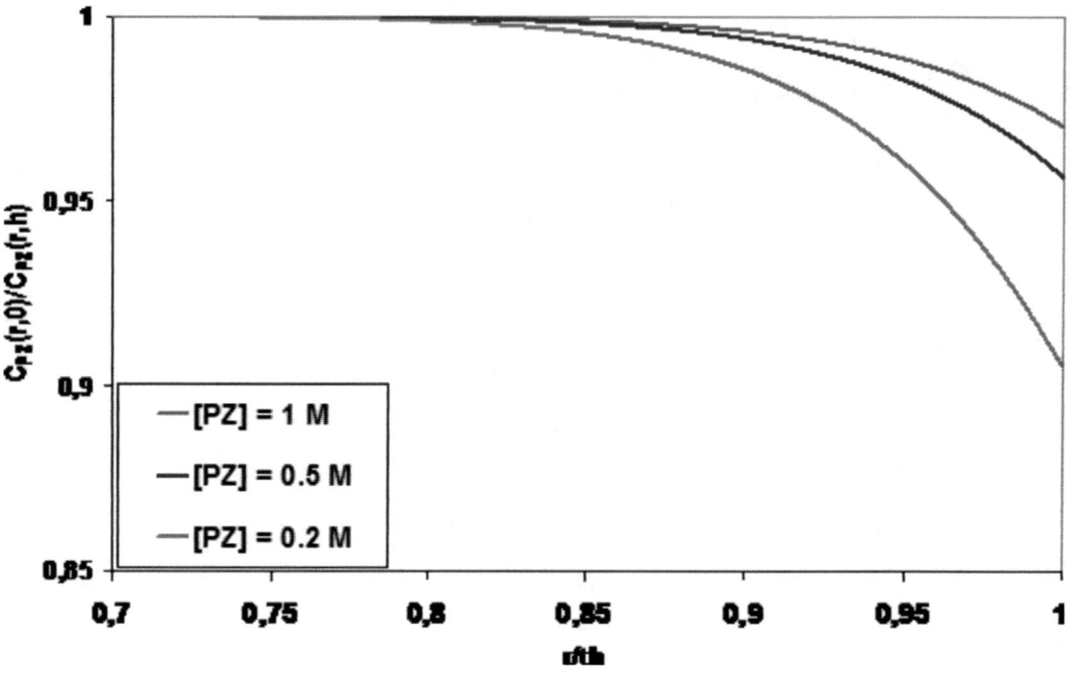

Figure 1 – Simulated normalized PZ concentration profiles in the liquid phase at the reactor exit at 60 °C.

The results presented in Figure 2 show a very good agreement between experimental and simulated CO_2 absorption fluxes, with an average absolute deviation of 3,7 %.

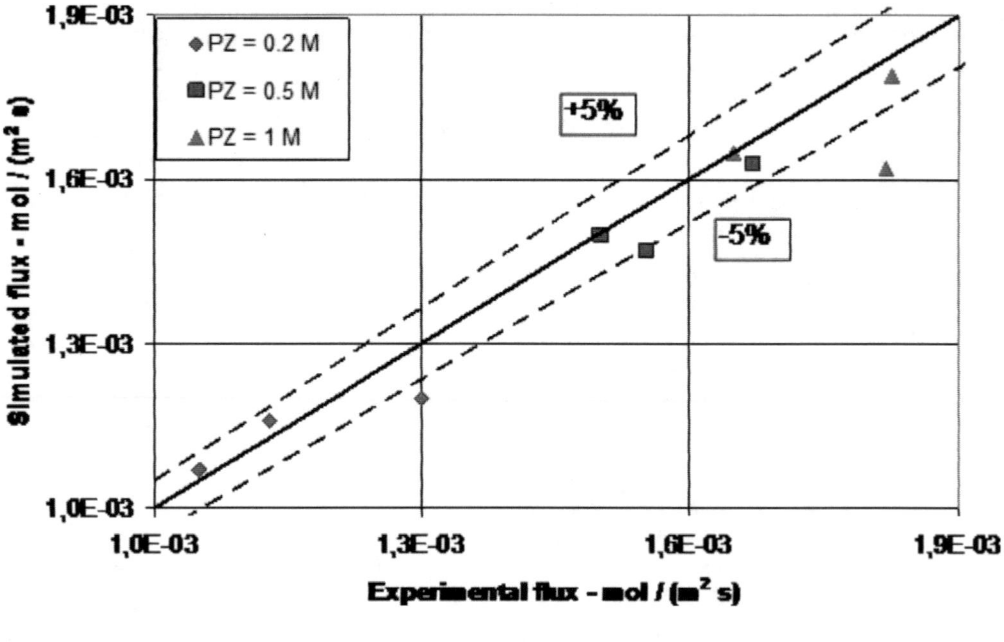

Figure 2 – Model validation with experimental data on unloaded PZ solutions.

5. Further work

CO_2 absorption experiments on loaded PZ solutions and on both unloaded and loaded blends of MDEA and PZ are being conducted. The objective is to confirm and quantify the synergy between both amines. Besides, the CO_2 absorption experiments on loaded solutions will provide information about the role of $PZCOO^-$ and its interaction with other species, especially with PZ, H_2O and MDEA. Finally, the model is expected to enable the prediction of CO_2 absorption in aqueous blends of PZ and MDEA in a wide range of operating conditions by implementing a complete reaction mechanism.

References

Bishnoi, S., Rochelle, G.T., (2002). Absorption of carbon dioxide in aqueous piperazine/methyldiethanolamine. Aiche Journal, 48, 2788-2799.

Chakravarty, T. et al. (1985). Reaction of Acid Gases with Mixtures of Amines. Chemical Engineering Progress, 81, 32-36.

Edali, M. et al. (2010). 1D and 2D absorption-rate/kinetic modeling and simulation of carbon dioxide absorption into mixed aqueous solutions of MDEA and PZ in a laminar jet apparatus. International Journal of Greenhouse Gas Control, 4, 143-151.

Samanta, A., Bandyopadhyay, S. S., (2011). Absorption of carbon dioxide into piperazine activated aqueous N-methyldiethanolamine. Chemical Engineering Journal, 171, 3, 734-741.

Zhang, X. et al. (2001). A kinetics study on the absorption of carbon dioxide into a mixed aqueous solution of methyldiethanolamine and piperazine. Industrial & Engineering Chemistry Research, 40, 3785-3791.

(772d) Aqueous Morpholine As a Potential Post-Combustion CO2 Absorption Solution

Krishna Mohan Jinka, Taylor Kuhn, Joseph E. Remias, James K. Neathery and Kunlei Liu, Center for Applied Energy Research, University of Kentucky, Lexington, KY

Due primarily to high cyclic capacity and high absorption rates compared to other methods, aqueous amine based CO_2 capture has gained great interest for application to post-combustion CO_2 capture. While aqueous amines have a long history for acid gas removal, challenges remain for the commercial application to power plant post-combustion CO_2 capture due to the high gas volume and low CO_2 gas pressure. These challenges require a solvent able to simultaneously deliver high mass transfer rate, high thermal and oxidative stability, low corrosion, and low energy of solvent regeneration to effectively reduce the cost of post-combustion CO_2 capture. Morpholine was identified to potentially meet the above needs. The absorption of carbon dioxide into aqueous morpholine solutions (35-53 wt%) using a wetted wall column were performed using simulated flue gas at atmospheric pressure and with 3-14 vol% of CO_2 to obtain the overall mass transfer coefficient. Temperatures between 30-40 °C were tested as a believed reasonable compromise between acceptable absorber tower temperatures and morpholine volatility.

(772e) Novel Liquid-Like Nanoscale Organic Hybrid Materials for CO2 Capture

Camille Petit, Department of Earth and Environmental Engineering and Department of Chemical Engineering, Columbia University, New York, NY, Youngjune Park, Earth and Environmental Engineering and Chemical Engineering, Columbia University, New York, NY, Kunyi Andrew Lin, Earth and Environmental Engineering, Columbia University, New York, NY and Ah-Hyung Alissa Park, Earth and Environmental Engineering & Chemical Engineering, Columbia University, New York, NY

New solvents referred to as Nanoparticle Organic Hybrid Materials (NOHMs) are being developed for various energy and environmental applications, with a particular focus on CO_2 capture. These nanoscale hybrid materials consist of an inorganic nanocore functionalized with a polymeric canopy, which acts as the solvating medium. Compared to the conventional amine-based solvents, NOHMs exhibit valuable properties such as negligible vapor pressure and high thermal stability. CO_2 capture by NOHMs is expected to arise from two properties: the tunable chemistry of the canopy via selection of task-specific functional groups, as well as the specific structural arrangement of the polymer chains, favored by decreases in entropy. In this study, different NOHMs were synthesized and they were characterized using thermogravimetric analysis, spectroscopic tools (i.e., NMR, Raman, ATR FT-IR), and viscosity measurement techniques. The CO_2 capture and CO_2-induced swelling behaviors of NOHMs were evaluated and the mechanisms of CO_2 absorption were investigated. The effects of the canopy structure (e.g., grafting density, polymer geometry), the canopy functionalization (e.g., presence of amine groups), and the presence of other flue gas components (e.g., N_2, SO_2, H_2O) on the overall CO_2 capture process were studied.

(772f) Development of Reaxff Force Field for Carbon Dioxide Capture with Ionic Liquids: A Combined First Principles and Classical Simulation

Bo Zhang, US DOE-National Energy Technology Laboratory, Pittsburgh, PA, Adri C.T. van Duin, Department of Mechanical and Nuclear Engineering, Pennsylvania State University, University Park, PA and Karl Johnson, Chemical and Petroleum Engineering, University of Pittsburgh, Pittsburgh, PA

Ionic liquid (IL) are salts in the liquid state below 100°C. The number of ions that can form IL is well over tens of thousands. ILs are considered as promising materials for CO_2 capture due to their wide liquid range, negligible vapor pressure, thermal and chemical stability, and the ability to tailor the interaction with CO_2 through choosing appropriate ions. We are exploring ILs that can capture CO_2 through chemical reactions and physical interactions via first principles and classical simulation techniques. Our goals are to understand the mechanism of IL and CO
to develop a new force field that will capture both the physical interactions and chemical reactions, and to calculate the thermodynamic and transport properties of bulk and confined IL with CO
simulations will be used to accelerate the design of new ILs that have tailored interactions and acceptable viscosities for pre-combustion CO_2 capture processes. We use the ReaxFF formalism for describing the physical and chemical interactions in the IL-CO_2 system. ReaxFF has been proven to be a very successful formalism for modeling chemical as well as physical interactions with an empirical approach. [P(C_4)
interactions between ILs and CO_2. Two possible interaction sites have been identified with first principles molecular dynamics (MD) simulations. One is the $-NH_2$ center and the other one is the $-CO_2^-$ center, both on the [Gly] anion. The calculated binding energy and CO_2 angle distribution indicate that the interaction with the amine center is a chemical type interaction and the interaction with carboxylic group appears to be a complicated weak chemical interaction. The reaction between CO_2 and $-NH_2$ has been identified as a two-step reaction mechanism. The first step involves a proton transfer with barrier height of about 1 kcal/mol, and the second step is mainly a libration motion, also leading to a proton transfer, with barrier height around 2.3 kcal/mol. The reaction pathways along with other data, including bond-energies, angle-energies, torsion angle-energies have been compiled into a training set to parameterize the ReaxFF force field. The force field is optimized against the calculated structure-energy relationships. With the optimized force field, the model IL [P(C_4)$_4$][Gly] interacting with CO_2 has been studied through large scale MD simulations. The thermodynamic and transport properties have been computed.

(772g) Vibrational Spectroscopy Characterization of CO2-Ionic Liquid Systems for Pre-Combustion CO2

Anita S. Lee and John R. Kitchin, Department of Chemical Engineering, Carnegie Mellon University, Pittsburgh, PA

Ionic liquids are attractive for pre-combustion CO_2 capture applications due to their low volatility, thermal stability, and non-aqueous environment.[1,2] Studies have shown that CO_2 solubility in ionic liquids can be modified through molecular changes in the anion and cation.[3–5] This is promising for the development of new ionic liquids. However, the larger number of possible cation and anion combinations suggest that a better understanding of molecular structure and CO_2 solubility is necessary guide the development of new ionic liquid CO2 solvents. Vibrational spectroscopy techniques can provide a picture at the molecular level of CO_2-ionic liquid interactions, aiding in understanding the mechanism of CO_2 capture and relationships of solubility to properties of the ionic liquid. In this work, we use Raman spectroscopy to study the interaction of CO_2 with a class of imidazolium ionic liquids a high pressure optical cell, we are able to study the vibrational spectra of solvated CO_2 from 1 to 40 bar, pressures relevant to pre-combustion capture condition. We observe peaks in the CO_2-IL spectra that correspond to the upper and lower band of the fermi-diad split of the CO_2 symmetric stretch vibration, suggesting that solvated CO_2 maintains a linear geometry. The bands are slightly shifted from the gas phase spectra of CO_2, indicating impact of the solvent environment. Additionally, we observe a pressure dependence on spectral intensity of the solvated CO_2 peaks that correlate with CO_2concentration.

1. Hasib-ur-Rahman, M., Siaj, M. & Larachi, F. Ionic liquids for CO2 capture—Development and progress. *Chemical Engineering and Processing: Process Intensification* **49**, 313-322 (2010).

2. Brennecke, J.F. & Gurkan, B.E. Ionic Liquids for CO 2 Capture and Emission Reduction. *Society*3459-3464 (2010).doi:10.1021/jz1014828

3. Bates, E.D., Mayton, R.D., Ntai, I. & Davis, J.H. CO 2 Capture by a Task-Specific Ionic Liquid. 2001-2002 (2002).

4. Wang, C. *et al.* Tuning the Basicity of Ionic Liquids for Equimolar CO 2 Capture **. *Communications*4918-4922 (2011).doi:10.1002/anie.201008151

5. Anthony, J.L., Anderson, J.L., Maginn, E.J. & Brennecke, J.F. Anion effects on gas solubility in ionic liquids. *The journal of physical chemistry. B* **109**, 6366-74 (2005).

(785a) Evaluation of Mercury Capture Technology Using Various Sorbents Injection

Xiaofei Wang, Department of Energy, Environmental & Chemical Engineering, Washington University in St Louis, St Louis, MO and Pratim Biswas, Department of Energy, Environmental and Chemical Engineering, Washington University, St Louis, MO

Coal combustion is a single largest known anthropogenic source for mercury (Hg) emission. Recently, US EPA has proposed Mercury and Air Toxics Standards (MATS) in order to regulate emissions of Hg and other pollutants. Many Hg control technologies have been developed, such as activated carbon injection, catalytic oxidation of Hg and metal oxide sorbent injections. However, no single technology can have effective and economical Hg removal performance for various coal combustion conditions and existing control devices for other pollutants. We evaluate the performance of the sorbent technologies and their combinations under different combustion conditions in both lab-scale (a drop-tube furnace) and pilot-scale (the facility capacity: 1 MW) coal combustors. These sorbents include titanium dioxide, selective catalytic reduction (SCR) catalysts (e.g. vanadium, molybdenum, tungsten oxides and zeolites), potassium iodide and modified carbon particles. Hg speciation in flue gas was also studied for a variety of combustion conditions and sorbent injection methods.

(785b) Overview on Mercury Control Options for Coal-Burning Power Plants

Evan J. Granite, National Energy Technology Laboratory, United States Department of Energy, Pittsburgh, PA and Henry W. Pennline, U.S. Department of Energy, National Energy Technology Laboratory, Pittsburgh, PA

With the USEPA issuing a national regulation requiring 91% removal of mercury, and many states promulgating their own regulations, the need exists for low-cost mercury removal techniques that can be applied to coal-burning power plants. The injection of powdered activated carbon into the ductwork upstream of the particulate control device is the most developed technology for mercury capture. Alternative techniques for mercury capture will also play a role in the near future because of the numerous configurations of air pollution control devices present within the power plants, as well as the many different coals being burned. These methods employ sorbents, catalysts, scrubber liquors, flue gas or coal additives, combustion modification, flue gas cooling, barrier discharges, and ultraviolet radiation for the removal of mercury from flue gas streams. The DOE Mercury Program has been a huge success, spurring development, demonstration, and commercialization of many technologies for the capture of mercury.

An overview of current and alternative technologies for mercury capture from coal-derived flue gas will be provided. In addition, six patent/patent pending methods for mercury as well as carbon dioxide control within coal-derived flue and fuel gases have been recently developed at NETL, and will be discussed. The future research needs for mercury control include improved sorbent-flue gas contact, development of poison-resistant sorbents and catalysts, new scrubber additives for retention of mercury within wet FGD systems, concrete-friendly activated carbons, new continuous measurement methods, byproducts research, and development of an ASTM standard lab test for sorbent activity for mercury capture.

The EPA's proposed Carbon Pollution Standard for New Power Plants will also be briefly discussed.

(785c) Palladium Sorbents for High Temperature Capture of Mercury, Arsenic, Selenium and Phosphorus from Fuel Gas

Evan J. Granite[1], Henry W. Pennline[2], Erik C. Rupp[3], Dennis Stanko[3], John P. Baltrus[2], Hugh Hamilton[4], Liz Rowsell[4], Stephen Poulston[4], Andrew Smith[4], Wilson Chu[5] and Tony Wu[6], (1)National Energy Technology Laboratory, United States Department of Energy, Pittsburgh, PA, (2)U.S. Department of Energy, National Energy Technology Laboratory, Pittsburgh, PA, (3)National Energy Technology Laboratory, Pittsburgh, PA, (4)Johnson Matthey Technology Centre, Reading, United Kingdom, (5)Johnson Matthey, West Chester, PA, (6)National Carbon Capture Center, Southern Company, Wilsonville, AL

In gasification for power generation, the removal of mercury by sorbents at elevated temperatures preserves the high thermal efficiency of the integrated gasification combined cycle system. Unfortunately, most sorbents will display poor capacity for elemental mercury at elevated temperatures. Previous experience with sorbents in flue gas has allowed for judicious selection of potential high temperature candidate sorbents. The capacities of many sorbents for elemental mercury from nitrogen, as well as from different simulated fuel gases at temperatures from 400 - 700°F, were determined. The simulated fuel gas compositions contain varying concentrations of mercury, arsine, hydrogen selenide, phosphine, carbon monoxide, hydrogen, carbon dioxide, moisture, and hydrogen sulfide. Palladium is an attractive sorbent candidate for the removal of mercury from fuel gases at elevated temperatures. Recent results suggest that palladium has excellent potential for arsenic, phosphorus and selenium capture from fuel gases, making it capable of multi-pollutant capture. A license agreement has been signed by the United States Department of Energy and Johnson Matthey for further development of the sorbents. The sorbents have recently removed nearly 100% of the mercury, arsenic and selenium from slipstreams of dirty syngas at 500F during several extended exposures at a pilot gasification facility. Future research areas and sorbent development for trace metal capture from fuel gases will be discussed.

AUTHOR INDEX

Aaggard, Per 165
Abdelsayed, Victor 244, 377
Adeyemo, Kehinde 14
Akbari, M. H. 219
Algadi, Otman 426
Allen, D. 33
Al-Muhtaseb, Shaheen A. 6
Al-Najem, Abdullah 351
Alquraishi, Abdulrahman A. 426
Anderson, Brian 166
Anugwom, Ikenna 413
Asaee, S. A. 219
Assabumrungrat, Suttichai 262
Atakan, V. 92, 93
Axelbaum, Richard L. 274
Azad, Abdul-Majeed 277
Azenkeng, Alexander 98
Azurdia, J. 92
Baliban, Richard C. 44
Baltrus, John P. 276, 377, 452
Baltus, Ruth E. 11
Banu, Laila 11
Barbieri, Giuseppe 326
Basile, Angelo 328
Bekemohammadi, Roxana 198
Berry, David 59, 68, 202
Besser, R. S. 236
Bhattacharyya, Debangsu 109, 303, 311
Bhown, Abhoyjit S. 19
Bhuwania, Nitesh 70
Biegler, Lorenz T. 45
Bielicki, Jeffrey 188
Birkholzer, Jens 415
Bischoff, Brian L. 12
Biswas, Pratim 450
Bitteto, M. A. 92
Boek, Edo 346
Booras, George 19
Borujeni, Ali Takbiri 350
Botte, Gerardine G. 175
Bouhlila, Rachida 172, 366
Bradley, Seth M. 350
Bromhal, Grant 168, 169, 347, 415
Brouwer, Jack 198
Brunet, Leopold 419
Brunetti, Adele 326
Buckley, Benjamin R. 108
Campanari, Stefano 329
Caravella, Alessio 326
Carneiro, Juliana S. A. 182
Carpenter, Stephen 296
Carr, Timothy 187
Carroll, Susan 185
Catalano, Jacopo 15
Chanburanasiri, Naruewan 262
Chapman, Emily 346
Chaudhari, Kiran 302
Chen, Liang 97
Chen, Liangyong 294, 295
Chen, Yongsheng 60
Cho, Su Hyun 173
Chong, Kok Lin 10
Choo, Teck Kwang 23
Chu, Deryn 55, 177
Chu, Wilson 452
Chung, Tai-Shung 9, 10
Cowan, Robert M. 98
Crandall, Dustin 168
Crawshaw, John 346
Czaun, Miklos 270
Czerepinski, J. 92
Da Rocha, Sandro R. P. 182
Damodharan, Shalini 324
Daramola, Damilola A. 175
Dariva, Cláudio 182
Datta, Arunabha 173
Datta, Ravindra 13
De Wit, Ellart 329
Decristofaro, Nick 92, 93
Deng, Hang 437
Deng, Xingyi 377
Deveau, Nicholas 13
Dhungel, Bhupesh 274
Dillich, Sara 313
Dillon, Des 35
Dilmore, Robert 33, 415
Dong, Junhang 345
Doucet, Frédéric J. 24
Dowell, Niall Mac 49
Dowling, Alexander W. 45
Drago, Francesca 328
Drioli, Enrico 326
Dufek, Eric 403
Dyer, Kathy 417
El-Hassan, Hilal 94
Elia, Josephine A. 44
Ellis, Brian R. 437
Elyassi, Bahman 8
Estejab, Ali 175
Eta, Valerie 413
Fan, Hongxia 385
Fan, Liang-Shih 208
Fang, Weijie 4
Feng, Jie 385
Feng, Li 378
Ferreira, Bianca M. S. 182
Fitts, Jeffrey 188
Floudas, Christodoulos A. 44
Fortuny, Montserrat 182
Franceschi, Elton 182

AUTHOR INDEX

Freeman, Benny 90
Frimpong, Reynolds A. 439
Frost, Lyman 207
Furler, Philipp 401
Gallagher Jr., Michael J 296
Gamadi, Talal 426
Gao, Peng 378
Garapati, Nagasree 166
Ge, Qingfeng 397
Gel, Aytekin 302
Ghezel-Ayagh, Hossein 37, 323
Gitari, Wilson M. 24
Godec, Michael L. 186
Goeppert, Alain 270
Gondle, Raj 169
Gor, Gennady 418
Grandjean, Julien 440
Granite, Evan J. 451, 452
Grant, Tim 21
Green Jr., William H. 57
Guazzone, Federico 15
Guenther, Chris 302, 304
Gundersen, Truls 46
Guo, Ruilan 90
Gupta, S. 92
Hamilton, Hugh 452
Hamza, Ali 71
Han, Chunli 294, 295
Hao, Yue 185, 417
Hasan, M. M. Faruque 44
Hasse, Hans 438
Hawley, Kyle 58
Haynes, Daniel J. 59, 68, 244
He, Jiajun 164
He, Xuezhong 46
Hedayati, Ali 277
Hedges, S. W. 33
Heebink, Loreal V. 98, 190
Horn, Norman R. 9
Houseworth, Jim 415
Howard, Bret H. 377
Hu, Liang 94
Hunt, Jennifer 37
Hunwick, Richard 30
Hussain, A. H. M. Shahadat 56
Iskhakov, Ruslan 167
Iulianelli, Adolfo 328
Jalan, Amrit 57
Jensen, Melanie D. 98, 190
Jeong, Hae-Kwon 87
Jhong, Molly 411
Jia, Yunlu 298
Jiang, Jianwen 4
Jin, Yi 272
Jinka, Krishna Mohan 446
Johnson, James 417
Johnson, Karl 448
Jolly, Stephen 37, 323
Junaedi, Christian 58
Kabir, Aslam M. 168
Kail, Brian 1
Karan, Kunal 52, 57
Karpyn, Zuleima 419
Kasule, Job 303
Kathe, Mandar 208
Kazantzis, Nikolas 15
Kelkar, Sharad 416
Keller, S. 196
Kenis, Paul J. A. 411
Keyvani, Majid 71
Kim, Bang Hee 173
Kim, Daejin 16
Kim, Jaeik 438
Kim, Sang Woo 173
Kim, Seok-Jhin 345
Kim, Seung-Hoon 173
Kitchin, John R. 50, 449
Kocherlakota, Lakshmi Suhasini 5
Kolb, Gunther 196
Koros, William J. 69, 70, 89, 91
Kraftschik, Brian E. 91
Krishnan, J. 92, 93
Kruger, Tim 31
Kuhn, Taylor 446
Kumfer, Benjamin M. 274
Kuncharam, Bhanu Vardhan Reddy 324
Kuramoto, Koji 209
Kutchko, Barbara 419
Kwon, Hyuk Taek 87
Laabidi, Ezzeddine 172, 366
Laloue, Nicolas 440
Laosiripojana, Navadol 262
Laumb, Jason D. 98
Lawrence, Richard 21
Lee, Anita S. 50, 449
Lee, Kwan-Young 173
Lee, Seung-Hwan 173
Leion, Henrik 277, 286
Lekse, Jonathan W. 276, 377, 400
Letvin, Peter A. 98
Lewis, James P. 400
Li, Fanxing 208
Li, Lei 178
Li, Li 419
Li, Pei 9
Li, Q. 92, 93
Li, Tingwen 302
Li, Wenying 385
Li, Xiaohong 385
Li, Yongdan 272

AUTHOR INDEX

Liese, Eric A. 109
Liguori, Simona....328
Lin, Jiefeng....315
Lin, Kunyi Andrew....447
Lindner, Peter J. 236
Lipscomb, Glenn....10
Lister, Tedd....403
Litster, Shawn....202
Liu, Changjun....397
Liu, Fang....294, 295
Liu, Kunlei....294, 295, 439, 446
Long, Richard....59
Lonkar, Sagar....87
Lu, Peng....165
Luebke, David....1
Luo, Siwei....208
Luo, Zhonglin....4
Lyngfelt, Anders....286
Ma, Jinliang....297
Ma, Juan....25
Ma, Sichao....411
Ma, Yi Hua....13, 15
Mahapatra, Priyadarshi....109
Mäki-Arvela, Päivi....413
Manzolini, Giampaolo....329
Mardilovich, Ivan....15
Margalef, Pere....198
Marina, Olga A. 37
Martin, Stephen M.....86
Matranga, Christopher....276, 377, 400
Mattisson, Tobias....277, 286
Matuszewski, Michael....18
Maurer, Gerd....438
May, Robert....270
McCandlish, L.....93
McCandlish, Larry E.....92
McGrath, James E. 90
McIlwain, Michael....403
McIntyre, Dustin....347, 426
Means, Nicholas C.....304
Meißner, Jan....61
Melzer, Steve....183
Middleton, Richard....188
Mikkola, Jyri-Pekka....413
Milan, Marta....21
Miller, David C.....45, 51
Miller, Duane....285
Miller, Eric....313
Moon, Dong-Ju....173
Morgan, Dave....21
Morinelly, Juan E. 51
Morrone, Pietro Paolo....328
Mozer, Thiago S. 201
Muriithi, Grace....24
Myers, Christina....1
Natesakhawat, Sittichai....276, 377
Neathery, James K. 294, 295, 439, 446
Nicolleti, Philip....302
Ning, Xue....69
Nulwala, Hunaid....1
O'Connell, M. 196
O'Connor, William....34, 168
Ohodnicki Jr., Paul R. 377
Olah, George....270
Oliver, Scott R. J. 55, 177
Ostwal, Mayur....14
Overney, René M.....5
Ozdogan, Ekin....2
Pakhare, Devendra....244
Park, Ah-Hyung Alissa....447
Park, Youngjune....447
Parmar, Rajesh D.....52, 57
Pasel, Joachim....61
Passos, Fabio B. 201
Patel, Dilip....37
Paul, Donald R.....9, 90
Pawar, Rajesh....415, 416
Pednekar, Pratik....311
Pei, Peng....190
Peng, Na....10
Pennline, Henry W. 451, 452
Peppley, Brant A. 52, 57
Peters, Catherine....188, 437
Peters, Ralf....61
Petit, Camille....447
Petrik, Leslie F.....24
Pham, Tiep Hoang....5
Pinacci, Pietro....328
Poe, Andrea....21
Pollak, Melisa....188
Poulston, Stephen....452
Powell, Lawrence E.....12
Prakash, Surya....270
Prevost, Jean-Herve....418
Rabie, Feras....86
Ramaraju, B.....243
Ramirez, Abelardo....417
Ramsey, John T. 296
Raymond, Laura J.....98
Reddy, P. Manoj Kumar....243
Remias, Joseph E.....439, 446
Renforth, Phil....31
Rengaswamy, Raghunathan....311
Ribeiro, Ana M. 262
Richards, George....296
Richardson, Carl F.....37
Riduan, Siti....398
Riman, R. 92, 93
Rochana, Panithita....2
Rode, Sabine....440

AUTHOR INDEX

Rodrigues, Alirio E. 262
Roizard, Christine 440
Romero, Elva 16
Roses, Leonardo 329
Rowsell, Liz 452
Roychoudhury, Subir 58
Rubel, Aurora M. 295
Rubin, Edward S. 17, 20
Rupp, Erik C. 164, 452
Rush, G. E. 34
Ryden, Magnus 277, 286
Samsun, Remzi Can 61
Sanders, David F. 90
Sanematsu, Paula 350
Santos, Alexandre F. 182
Sarker, Md Rakibul 347
Schaap, Marcel 348
Scheffe, Jonathan 401, 402
Schoff, Ronald 19
Schuerer, Jochen 196
Seabaugh, Matthew M. 59
Servia, Alberto 440
Shah, Nilay 49
Shahnam, Mehrdad 302
Shaik, Mahammadunnisa 243
Shao, Yixin 94
Shekhawat, Dushyant 57, 59, 68, 202, 244
Shen, Yijie 350
Shi, Wei 1
Siddiqui, Shameem 347, 351, 426
Siefert, Nicholas 202
Siriwardane, Hema 169
Siriwardane, Ranjani V. 285
Sivaniah, Easan 6
Skliar, Mikhail 298
Slone, Jeffrey 295
Smith, Andrew 452
Smith, Daniel 315
Smith, Mark W. 59, 68
Smith, Megan 185
Smith, Zachary P. 90
Snyder, Anthony C. 190
Soliman, Mohamed 347, 351
Song, Chunshan 60
Song, Qilei 6
Soong, Yee 33
Spivey, James J. 59, 68, 244
St John, Jessica 55, 177
Stanko, Dennis 452
Stauffer, Philip 416
Steen, William A. 37
Steinberg, Meyer 273
Steinfeld, Aldo 401, 402
Stolten, Detlef 61
Strazisar, Brian 419
Studer, Sarah 313
Sun, Yuhan 378
Surbahmanyam, Ch. 243
Sutherland, Erika 313
Tang, L. 92, 93
Tatarchuk, Bruce J. 56
Tchelepi, Hamdi 167
Thompson, Karsten E 350
Thompson, Robert L. 1
Tiemann, David 196
Towe, Glenn 71
Trabold, Thomas 315
Tran, Dat T. 55, 177
Trentham, Robert 183
Tsapatsis, Michael 8
Turton, Richard 109, 302, 303, 311
Tyagi, Mayank 350
Underwood, M. Kylee 400
Vakifahmetoglu, Ç. 92, 93
Valenstein, Jason 21
Van Duin, Adri C. T. 448
Vance, David 183
Vanessendelft, Dirk 302
Velaga, Srinath 166
Verba, Circe 34
Vilekar, Saurabh A. 58
Virtanen, Pasi 413
Viswanathan, Hari 416
Von Harbou, Inga 438
Voskov, Denis 167
Wachsman, Eric D. 325
Wagner, Michael 438
Wainwright, Haruko 415
Walluk, Mark 315
Walsh, Dennis 58
Walter, Michael 329
Wang, Feng 178
Wang, Gaowei 272
Wang, Qian 385
Wang, Xiaofei 450
Warzinski, Robert P. 168
Way, J. Douglas 14
Wei, Wei 178, 378
Weibel, David 402
Weiland, N. 302, 304
Wen, Xia 178
White, Donald J. 417
Wichert, Martin 196
Wickramanayake, Shan 1
Wilcox, Jennifer 2, 164
Wildenschild, Dorthe 348
Wilhite, Benjamin 16, 324
Willson, Clinton S. 350
Wilson, Elizabeth 188
Wimer, John 22

AUTHOR INDEX

Wolden, Colin A. 14
Woods, Mark 22
Worhach, Paul 22
Wu, Tony 452
Xia, Fei 274
Xia, Jianzhong 9
Xiao, Fukui 178, 378
Xie, Chao 60
Yang, Hongyun 56
Yang, Jianhui 346
Yang, Shaowei 345
Ye, Jingyun 397
Yin, Hang 10
Ying, Jackie Y. 398
Yoon, Roe-Hoan 25
Yu, Dingyi 398
Zeng, Liang 208
Zhang, Bo 448
Zhang, Chen 89
Zhang, Lian 23
Zhang, Shujiao 385
Zhang, Xiangping 46
Zhang, Xueyi 8
Zhang, Yi 294, 295
Zhang, Yugen 398
Zhao, Ning 178, 378
Zhu, Chen 165
Zitney, Stephen E. 109, 297, 303
Zondlo, John 68
Zyvoloski, George A. 416

American Institute of Chemical Engineers
3 Park Avenue
New York, New York 10016-5991

CURRAN ASSOCIATES INC
proceedings
.com